Sergej Hloch · Dagmar Klichová ·
Frank Pude · Grzegorz M. Krolczyk ·
Somnath Chattopadhyaya
Editors

Advances in Manufacturing Engineering and Materials II

Proceedings of the International Conference on Manufacturing Engineering and Materials (ICMEM 2020), 21–25 June, 2021, Nový Smokovec, Slovakia

Springer

Editors
Sergej Hloch
Faculty of Manufacturing Technologies
TUKE
Bayerova 1, Slovakia

Frank Pude
High-Pressure Waterjet Technology
Steinbeis Consulting Center
Horgau, Germany

Somnath Chattopadhyaya
Indian Institute of Technology, Indian
School of Mines
Dhanbad, India

Dagmar Klichová
Institute of Geonics of the CAS
Ostrava, Czech Republic

Grzegorz M. Krolczyk
Opole University of Technology
Faculty of Mechanical Engineering
Opole, Poland

ISSN 2195-4356 ISSN 2195-4364 (electronic)
Lecture Notes in Mechanical Engineering
ISBN 978-3-030-71955-5 ISBN 978-3-030-71956-2 (eBook)
https://doi.org/10.1007/978-3-030-71956-2

This Springer imprint is published by the registered company Springer Nature Switzerland AG
The registered company address is: Gewerbestrasse 11, 6330 Cham, Switzerland

Lecture Notes in Mechanical Engineering

Lecture Notes in Mechanical Engineering (LNME) publishes the latest developments in Mechanical Engineering—quickly, informally and with high quality. Original research reported in proceedings and post-proceedings represents the core of LNME. Volumes published in LNME embrace all aspects, subfields and new challenges of mechanical engineering. Topics in the series include:

- Engineering Design
- Machinery and Machine Elements
- Mechanical Structures and Stress Analysis
- Automotive Engineering
- Engine Technology
- Aerospace Technology and Astronautics
- Nanotechnology and Microengineering
- Control, Robotics, Mechatronics
- MEMS
- Theoretical and Applied Mechanics
- Dynamical Systems, Control
- Fluid Mechanics
- Engineering Thermodynamics, Heat and Mass Transfer
- Manufacturing
- Precision Engineering, Instrumentation, Measurement
- Materials Engineering
- Tribology and Surface Technology

To submit a proposal or request further information, please contact the Springer Editor of your location:

China: Dr. Mengchu Huang at mengchu.huang@springer.com
India: Priya Vyas at priya.vyas@springer.com
Rest of Asia, Australia, New Zealand: Swati Meherishi at swati.meherishi@springer.com
All other countries: Dr. Leontina Di Cecco at Leontina.dicecco@springer.com

To submit a proposal for a monograph, please check our Springer Tracts in Mechanical Engineering at http://www.springer.com/series/11693 or contact Leontina.dicecco@springer.com

Indexed by SCOPUS. All books published in the series are submitted for consideration in Web of Science.

More information about this series at http://www.springer.com/series/11236

Preface

It is a great delight to write the introduction to this volume of Lecture Notes of Mechanical Engineering, subtitled "Advances in Manufacturing Engineering and Materials II". This volume comprises the accepted manuscripts for the International Conference on Manufacturing Engineering and Materials II (icmem.eu). The conference is organized under the auspices of Dean of the Faculty of Manufacturing Technologies TUKE in Prešov; under the auspices of Dean of the Faculty Dr. h. c. Prof. Ing. Jozef Zajac, PhD.; in collaboration with the Institute of Geonics of the CAS in Ostrava-Poruba under the auspices of its Director Ing. Josef Foldyna, CSc.; the Indian Institute of Technology (Indian School of Mines), Dhanbad, India; the Faculty of Mechanical Engineering at the Opole University of Technology; the Mechanical Engineering Faculty in Slavonski Brod under the auspices of Dean Prof. Dr. Sc. Ivan Samardzić; and the TEAM International Society, under the auspices of President Prof. Dr. sc. Dražan Kozak with the further cooperation of Baden-Wuerttemberg Cooperative State University; the Department of Industrial Engineering, University of Salerno; the Mechanical Engineering Faculty of VŠB-TU Ostrava; the Faculty of Engineering, Czech University of Life Sciences, Prague; and Istanbul Medipol University, supported by Technical Diagnostics, Ltd. in Prešov and Steinbeis Consulting Center for High Pressure Water Jet Technology.

This proceedings was created at a more complicated time, when human beings were affected by a pandemic situation caused by the presence of a novel coronavirus—Coronavirus SARS-CoV-2. Some studies have pointed to this phenomenon before, and as a result of human activity, this phenomenon has now significantly affected our lives, globally. The virus showed us that our current living style is not compatible with the environment. The current destruction and metamorphosis of the environment are associated with anthropogenic activity. In addition to its benefits, this activity has been threatening the very essence of life. Recently, as humankind, we managed to solve the problem of depletion of the ozone layer. Currently, humankind is facing a much more serious problem—global warming. This phenomenon is mostly related to the rapid conversion of fossil fuels into energy, which has been driving technological developments. As these technologies are helping us in our daily lives and to reach milestones in science and progress, to successfully handle this difficult situation, we will have to change our mindsets. Changes must be led by the current generation, which can no longer ignore the current global environmental situation. We are, with high probability, the last generation that can derive responsibility from knowledge. Following the pandemic coincidence and relating to the aforementioned serious problems, we should become active in rebuilding the future. Therefore, the "motto" of the International Conference on Manufacturing Engineering and Materials (ICMEM2021) is "Nature First":

- How can we generate new knowledge supporting both the environment and sustainable development?
- How can we quickly transfer knowledge to education and daily life?
- How can we translate this into the development of new technologies, with reduced energy consumption?
- How can we translate this into the development of CO_2-neutral technologies?

Since the pandemic coincidence highlighted the need for sustainable production for this purpose, some experts in the field were invited to give their keynote lectures related to sustainable machining.

A member of the Federal Association of German Management Consultants and ambassador of the Global Ethic Institute will deliver an interesting invited lecture entitled "Only Fools Stretch the Rules? Why Might People Comply with Rules or Ignore Them?". As Entrepreneur of the Year and successful Managing Director of mid-size enterprises, Dr. Jan-Arne Gewert has many years of practical experience in the business. Since 2015, he has consulted managers in the economy and researched questions of business as well as strategy development. The next interesting keynote on the environmental footprint of distributed manufacturing using 3D printers will be introduced by Prof. Dr. Jan Michael Olaf from Baden-Wuerttemberg Cooperative State University, Loerrach. He will introduce options in reducing logistics efforts by investing in secure IT solutions.

Currently, the wide and various applications of exotic materials and their alloys are related to the necessity to manufacture machine parts while meeting adequate construction requirements. Nowadays, it is important that the manufacturing process should also incorporate aspects of sustainable machining, taking into consideration questions of economy, the environment, and social implications related to production. Those requirements determine the technological approach to the process of machining. Related problems regarding the sustainable machining of titanium and its alloys will be introduced by Professor Witold Habrat from Rzesow University of Technology. Sustainability issues are trending due to the environmental problems caused by the unbridled use of environmental resources. Governments are making efforts to increase industrial sensibility to environmental problems, promoting actions to reduce the impact of their activities. Processes and manufacturing technologies are involved in finding solutions with a lower environmental impact. A method to evaluate the sustainability of micro-electrical discharge machining (EDM), using quantitative and qualitative indicators, was developed taking into account the critical aspects of the technology. This method can support in the finding of the process conditions that minimize environmental impact. The abovementioned critical aspects are included in the invited lecture by Prof. Chiara Ravasio, entitled "Micro-EDM Process Sustainability Aspects".

Prof. Dr. Julia Rauenberger from the University of Applied Sciences and Arts introduces a lecture entitled "How to Strengthen Today's Math Skills of Tomorrow's Engineers—Practical Experiences with Agile Approaches to Innovative University Math Lectures". Other interesting invited lectures are also promised which will be presented by respected scientists: Prof. Gabriëlle Tuijthof from Maastricht University & Amsterdam UMC will present "Healing Water—Application of Water Jet Drilling for Orthopaedic Surgery", Dr. Jaroslava Svobodová will present "Fractography of Al and Fe Alloys—Air and Other Disasters", and Prof. Alessandro Ruggiero will present "Tribology of Environmentally Friendly Vegetal Lubricants: Some Recent Scientific Results".

The book gathers the proceedings of ICMEM 2020. While the conference date had to be shifted to 2021 (icmem.eu) because of the pandemic, the preparation of these proceedings has continued. Papers accepted to the conference upon a rigorously double-peer-reviewed process represent the core of this book, which covers research and applications that are expected to foster advancements in the field of manufacturing materials, specifically in the area of sustainable manufacturing. The content of the manuscripts includes various aspects of progress in (un)conventional manufacturing technologies mainly focusing on environmentally friendly technology, such as water jet technology and its technological modifications for industrial and medical applications. The authors' efforts were focused on contributing to increased competitiveness of industry with regard to sustainable development and preservation of the environment and natural resources.

For their efforts, I would like to thank the organization staff, the members of the scientific committee, and the reviewers for their time and energy in rigorously reviewing papers and providing valuable remarks to improve the authors' work. At this point, I want to thank the authors who, despite the problems due to the pandemic situation, were able to finish and refine their articles. In particular, I would like to thank Senior Editor for Engineering/Applied Sciences from Springer Verlag GmbH, Leontina Di Cecco, and Project Coordinator from Springer Nature, Ms. Divya Meiyazhagan, for their endless patience.

I firmly believe that many readers will find in this book interesting information that will support their research and foster their further—both academic and professional—creative activity.

Sergej Hloch

Organization

Scientific Board

Samardžić Ivan	J. J. Strossmayer University of Osijek, Croatia
Schlatter Manfred	Baden-Wuerttemberg Cooperative State University, Germany
Sedmak Aleksandar	MF UNI Belgrade, Serbia
Sennaroğlu Bahar	Marmara University, Istanbul, Turkey
Sergi Vincenzo	University of Salerno, Italy
Sitek Libor	Institute of Geonics of the CAS, The Czech Republic
Stančeková Dana	University of Žilina, Slovakia
Surzenkov Andrei	Tallinn University of Technology, Estonia
Šofer Michal	VŠB-TU Ostrava, The Czech Republic
Stoić Antun	J. J. Strossmayer University of Osijek, Croatia
Tozan Hakan	Medipol University Istanbul, Turkey
Traussnigg Udo	CAMPUS o2, Graz, Austria
Tropša Vlado	University North Varaždin, Croatia
Tuijthof Gabrielle	Maastricht University, The Netherlands
Turcin Ioan	CAMPUS o2, Graz, Austria
Ungureanu Nicolae	NUC OF Baia Mare, Romania
Valášek Petr	Czech University of Life Science, The Czech Republic
Valentinčič Joško	University of Ljubljana, Slovenia
Vayvay Özalp	Marmara University, Istanbul, Turkey
Vinay Sharma	Birla Institute of Technology, India
Weise Dieter	Fraunhofer Institute, Germany
Wieczorowski Michał	Poznan University of Technology, Poland
Yagimli Mustafa	Istanbul Gedik University, Turkey
Wojciechowski Szymon	Poznan University of Technology, Poland
Yashar Javadi	Islamic Azad University, Iran
Natarajan Yuvaraj	R&D Institute of Science and Technology, India
Zajac Jozef	Technical University of Košice, Slovakia

Reviewers

Steven Dunnen	TU Delft
Ashish Kumar Srivastava	Bajaj Institute of Technology and Management
Madhulika Srivastava	Amrita School of Engineering, Chennai
Somnath Chattopadhyaya	Indian Institute of Technology, Dhanbad
Miroslav Muller	University of Life Science, Prague
Dagmar Klichová	Institute of Geonics of the Czech Academy of Sciences
Petr Hlaváček	Institute of Geonics of the Czech Academy of Sciences
Jiří Klich	Institute of Geonics of the Czech Academy of Sciences

Libor Sitek	Institute of Geonics of the Czech Academy of Sciences
lessandro Ruggiero	University of Salerno, Italy
Dominika Lehocká	Faculty of Manufacturing Technologies TUKE with a seat in Prešov
Andrzej Perec	The Jacob of Paradies University
Sachindar Shankar	Indian Institute of Technology, Dhanbad
Frank Pude	ETH Zurich
Pavol Hvizdos	Institute of Material Sciences, SASKE, Košice
Hakan Tozan	Istanbul Medipol University, Istanbul

Organizing Committee

Sergej Hloch	Faculty of Manufacturing Technologies TUKE with a seat in Prešov
Ľubica Siváková	Faculty of Manufacturing Technologies TUKE with a seat in Prešov
Dagmar Klichová	Institute of Geonics of the Czech Academy of Sciences
Frank Pude	ETH Zurich
Jana Petrů	VŠB – TU Ostrava
Michaela Žembová	Atrium Hotel
Alessandro Ruggiero	Salerno University
Hakan Tozan	Istanbul Medipol University, Istanbul
Madhulika Srivastava	Amrita School of Engineering, Chennai
Akash Nag	Indian Institute of Technology, Dhanbad
Dominika Lehocká	Faculty of Manufacturing Technologies TUKE with a seat in Prešov
Ladislav Morovič	Slovak University of Technology in Bratislava

Contents

Keynote Lectures

How to Strengthen Today's Math Skills of Tomorrow's Engineers – Practical Experiences with Agile Approaches to Innovative University Math Lectures 3
Julia Rausenberger, Lilian Gilgen, Oliver Mülken, Stefanie Feiler, Roger Burkhard, Nico Erb, Anna Luther, Meike Hölscher, Silke Bock, and Frank Pude

Fractography of Al and Fe Alloys – Aircraft Accident and the Other Disasters 12
Jaroslava Svobodova

Micro-EDM Process Sustainability Aspects 23
Chiara Ravasio, Giancarlo Maccarini, and Giuseppe Pellegrini

Water Jet Session

Drilling Difficult-to-Machine Al_2O_3 Ceramics Using an Injection Abrasive Water Jet 41
Jiri Klich, Dagmar Klichova, Guoyi Peng, and Libor Sitek

Multiple Criteria Optimization of Abrasive Water Jet Cutting Using Entropy-VIKOR Approach 50
Andrzej Perec and Wojciech Musial

Identification of Local Microplasticity on Ti6Al4V After Impingement of Periodically Acting Water Clusters 63
Jakub Poloprudský, Alice Chlupová, Tomáš Kruml, and Sergej Hloch

Preliminary Testing of Production Thin-Walled Ribs into Aluminum Alloy AW6060 Using the Abrasive Water Jet 75
Frantisek Botko, Dominika Botkova, Jiri Klich, Dagmar Klichova, and Vladimir Simkulet

Material Reuse of Waste Abrasive Particles from Abrasive Water Jet Technology in the Field of Polymer Particle Composite Systems ... 87
Miroslav Muller, Martin Tichý, Monika Hromasová, Jan Novotný, Anna Rudawska, and Viktor Kolář

On-Line Monitoring of In-Vitro Application of PWJ for Bone Cement Disintegration ... 100
Akash Nag, Sergej Hloch, and Amit Rai Dixit

Effect of Surface Modification and Standoff Distance on Hydrodynamic Ductile Erosion ... 111
Ashokkumar Ravichandiran, Peter Pavol Monka, Katarína Monková, and Monika Hromasova

Effect of Acoustic Chamber Length on Disintegration of Ductile Material with Pulsating Water Jet ... 120
Akash Nag, Sergej Hloch, Avinash Babu Ramesh Babu, Monika Hromasova, and Amit Rai Dixit

Investigation of Different Abrasives Mixtures Ratio in Abrasive Water Jet Drilling of SS304 ... 132
Sounder Rajan Rajendran, Balaji Kuppusamy, Yuvaraj Natarajan, and Sivakumar Mahalingam

Effect of Water Flow Rate on Operating Frequency and Power During Acoustic Chamber Tuning ... 142
Akash Nag, Gabriel Stolárik, Brano Svehla, and Sergej Hloch

The Study of Method of Acoustic Generation of Pulses for Creating Pulsating Water Jet ... 155
Piush Raj, Somnath Chattopadhyaya, and Amitava Mondala

Abrasive Water Jet Piercing of Superalloys: A Study of Small Diameter Deep Holes ... 183
Natarajan Yuvaraj, Chinnakulanthai Sampath Shamli, Mohan Mughilvalavan, and Radhakrishnan Muruganandhan

Manufacturing Engineering Session

Sustainability of the Production Process by Applying Lean Manufacturing Through the PDCA Cycle – A Case Study in the Machinery Industry ... 199
Marko Milosevic, Marko Djapan, Roberto D'Amato, Nicolae Ungureanu, and Alessandro Ruggiero

Areal Decomposition Methodology for a 5-Axis Milled Surface ... 212
Chun Chan and Karl Walton

Tool Wear of Carbide Cutting Inserts Coated with TiAlN and AlTiSiN in Finish Turning of Inconel 718 226
Joanna Lisowicz, Witold Habrat, Krzysztof Krupa, and Mateusz Fiedeń

The Effect of Magnetic Assisted Ball Burnishing on the Corrosion Resistance of C45 Steel 236
Zsolt Ferenc Kovács, Zsolt János Viharos, and János Kodácsy

Characterization of Spheroidized Alumina Powders for Ceramic 3D Printing Application 249
Ram Krishna, Mohan Sreenivasan, Mykola Lyutyk, and Mykola Barosh

Layered Composite Materials and the Use of Dynamic Model Analysis 258
Gabriela Kuchtová

A Comparative Simulation Analysis of Conventional and Composite Leaf Spring 270
Dinesh Kumar Madheswaran, Mohanraj Thangamuthu, Suresh Gopi, Ram Krishna, Sreenivasan Mohan, Arunkumar Jayakumar, Yogesh Chinnampalli, and Shaik Tappa Nayab Hussain

Research of the Effect of Macrogeometric Structures on the Melt Front Using Simulation 282
Krisztian Kun and Zoltan Weltsch

Replication of Microstructures Formed by Femtosecond Laser During Injection Moulding 290
Krisztian Kun, Janos Líska, and Zoltan Weltsch

Design of a Rig to Assess the Structural Performance of Rotors for High Speed Electric Machines 301
Levi Mallin

A Commentary on the Role of Noise Measurement Uncertainty in the Industrial Context 324
Consolatina Liguori, Alessandro Ruggiero, Domenico Russo, and Paolo Sommella

Attendants at MMC Composite Milling 332
János Líska, Krisztián Kun, and Zsolt Kovács

Design and Verification of an Acoustic Mitigation Intervention for an Air-Cooled Inverter Heat Pump: A Case Study 343
Alessandro Ruggiero, Domenico Russo, Roberto D'Amato, and Nicolae Ungureanu

Effects of Tool Rotation Speed on Friction Stirred Material Flow Behavior of AA6061-T6 Aluminum Alloys 355
Anil Kumar Maurya, Anuj Jain, and Akshoy Ranjan Paul

Enhancement of Surface Properties of High Speed Steel Using Powder Mixed Micro-electrical Discharge Process 361
Shalini Mohanty, Partha Bhattacharyya, Alok Kumar Das, and Amit Rai Dixit

Friction Comparison on Tribological Pairs HXLPE/Ti6Al4V Alloy and VE-HXLPE/Ti6Al4V Alloy Under Dry and Physiological Lubricated Conditions 371
Alessandro Ruggiero, Roberto D'Amato, Nicolae Ungureanu, and Saverio Affatato

Computational Study of Metal Flow Behavior of AA 6061 in Friction Stir Processing with Twin Tools 386
Vineet Kumar Tiwari, Akshoy Ranjan Paul, Anuj Jain, Somnath Chattopadhyaya, Sachindra Shankar, and Pedro Vilaca

Nano Particles as Lubricant Additive: A Concise Review 392
Roberto D'Amato, Alessandro Ruggiero, Petr Valašek, and Nicolae Ungureanu

Automatic Generation of the Machining Processes for Turned Components in CAD/CAM System 404
Tomáš Dodok, Nadežda Čuboňová, Miroslav Císar, and Vitalii Ivanov

Author Index 411

Keynote Lectures

How to Strengthen Today's Math Skills of Tomorrow's Engineers – Practical Experiences with Agile Approaches to Innovative University Math Lectures

Julia Rausenberger[1(✉)], Lilian Gilgen[1], Oliver Mülken[1], Stefanie Feiler[1], Roger Burkhard[2], Nico Erb[2], Anna Luther[3], Meike Hölscher[4], Silke Bock[4], and Frank Pude[1]

[1] School of Life Sciences, University of Applied Sciences and Arts Northwestern Switzerland, Hofackerstrasse 30, 4132 Muttenz, Switzerland
julia.rausenberger@fhnw.ch
[2] School of Applied Psychology, University of Applied Sciences and Arts Northwestern Switzerland, Riggenbachstrasse 16, 4600 Olten, Switzerland
[3] Faculty of Mechanical Engineering, University of Applied Sciences Mannheim, Paul-Wittsack-Strasse 10, 68163 Mannheim, Germany
[4] Centre for Collaborative Teaching and Learning, Technische Hochschule Mittelhessen University of Applied Sciences, Eichgärtenallee 6, 35390 Giessen, Germany

Abstract. How can math lectures within the life sciences curriculum take into account student heterogeneity in terms of prior mathematical knowledge and learning pace? And how can they do this while combining the achievement of learning goals with elements of agile working, such as self-organization in heterogeneous teams or promotion of creativity and motivation? At the start of our new "BSc In Life Sciences" curriculum, the focus was on two approaches to address student heterogeneity – eduScrum as an undergraduate math learning framework and the qualification of highly motivated students as tutors. This paper reports on the motivation and development process to adapt teaching settings and presents first insights into the acceptance and impact of both approaches. In addition to achieving the learning objectives, both the eduScrum framework and the qualification of tutors promote skills such as collaboration, communication, creativity, IT skills and critical thinking - requirements that tomorrows' employees will encounter in their carriers in the twenty-first-century.

Keywords: Mathematical skills · eduScrum · Agile university didactic · Qualification of tutors · Empowerment · Academic staff development

1 Introduction/Background

For the BSc in Life Sciences degree at the School of Life Sciences FHNW, a basic mathematical education is mandatory for all students. The different fields of study within the

S. Hloch et al. (Eds.): ICMEM 2021, LNME, pp. 3–11, 2021.
https://doi.org/10.1007/978-3-030-71956-2_1

BSc "Life Sciences" determine the depth and scope of the mathematical study contents. For the engineers of tomorrow, rigorous math skills are mandatory.

In recent years, the School of Life Sciences FHNW has successfully launched a number of special courses to address the heterogeneity of basic mathematical knowledge, especially for first-year students. Providing a math app, regular consultation hours and a wide range of online material for self-study ensured instantaneous support. As a result, the dropout rate was already effectively reduced. By introducing new teaching formats such as eduScrum and implementing tutorial work, students are encouraged to take ownership of their learning process as well as develop their personality.

Different educational paths and histories lead to significant heterogeneity in the basic mathematical knowledge of first-year students. Traditional, lecturer-centred courses as a means of "one-directional" knowledge transfer contradict the students' heterogeneity. Students with moderate prior knowledge may be bored, while students with low prior knowledge are overloaded and might have to abandon studying early due to insufficient grades.

The labour market requires not only educated workers but also highly motivated thinkers and leaders [1]. Apart from supporting heterogeneous ways of learning, a program to qualify academically advanced, socially competent and motivated students to become tutors was started. To improve their skills they trained as peer-teachers and co-developers of learning settings of university courses along with their lecturers. Using an agile approach to qualify tutors increased their autonomy and involvement in the learning processes. By empowering tutors to co-create and reflect didactic designs and methods as well as to test and subsequently modify them, there is a better chance that they can motivate their peers in the tutorial setting. Qualified tutors can therefore act as role models for independent proactive learning, especially in heterogeneous university classes.

2 Different Approaches to Heterogeneity

2.1 Dealing with Heterogeneous Prior Knowledge

eduScrum. The eduScrum framework is derived from Scrum [2, 3] a framework for the development of complex products in project and product management. eduScrum transfers ideas of agile working methods to education [1] and therefore provides an active, collaborative and co-creative educational approach based on mutual trust between teachers and students. The lecturer ("eduScrum Master" and "product owner") determines *what* the team should learn during the learning unit and what the learning objectives are. Fig. 1. Visualizes the iterative structure of the working and learning environment with eduScrum: The lecturer also justifies the learning objectives (gives the "*why*") and explains what the learning objectives are useful for. The students in turn work in heterogeneously composed, self-organised teams. They decide *how* to achieve the learning objectives within the learning unit, the so-called "sprint". A sprint lasts for two to four weeks. At the end of the sprint the students demonstrate in a "review" that they own a usable "product" – they have reached the learning objectives. Afterwards the students reflect in a "retro-perspective" on their own contribution and the cooperation within the team. They come to an agreement on teamwork improvements for the next sprint. In summary, the lecturer-centred teaching format of a traditional lecture is broken up in favour of student-oriented, self-organised teamwork accompanied by input, coaching and review phases of the lecturer.

Fig. 1. The iterative structure of the working and learning environment with eduScrum.

Basic mathematical understanding and the corresponding tools of differential and integral calculus, linear algebra, statistics and differential equations form one of the pillars of life sciences and engineering. In addition to training in analytical solution methods, application-oriented life science and chemical engineering examples, as well as mathematical simulation programs such as Matlab, have become an integral part of the mathematical curriculum. The use of numerical tools requires an increased understanding of the underlying processes, as well as critical questioning and classification of numerical results. The benefit of being able to deal with mathematical content actively is that students can simulate technical processes in advance and thus minimise experimental effort, in terms of both money and time. Tomorrow's engineers will be able to use this competence profitably in their professional lives.

Learning within the eduScrum framework results in a twofold benefit for the students. On the one hand, they learn the contents of the math course, in a self-organized way; on the other hand, they obtain valuable additional skills in terms of team working experience, self-organisation, and, to a certain extent, having to search themselves for appropriate material. Students will benefit from these additional skills not only in the further progress of their studies but also in later professional life. The ability to search for and obtain the right material for understanding a new subject and to efficiently self-organise the studies is an important pre-requisite for later scientific studies, in particular for theses. Working in heterogeneous teams where people with different personalities are contributing is a preparation for teamwork situations in professional life, and the team-leading qualities developed as a scrum captain facilitate a career in higher positions.

The University of Applied Sciences Mannheim has introduced learning mathematics with eduScrum at tertiary education level [4] already in autumn 2017/18. The introduction took place in parallel with traditional lectures, whereby control groups helped to compare the results. The focus was not on achieving better exam results but rather on making the learning process more sustainable. All math courses for future engineers were successively converted from a classical teaching scenario to eduScrum over five semesters [5] and cooperation with other faculties e.g. Faculty of Electrical Engineering was sought. Overall, students' acceptance of working with the eduScrum framework increases as they move on to more advanced semesters. Students in their second semester are already largely able to work on topics not yet familiar to them from the high school context using provided or self-organized materials. After the third semester, the students have established an independent way of working and many of them are ahead of schedule. Adapting the schedule to the speed of the students and offering new topics ahead of time lead lecturers to work in an agile way, to assume the role of an idea generator and coach instead of a traditional lecturer. The students take ownership of their learning pace and process.

In a first pilot study in autumn 2019/20 at the School of Life Sciences FHNW, the focus was set on the mathematical success rather than on the secondary soft skills that are achieved by working with eduScrum. The introductory mathematical course was run in parallel: one eduScrum course with 40 students and three classical lectures, each with 70 students. The students of all courses worked on the same exercises during the semester and had to pass the same final exam. For data analysis, only those students were considered who participated in the pre- and post-survey ($n = 146$ valid out of 250 possible). The first results showed that academic success, i.e. passing the final exam, did not significantly depend on the framework. Figure 2 summarizes the passing rate of the introductory math course of eduScrum students and students in the traditional setting. Students working within the traditional or eduScrum framework achieved similar grades, cf. Fig. 3. Thus, it can be concluded that students managed to achieve their personal level of knowledge either way. However, beyond the scientific training, the eduScrum students could enhance their soft skills.

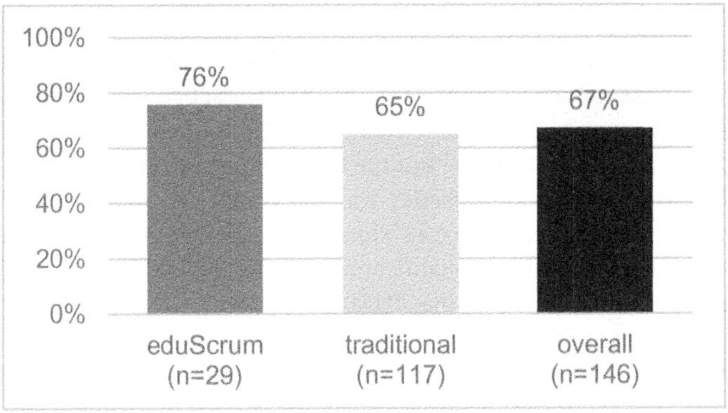

Fig. 2. Comparison of exam passing rate when working with eduScrum or in the traditional setting.

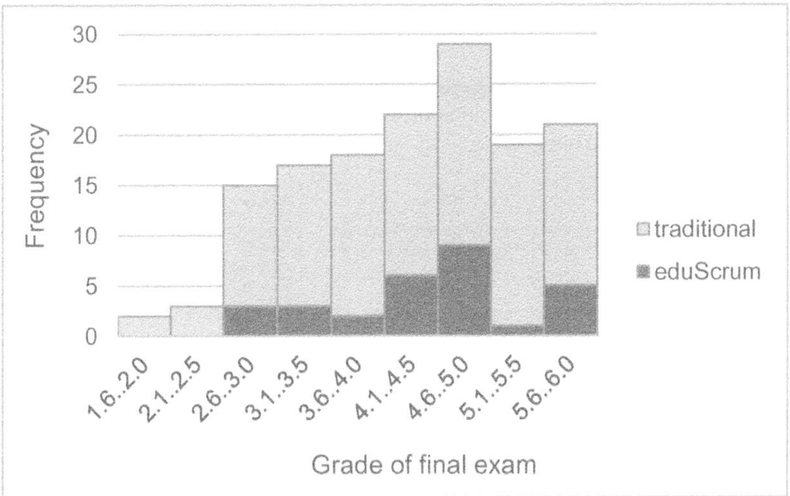

Fig. 3. Histogram of final grades of introductory math course indicating the traditional or eduScrum setting measured in the Swiss Grading System (4.0 = pass; 6.0 = optimum).

2.2 How to Develop Advanced and Motivated Students

Involving students in tutorial work provides peer-level support for their learning processes and offers the student-tutors qualified training and further development in an academic environment [6]. Especially for large courses at beginner level, tutorial work is suitable for reflecting on and deepening the learning objectives of the associated lectures. Nevertheless, close interaction and communication with the corresponding lecturer is indispensable for deriving benefit from the course.

Qualification of Tutors. In collaboration with the TH Mittelhessen University of Applied Sciences (THM) and its Center for Cooperative Teaching and Learning (ZekoLL), a one-week basic qualification for future tutors was developed. Based on more than ten years' experience in developing agile teaching approaches in the THM tutoring program and on having successfully qualified more than 1200 tutors for their jobs as peer-teachers in the last eight years, the good-practice components of the THM workshop modules were jointly transferred, according to the needs and contexts of the School of Life Sciences FHNW. Within the developmental process, the experience of THM in the field of laboratory didactics was integrated into the new concept, particularly with regard to the importance of simulations for learning how to teach [7], cf Table 1 for the agenda of the basic qualification. The main differences between the basic qualification week at the HLS and the THM tutor qualifications [8] are the following: provision of more time (5 days instead of half a day), the consequent integration of three experienced THM-tutors as co-coaches as well as a full day for a hands-on simulation of the tutorial settings. These preconditions proved successful for a highly agile workshop design as team-building and learning processes could be intensified by iterated reflection phases. The tutorial simulations helped the future tutors to experience their didactic implementations in a direct way, receive instant feedback from their peers and teaching staff and adapt their tutorials appropriately.

Table 1. Agenda of the basic qualification of tutors at the School of Life Sciences FHNW. (* involvement of experienced tutors).

Monday	Tuesday	Wednesday	Thursday	Friday
Teaching and learning at tertiary level/role-models/basic principles of tutorial teaching	Planning of processes to support learning/communication skills	Reflection of learning and teaching processes*/feedback culture	Micro-teaching-unit: Simulation of different tutorial settings: mathematics, laboratory	Discussion and reflection of micro-teaching unit
Challenges and expectations of tutors*/students' motivation	Principle of minimal help*/Dealing with challenging situations*	Preparation of micro-teaching-unit		Peer-tutor-shadowing/Tutor's checklist/

Most of the 13 tutors lacked teaching experience, but all were highly motivated and actively involved in co-creating the qualification workshop. This became very clear during the first phase of the workshop, when an intensive socializing and getting-to-know each other activity ("Triangle of common interests") helped to bring out the various pre-existing methodological and social competences of the student participants.

In contrast to a classical qualification of tutors at the THM, the math and physics lecturers of the HLS and the future corresponding lecturer of the tutors were present during the qualification. Their participation proved to be very beneficial as the future tutors could align expectations on tutorial work, responsibilities and role models of tutors and lectures instantaneously.

In addition to the THM's main coach, three experienced THM tutors participated in and stimulated the group work and reflection processes with the future tutors. Their presence allowed iterative adjustment of defining general tutorial settings and appropriate HLS-specific ways to support heterogeneous groups of first year students. The THM tutors acted as role-models for their peer-tutors by presenting different opinions on co-creating learning processes, e.g. short presentations on learning techniques or sharing experiences "the principle of minimal help" [9], as well as testing, e.g. simulations. This approach stressed the importance of collaboration between lecturers and tutors as well as between tutors and the students [10].

The simulation of the math and physics tutorials as well as reflection on that simulation proved to be very effective. The future tutors experienced themselves as designers and teachers/tutors in action. In addition, they obtained detailed feedback from all parties involved, their math and physics lecturers, peers and the experienced tutors. The high level of involvement and empowerment of the tutors was used to transfer the processes and methods of the qualification week into self-designed rules for future tutorial settings and project-plans for cooperation and collaboration with their math and physics lecturers.

Although the strong involvement of the future tutors from then on changed the course of the week, these "presence didactics" [11] enabled us to respond to the needs of the tutors. In addition, by adapting the program, the tutors became more independent and self-directed in their teaching experience and could take ownership of their learning process.

2.3 Qualitative Evaluations of Both Settings

To get a first impression of the two approaches with regard to acceptance of the interventions, qualitative evaluations were performed, in the form of interviews for the eduScrum setting and evaluation forms for the tutor qualification week.

eduScrum. The development of the eduScrum framework at the HLS FHNW was based on Human Centered Design (HCD) principles. This approach, also described in the DIN standard [12], is a procedure for the development and evaluation of products, services and processes and is therefore not limited to interactive systems. This implies that the eduScrum framework was evaluated at different points in time with students and lecturers in order to identify optimization possibilities and to check effectiveness. To address the question of optimization, two pluralistic walkthroughs (May/June 2019) were conducted with a "prototype" of the eduScrum framework before the pilot implementation and two qualitative guideline-based interviews were carried out (October/November 2019 with a total of 8 students) during the pilot implementation. Here, a selection of feedback from the interviews with students of the eduScrum group is presented:

- Strengths and weaknesses of the group members could be compensated during teamwork learning.
- The students were more engaged to reach the learning objectives. They reported the impression that learning was deeper and more long-term.
- The involvement and cooperation during the teamwork was evaluated more positively compared to the classical lecture setting.
- Compared to other lectures, learning with eduScrum required a relatively higher initial workload during the semester.
- eduScrum required much more self-discipline than a classical setting.

In order to evaluate the effectiveness of the teaching format and the learning experience by statistical means, a larger sample over a longer period has to be taken into account. The sustainability of the learning effect should be also assessed a year after the final exam.

Qualification of Tutors. The one-week basic qualification for future tutors was carried out in February 2020 with 13 participating students. Every day the students were asked to assess their experiences of that day. The answers to the standardized questions provided a detailed account of the successes and limitations of the approach. Below, a summary of the most prominent aspects of the evaluations is presented:

- During the basic qualification the role, the responsibilities and the duties of future tutors were clarified.
- The principle of and the necessity for "minimal help" in a tutorial setting was made very clear. Among other didactic methods, the students tried to implement "the principle of minimal help" in their simulations.
- In preparing and while conducting their respective simulations, the students experienced the importance of time-management and the differentiation of the course material for different levels of personal knowledge of the prospective students in the tutorial classes.

After the basic qualification week, the students were asked to give feedback on the entire week. There are three prominent positive outcomes of this evaluation:

- The involvement of the lecturers during the whole week was highly appreciated.
- The active participation of the three experienced tutors from THM significantly increased the learning success.
- A trusting learning environment resulted in active participation of the students.

The participating students suggested further adjustments and modifications of the concept:

- To introduce other didactic methods, e.g., buzz groups, think-pair-share, or speed-dating in addition to the "principle of minimal help",
- To include more practical exercises similar in form to the simulations they performed at the end of the week.

3 Conclusions

If one wants to address the heterogeneity of undergraduates, it is essential to consider the "mathematically weaker" students as well as the "mathematically stronger" and highly motivated students.

Here, first insights into two approaches are presented that have been further developed to strengthen the math skills of tomorrows' engineers and life scientists. The major outcome of this pilot project can be concluded as follows:

- Both approaches need to be continued for at least three successive years.

- According to the evaluation of the eduScrum interviews, some of the first year students doubt that eduScrum is an adequate teaching format for beginners. However, students have to change their consumer attitude to a more self-directed and independent attitude of learning.
- In order to increase the acceptance of the seemingly unfamiliar teaching setting already during the first year of study, an integration of experienced eduScrum-students into the future implementations of the module on a selective and recurrent basis is planned.

4 Outlook

As described, similar projects are being carried out at different educational institutions in Switzerland and abroad. It will be interesting to compare the general conditions or differences in cultures – due to origin, field of specialization or educational system.

For the long-term perspective, a combination of both approaches is desirable, i.e. math tutors support undergraduate lectures within an eduScrum setting. While the eduScrum setting fosters twenty-first-century competences such as collaboration or communication, the qualification of tutors contributes to personal and academic staff development.

References

1. Wijnands, W., Stolze, A.: Transforming education with eduScrum. In: MacCallum, D., Parson, K. (eds.) Agile and Lean Concepts for Teaching. Springer Verlag (2019)
2. Sutherland, J., Sutherland, J.J.: Scrum: The Art of Doing Twice the Work in Half the Time. Crown Business, New York (2013)
3. Schwaber, K., Sutherland, J.: The Scrum guide: the definitive guide to Scrum (2017). https://www.scrumguides.org/docs/scrumguide/v2017/2017-Scrum-Guide-German.pdf
4. Raab, P., Werft, W., Pinkernell, G., Luther, A.: eduScrum–ein methodischer Rahmen in einer Mathematik-Vorlesung in der Grundausbildung von Ingenieuren. Fachgruppe Didaktik der Mathematik der Universität Paderborn, pp. 1439–1442. WTM-Verlag, Münster (2018)
5. Čukić, P., Pinkernell, G., Werft, W., Luther, A.: Mathematikvorlesungen für Maschinenbaustudierende als Projektmanagement. die Hochschullehre Praxis (2020)
6. Kröpke, H., Ladwig, A.: Tutorienarbeit im Diskurs – Qualifizierung für die Zukunft. LIT-Verlag (2013)
7. Behr, I., Bock, S., Weimar, P.: Didaktik im Labor - Eine Kunst für sich. In: Berendt, B., Voss, H.-P., Wildt, J. (eds.) Neues Handbuch Hochschullehre (E.5.1.). Raabe Verlag, Stuttgart (2009)
8. Kröpke, H., Szabo-Batancs, M., Bock, S.: Investition in die Qualität der Lehre. Einsatz studentischer Tutorinnen und Tutoren am Beispiel zweier Fachhochschulen in Deutschland. In: Zimmermann, T., Zellweger, F. (eds.) Lernendenorientierung. Studierende im Fokus. Forum Hochschuldidaktik und Erwachsenenbildung. Band 3. hep Verlag, Bern (2012)
9. Aebli, H.: Zwölf Grundformen des Lehrens: eine allgemeine Didaktik auf psychologischer Grundlage: Medien und Inhalte didaktischer Kommunikation, der Lernzyklus (15. ed.). Klett-Cotta, Stuttgart (2019)
10. Rohr, D., den Ouden, H., Rottlaender, E.-M.: Hochschuldidaktik im Fokus von Peer Learning und Beratung. Weinheim und Basel: Beltz Juventa (2016)
11. Arn, C.: Agile Hochschuldidaktik. Beltz-Juventa (2017)
12. DIN EN ISO 9241-210. Prozess zur Gestaltung gebrauchstauglicher Interaktiver Systeme. Beuth Verlag GmbH, Berlin (2010)

Fractography of Al and Fe Alloys – Aircraft Accident and the Other Disasters

Jaroslava Svobodova(✉)

The University of J.E. Purkyně, Ústí nad Labem, Pasteurova 7, 40096 Ústí nad Labem, Czech Republic
jaroslava.svobodova@ujep.cz

Abstract. Machines and constructions fail in operation without warning, especially in the case of fatigue cracks and often with catastrophic consequences. The collapse of bridges and buildings, the massive breakage of ships and tankers, the explosions in chemical industry factories, the disintegration of aircraft and space rockets in the past, caused by various causes, are an unforgivable and unforgettable failures of materials but also a human factor, who designed the wrong material, or improperly selected the structure, regardless of the conditions that surround the structure and the environment in which the structure will operate. The lecture is focused on the fractography of Al and Fe alloys in specific cases of aircraft accident and the other disasters. Fatigue fractures are responsible for many disasters e.g. aircraft, ships, bridges accident, and other structures, and their greatest danger lies in the fact that occurs suddenly.

Keywords: Fractography · Fracture · Aluminium alloys · Steels · Fatigue fracture · Failure analyses

1 Introduction

Fractography is a term by which we understand fracture surface analysis. The term comes from the Latin word fractus. Fractography evaluates the quantitative and qualitative fracture area and determines the cause of the fracture.

The fracture area characterizes the relevant way of material failure. It is analysed by macroscopic, microscopic or electron scanning microscope (SEM) to determine the exact cause of the failure. Figure 1 illustrates the fracture area of the AlSi8Cu3 alloy in the region of the primarily excluded unmodified eutectic with platy excluded Si particles (yellow irregular circles) and intact sharp-edged Si particles (red arrow).

The fracture line characterizes the appropriate form of material failure concerning the local structure, Fig. 1.

Morphology (appearance) - each fracture surface has its specific appearance - morphology. By its analysis, we can sometimes find out the exact cause of the fracture surface and propose measures that would prevent material damage.

S. Hloch et al. (Eds.): ICMEM 2021, LNME, pp. 12–22, 2021.
https://doi.org/10.1007/978-3-030-71956-2_2

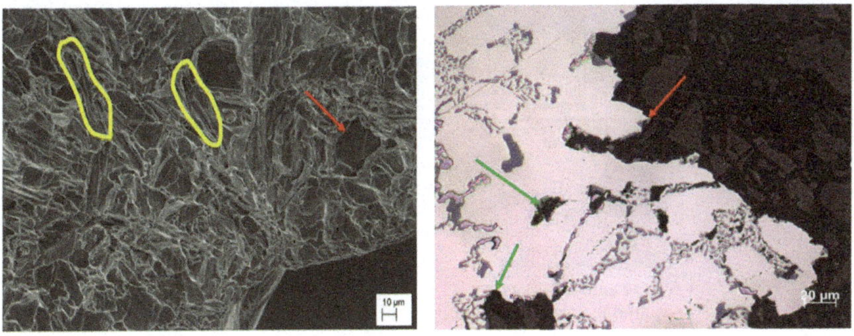

Fig. 1. SEM of the fracture surface AlSi8Cu3 and the fracture line of the cracked casting, AlSi8Cu3 alloy.

1.1 Criteria for the Evaluation of the Fractures

According to the Consumed Energy. According to this criterion, we divide the fractures into tough (ductile - the time of fracture surface formation is longer) and brittle (the time of fracture surface formation is short). The explanation is characterized by Fig. 2 of the static tensile test record at room temperature and Fig. 3 of the fracture surface documentation. The energy of both types of fracture formation is different. While, for the brittle fracture formation, the lower amount of energy is needed (lower deformation, ε), for the ductile fracture formation is needed a higher amount of energy and bigger deformation ε. The scale (for both types of fracture formation) is absorbed energy consumed to deformation and fracture of the testing rod. This absorbed energy is proportional to area, which lies under the working diagram curve (brittle fracture the area is smaller, ductile fracture the area is larger – hatched areas, Fig. 2). Brittle fracture is spreading with at a high speed with low consumption of energy and a slight ductile deformation ε. Material resistance against the brittle fracture indicates the toughness of the material.

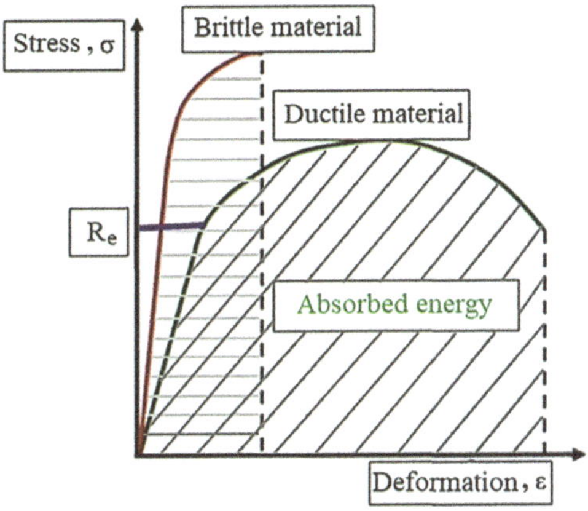

Fig. 2. Energy consumed for ductile and brittle fracture (hatched areas) [1].

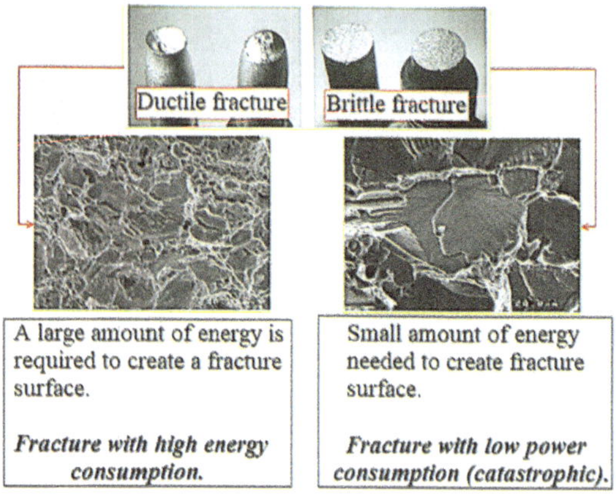

Fig. 3. Character of fracture surface at brittle and ductile fracture [1].

According to the Type of Stress. According to this criterion, we know two kinds of fractures. *Force fractures* arise from slow, continuously increasing load or dynamic disposable load. *Fatigue fractures* occur under cyclic, repeated loading. They have the characteristic morphology (shape) of the fracture surface, the schema is in Fig. 4 and the fatigue fracture of the shaft.

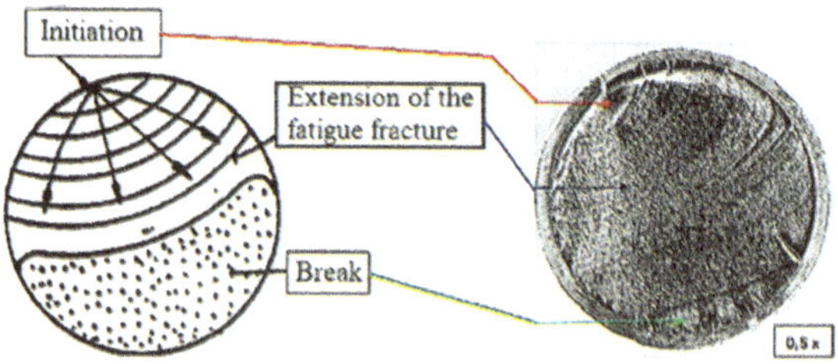

Fig. 4. Scheme of the fatigue fracture formation [2].

According to the Process of the Crack Transition (Fracture Surface) Through the Structure. This may be either a *transcrystalline* or an *intercrystalline* fracture. In connection with the energy criterion, these are fractures showed in the scheme in Fig. 5.

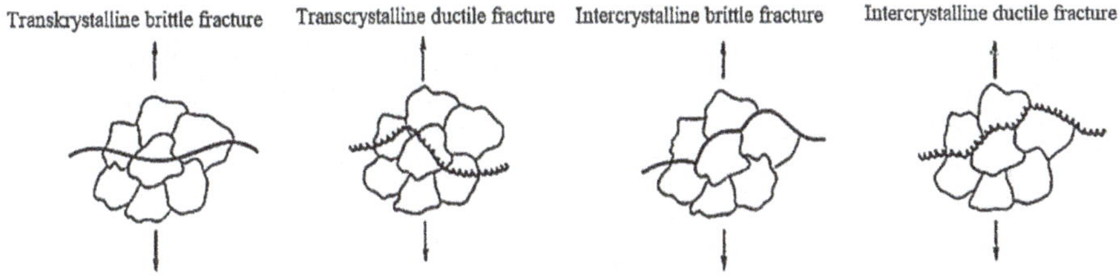

Fig. 5. Scheme of fractures crack pass through grains [2].

1.2 Theoretical Knowledge of Fatigue Failure

Fatigue fracture occurs at cyclic loading lower than the limit of elasticity R_e of the material, Fig. 7. Fatigue fracture was found in 90% of metal materials failure (e.g. bridges, aircraft, dynamically stressed machine components, components of various equipment) and has often catastrophic aftermath. The fatigue fracture has a characteristic macroscopic appearance (morphology) unmistakable with other failures. It is possible to observe the irregular cracks on the surface of the Al-Si alloys castings as shows Fig. 6.

Fig. 6. Propagation of the fatigue cracks on the surface of the AlSi9Cu3 after 1.2×10^9 cycles.

When analysing the fracture surface, we try to determine the initiation point, Fig. 4. We know three stages of fatigue fracture:

1. The concentration of stress at the initiation site and crack formation – Stage I.
2. Fatigue crack propagation – Stage II.
3. Break, catastrophic material failure – Stage III.

The *fatigue strength* is defined by the stress ***S*** at which the material breaks for a defined number of cycles (e.g. 10^7). The *fatigue life* is defined by the number of cycles ***N*** at which the material breaks at a given stress. Figure 7 shows the fatigue strength and fatigue life schemes, as well as an example of the fatigue curve of 12 040 steel and $AlCu_4Mg_1$ alloy.

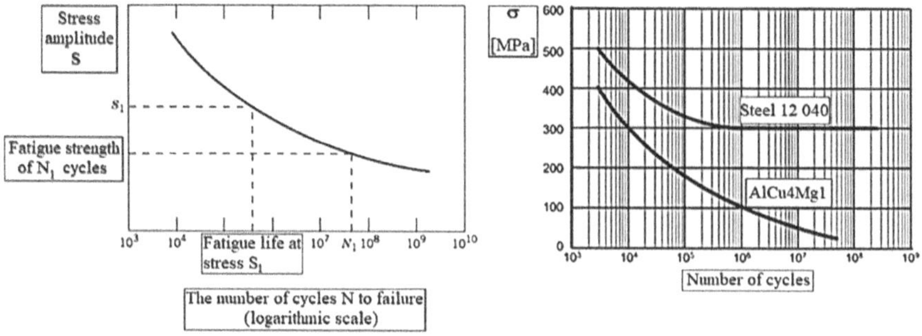

Fig. 7. Fatigue strength and fatigue life.

2 Aircraft Accident and the Other Disasters

2.1 Fatigue Fracture Causes

Factors that cause crack formation (respectively fracture) are mainly: **Faulty design solutions, Wrong chosen conditions of material production technology, Local environment and service conditions.**

2.2 Examples of Specific Cases of Material Failures

Tay Bride, Scotland (December 28, 1879). The classic and oldest known example of miscalculation is the collapse of Tay Bridge in Scotland in 1879, Fig. 8. The bridge collapsed during a storm when the train passed across the bridge, 75 passengers perished. This was due to poor structural calculations of the load during the whirlwind (0.5 kPa) instead of (2.5 kPa), which causes fatigue fracture.

Fig. 8. Tay bridge fatigue [7].

Broken Ship Liberty (April 29, 1943). On April 29 1943, it occurs to break of the newly built ship Liberty in the USA which was intended for an importing materials convoy during the World War II to Great Britain, Fig. 9. In the analysis of the steel at the failure locality, a characteristic brittle fracture morphology was found, which made it possible to identify the cause of the accident, Fig. 9. The cause of the fracture was low-temperature embrittlement of the steel. The temperatures of the seawater in the North Atlantic can be very low. The steel was exposed to low temperatures changed its properties, that means, reach the critical point at which ductile steel becomes brittle and the cracks are formed easily in these conditions. Unlike aluminium alloys, steels (other than austenitic steels, K12 lattice) may be brittle or ductile depending on temperature.

Fig. 9. Broken ship Liberty 29. 4. 1943 and the fracture surface of the steel [8].

Flight United Airlines 232. Fatigue fractures due to material defects have disastrous consequences in aviation. An example would be United Airlines 232 flight to Chicago, Illinois on July 19, 1989. The tail engine exploded, as a result, the tail surfaces of the aircraft were damaged and all hydraulics were discarded. The cause of the accident was the fatigue fracture of the rotary disc of the aircraft, which initiated a material defect, destroyed the hydraulic system of the aircraft that became uncontrollable at the landing and disintegrated into four pieces. The accident claimed 111 victims, Fig. 10.

Fig. 10. Part of the aircraft Flight 232 to Chicago and the Ti alloy wheel and schematic of damaged hydraulic system [9].

Japan Airlines Flight 123. An extremely large number of victims - 520 was due to the fatigue fracture of horizontal stabilizer in flight Japan Airlines Flight 123 in 1985, Fig. 11, 12. The initiation has not been determined. We can observe fatigue lines on the detail of the fracture surface, which proves that it is a fatigue fracture, Fig. 12.

Fig. 11. An amateur shot of an aircraft with a stabilizer broken off before crashing and the broken stabilizer caught from the sea [11].

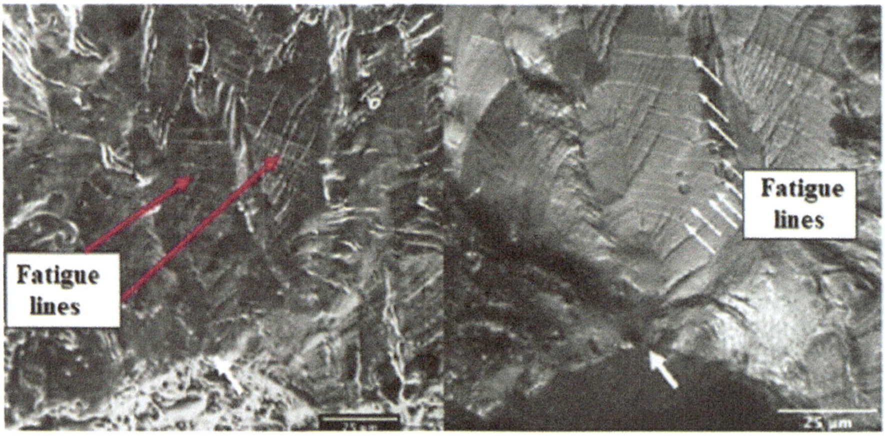

Fig. 12. Detail of fracture surface of broken stabilizer [11].

American Airlines Flight 587. The presence of oxidic films in the Al alloy 7075, initiated a fatigue fracture that was the cause of the American Airlines Flight 587 12 November 2001 disaster with 265 victims, Figs. 13, 14. The disaster was caused because of the vertical aircraft stabilizer that broke off during the flight.

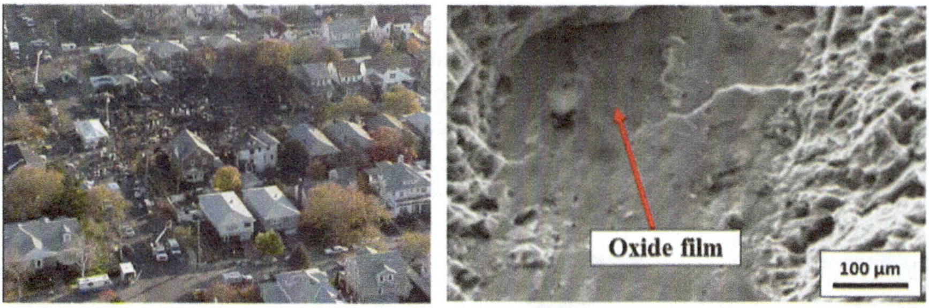

Fig. 13. Aircraft impact site and the oxidic membrane, initiation of fracture, on fracture surface of broken stabilizer [12].

Fig. 14. Detail of fracture surface [12].

Chalk's Ocean Airways Flight 101. On December 19 2005, an aircraft crashed in the USA (the wing broke off) by Chalks Ocean Airways, which did not survive 20 passengers. Documentation is in Fig. 15. The cause of the disaster was the fatigue fracture of the 7075 alloy wing.

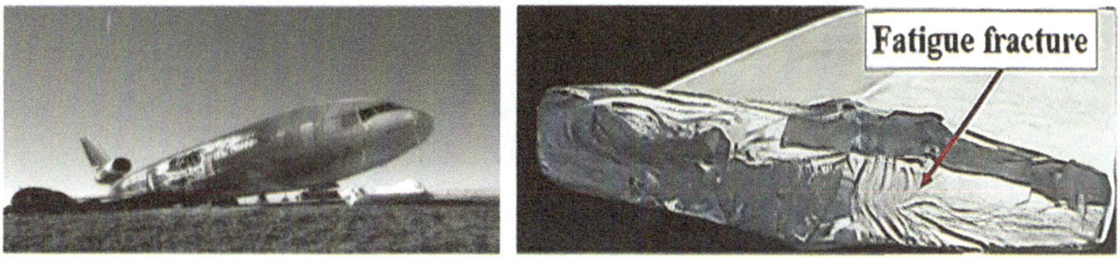

Fig. 15. Aircraft without right wing and the Fatigue fracture of Al alloy wing 7075 [13].

Fatigue Fracture of the Freight Wagon Shaft. A typical fatigue fracture of a railway freight wagon shaft initiated by the presence of a non-metallic inclusion is documented in Figs. 16 and 17. The macroscopic appearance of the fracture surface of the railcar shaft does not always have clear morphological features of the fatigue fracture, Fig. 16 b.

For accurate identification, what kind of fracture it is, it is necessary to carry out a substantial metallographic analysis. Preferably, the fracture surface is observed and documented using an electron scanning microscope (SEM). The presence of striations on the fracture surface of Fig. 16 b was observed and documented by SEM of Fig. 17. Their presence is evidence that the fracture was initiated by the fatigue process.

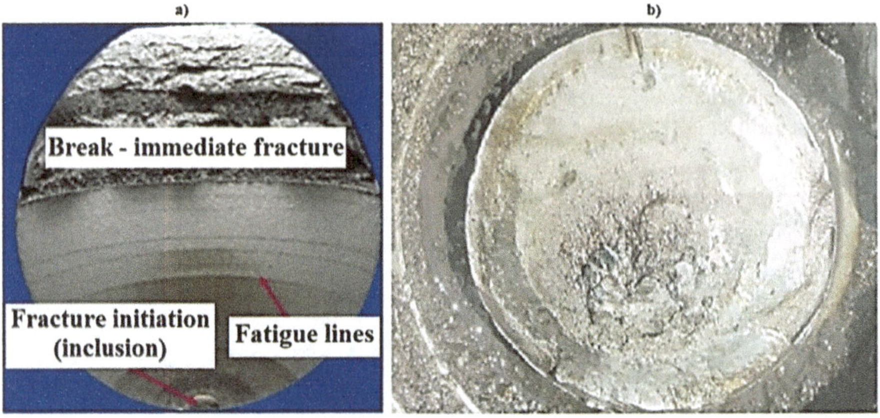

Fig. 16. Typical and atypical fatigue fracture of the rail freight wagon shaft [15].

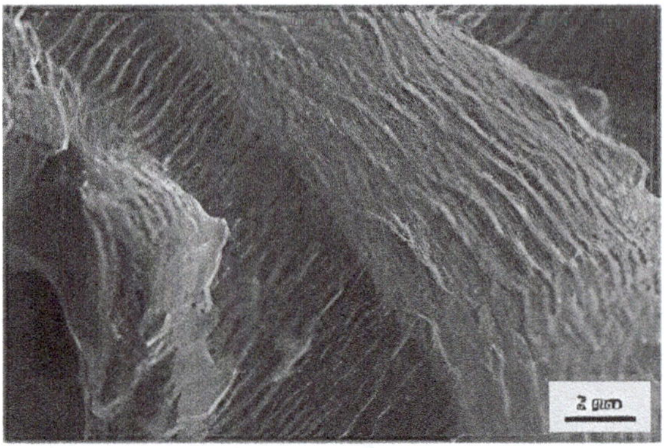

Fig. 17. Striations on fatigue fracture surface from picture 48 b. [15].

3 Conclusion

A typical feature of fatigue fractures is that they occur suddenly, without previous material macro deformation, even in the case of tough materials which fail in a static tensile test only after significant plastic deformation. They are insidious in time. The propagation of fatigue fracture, which has usually transcrystalline character, results in a weakening of the cross-section of the material. If the fracture surface reaches a critical value, a sudden fracture of the remaining cross-section will occur.

In most cases, the macroscopic appearance of the fatigue fracture can be identified based on the presence of three regions: the fatigue fracture initiation region, the fatigue damage region and the residual fracture region.

The objective of any accident investigation is to identify not only the failure mechanism but also the main causes, which may be related to the defective application of the structure in connection with material properties, design or production parameters.

Despite developments and progress in solving the various causes and types of fractures, there are still many unresolved questions that must be answered by fractography and the other methods. Below are some questions that need to be clarified:

1. Explanation, why some planes and directions in the crystallographic structure of materials are preferred especially for brittle (quasi-brittle) fractures?
2. What are minimum stress rates for the localized plasticity/coalescence process?
3. Why there is often a minimum distance between sections below a certain ΔK (one of the components of crack growth - cyclic load amplitude) due to different materials and environments?
4. Explanation of fracture propagation rates for different fracture types due to the surrounding environment – mechanism and controlling processes.

In addition to the issues outlined above, there will be a need to respond to new materials in different industries in the future.

References

1. Udomphol, T.: Fracture. Suranaree University of Technology (2007)
2. Veles, P.: Mechanické vlastnosti a skúšanie. 2. vydání. Bratislava: Alfa, 408 (1989)
3. Lados, D.A.: Fatigue Crack Growth Mechanisms in Al-Si-Mg Alloys.Polytechnic Institute, Ph.D. thesis (2004)
4. Jiang, X., et al.: Microstructure-based analysis of fatigue behaviour of Al-Si-Mg Alloy. Trans. Nonferrous Met. Soc. China **21**, 443–448 (2011)
5. Boromei, I., et al.: Influence of the solidification microstructure on porosity and the fatigue strength of Al-Si-Mg Alloys. Metall. Sci. Technol. **28**(2), 18–24 (2010)
6. Alexopoulos, N.D.: Impact Properties of the Aircraft Cast Aluminium Alloy AlSi7Mg0.6 (A357). https://www.epi-conferences.org
7. Anderson, T.L.: Fracture Mechanics, Fundamentals and Applications. Taylor&Francis Group. 630 (2017)
8. Harris, M.D., Grogg, W.J., Akoma, A., Hayes, B.J., Reidy, R.F., Imhoff, E.F., Collins, P.C.: Revisiting (some of) the lasting impacts of the liberty ships via a metallurgical analysis of rivets from the SS "John W. Brown". JOM **67**(12), 2965–2975 (2015)
9. Matlock, D.K.: Learning from the Past? Fatigue Failures in Engineering Systems. The Hatfield Memorial Lecture, Colorado School of Mines (2014)
10. Lukáč, I.: Fractography, US Steel s.r.o. Košice (2018)
11. Japan Airlines Flight 123. https://en.wikipedia.org/wiki/Japan_Airlines_Flight_123
12. Locovei, C., et al.: Analysis of fatigue fracture of tank wagon railway axles. In: Proceeding of the 3rd WSEAS International Conference, pp. 219–223 (2015)

13. American Airlines Flight 587. https://en.wikipedia.org/wiki/American_Airlines_Flight_587
14. Aircraft Accident Report, NTSB/AAR-07/07 (2007)
15. Railway Investigation Report, R 10T0035, Canada (2010)

Micro-EDM Process Sustainability Aspects

Chiara Ravasio(✉), Giancarlo Maccarini, and Giuseppe Pellegrini

Department of Management, Information and Production Engineering, University of Bergamo, viale Marconi, 5, 24044 Dalmine, BG, Italy
chiara.ravasio@unibg.it

Abstract. Nowadays the sustainability issues are trendy considering the environmental problems caused by an unbridled use of the resources. Governments are making efforts aiming to increase the industrial sensibility to environmental problems promoting actions to reduce the impact of their activities. The manufacturing technologies are involved in this scenario too. Electrical Discharge Machining (EDM) is a technology that finds a wide use in the micromachining sector for a variety of drilling applications in automotive, aerospace, biomedical sectors on different materials such as stainless steels, titanium alloys and others metals regardless of their mechanical properties. In this paper, a method to evaluate the sustainability of this process is proposed, taking into account the consumption of the electrical energy, the wear of the tool electrode during the machining, the consumption of the dielectric fluid, the environmental impact and part conformance. This method was applied for the drilling of a component in biomedical sector such as a bone plates in titanium alloy for fractures of small body parts. The micro holes were executed in different conditions as regards both electrode material and dielectric fluid. The sustainability of the machining was evaluated for all the tested conditions. In this way, the best operative conditions to minimize the environmental impact by using this technology were found. Moreover, the more critical aspects were identified. This allows targeted actions to be taken to diminish the impact of these aspects.

Keywords: Micro-EDM · Sustainability · Micro drilling

1 Introduction

In the last years, the miniaturization of the components took place in many industrial sectors such as automotive, aerospace, biomedical. Significate industrial efforts were made to support this new manufacturing trend in order to propose effective solutions. In this context, Electrical Discharge Machining (EDM) finds use for a variety of applications in automotive, aerospace, biomedical sectors on different materials such as stainless steels, titanium alloys; in fact, this manufacturing technology is insensitive to work strength. The material removal occurs thorough electrical discharges between the electrode tool and the workpiece in a dielectric fluid, thus dielectric, electrode and workpiece properties can affect the discharge behaviour.

EDM can be used to manufacture small features (with characteristic size of less than 1 mm), in this case the term "micro EDM" is often used. Compared to traditional

S. Hloch et al. (Eds.): ICMEM 2021, LNME, pp. 23–37, 2021.
https://doi.org/10.1007/978-3-030-71956-2_3

EDM, in Micro EDM sparks are shorter and less energetic; discharge frequency is higher and consequently pulse on times are shorter. Because of these features, micro EDM equipment differs from macro EDM machines, especially regarding the generator architecture. Micro EDM can be exploited to perform different tasks, namely micro die sink and wire EDM (acting like their macro-scale counterparts), micro ED drilling, micro ED milling, micro ED grinding and micro wire ED grinding (micro WEDG).

The effectiveness of EDM has often been evaluated through a number of indicators, such as Material Removal rate MRR, Tool Wear TW (or Tool Wear Ratio TWR, ratio between the amounts of material removed from the tool and from the workpiece) and quality parameters. Part quality can be assessed through several measures of overcut OC that is an indicator of overall spark length; other indexes of geometrical accuracy can be used, as an example, in drilling through holes, taper as the difference of diameters at the entrance and at the exit side. Roughness measures had been considered as well [1–3], although such measurement may be difficult on inner surfaces, as for the case of micro holes.

Recently, sustainability issues are becoming more and more important. Unlike part quality, process productivity and cost, for which well established techniques are available, sustainability evaluation is still matter of research. The concept of 'sustainability' itself is still open to different interpretations. When dealing with a manufacturing process, sustainability can be related to resource consumption (materials and energy), waste production and impact on health and safety [4]. Basing on the above mentioned aspects, sustainability can be assessed by means of indicators [5], otherwise a more complex method of life cycle assessment has been proposed by international standards [6].

Sustainable production through electro discharge machining (EDM) has been studied by some authors [7–9], as an example. Although a relatively small number of parts are machined via EDM, when compared to conventional operations, EDM specific energy consumption is one of the highest among machining processes [10]. Similar remarks can be made for micro EDM too, by which small but increasing amounts are produced and the power consumption per unit volume is higher [11, 12].

The energy used for the dielectric management (circulation and filtering) is in general a significant part of the overall power consumption [12], thus affecting both sustainability and process cost. It is worth noting that energy consumption in EDM is the most important source of both cost and sustainability issues. Most energy consumption is independent from electrical process parameters [13] thus, the machining time affects both cost and sustainability; thus, MRR considerations play a major role.

Dielectric management is the main source of health and safety issues as well. During the spark on time, vapours and fumes may arise. Also, they in general require a treatment (filtering, deionizing) and, finally, they become exhausted. A quantitative assessment of health impact requires more information on the chemical composition of pollutants that is heavily dependent on dielectric type [14, 15].

Although studies on the critical aspects of the process related to the environmental impact of the EDM process are present in the literature, a method for the evaluation of its sustainability is missing. In this paper, a method to evaluate the sustainability of this process is proposed taking into account the consumption of the electrical energy, the wear of the tool electrode during the machining, the consumption of the dielectric fluid, the environmental impact and part conformance. This method was applied for the drilling of a component in biomedical sector. The experimental campaign was executed varying both electrode material and dielectric fluid.

2 Evaluation of the Micro-EDM Process Sustainability

In the micro-EDM process, the input parameters are the workpiece material, the characteristics of the electrode (material, type and diameter), the dielectric fluid and then the selected electrical process parameters. The output is the workpiece and the amount of undesirable wastes and emissions (Fig. 1).

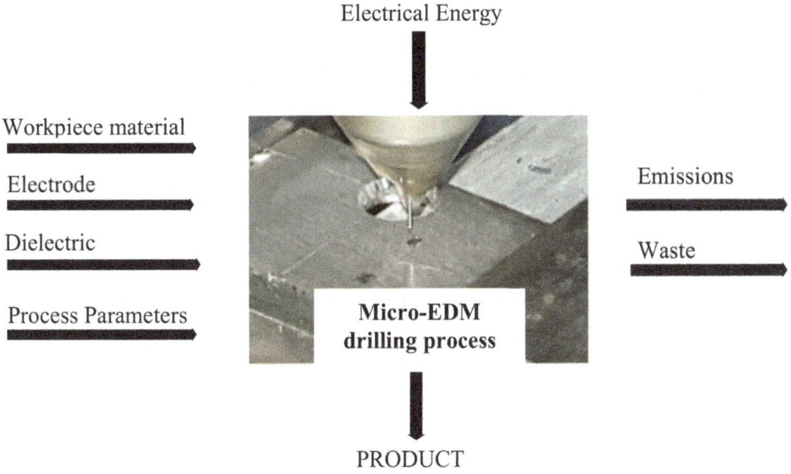

Fig. 1. Inputs and outputs of the micro-EDM drilling process.

From the point of view of the sustainability, the main aspects related the EDM machining are the consumption of the electrical energy, the wear of the tool electrode during the machining, the consumption of the dielectric fluid, the environmental impact and part conformance (see Fig. 2). In order to evaluate the level of the sustainability of the machining, it was chosen to link the economic value of the used resource with the environmental impact [16]. Under this assumption, increasing the consumed resources by the process, its sustainability gets worse. In other words, a process is evaluated as sustainable if it consumes few resources.

Sustainability Index is composed by five sub-indexes related to:

- *energy consumption*: the impact of the energy consumption is easily evaluated by measuring the energy adsorbed during the machining and fixing the unitary cost of electricity;
- *electrode consumption*: the measurement of the consumed tool is made through a touching operation at the end of the machining in a reference point on the sheet. Then the wear can be estimated by the difference between the start and final length of the

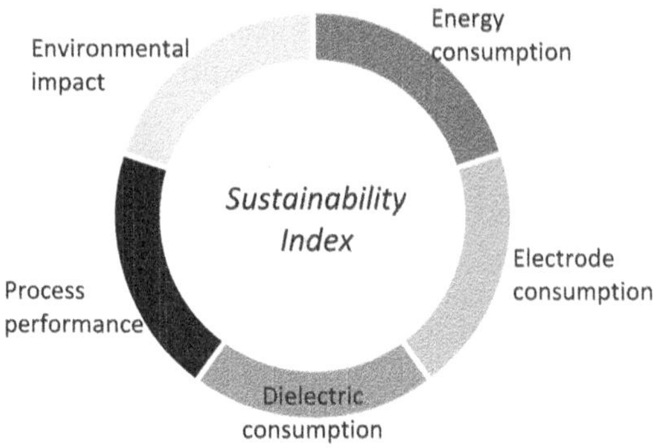

Fig. 2. Critical sustainability aspects of the micro-EDM drilling process.

tool. Noted the unit cost per length of tool, the economic value of the electrode used is evaluated;

- *consumption of the dielectric*: the purchase unitary costs of the dielectric liquid and the filters have to be taken into account. Considering that at the end of theirs life, they have to be disposed, the disposal costs were considered. Under the hypothesis of a certain lifetime of the dielectric and of the filters and assuming to know the unitary cost of the dielectric, the unitary cost of the filter and the unitary dielectric dismantling cost, the consume of the dielectric is proportional to the machining time;
- *environmental impact*: it is mainly due to the dielectric. In fact during the machining fumes, vapors and dust are generated that can cause respiratory problems to the operator. For some dielectrics, if they come into contact with the human skin, they could cause dermatitis. Moreover it useful to take into account the possibility to either recycle or dispose the dielectric. It is extremely difficult to realize an objective measurement of the entity of these elements. For this reason, they were estimated based on a subjective evaluation as suggested [10, 17]. The environmental impact of the machining is proportional to the machining duration;
- *part conformance*: it is related to the accuracy of the machining. It was evaluated calculating the variability in term of the diameter of the hole (standard deviation). This value can be associated to the predictability of the process so affecting scrap production. The sustainability is inversely proportional to the variability of the diameter.
  The probability to machine a non-conforming item is calculated under the hypothesis of normal distribution of the diameter. The standard deviation of the diameter was estimated using residuals from a linear regression model on the experimental data. Once a tolerance of the diameter, the scrap probability can be evaluated and then knowing the economical value of the used resources, this penalty can be computed.

In this way, an economical value was assigned to each element that should be taken into account for the evaluation of the sustainability of the micro-EDM process. The index proposed has the advantage to be simple and easily implementing to real application cases as recommended by [16]. An alternative of the developed index could be made through the estimation of the produced CO_2 of all the aspects taken into account. Anyway, the results in term of the rating of the sustainability at different experimental conditions should not change significantly and moreover the approach appears more difficult to be implemented.

3 Application of the Proposed Index for a Case Study

The biomedical sector is an important user of the micromachined parts, in particular to produce components to be implanted on the human body. The selected case study consists of the bone plates for fractures of small body parts, such as hands and feet and fractures of the skull. This component is realized in titanium alloy Ti-6Al-4V, having a depth of 0.5 mm and a dimension of 60 × 30 mm. The number of micro holes on the plates varies as a function of the area of the body that they have to be installed. The selected case study has 4 micro holes as reported in Fig. 3.

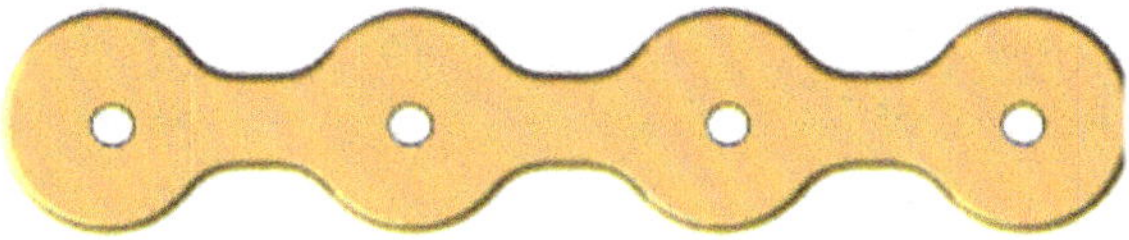

Fig. 3. Plate on titanium for biomedical application.

The micro holes were executed using a Sarix SX-200 machine available at the microtechnologies laboratory of University of Bergamo. The scheme of the machining system is shown in Fig. 4.

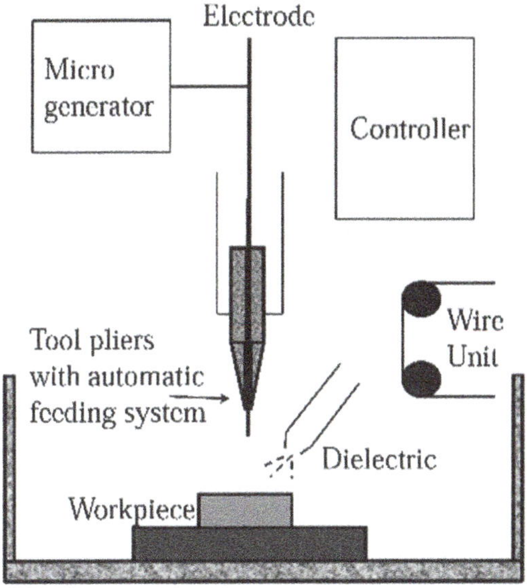

Fig. 4. Scheme of the Sarix SX-200 machining system [18].

Table 1. Variable process parameters.

	Brass		*WC*	
	Current [-]	Voltage [-]	Current [-]	Voltage [-]
H_2O	55	85	55	85
	49	100	49	100
	70	79	70	79
	55	115	55	115
	70	100	70	100
	85	85	85	85
	70	121	70	121
	91	100	91	100
	85	115	85	115
Ke	55	85	55	85
	49	100	49	100
	70	79	70	79
	55	115	55	115
	70	100	70	100
	85	85	85	85
	70	121	70	121
	91	100	91	100
	85	115	85	115
Soya Oil	55	85	80	100
	49	100	78	110
	70	79	90	98
	55	115	80	120
	70	100	90	110
	85	85	100	100
	70	121	90	122
	91	100	102	110
	85	115	100	120

Three different dielectric fluids were adopted, both conventional and unconventional: demineralized water, kerosene (HEDMA 111) and vegetable oil (soya oil). Two types of electrode material were used, brass and tungsten carbide, both tubular having an external diameter of 0.3mm and the internal one 0.12 mm. At the end of each hole, the tool wear was measured and then, the electrode was cut using the wire EDM unit in order to assure the same starting shape of the electrode tip for each hole. The absorbed energy by the machine during the machining was measured through a watt meter with an acquisition rate of 0.2 Hz. A Zeiss O-Inspect 543 CMS (Coordinate Measuring Machine) measured the diameter of the holes. Some electrical process parameters were varied such as the current and the voltage. As regards the process parameters, only peak current and voltage were varied and the tests were planed using a central composite design (CCD) 2^k with central point; the central point was replicated 10 times while the others points twice. In total, 156 micro holes were executed. The polarity was negative, pulse on time 3 μs, frequency 150 kHz and a roughness energy shape was used. The variable process parameters are reported in Table 1.

3.1 Machining Time and Tool Wear

The main variables used in the proposed index are the machining time, the tool wear and the scrapping probability. In this section, these data obtained during the experimental campaign are presented and discussed. As criteria to characterize the electrical process parameters, the product between peak current and voltage was chosen. In general, this product evaluates the power of the discharge. In this case, it is only representative because the value of current is not expressed in Ampere but in conventional unit defined by the machine tool manufacturer. The graphs report the mean value of the repetition. The values are related to the machining of 4 micro holes of the case study. In Fig. 5 and 6, machining time as a function of the electrical process parameters varying the dielectric fluid for brass and tungsten carbide electrode respectively was shown. The product of the electrical process parameters affects the machining time only in some conditions; otherwise its effect is almost negligible. In particular, the performance using vegetable oil needs an accurate selection of the process parameters. This consideration holds for kerosene using tungsten carbide electrode too. Comparing electrode materials, brass obtains a lower machining time than tungsten carbide electrode. This behaviour can be explained taking into account the electrical resistivity of the electrode material [1]; as this property decreases, the machining occurs in faster way because each spark delivers to the target a higher amount of its energy.

As regards dielectric, machining time using water is always lower than the oil based dielectrics. The effect of the dielectric fluid on the performance may be related to both physical and chemical phenomena: the viscosity, the electrical conductivity and the dielectric strength of the dielectric fluid seem to be the most important properties. It can be noted that these properties act in different way and very often, their increment has both positive and negative aspects. Also in literature, the overall effect of viscosity is still debated. Another important aspect is the presence of reaction products: water based dielectric fluids generally lead to oxidised products, whereas organic fluids may cause carbon deposition [19]. Oxides and carbides have different physical properties especially their melting point. This can justify the faster removal rates when using water compared with the others fluids.

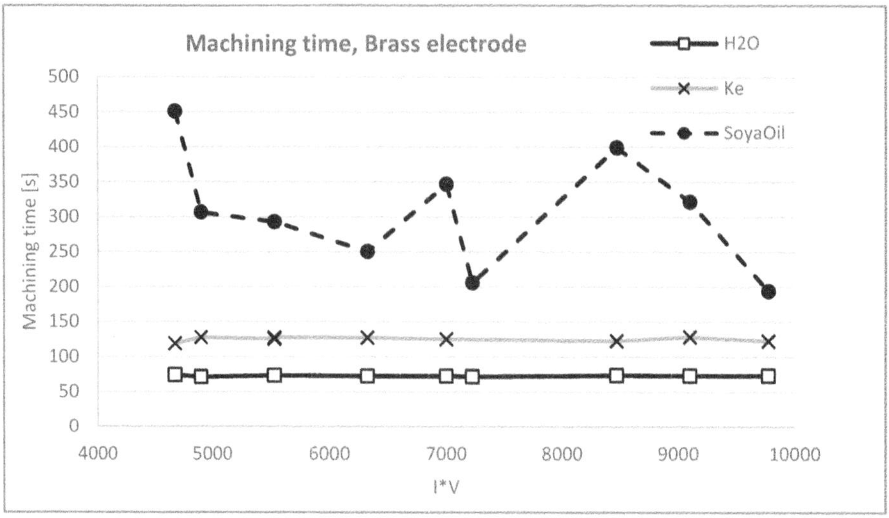

Fig. 5. Machining time with brass electrode as a function of the electrical parameters varying the dielectric.

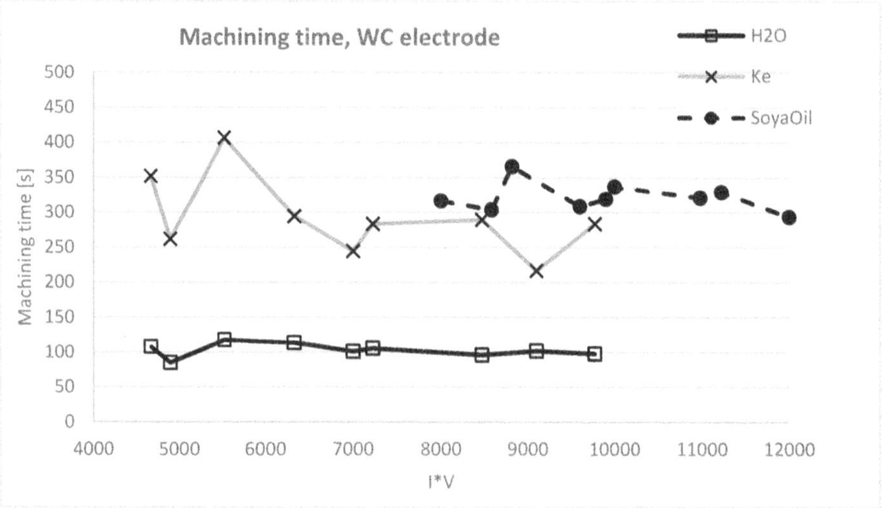

Fig. 6. Machining time with tungsten carbide electrode as a function of the electrical parameters varying the dielectric.

Figure 7 and 8 show the wear of the brass and tungsten carbide electrode as a function of the process parameters with the tree types of dielectric. It is useful to remember that these data represent the wear due to the machining of 4 micro holes on the titanium plate.

Except for the case of soya oil with brass electrode, also for the electrode wear the effect of the process parameters is almost negligible. Water is the dielectric that causes the lowest wear in all the tested conditions. As regards the effect of the electrode material, using brass the electrode wear is higher than tungsten carbide one. In literature the erosion resistant index is proposed to evaluate how well a material can be removed by the EDM [20]. The properties of the two electrode materials are significantly different and the erosion resistant index assumes different values for the two electrodes. This index is calculated as:

$$ER = c \cdot k \cdot T^2 \qquad (1)$$

where c is the specific heat, k is the thermal conductivity and T is the melting point of the material. For low value the electrode material is subjected to a higher wear while for high value the erosion occurs in more difficult way.

In general, water as dielectric grants both a good removal rate a low tool wear.

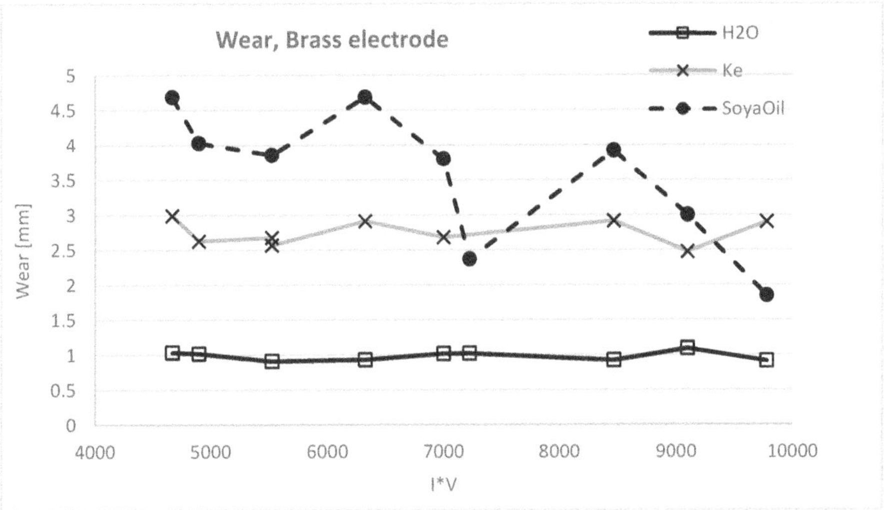

Fig. 7. Electrode wear of brass electrode as a function of the electrical parameters varying the dielectric.

Another aspect that should be taken into account is the geometrical characteristics of the micro holes. Under the hypothesis that the quality requirement regards only the top diameter, a tolerance of 0.05 mm was fixed. The probability of producing a non-conforming hole was linked to the standard deviation alone since the diameter mean value can be controlled through other techniques (i.e. by adjusting the electrode size). The probability of producing a hole out of tolerance is higher for brass electrode and depends on the dielectric fluid as reported in Table 2. Brass permits to obtain faster machining to the detriment of high probability to obtain a hole out of tolerance. The accuracy of the holes improves when tungsten carbide electrode is adopted especially with kerosene as dielectric.

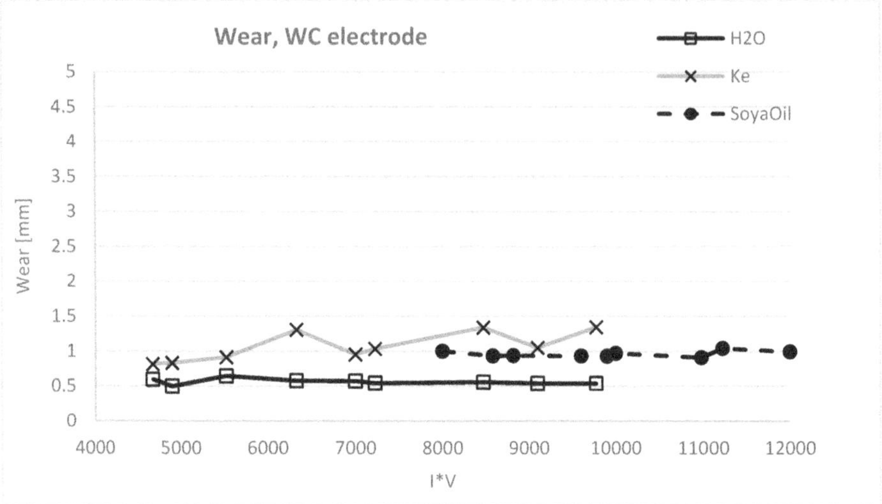

Fig. 8. Electrode wear of tungsten carbide electrode as a function of the electrical parameters varying the dielectric.

Table 2. Probability of producing a defective hole.

	Brass	WC
Water	10.5%	0.11%
Kerosene	10.7%	0.05%
Soya Oil	7.58%	0.35%

3.2 Sustainability Index

The coefficients used to calculate the sustainability index are reported in Table 3.

The Sustainability Index obtained for the tested conditions using brass and tungsten carbide electrode are reported in Figs. 9 and 10 respectively. For both electrode materials, water is the dielectric most sustainable, followed by the oil based dielectrics. Also for the sustainability index, in most cases, the effect of the process parameters is almost negligible. With brass electrode, soya oil shows the worst index but for some condition in terms of current and voltage, it reaches value similar than kerosene. With tungsten carbide electrode, the vegetable oil is comparable to kerosene under the point of view of the sustainability. Considering that the use of the vegetable oils in the EDM process is still under investigation, some degree of improvement of its performance can be reasonable expected. For water and kerosene, a light difference of the index varying the electrode material was found.

It useful to analyze how the sustainability index is composed in order to highlight the critical aspects for each condition in terms of dielectric and electrode adopted. Figure 11

Table 3. Coefficients to calculate Sustainability Index.

Cost per unit of electricity [€/kWh]	0.156
Cost per volumetric unit of tool [€/mm^3]	Brass: 0.024362
	WC: 0.1054
Price per litre of the dielectric [€/l]	Kerosene: 9.63
	H_2O: 0.25
	Vegetable oil: 1.4
Dielectric tank volume [l]	25
Lifetime of the dielectric [h]	1000
Cost of the filter [€]	117
Lifetime of the filter [h]	1000
Unitary dielectric dismantling cost [€/l]	Kerosene: 0.215
Cost per hour of the micro-EDM machine [€/h]	40

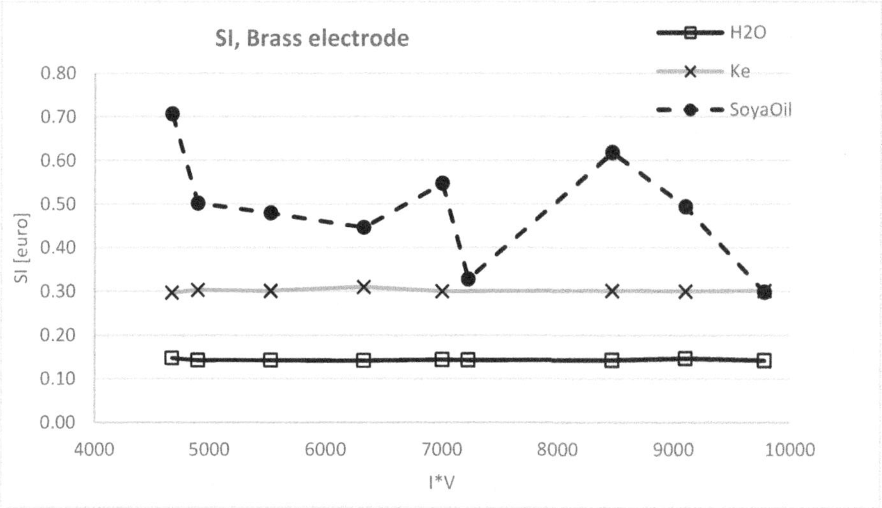

Fig. 9. Sustainability Index as a function of the electrical parameters using brass electrode varying the dielectric.

shows the average percentage composition of the sustainability index varying the electrode material and the dielectric fluid. In all the conditions, the contribution of terms related to the energy and dielectric consumption is low. Anyway, kerosene is the dielectric that has the highest impact as regards the term related to the dielectric. The sustainability related to the part conformance (performance) affects deeply the global index only for the cases where brass electrode is used. Using tungsten carbide electrode the consumption of the electrode and the environmental aspects represent the most important

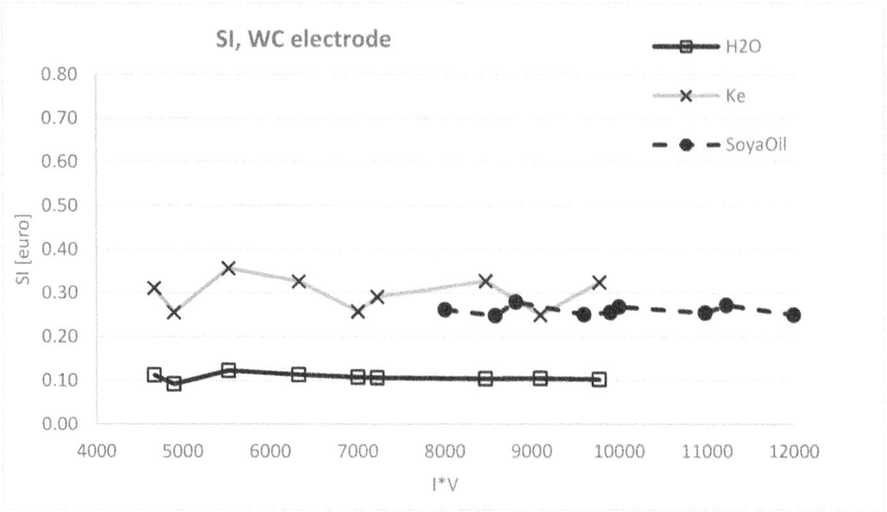

Fig. 10. Sustainability Index as a function of the electrical parameters using tungsten carbide electrode varying the dielectric.

contributions in the formation of the index. In all the conditions, the terms related to the environmental impact and the electrode wear have the same weight on the sustainability index. This graph can help to find solutions to decrease the more important causes affected the sustainability of the machining for each experimental conditions.

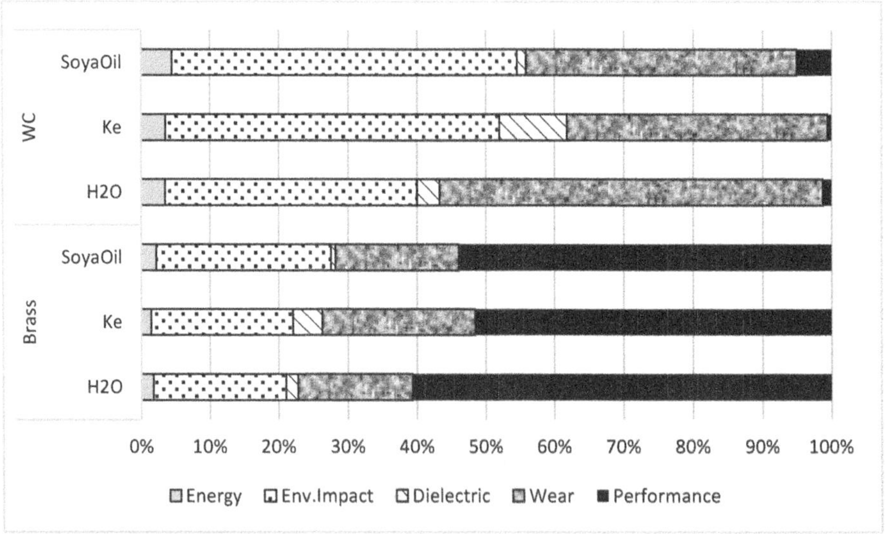

Fig. 11. Average percentage composition of the sustainability index varying electrode material and dielectric fluid.

Finally, the minimum values of the sustainability index for each experimental condition tested, subdivided in the five contributions, are reported in Fig. 12. The solution using water as dielectric and tungsten carbide as electrode material is the best solution to minimize the environmental impact due to the machining. In general, brass electrode penalizes deeply the performance in term of accuracy of the hole while tungsten carbide electrode penalizes the terms related to both the consumption of the electrode and the environmental aspects overall using oil-based dielectrics.

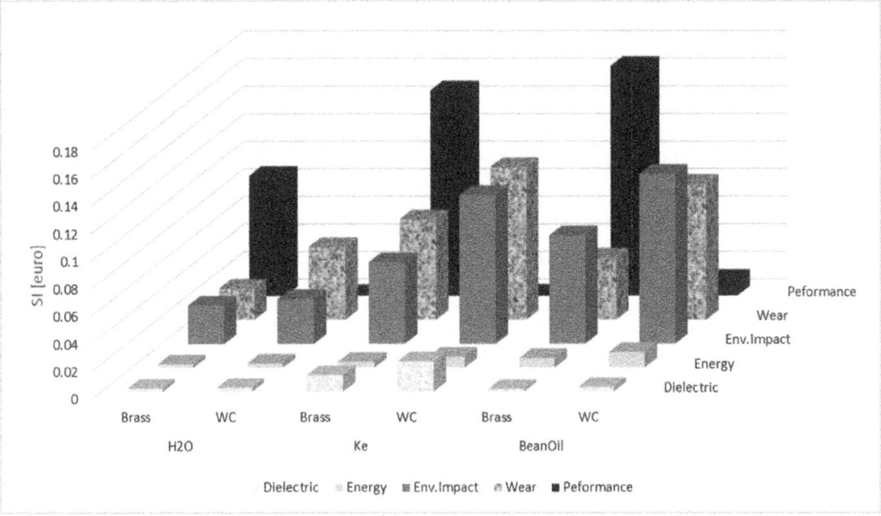

Fig. 12. The minimum index values varying the electrode and the dielectric fluid.

The regression equation of the sustainability index as a function of the most important variables (machining time "t", electrode wear "Wear" and the probability to obtain a defective hole "P") is here reported:

$$\text{Stot}\,[\text{euro}] = -0.0262 + 0.000919\,\text{t}\,[\text{s}] + 0.0548\,\text{Wear}\,[\text{mm}] + 0.517\,\text{P}$$

All the three variables are significant (p-values null) and the correlation index is higher than 97%.

4 Conclusion

Aim of this paper was to give a method to evaluate the sustainability of the micro-EDM drilling process. This method is based on the assumption that there is a correlation between the environmental impact of the process and the quantity of the consumed resources and the pollution effects. In fact, considering that to make the method easily implementable even in industrial cases, this must be very simple and intuitive.

Under this assumption, the sustainability of the process is evaluated taking into account the energy consumption, the dielectric consumption, the wear of the electrode, the environmental impact and the part conformance.

The proposed method was applied to a case study, the drilling of a component in titanium alloy for the biomedical sector. In order to find the best operative conditions to maximize the sustainability of the process, micro holes were executed in different conditions as regards both electrode material (brass and tungsten carbide) and dielectric fluid (kerosene, water and vegetable oil).

In the case study the use of water as dielectric and tungsten carbide as electrode material is the best solution to minimize the environmental impact due to the machining obtaining a sustainability index approximately equal to 0.1 €. Kerosene and vegetable oils associated with tungsten carbide electrode have a greater consuming of resources than water as dielectric and the sustainability index is around the triple. Using brass, the sustainability is overall worst when vegetable oil is used while using water and kerosene the sustainability is lightly worse than that obtained using tungsten carbide. From the analysis of the components of the sustainability index, a general trend for the use of the two electrode materials was found. Brass electrode penalizes deeply the performance in term of accuracy of the hole, more than 50% of the total index. Tungsten carbide electrode penalizes the terms related to both the consumption of the electrode and the environmental aspects overall using oil-based dielectrics. The sum of the contributions of these terms is higher than 80% of the sustainability index.

Moreover, the critical aspects for each experimented test are highlighted as follows. In general, the effects of the energy and dielectric consumption are lower than 10% but in most cases are lower than 5%. Among all cases, kerosene is the dielectric that has the highest impact, around 5–10 times higher respect to the other ones.

The proposed method permitted to deeply analyze the process form the point of view of its sustainability. Both the critical aspects and the strengths of different electrodes and dielectrics have been pointed out; such knowledge may be exploited for either optimizing process conditions or promoting actions that minimize the environmental impact. A similar approach could be implemented to study sustainability issues for others machining operations.

References

1. Bhaumik, M., Maity, K.: Effect of electrode materials on different EDM aspects of titanium alloy. Silicon **11**, 187–196 (2019)
2. Rahul, D., Mishra, S., Masanta, M.: Effects of tool electrode on EDM performance of Ti-6Al-4V. Silicon **10**(5), 2263–2277 (2018)
3. Chakraborty, S., Dey, V., Ghosh, S.K.: A review on the use of dielectric fluids and their effects in electrical discharge machining characteristics. Precis. Eng. **40**, 1–6 (2015)
4. Zhang, Y., Liu, Y., Shen, Y., Ji, R., Li, Z., Zheng, C.: Investigation on the influence of the dielectrics on the material removal characteristics of EDM. J. Mater. Process. Technol. **214**, 1052–1061 (2014)
5. Singh, R.K., Murty, H.R., Gupta, S.K., Dikshit, A.K.: An overview of sustainability assessment methodologies. Ecol. Ind. **9**, 189–212 (2009)
6. Priarone, P.: Quality-conscious optimization of energy consumption in a grinding process applying sustainability indicators. Int. J. Adv. Manuf. Technol. **86**, 2107–2117 (2016)
7. Tan, X.C., Liu, F., Cao, H.J., Zhang, H.: A decision-making framework model of cutting fluid selection for green manufacturing and a case study. J. Mater. Process. Technol. **129**, 467–470 (2002)

8. Valaki, J.B., Rathod, P.P., Khatri, B.C.: Environmental impact, personnel health and operational safety aspects of electrical discharge machining: a review. J. Eng. Manuf. **229**(9), 1481–1491 (2015)
9. Valaki, J.B., Rathod, P.P., Sankhavara, C.D.: Investigations on technical feasibility of Jatropha curcas oil based bio dielectric fluid for sustainable electric discharge machining (EDM). J. Manuf. Process. **22**, 151–160 (2016)
10. Wang, X., Chen, L., Dan, B., Wang, F.: Evaluation of EDM process for green manufacturing. Int. J. Adv. Manuf. Technol. **94**, 633–641 (2018)
11. Yi, Q.: Micromanufacturing Engineering and Technology, 2nd edn. Elsevier (2015)
12. Gutowski, T., Dahmus, J., Thiriez, A.: Electrical energy requirements for manufacturing processes. In: 13th CIRP International Conference of Life Cycle Engineering, Leuven, Belgium (2006)
13. Marashi, H., Jafarlou, D.M., Sarhan, A.A.D., Hamdi, M.: State of the art in powder mixed dielectric for EDM applications. Precis. Eng. **46**, 11–33 (2016)
14. Niamat, M., Sarfraz, S., Aziz, H., Jahanzaib, M., Shehab, E., Ahmad, W., Hussain, S.: Effect of different dielectrics on material removal rate, electrode wear rate and microstructure in EDM. Procedia CIRP Des. **60**, 2–7 (2017)
15. Srinivas, V., Ramanujam, R., Rajyalakshmi, G.: A review of research scope on sustainable and eco-friendly electrical discharge machining (E-EDM). Mater. Today Proc. **5**, 12525–12533 (2018)
16. Dickinson, D.A., Caudill, R.J.: Sustainable product and material end-of-life management: an approach for evaluating alternatives. In: Proceedings of the IEEE International Symposium on Electronics and the Environment, Boston, pp. 153–158 (2003)
17. Schuch Bork, C.A., de Oliveira Gmoes, J., Venancio Pappetti Canhete, V., De Barba Jr., D.J.: Methodological tools for assessing the sustainability index (SI) of industrial production processes. Int. J. Adv. Manuf. Technol. **87**, 1313–1325 (2016)
18. D'Urso, G., Maccarini, G., Ravasio, C.: Process performance of micro-EDM drilling of stainless steel. Int. J. Adv. Manuf. Technol. **72**, 1287–1298 (2014)
19. Kumar, A., Chattopadhyay, S., Prakash, V.: Micro-EDM performance using different dielectrics. In: Kibria, G., Jahan, M.P., Bhattacharyya, B. (eds.) Micro-Electrical Discharge Machining Processes, pp. 125–136. Springer (2019)
20. Reynaerts, D., Van Brussel, H.: Microstructuring of silicon by electro-discharge machining (EDM) – part I: theory. Sens. Actuators A **60**(1), 212–218 (1997)

Water Jet Session

Drilling Difficult-to-Machine Al_2O_3 Ceramics Using an Injection Abrasive Water Jet

Jiri Klich[1(✉)], Dagmar Klichova[1], Guoyi Peng[2], and Libor Sitek[1]

[1] Institute of Geonics, Czech Academy of Sciences, Ostrava-Poruba, Czech Republic
jiri.klich@ugn.cas.cz

[2] College of Engineering, Nihon University, Koriyama, Fukushima, Japan

Abstract. The article studies the process of drilling with an Injection Abrasive Water Jet (IAWJ). Two different aluminium-based materials with various properties were chosen as test specimens for drilling: (i) very hard and abrasion-resistant alumina ceramics and (ii) soft and easy-to-machine aluminium alloy. The development of drilled holes was observed as a function of the exposure time of the jet acting on a drilled material surface. The exposure time was varying from 0.1 s to 102.4 s. The inlet diameter of the drilled hole, hole depth, volume of material removed from the hole and hole shape were continuously observed. It has been found that the drilled holes differ significantly depending on the type of material. While the soft aluminium alloy can be drilled easily and the hole was drilled through the material in a short time (20 mm thick sample in 24 s), the drilling process was significantly slowed down due to the strength and abrasion resistance of ceramics (12.7 mm deep hole in 102.4 s). The drilled hole shape is flatter in ceramics than in aluminium alloys.

Keywords: Drilling · Injection abrasive water jet · Al_2O_3 ceramics

1 Introduction

Drilling of small holes represents a relatively complicated operation of machining workpieces made of hard and abrasion resistant materials, such as hardened steels, titanium alloys, metal matrix composites or ceramics. The drilling process is often time-consuming, regarding the early wear of a drilling tool and consequent need to change it, as well as the requirements on the expected quality of a drilled hole. The Injection Abrasive Water Jet (IAWJ) [1] technology is one of few production technologies that can rather easily drill these materials. This technology has unique properties, such as the absence of a heat-affected zone near the cut (maximum temperature in the place of cut is about 70 °C), low forces on the workpiece (units up to tens of N), zero wear of the cutting tool and its negligible vibration [2].

S. Hloch et al. (Eds.): ICMEM 2021, LNME, pp. 41–49, 2021.
https://doi.org/10.1007/978-3-030-71956-2_4

Due to some ambiguity in terminology, it should be emphasized that the process described in the paper as "drilling with IAWJ" means a technological operation where the cutting head remains in a fixed position while the abrasive water jet works its way through the material. Opposed to the drill bit mechanism, the IAWJ drilling mechanism has its proper specifics. First, the impacting abrasive water jet removes an appropriate volume of material. Then, the reflected water flows out around the continuous incoming water jet, thus widening the inlet opening of the drilled hole compared to its bottom (see Fig. 1). In depths of several milimeters, the drilling process usually slows down due to the interference of the incoming and outgoing water flows. Backflow of the water jet gradually widens the hole, thus making space for the incoming jet.

One of the reasons for the realization of the presented preliminary study is the research of technological differences of drilling using IAWJ with respect to drilling with mechanical tools, another is to find out mainly dimensional and qualitative properties of hole drilled by IAWJ. This can be useful for the so-called predrilling of the cut material during the start of cutting with IAWJ technology.

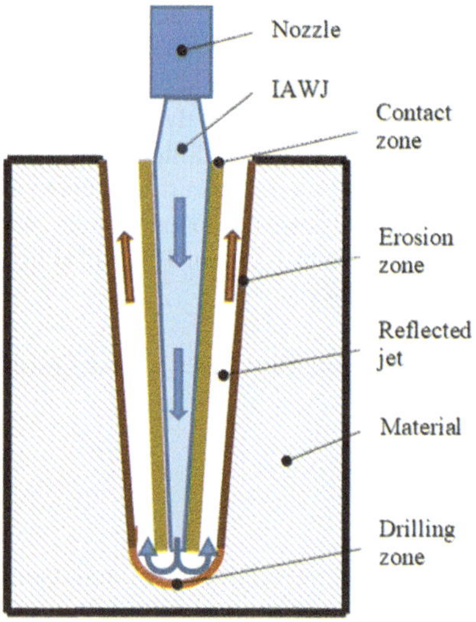

Fig. 1. IAWJ drilling scheme.

Several studies were published on drilling or piercing of materials using the abrasive water jets. Orbanic studied the change in geometric parameters of a hole over time on three types of metallic materials [3]. Kowsari investigated drilling with a micro-slurry jet on borosilicate glass [4]. Akkurt studied the effect of metal material type and plate thickness on the drilling time [5]. Fredin investigated various piercing methods (stationary, linear, repeated linear, circular) depending on the time [6]. Ohlsson made some important optimisations of the piercing or drilling mechanism [7]. Although Changshui et al. drilled thin TBC ceramic layers with an abrasive water jet [8], none of published articles focused on drilling deep holes (over 10 mm) of a small diameter (up to about 2 mm) into ceramics.

Typical representative of hard ceramics is the aluminium oxide Al_2O_3. Alumina ceramics show a very good hardness and abrasion resistance. This material is very good chemically and high-temperature resistant in aggressive environments. In industry, it is widely used in many applications: production of refractory "boats" used in metallurgy, sealing elements, ceramic insulators, abrading tools, container tiles, conveyors and other components with special requirements on heat and wear resistance. Alumina ceramics were thus selected as a test material for experiments, aiming at the investigation of possibilities of drilling ceramics by means of the IAWJ, specification of the ceramics disintegration mechanism during drilling and comparison of the obtained results with the results of aluminium alloy disintegrated under similar conditions.

2 Materials and Methods

Alumina ceramics with 92% of Al_2O_3 (see above) and the EN AW 6061 T651 aluminium alloy (a high-strength steel for highly-loaded structural applications, such as the production of scaffolding elements, rail coach parts, containers, machine components, aerospace parts, etc.) were selected as test materials for drilling by IAWJ. The test specimens had the form of solid blocks with a thickness of 20 mm. Chemical composition and physical properties of the materials used are presented in Table 1. Operating parameters of the IAWJ used for drilling are specified in Table 2.

Table 1. Chemical and physical properties of sample materials.

Material	Density [g/cm^3]	Hardness	Modulus of elasticity [GPa]	Tensile strength *Rm* [MPa]	Melting range [°C]
Al_2O_3 (92%)	3.8-4	HV1.0 900-1200	340-400	400	2000-2090
EN AW 6061 T651	2.7	HB 85	70	260	585-640

Chemical composition of EN AW 6061 T651						
Si	Fe	Cu	Mn	Cr	Zn	Ti
0.4-0.8	Max. 0.7	0.15-0.4	Max 0.8-1.2	0.04-0.35	Max 0.25	Max 0.15

Table 2. Technological parameters of IAWJ.

Water pressure	300 MPa	Standoff distance	4 mm
Abrasive mass flow rate	250 g/min	Water nozzle diameter	0.33 mm
Abrasive	Australian garnet GMA 80 MESH	Focusing tube diameter	0.9 mm

The jet was directed in the free space for a few seconds in order to stabilize the high-speed abrasive water jet before drilling and to fill particular parts of the abrasive cutting head with abrasive particles. After this procedure, the drilling process of holes could started. Several holes were gradually drilled in each of the tested materials under the same conditions (Fig. 2). The only changing parameter was the exposure time of the jet, varying from 0.1 s to 102.4 s. Each time periods of the jet action was twice as long as the previous time period. This method was chosen on the basis of our own experience with the abrasive water jet drilling in various materials. Subsequently, the properties of individual holes were determined using a VHX5000 Keyence digital microscope – i.e. diameter of a hole opening *D*, hole depth *H*, volume of material removed *V* and shape of a drilled hole. During the experiments, 3 holes were always created under the same condition. By successively measuring all parameters (*D*, *H* and *V*) in each of these holes, an average value was obtained for the holes created under the same condition. These parameters are then plotted in graphs (Figs 3, 4 and 5). The setting of parameters was strictly observed (water pressure, smoothness of abra-sive supply, etc.) during the experiments, so there was no significant variability in the results obtained.

Fig. 2. One series of holes in ceramics drilled by IAWJ.

3 Results

Results of the experiments are presented in graphical form (Figs. 3, 4, 5, 6 and 7). The diameter of the hole opening *D* in the alumina ceramics increased from 1.24 mm (after 0.1 s of drilling) to 2.02 mm (after 102.4 s of drilling) – Fig. 3. Accordingly, the hole depth *H* increased from 0.41 mm to 12.73 mm, keeping the same time intervals (Fig. 4). The volume of material removed from holes was $V = 0.28\ \text{mm}^3$ to $15.08\ \text{mm}^3$ (see Fig. 5).

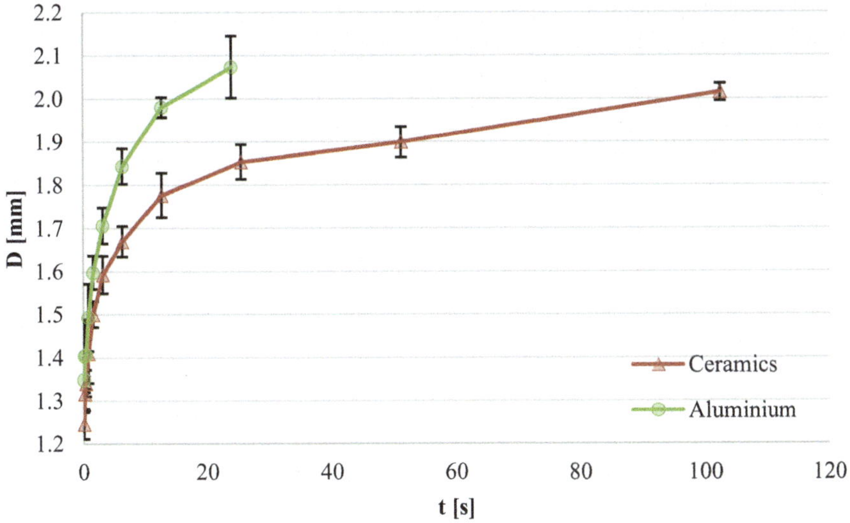

Fig. 3. Dependence of the diameter *D* [mm] of the hole opening on the drilling time *t* [s].

The parameters of a drilled hole in the alumina alloy were changing similarly. However, due to the properties of the aluminium alloy, while keeping the same time intervals during drilling, the test specimen was drilled through earlier than the specimen of alumina ceramics (i.e. in approximately 24 s). Therefore, the investigated parameters for this material were related to the maximum drilling time of 24 s, when the material was

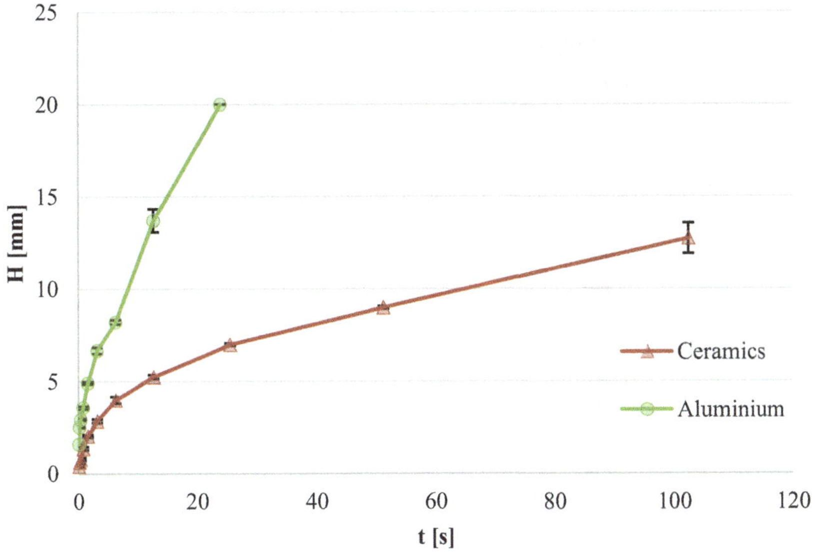

Fig. 4. Dependence of the hole depth *H* [mm] on the drilling time *t* [s].

drilled through. The parameter D thus varied from 1.35 mm to 2.07 mm, H from 1.59 mm to 20 mm and V from 1.45 mm^3 to 36.73 mm^3 (see Figs. 3, 4 and 5).

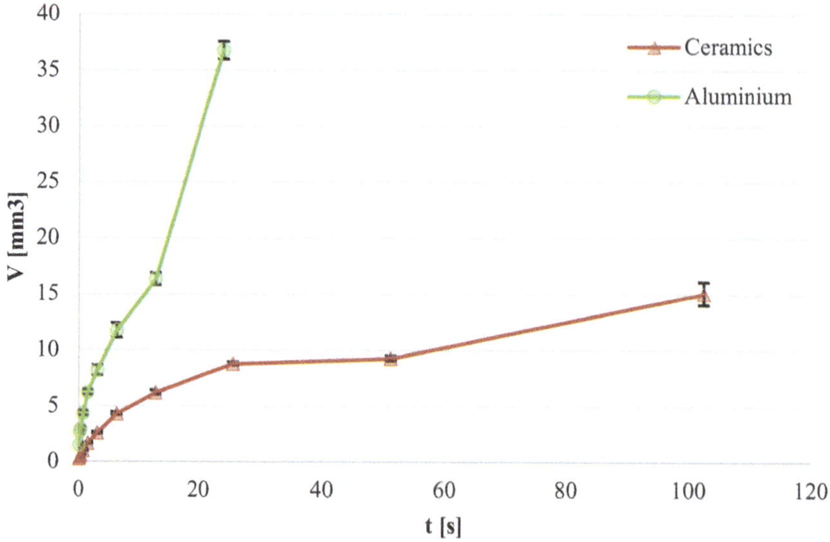

Fig. 5. Dependence of the volume V [mm^3] of material removed from holes on the drilling time t [s].

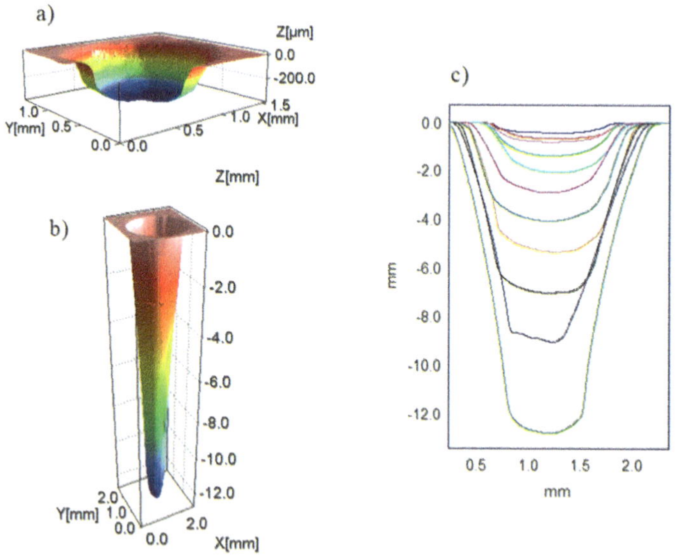

Fig. 6. Reconstruction of drilled holes created in alumina ceramics a) $t = 0.1$ s, b) $t = 102.4$ s, c) 2D profiles of holes at different drilling times (from 0.1 s to 102.4 s).

Figures 6 and 7 show shapes of the resulting drilled holes in both tested materials at minimum drilling time of 0.1 s (a), at drilling time of 102.4 s or 6.66 s, respectively (b), and a comparison of profiles of drilled holes (c).

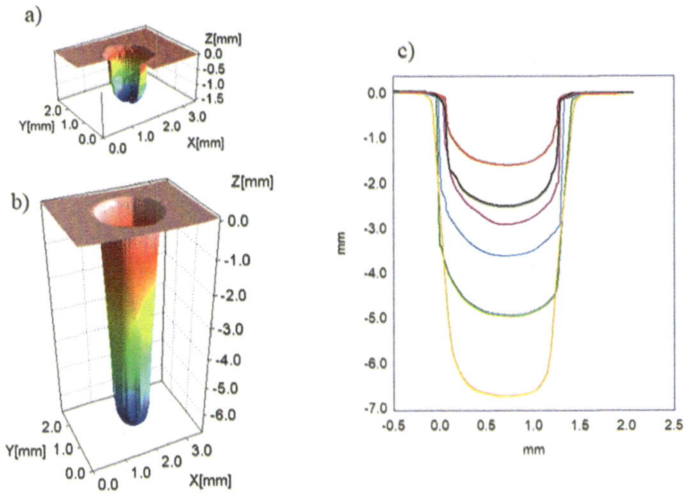

Fig. 7. Reconstruction of drilled holes created in aluminium alloy a) $t = 0.1$ s, b) $t = 3.2$ s, c) 2D profiles of holes at different drilling times (from 0.1 s to 6.66 s).

4 Discussion

The results obtained have proved the influence of a drilled material on the drilling performance. Character and shape of the drilled holes are similar in both tested materials and are given by the mechanism of drilling by means of IAWJ. The opening of a hole is always wider than the hole bottom (see Figs. 6 and 7) [11]. However, the profile of drilled holes in aluminium alloy is narrower than the profiles in hard ceramics (circa 0° to 5° from the hole axis for aluminium alloy and 5° to 10° for ceramics). This is due to a lower level of disintegration of hole walls in alumina ceramics by the reflected jet out-coming from the drilled hole. When drilling alumina ceramics, the hole depth increases very slowly and slows down with the increasing drilling time. Drilling is thus a very slow process, especially due to high levels of strength, hardness and resistance of the selected ceramic material against abrasive wear. The backflow of the reflected jet widens the hole only very slowly (see e.g. Fig. 5). The incoming jet lacks sufficient space for penetration and with the increasing depth, the water jet energy is getting lost in the fluid damping phenomena. The ability of the jet to remove more material from the bottom of the hole to be drilled thus further decreases [3].

However, when drilling aluminium alloy, the hole depth increases rather linearly with the drilling time and the drilling efficiency of the jet is not decreased by the backflow of the reflected jet to such an extent. The hole is drilled through after 24 s.

Due to the hole diameter varying in relation to the depth, deep holes drilled by a stationary IAWJ are more suitable for rough drilling or pre-drilling used for the consequent machining with mechanical tools [7].The drilling efficiency of the jet can be increased, and the drilling time reduced, by the application of the alternative drilling method, e.g. circular drilling [9]. The water jet makes small circular motion over the borehole, thus improving the backflow of the reflected jet and significantly saving the drilling time [7]. It can be reduced up to 10 times [9]. However, the stationary IAWJ technology is often very effective for drilling thin and layered materials [8, 10].

5 Conclusion

The technology of deep hole drilling in hard and abrasion-resistant ceramic materials using IAWJ is a relatively time-consuming process which is, in special cases, suitable for the creation of rough holes for a consequent technological operation. On the contrary, deep hole drilling in soft materials is rather easy and relatively fast. The results of the experiment show that:

- Geometric parameters of a drilled hole and the drilling speed are strongly influenced by the type of a material to be drilled. It is evident that the drilling in hard materials is slower (12.7 mm deep hole in the ceramic was drilled in 102.4 s) and the profile of the resulting hole is wider (the diameter of the hole at the entry of the jet into the ceramic is about 4 times larger than the diameter of the hole at the bottom).
- In contrast, drilling an aluminum alloy did not cause any problems for the jet, and the 20 mm thick test specimen was drilled quickly (in 24 s) compared to the ceramic.
- The drilling time could be reduced by the application of a circular drilling method which is based on a circular motion added to the cutting head with better backflow of the reflected jet consisting of water and remains of abrasive particles, as well as particles of the material to be drilled. According to [9], the drilling time can be reduced up to 10 times.

Further research will be primarily aimed at finding a balance between the drilling time, required quality of the resulting hole and economic justification of the drilling process.

Acknowledgements. This work was supported by the project "Programme for research and mobility support of starting researchers" MSM100862001. The authors are also very grateful for the support of the Nihon University.

References

1. Zhang, F.L.: Machining mechanism of abrasive water jet on ceramics. In: Zuo, D.W. (ed.) 3rd International Conference on Engineering and Technologies and Ceeusro, ICETC 2009, Changzhou, vol. 426–427, pp. 212–215 (2009)
2. Hashish, M.: A model for abrasive-waterjet (AWJ) machining. J. Eng. Mater. Technol. **111**, 154–162 (1989)

3. Orbanic, H., Junkar, M.: An experimental study of drilling small and deep blind holes with an abrasive water jet. Eng. Manuf. **218**, 1–6 (2004)
4. Kowsari, K., Nouraei, H., James, D.F., Spelt, J.K., Papini, M.: Abrasive slurry jet micro-machining of holes in brittle and ductile materials. J. Mater. Process. Technol. **214**(9), 1909–1920 (2014)
5. Akkurt, A.: The effect of material type and plate thickness on drilling time of abrasive water jet drilling process. Mater. Des. **30**, 810–815 (2008)
6. Fredin, J., Jonsson, A.: Experimentation on piercing with abrasive waterjet. Int. J. Mech. Aerosp. Ind. Mechatron. Manuf. Eng. **5**(11), 2400–2406 (2011)
7. Ohlsson, L., Powell, J., Ivarson, A., Magnusson, C.: Optimisatoin of the piercing or drilling mechanism of abrasive water jets. In: Proceedings of the 11th International Conference on Jet Cutting Technology, St Andrews, pp. 359–370 (1992)
8. Changshui, G., Zhuang, L., Kai, Z., Chao, G.: Abrasive water jet drilling of ceramic thermal barrier coatings. In: 19th CIRP Conference on Electro Physical and Chemical Machining, Bilbao, Spain, 23–27 April 2018 (2018). https://doi.org/10.1016/j.procir.2017.12.106
9. Klich, J., Klichova, D., Peng, G.: Submerged abrasive water jet piercing/drilling: preliminary tests. In: Advances in Water Jetting. International Conference on Water Jet 2019 - Research, Development, Applications, Čeladná, Czech Republic, 20–22 November 2019 (2020). https://doi.org/10.1007/978-3-030-53491-2_10
10. Hlaváček, P., Zlámal, T., Sitek, L.: Abrasive water jet drilling of cooling holes in aeroengines: preliminary experimental study. MM Sci. J. (1), 2218–2222 (2018). ISSN 1803-1269
11. Lenin, S., Rajadurai, A.: Experimental study on deep-hole making in Ti-6Al-4V by abrasive water jet machining. Mater. Res. Exp. **6**(6), 066532 (2019). https://doi.org/10.1088/2053-1591/ab0c35

Multiple Criteria Optimization of Abrasive Water Jet Cutting Using Entropy-VIKOR Approach

Andrzej Perec[1](✉) and Wojciech Musial[2,3]

[1] Faculty of Technology, Jacob of Paradies University, Chopina 52, 66-400 Gorzów Wielkopolski, Poland
aperec@ajp.edu.pl

[2] Faculty of Mechanical Engineering, Koszalin University of Technology, Raclawicka 15-17, 75-620 Koszalin, Poland

[3] Regional Research and Development Center in Walcz, Wojska Polskiego 99, 78-600 Walcz, Poland

Abstract. The paper shows multiple performance optimisation of Hardox steel machining effects by the Abrasive Water Jet (AWJ). An innovative Entropy/VIKOR approach is implemented for the optimisation of parameters in cutting like pressure, feed rate, and abrasive flow rate over cutting depth, cut surface roughness, and angle of cut kerf was conducted. For determine the best set of the process control parameters VIKOR approach with entropy-based weights determining was done. To establish the weights of the initial parameters, it was proposed to calculate individual weights based on entropy. For the multiple criteria optimisation VIKOR approach was assumed.

Keywords: High-pressure abrasive water jet · Cutting · Hardox steel · Multi-objective optimization · VIKOR

1 Introduction

Usual cutting methods for heavy to machining metals do not provide enough effectiveness and accuracy. Abrasive water jet [1, 2] and pulsating water jet [3, 4] treatment are devoid of such disadvantages. An additional advantage of the water jet technology is its environmental friendliness [5, 6].

To optimisation of AWJ cutting method a VIKOR approach was chosen. The VIKOR technique was formed by Opricovic [7] for complete multiple choice questions that imply discrepant criteria by considering the compromise as satisfactory. The solution is grounded on "maximum group utility" for the majority cand "mini-mum individual regret" in the opposite case. This technique uses the VIKOR factor grounded on its "proximity" to an "ideal" solution. Furthermore, this manner provides both a real solution, close to the ideal and farthermost from the weakest solution.

S. Hloch et al. (Eds.): ICMEM 2021, LNME, pp. 50–62, 2021.
https://doi.org/10.1007/978-3-030-71956-2_5

The VIKOR approach can be merged with the DoE method, which is often used in research on AWJ. For example, Tavodova et al. [8] presented the using of DoE with a Taguchi table, during estimation of aluminum alloy area properties after AWJ machining. Grounded on the effects of the experiments the top kerf quality cutting parameters were recognized.

Also, Michalska-Pozoga et al. [9] published the research on selected mechanical and physical properties of the composites using the DoE/Taguchi method to find optimal extrusion process parameters and combinations of components, considering the best feasible properties of the extrudate. Other DoE method – response surface method in modeling of Abrasive Water Suspension Jet cutting efficiency of limestone was presented by Perec [10] and aluminum alloy published by Radomska-Zalas et al. [11].

The VIKOR approach was used for multiple criteria optimisation in bore machining of A356 - TiB2/TiC material, fabricated by stir casting route [12]. Experiments were carried out under Taguchi L27 orthogonal array and all output parameters was transformed to an individual VIKOR factor to establish its impact for drilling. In result, verification experiment was executed and checked the regression equation to make the efficacy of this method.

Gangil et al. [13] published the research of EDM parameters optimisation for maximum efficiency and minimal surface reliability using joined RSM - VIKOR technique. The research was carried out on Titanium using Central Composite Design. The optimum circumstance of the tests was found by VIKOR technique. The optimisation result was confirmed by the test. The research effects confirm higher MRR and better machining accuracy under optimal conditions.

Rajesh Kumar Bhuyan et al. [14] published the effects of optimizing the process parameters of the Electrical Discharge Machining (EDM) of metal matrix composite (MMC) by a VIKOR and Entropy weight measurement method. The Entropy method estimates the separate weights effect and utilizing the VIKOR approach, the output parameters was computed to obtain a specific VIKOR factor. The VIKOR factor was validated by confirmation test using the linear equation getting by RSM to recognize the proposed approach efficiency.

Results of optimizing various process parameters of AWJM of stainless steel AISI 304 for obtaining the better machining characteristics presented Chaturvedi et al. [15]. In the tests considered process parameters as follows feed rate, standoff distance, abrasive mass rate, and pressure. Output parameters were removal rate and surface roughness. The MRR is classified as "higher is better" and the SR is classified as "lower is better". After the determination of the VIKOR factor, an analysis was performed applying the signal/noise (S/N) ratio and predicted the set of the optimum parameters for a greater material removal rate and lower surface roughness. Verification was done by confirmatory tests under optimal parameters. It was confirmed that the optimal set of control parameters provides the smallest VIKOR factor.

Gupta et al. [16] researched the cut slot properties after AWJ machining of marble. The effect of pressure, abrasive feed, and feed rate were tested were conducted aptly to appropriate orthogonal table. The results effects showed that the feed rate is the major parameter influencing on the slot size and the slot shape.

AWJ cutting research of CFRP to achieve the cut slot best characteristic in high performance conditions using the multiple pass process presented Xiao et al. [17]. It was procured on basis of test results that the slow traverse velocity in the first pass of process was better than in the second pass.

The cut kerf surface qualities estimation after PWJ machining published Lehocka et al. [18]. Cut texture was assessed utilizing optical profilometry and Vickers hardness tester was used to measure of surface hardness. Measurement suggest a minor rise in strain hardening undercutting region.

Babu et al. [19] published the experimental research of the feed rate, pressure, standoff distance, and abrasive output effect on the kerf roughness and kerf-angle of AISI1018 steel. The tests were done grounded on the orthogonal array. The best set of parameters that significantly entailed a reduction of the roughness and angle was got by RSM. Machined surfaces for a thorough analysis on its quality were subjected to research with electron microscopy and scanning force microscopy.

2 Experiments

2.1 Cut Materials and Abrasives

Hardox steel is high wear-resistant material, manufactured by the company SSAB [20]. This steel wears slowly, even under big mechanical loads as well as is typified by its constant evenness, high hardness, and high strength. HARDOX wear sheet does not drop its shock resistance, even at low temperatures. HARDOX sheets are a quenched and tempered. These outstanding properties can be lost if steel will be machined at a high-temperature level. Therefore, water jet treatment is recommended [21], because it does not increase the temperature in the cutting zone above 65 °C, even in the case of extreme working pressures [22, 23]. The basic properties of HARDOX steel presents Table 1.

Table 1. Basic properties of Hardox 500 steel [21].

Chemical composition									
C	Si	Mn	P	S	Cr	Ni	Mo	B	
Max %	Max %	Max %	Max %	Max %	Max %	Max %	Max %	Max %	
0.30	0.70	1.60	0.020	0.010	1.50	1.5	0.60	0.005	

Mechanical properties			
Grade	Thickness (mm)	Hardness (HBW)	Typical yield strength (MPa)
Hardox 500 sheet	2.0–7.0	470–530	1400
Hardox 500 plate	4.0–32.0	470–530	1400
Hardox 500 plate	32.1–103.0	450–540	1400

2.2 Material

A garnet from a Chinese deposit in Jinhong was utilized for the research. Fig. 1a presents view of grains, Fig. 2b, view of single grain and Fig. 1c shows grain size distribution of this abrasive.

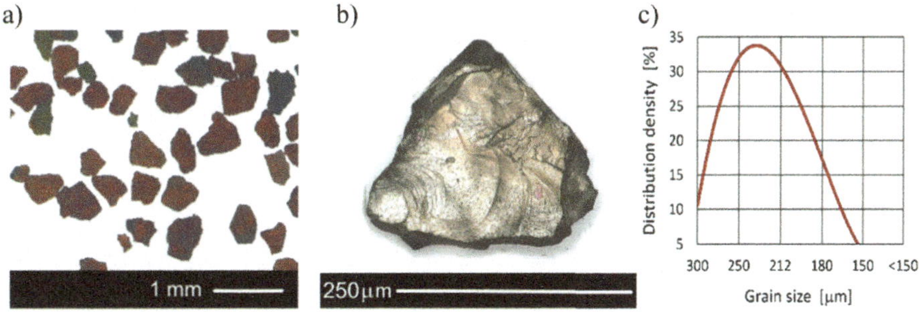

Fig. 1. J80A garnet: a) grains view, b) details of single grain, c) grains distribution.

The form of grains is almost isometric. The abrasive particle is typified by clear edges and pointed angles. Abrasive particles are most often red violet, less often dark brown to black. Table 2 presents most important properties of J80A abrasive.

Table 2. Chosen properties of J80A garnet [24].

Abrasive grain sizes	80 mesh
Color	Red
Hardness	7.5–8.0 Mohs Scale
Bulk density	1960–2150 kg/m^3
Specific density	3960–4200 kg/m^3
Grain shape	Sub angular

2.3 Test Stand

Cut researches were conducted on the OMAX 60120 machine according to L27 orthogonal array design model with the following factors: pressure, feed rate, and garnet amount (Table 3). Factors selected based on their own previous research works [25–28], and other scientific papers [29, 30].

2.4 Research Procedure

To numerically appraise tests results, a cutting depth, roughness of the kerf surface and taper kerf angle, was measurement. The cutting depth was measured on digital caliper altimeter Digimatic Height Gage. Surface roughness was determined with the Olympus LEXT OLS5000 laser scanning confocal microscope. The taper kerf was calculated on basis measuring the kerf wall inclination (W_t – W_b) on digital optic microscope from the top and bottom kerf edge (Fig. 1) as follow:

$$\delta = \arctan\left(\frac{W_t - W_b}{2h_c}\right) \quad (1)$$

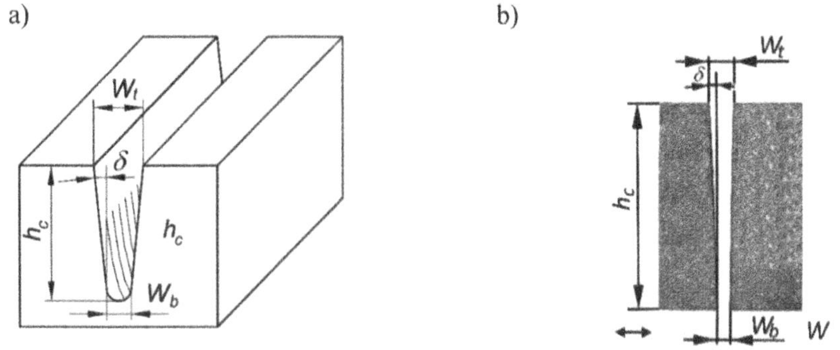

Fig. 2. Details of evaluate parameters of kerf: a) schematic, b) real view of sample cross section W_t – top width, W_b – bottom width, h_c – cutting depth, δ – kerf taper angle.

2.5 Entropy Weight Determination

In the entropy technique, the amount of choices, and different criteria get to appraise multiple criteria optimisation on basis establishing a comparative decision matrix. If the number of choices (garnet rate, pressure, and feed rate) getting as 'M', and the numbers of conditions are cutting depth surface rough-ness and angle of cut kerf get as 'N' then relative decision matrix having a dimension of M × N.

$$D = \begin{bmatrix} x_{1,1} & x_{1,2} & x_{1,n} \\ x_{2,1} & x_{2,2} & x_{2,3} \\ x_{m,1} & x_{m,2} & x_{m,n} \end{bmatrix} \tag{2}$$

The decision matrix must be standardized through maximal levels for a beneficial element:

$$r_{i,j} = \frac{x_{i,j} - \min x_{i,j}}{\max x_{i,j} - \min x_{i,j}} \tag{3}$$

Where:
$R_{i,j}$ is normalized decision matrix element.
$x_{i,j}$ is relative decision matrix element.
and minimal values for non-beneficial (cost) elements:

$$r_{i,j} = \frac{\max x_{i,j} - x_{i,j}}{\max x_{i,j} - \min x_{i,j}} \tag{4}$$

The various criteria determined on basis the entropy manner projection factors ($T_{i,j}$) was appraise as follow:

$$T_{i,j} = \frac{r_{i,j}}{\sum_{i=1}^{m} r_{i,j}} \tag{5}$$

To establish the entropy factor ($e_{i,j}$) exploiting the projection value of the alternative, the equation as follow:

$$e_{i,j} = -\frac{1}{\ln m}\sum_{i=1}^{n} T_{i,j} \ln T_{i,j} \tag{6}$$

and entropy weight of the j-th index is determined by equation:

$$w_i = \frac{1 - e_{i,j}}{\sum_{i=1}^{n}\left(1 - e_{i,j}\right)} \tag{7}$$

2.6 VIKOR Multi-optimisation Approach

The VIKOR technique is relevant for multiple criteria optimisation for achieves a ultimate solution. In the initial stage, was determining the matrix D, which is of course the same as when determining the weights using the entropy method (Eq. 1). In the next step, the normalized matrix should be defined on the base following equation:

$$f_{i,j} = \frac{x_{i,j}}{\sqrt{\sum_{i=1}^{m} x_{i,j}^2}} \tag{8}$$

Where:
$f_{i,j}$ is normalized matrix element,
$x_{i,j}$ is relative decision matrix element.
Next, calculation of utility measure S_i following equations for beneficial attribute:

$$S_i = \sum_{i=1}^{n} w_i \left[\frac{\max f_{i,j} - f_{i,j}}{\max f_{i,j} - \min f_{i,j}} \right] \tag{9}$$

Where:
w_i is weight of attribute.
and for non-beneficial (cost) attribute:

$$S_i = \sum_{i=1}^{n} w_i \left[\frac{f_{i,j} - \min f_{i,j}}{\max f_{i,j} - \min f_{i,j}} \right] \tag{10}$$

Calculation of regret measure R_i following equation for beneficial attribute:

$$R_i = \max \left\{ w_i \left[\frac{\max f_{i,j} - f_{i,j}}{\max f_{i,j} - \min f_{i,j}} \right] \right\} \tag{11}$$

and for non-beneficial (cost) attribute:

$$R_i = \max \left\{ w_i \left[\frac{f_{i,j} - \min f_{i,j}}{\max f_{i,j} - \min f_{i,j}} \right] \right\} \tag{12}$$

In the last step, the VIKOR factor Q_i is determined on the equation:

$$Q_i = \nu \left[\frac{S_i - \min S_i}{\max S_i - \min S_i} \right] (1 - \nu) \left[\frac{R_i - \min R_i}{\max R_i - \min R_i} \right] \tag{13}$$

Where:
ν is the influence factor, which is held as 0.5. This factor can take any value from 0 to 1 [31].

3 Results and Discussion

Variable parameters of the cutting process and impacts of the laboratory research presents Table 3. The remaining process control parameters were kept constant, such as water nozzle diameter = 0.3 mm, focusing tube diameter = 0.76 mm, and standoff distance = 4 mm.

Table 3. The cutting process parameters and effects.

No	Machining factors			Outputs		
	Garnet rate	Pressure	Feed rate	Cutting depth h_c	Roughness S_a	Kerf angle δ
	[g/min]	[MPa]	[mm/min]	[mm]	[μm]	[deg]
1	250	350	100	17.09	4.91	2.20
2	250	350	200	10.74	6.57	1.94
3	250	350	300	7.86	8.18	1.65
4	250	375	100	19.09	5.15	2.44
5	250	375	200	11.63	7.11	2.28
6	250	375	300	8.04	8.73	1.95
7	250	400	100	20.30	3.83	2.70
8	250	400	200	12.89	7.41	2.53
9	250	400	300	8.9	10.56	2.10
10	350	350	100	18.21	3.86	2.61
11	350	350	200	11.87	6.01	2.37
12	350	350	300	8.37	10.2	2.04
13	350	375	100	19.66	4.09	2.13
14	350	375	200	12.30	5.29	2.31
15	350	375	300	9.28	7.32	2.29
16	350	400	100	21.06	4.21	2.80
17	350	400	200	13.75	7.02	2.22
18	350	400	300	9.98	9.48	2.16
19	450	350	100	19.81	4.23	2.62
20	450	350	200	12.16	4.88	1.96
21	450	350	300	9.08	5.56	2.24
22	450	375	100	20.8	4.14	2.53
23	450	375	200	13.07	5.78	1.99
24	450	375	300	9.96	6.88	1.92
25	450	400	100	21.14	5.17	2.55
26	450	400	200	14.53	7.98	2.39
27	450	400	300	11.05	10.64	2.02

Table 4 shows the calculated entropy indexed values, dispersion values and weight vector on basis Eq. (5–7).

Table 4. Entropy, dispersion, and weight vector values.

Criteria	Depth of cut	Surface roughness S_a	Taper kerf angle
Entropy projection $T_{i,j}$	0.9840	0.9849	0.9977
Dispersion value $e_{i,j}$	0.0160	0.0151	0.0023
Weight w_i	***0.4803***	***0.4519***	***0.0678***

Table 5. VIKOR approach coefficients: utility measures, regret measures, VIKOR factors, and its rank.

No	Sij (nb)	Sij (b)	Sij	R_i	Q_i	Rank
1	0.1041	0.1465	0.2506	0.1465	0.2086	9
2	0.1992	0.3761	0.5753	0.3761	0.6730	18
3	0.2887	0.4803	0.7689	0.4803	0.9116	23
4	0.1343	0.0741	0.2084	0.1343	0.1690	8
5	0.2548	0.3439	0.5987	0.3439	0.6494	17
6	0.3427	0.4737	0.8165	0.4737	0.9325	24
7	0.0620	0.0304	0.0924	0.0620	0.0145	2
8	0.2895	0.2984	0.5878	0.2984	0.5895	14
9	0.4733	0.4426	0.9160	0.4733	0.9919	27
10	0.0589	0.1060	0.1649	0.1060	0.1096	6
11	0.1865	0.3352	0.5217	0.3352	0.5929	15
12	0.4458	0.4618	0.9076	0.4618	0.9733	26
13	0.0458	0.0535	0.0993	0.0535	0.0087	1
14	0.1356	0.3197	0.4553	0.3197	0.5348	13
15	0.2692	0.4289	0.6981	0.4289	0.8088	21
16	0.0930	0.0029	0.0959	0.0930	0.0530	4
17	0.2451	0.2673	0.5123	0.2673	0.5076	11
18	0.4053	0.4036	0.8089	0.4053	0.8478	22
19	0.0839	0.0481	0.1320	0.0839	0.0641	5
20	0.0881	0.3248	0.4129	0.3248	0.5152	12
21	0.1496	0.4361	0.5857	0.4361	0.7496	20
22	0.0725	0.0123	0.0848	0.0725	0.0222	3
23	0.1495	0.2918	0.4413	0.2918	0.4937	10
24	0.2181	0.4043	0.6224	0.4043	0.7345	19
25	0.1418	0.0000	0.1418	0.1418	0.1377	7
26	0.3193	0.2390	0.5583	0.3193	0.5963	16
27	0.4738	0.3649	0.8387	0.4738	0.9460	25

The utility measure S_i for non-beneficial (nb) and beneficial (b) factors, regret measure R_i and VIKOR factor, calculated on basis Eqs. (9–13) with the VIKOR factor rank presents Table 5.

The estimated VIKOR factor varies from 0 to 1 and the largest value fetches it closely approximate to the ideal one. Out of all the combinations of process parameters employed, one that has got the biggest VIKOR factor is considered as the best set of parameters and thus recommended. Musani et al. [32] and Shemshadi et al. [33] introduced the subsequent condition to verify whether the best option gives the compromise solution. Satisfactory advantage: $Q_i \geq DQ$, where DQ is the threshold and was determined on basis following equation:

$$DQ = \frac{1}{p-1} \tag{14}$$

Where p is amount of measurements. The threshold value of DQ is equal:

$$DQ = \frac{1}{27 - 1} <=> DQ = 0.0385$$

The VIKOR factor value for the first three alternatives is less than 0.0385. The fourth best alternative exceeds the threshold, because $0.0530 \geq 0.0385$ and is the acceptable advantage. The fourth best combination of control parameters is AFR2-P3-TS1, which corresponds to garnet rate 350 g/min, pressure 400 MPa and feed rate 100 mm/min. Sample microscopic view presents Fig. 3 and 3D surface view, cut at optimal condition, presents Fig. 4.

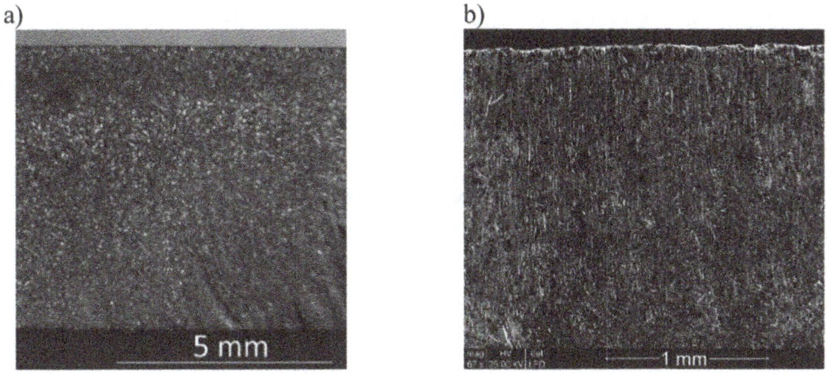

Fig. 3. Machined surface exemplary view: a) optical microscope, b) EDX microscope.

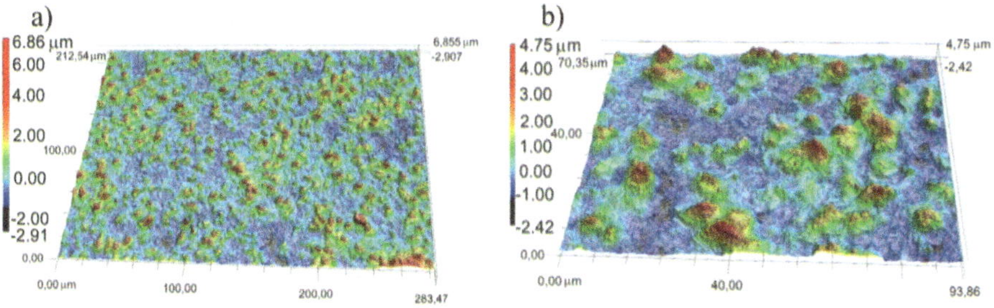

Fig. 4. Exemplary 3D roughness map: a) magnification 50x, b) magnification 150x.

4 Conclusions

VIKOR approach turned out to be an effective method of optimisation of problems related to the AWJ process. This approach of the multiple characteristics vitally reduces the computation and determines the positions from computational results by optimal control parameters combination. The Entropy/VIKOR method used in this research efficiently copes with conflicting responses. This technique simplifies manifold cut composite responses by obtaining one individual response as the VIKOR factor.

The best possible set of the control parameters is garnet rate = 350 g/min, pressure = 400 MPa, feed rate = 100 mm/min. The variation in feed rate affects the biggest changes in results, next by the garnet rate, and the lowest influence for the pressure was observed.

References

1. Perec, A.: Experimental research into alternative abrasive material for the abrasive water jet cutting of titanium. Int. J. Adv. Manuf. Technol. **97**, 1529–1540 (2018). https://doi.org/10.1007/s00170-018-1957-2
2. Radomska-Zalas, A., Perec, A., Fajdek-Bieda, A.: IT support for optimisation of abrasive water cutting process using the TOPSIS method. In: IOP Conference Series: Materials Science and Engineering, vol. 710, p. 012008 (2019). https://doi.org/10.1088/1757-899X/710/1/012008
3. Lehocka, D., Klich, J., Pitel, J., Krejci, L., Storkan, Z., Duplakova, D., Schindlerova, V., Sajdlerova, I.: Analysis of the pulsating water jet maximum erosive effect on stainless steel. In: Gapiński, B., Szostak, M., Ivanov, V. (eds.) Advances in Manufacturing II, pp. 233–241. Springer, Cham (2019). https://doi.org/10.1007/978-3-030-16943-5_21
4. Lehocka, D., Klich, J., Botko, F., Simkulet, V., Foldyna, J., Krejci, L., Storkan, Z., Kepic, J., Hatala, M.: Comparison of ultrasonically enhanced pulsating water jet erosion efficiency on mechanical surface treatment on the surface of aluminum alloy and stainless steel. Int. J. Adv. Manuf. Technol. **103**, 1647–1656 (2019). https://doi.org/10.1007/s00170-019-03680-8
5. Kukielka, K.: Ecological aspects of the implementation of new technologies processing for machinery parts. Rocznik Ochrona Srodowiska – Ann. Set Environ. Protect. **18**, 137–157 (2016)
6. Perec, A.: Environmental aspects of abrasive water jet cutting. Ann. Set Environ. Protect. - Rocznik Ochrona Srodowiska **20**, 0258–0274 (2018)
7. Opricovic, S., Tzeng, G.-H.: Compromise solution by MCDM methods: a comparative analysis of VIKOR and TOPSIS. Eur. J. Oper. Res. **156**, 445–455 (2004). https://doi.org/10.1016/S0377-2217(03)00020-1
8. Cierna, H., Tavodova, M.: Using the design of experiment method to evaluate quality of cuts after cutting aluminum alloy by AWJ. Manuf. Technol. **13**, 303–307 (2013)
9. Michalska-Pozoga, I., Wegrzyk, S., Rydzkowski, T.: Assessment of influence of extrusion method on selected properties of wood-polymer composites using Taguchi method of experiment planning. Polimery **62**, 686–692 (2017). https://doi.org/10.14314/polimery.2017.686
10. Perec, A.: Investigation of limestone cutting efficiency by the abrasive water suspension jet. In: Hloch, S., Klichová, D., Krolczyk, G.M., Chattopadhyaya, S., Ruppenthalová, L. (eds.) Advances in Manufacturing Engineering and Materials, pp. 124–134. Springer, Cham (2019). https://doi.org/10.1007/978-3-319-99353-9_14
11. Radomska-Zalas, A., Perec, A.: Modeling of abrasive water suspension jet cutting process using response surface method. In: AIP Conference Proceedings, pp. 0200511–0200518. AIP Publishing, Melville (2019). https://doi.org/10.1063/1.5092054
12. Kakaravada, I., Mahamani, A., Pandurangadu, V.: Optimization of machining parameters using Entropy-VIKOR method in drilling of A356-TiB 2 /TiC in-situ composites. In: IOP Conference on Series: Materials Science and Engineering, vol. 390, p. 012099 (2018). https://doi.org/10.1088/1757-899X/390/1/012099

13. Gangil, M., Pradhan, M.K.: Optimization the machining parameters by using VIKOR Method during EDM process of Titanium alloy. Mater. Today: Proc. **5**, 7486–7495 (2018). https://doi.org/10.1016/j.matpr.2017.11.420
14. Bhuyan, R.K., Routara, B.C.: Optimization the machining parameters by using VIKOR and entropy Weight method during EDM process of Al–18% SiCp Metal matrix composite, pp. 269–282 (2016). https://doi.org/10.5267/j.dsl.2015.11.001
15. Chaturvedi, V., Singh, D.: Multi response optimization of process parameters of abrasive water jet machining for stainless steel AISI 304 using VIKOR approach coupled with signal to noise ratio methodology. J. Adv. Manuf. Syst. **14**, 107–121 (2015). https://doi.org/10.1142/S0219686715500080
16. Gupta, V., Pandey, P.M., Garg, M.P., Khanna, R., Batra, N.K.: Minimization of Kerf taper angle and kerf width using taguchi's method in abrasive water jet machining of marble. Proc. Mater. Sci. **6**, 140–149 (2014). https://doi.org/10.1016/j.mspro.2014.07.017
17. Xiao, S., Wang, P., Gao, H., Soulat, D.: A study of abrasive waterjet multi-pass cutting on kerf quality of carbon fiber-reinforced plastics. Int. J. Adv. Manuf. Technol. **105**, 4527–4537 (2019). https://doi.org/10.1007/s00170-018-3177-1
18. Lehocka, D., Simkulet, V., Legutko, S.: Assessment of deformation characteristics on CW004A copper influenced by acoustically enhanced water jet. In: Hamrol, A., Ciszak, O., Legutko, S., Jurczyk, M. (eds.) Advances in Manufacturing, pp. 717–724. Springer, Cham (2018). https://doi.org/10.1007/978-3-319-68619-6_69
19. Babu, M.N., Muthukrishnan, N.: Exploration on Kerf-angle and surface roughness in abrasive waterjet machining using response surface method. J. Inst. Eng. India Ser. C 1–12 (2017). https://doi.org/10.1007/s40032-017-0366-x
20. SSAB Homepage. https://www.ssab.com/products/brands/hardox/products/hardox-500. Accessed 10 Oct 2019
21. Krolczyk, G.M., Krolczyk, J.B., Maruda, R.W., Legutko, S., Tomaszewski, M.: Metrological changes in surface morphology of high-strength steels in manufac-turing processes. Measurement **88**, 176–185 (2016). https://doi.org/10.1016/j.measurement.2016.03.055
22. Perec, A., Trieb, F., Pude, F.: Some investigations into 1,000 mpa pure waterjet cutting. In: Klichová, D., Sitek, L., Hloch, S., Valentinčič, J. (eds.) Advances in Water Jetting, pp. 155–163. Springer, Cham (2021). https://doi.org/10.1007/978-3-030-53491-2_17
23. Wessels, V., Grigoryev, A., Dold, C., Wyen, C.-F., Roth, R., Weingaertner, E., Pude, F., Wegener, K., Loeffler, J.F.: Abrasive waterjet machining of three-dimensional structures from bulk metallic glasses and comparison with other techniques. J. Mater. Res. **27**, 1187–1192 (2012). https://doi.org/10.1557/jmr.2012.36
24. 80 Mesh waterjet Abrasive Garnet | Lianyungang Jinhong Mining Co., Ltd. https://www.jhmining.cn/75.html. Accessed 23 Aug 2020
25. Perec, A., Pude, F., Grigoryev, A., Kaufeld, M., Wegener, K.: A study of wear on focusing tubes exposed to corundum-based abrasives in the waterjet cutting process. Int. J. Adv. Manuf. Technol. **104**, 2415–2427 (2019). https://doi.org/10.1007/s00170-019-03971-0
26. Perec, A.: Disintegration and recycling possibility of selected abrasives for water jet cutting. DYNA **84**, 249–256 (2017). https://doi.org/10.15446/dyna.v84n203.62592
27. Perec, A., Musial, W., Prazmo, J., Sobczak, R., Radomska-Zalas, A., Fajdek-Bieda, A., Nagnajewicz, S., Pude, F.: Multi-criteria optimization of the abrasive waterjet cutting process for the high-strength and wear-resistant steel Hardox®500. In: Klichová, D., Sitek, L., Hloch, S., Valentinčič, J. (eds.) Advances in Water Jetting, pp. 145–154. Springer, Cham (2021). https://doi.org/10.1007/978-3-030-53491-2_16
28. Perec, A., Pude, F., Kaufeld, M., Wegener, K.: Obtaining the selected surface roughness by means of mathematical model based parameter optimization in abrasive waterjet cutting. Strojniški vestnik – J. Mech. Eng. **63**, 606–613 (2017). https://doi.org/10.5545/sv-jme.2017.4463

29. Filip, A.C., Morariu, C.O., Mihail, L.A., Oancea, G.: Research on surface roughness of hardox steels parts machined by abrasive waterjet. Strojniški – vestnik J. Mech. Eng. **65**, 230–237 (2019). https://doi.org/10.5545/sv-jme.2019.5998
30. Spadło, S., Krajcarz, D.: Study of the geometrical structure of copper surface after abrasive waterjet cutting. In: IOP Conference Series: Materials Science and Engineering, vol. 461, p. 012045 (2018). https://doi.org/10.1088/1757-899X/461/1/012045
31. San Cristóbal, J.R.: Multi-criteria decision-making in the selection of a renewable energy project in Spain: the Vikor method. Renew. Energy **36**, 498–502 (2011). https://doi.org/10.1016/j.renene.2010.07.031
32. Musani, S., Jemain, A.A.: Ranking schools' academic performance using a fuzzy VIKOR. J. Phys.: Conf. Ser. **622**, 012036 (2015). https://doi.org/10.1088/1742-6596/622/1/012036
33. Shemshadi, A., Shirazi, H., Toreihi, M., Tarokh, M.J.: A fuzzy VIKOR method for supplier selection based on entropy measure for objective weighting. Expert Syst. Appl. **38**, 12160–12167 (2011). https://doi.org/10.1016/j.eswa.2011.03.027

Identification of Local Microplasticity on Ti6Al4V After Impingement of Periodically Acting Water Clusters

Jakub Poloprudský[1(✉)], Alice Chlupová[1], Tomáš Kruml[2], and Sergej Hloch[3]

[1] Institute of Physics of Materials, Czech Academy of Sciences, Brno, Czech Republic
poloprudsky@ipm.cz

[2] Institute of Physics of Materials, CEITEC IPM, Czech Academy of Sciences, Brno, Czech Republic

[3] Institute of Geonics, Czech Academy of Sciences, Ostrava-Poruba, Czech Republic

Abstract. Effect of the pulsating water jet on the plastic deformations of surface of Ti6Al4V was examined. Existing research papers investigating the interaction of Ti6Al4V with pulsating water and water impact are focused on culmination erosion stage of the impacted area. In this article, the initial stages of erosion were studied in term of the increasing but relatively low number of impingement on the microplastic deformation. The number of impacts varied from 1050 to 63000 impacts. The PWJ was positioned at the standoff distance of 5.5 mm above material surface. The aim of this research was to study the influence of early erosion stage, consisting of plastic deformation and material removal in preferential sites, on the material structure. The determination of early erosion stages can be used for selecting the conditions of PWJ leading to the surface strengthening, analogical to shot peening. Sites with significance to the surface strengthening were observed and discussed. Interphase boundaries and other structural characteristic play significant role in the earliest stages of erosion.

Keywords: Pulsating water jet · Ti6Al4V alloy · Early erosion stage · Initial damage · Erosion · Water clusters · Microplasticity

1 Introduction

Pulsating water jet (PWJ) is a method developed in order to enhance the effectivity of the cutting process [1], i.e. to facilitate separation of material using lower pressure compared to continuous water jet (CWJ). During the PWJ process, the surface of the material is impinged with periodically repeated clusters of droplets of water. When the front of the cluster hits the surface the transient increase in pressure called the water hammer effect is created [1]. When PWJ clusters are interacting with the surface of the material three distinguished stages of damage can be observed [2]. The initial stage creates a plastic deformation of the affected surface. The second stage is characterized by the appearance of micro-voids or pits and their merging into a single crater. The third stage is characterized by an increase in the depth of the erosion crater [2]. The erosion stages

S. Hloch et al. (Eds.): ICMEM 2021, LNME, pp. 63–74, 2021.
https://doi.org/10.1007/978-3-030-71956-2_6

were further divided by Hloch et al. [3] based on PWJ morphology and standoff distance. When mild conditions of PWJ are applied, i.e. conditions when no loss of material is reached but only the plastic deformation appears, the procedure can serve as a method of surface treatment. Impingement of water clusters on the surface induces deformation of the surface layer and creates compressive stresses in it [4]. While the application of continuous water jet as a peening method has been studied [5, 6] peening using PWJ is recently developed technology. Srivastava et al. [7] observed an increase in compressive residual stress from −67 MPa to −332 MPa in the welded region of stainless steel of grade SS304 after PWJ procedure. Along with the induction of compressive stresses, the increase of hardness in an impinged area was also observed [7].These findings favour use of PWJ process as a surface treatment method comparable to shot peening. Hlaváček et al. [8] studied the effect of PWJ peening on the fatigue life of AISI 316L. It has been concluded that with suitable process parameters of PWJ an increase in fatigue life in the area of 10^5 cycles can be reached. Arola et al. [9] observed an increase in fatigue properties after application of abrasive water jet on AISI 304 and Ti6Al4V. In their work the increase in fatigue life was observed, however number of abrasive particles was embedded into the surface of the material after the process. Advantage of using PWJ in comparison to abrasive water jet is that the impingement surface is free of abrasive particles.

For enhancing fatigue properties namely in high cycle area it is crucial to increase compressive residual stresses while keeping surface roughness minimal, analogically to shot peening [10]. Surface roughening caused by erosion damage needs to be limited in order to use PWJ as surface strengthening method [11]. It is, therefore, necessary to minimalize treated surface irregularities. From this point of view, the first erosion stage as defined by Foldyna [2] is the most important for surface strengthening. The first erosion stage, however, means considerable difficulties when using the same observation methods as for third erosion stage. It is mainly because of too shallow depth profile in the first stage compared to the depression/elevation (pile-up) created during the second and the third erosion stage.

In this article, the first erosion stage created on the surface of the Ti6Al4V was observed after application of PWJ for relatively low effective time (from 0.05 to 3 s) corresponding to the low number of impingements (from 1050 up to 63000). This article was focused mainly on the initial period of erosion (plastic deformation) and creation of first pits in Ti6Al4V during appropriate conditions of PWJ which should ensure the strengthening of the surface layer without damaging or material removal. From the aforementioned review, the focus on third erosion stage in the literature is clear. For successful use of PWJ peening, it is necessary to induce several plastic deformation with minimal changes in morphology, as these could act as stress concentrators. In case of Ti6Al4V local changes in morphology are influenced by the material duplex structure. It is therefore first necessary to understand effect of the material structure on the effect of PWJ.

The erosion mechanism in the same type of material, i.e. rolled Ti6Al4V, was observed by Kamkar et al. [12]. Focus of his study was on the stage of erosion with significant removal of the material, caused by CWJ after 300 000 of impingements in relevance to local microstructure and the mechanism of the erosion and material removal was analysed. Although Huang et al. [13] observed the initial stages of erosion in Ti6Al4V, they used also CWJ with one order higher jet pressures (138–207 MPa) than in this paper (15 MPa). The surface shown in this paper is comparable to results of Huang et al. [13], after one pass of CWJ.

Surface roughening caused by erosion damage needs to be limited in order to use PWJ as surface strengthening method [11]. Especially for fatigue enhancement in high cycle area [14]. It is, therefore, necessary to minimalize treated surface irregularities.

Aim of this paper is to study and understand local microplasticity which takes place in the affected area after very low number of periodic impingements. Secondary aim of this paper is to pinpoint places for defect initiation caused by PWJ at given process parameters, while taking into account the structure of the material. The material used in this study is alpha-beta phase titanium alloy. The local changes in morphology are influenced by the duplex structure of material.

2 Materials and Methods

2.1 Material

Ti6Al4V was used as the experimental material. This alloy consists of equiaxed grains of the alpha phase with the hexagonal close-packed (HCP) lattice and the beta phase with the body-centred cubic (BCC) lattice at the grain boundaries. The average grain size was evaluated using a linear interception method as 7.2 ± 0.8 μm.

The microstructure of the experimental material is shown in Fig. 1a in BSE contrast in order to enhance the visibility of individual phases. Fig. 1b shows the results of the EBSD analysis. According to the phase analysis, there is approximately 86% of the alpha phase and 4% of the beta phase, whereas 10% belongs to the zero solution i.e. grain or phase boundaries, where it is not possible to evaluate the type of lattice.

The specimens for the metallographic observation were rectangular with dimensions 10×10 mm and thickness 4 mm. The specimens were mechanically ground using SiC papers, polished using 1 μm diamond paste and ethanol as a lubricant and chemical-mechanically polished using colloidal silica suspension with the particle size of 0.04 μm.

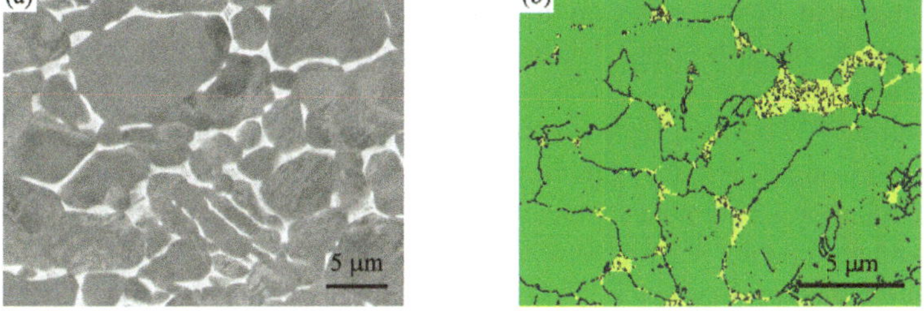

Fig. 1. a) SEM micrograph of initial microstructure in BSE contrast, b) the EBSD analysis of phases (green = alpha, yellow = beta) together with grain boundaries.

The Vickers hardness value of the material before PWJ treatment, i.e. in the as-received state, measured using ZHV30 Vickers ZwickRoell reached the value of 372 ±

6 HV0.2. The tensile strength of the material under investigation evaluated using MTS 809 servo-hydraulic testing machine was 1036 MPa.

2.2 Experiments

The surface was treated using the ultrasonic pulsating water jet technology at the Institute of Geonics of the CAS. The experimental apparatus for creating pressurized liquid was the Hammelmann HDP 253 high-pressure pump with operating pressure up to 160 MPa and flow up to 67 l/min. PWJ head was used with a circular nozzle Stone Age with a diameter of 0.4 mm. The pressure was 15 MPa. A robot ABB URB 6640–180//2.55 handling PWJ cutting head was used in the study. The pulses were created by ultrasonic generator Ecoson WJ-UG 630–40 connected to the piezoceramics with power output up to 800W. The frequency of a sonotrode was set to the frequency of 21 kHz. After setting up an optimal standoff distance $z = 5.5$ mm the experiment runs were realised according to Table 1. Polished Ti6Al4V surface was exposed to periodical impingements of PWJ stationary positioned above collision epicentre with time exposures. Set-up of experiment is graphically expressed in Fig. 2.

Table 1. Parameters of the PWJ process during the experiment.

Effective time	No. of impacts	Standoff distance	Pressure of water	Frequency sonotrode	Chamber length	Nozzle diameter	Traverse speed	Liquid density
[s]	[-]	[mm]	[MPa]	[kHz]	[mm]	[mm]	[mm/s]	[kg/m^3]
0.05 ↓ 3	1050 ↓ 63000	5.5	15	21	12	0.4	0	998

2.3 Measurement

The impingement region was examined by the digital optical microscope Olympus DSX1000 at low magnifications. Differential interference contrast (DiffIC), also known as Nomarski contrast, was used in order to enhance the visibility of surface irregularities caused by PWJ. The affected surface was studied also using Tescan Lyra3 XMU scanning electron microscopy with the aim to observe microstructural changes at grain level. The roughness parameters were measured using confocal microscope LEXT OLS 3100. Roughness parameter was measured using magnification 240× (view field 1280 × 960 μm) and cut-of frequency 1/50 (2.5 μm). Measured area was 100 × 100 μm and the measurement was done 8 times at randomly selected places in the impacted area.

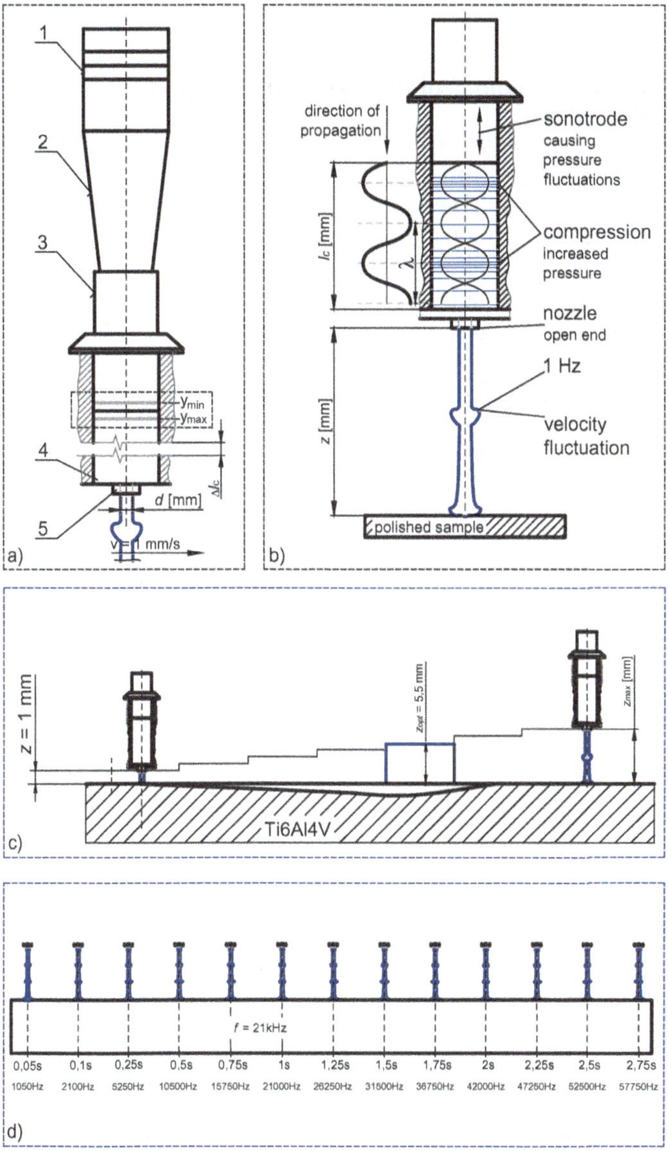

Fig. 2. The scheme of experiment apparatus and test set-up: a) PWJ head with a circular nozzle of the diameter of 0.4 mm, b) detail of a sonotrode causing water pressure fluctuations, c) setting of the optimal standoff distance $z = 5.5$ mm, d) test set-up for increasing time and a corresponding number of impingements.

3 Results

3.1 Topographic Observation of PWJ Treated Surfaces

The optical micrographs (OM) of three selected areas impacted with PWJ for 0.05, 0.25 and 0.5 s are in Fig. 3. This effective time can be recalculated to 1050, 5250 and 10500 impacts, respectively. The micrographs were taken using DiffIC technique. It is evident from Fig. 3a, that after 1050 impingements, only few surface irregularities were created with no signs of plastic deformation. After 5250 impacts (see Fig. 3b) the amount of surface irregularities increased and they were arranged into crescent-like shape. This shape was with high probability caused by the angle between the PWJ and the surface of the impacted area, when not a perfect perpendicular position of nozzle and specimen surface was kept. Some signs of plastic deformation can be seen in that area. The impacted area after 10500 impacts is shown in Fig. 3c. In Fig. 3d and Fig. 3e there are two SEM micrographs of samples impacted with PWJ with 1050 and 5250 impacts respectively. These micrographs show the same area as in Fig. 3a, Fig. 3b at the similar magnification indicating challenges of using the SEM technique. In Fig. 3f there is the optical micrograph of the sample from Fig. 3c but in the Dark Field (DF). From the comparison of these two micrographs is evident that DiffIC contrast exhibits that amount of surface irregularities increased, nevertheless the visibility of the whole shape of the crescent area is enhanced even further in DF.

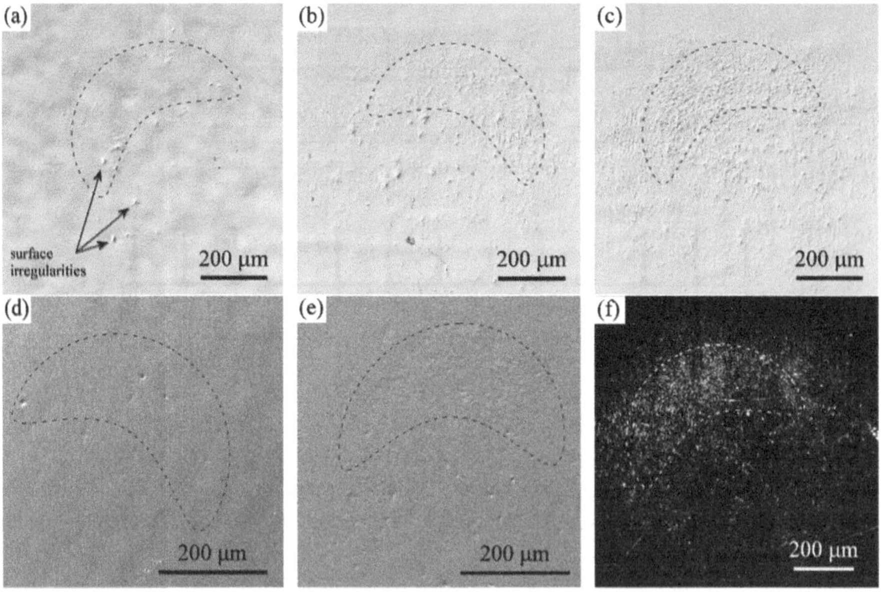

Fig. 3. OM micrographs of impacted area in DiffIC contrast after (a) 0.05 s/1050 impacts, (b) 0.25 s/5250 impacts and (c) 0.5 s/10500 impacts, (d) SEM micrographs after 0.05s/1050 impacts and (e) 0.25 s/5250 impacts and (f) OM after 0.5 s/10500 impacts in dark field.

3.2 Roughness Measurement

The roughness was measured at areas impacted by PWJ for chosen effective times. Results are in Table 2. The measurement was carried under conditions described above. The measured parameters are SRa, SRz and SRq. The effect of number of impingements (calculated from time when pulses were acting) on arithmetical mean height of the surface (SRa) is visible in the graph in Fig. 4. The other measured parameters listed in Table 2 are SRz and SRq.

Table 2. Surface roughness values.

Effective time [s]	0	0.05	0.1	0.5	1	1.5	2
Number of impingement	0	1050	2100	10500	21000	31500	42000
SRa [μm]	0.044 ± 0.007	0.082 ± 0.04	0.113 ± 0.01	0.241 ± 0.048	0.285 ± 0.094	0.379 ± 0.19	0.360 ± 0.05
SRz [μm]	0.569 ± 0.276	2.873 ± 2.870	2.561 ± 0.631	5.019 ± 1.437	4.872 ± 2.384	5.951 ± 2.877	5.22 ± 2.066
SRq [μm]	0.056 ± 0.010	0.145 ± 0.108	0.165 ± 0.021	0.357 ± 0.068	0.392 ± 0.140	0.520 ± 0.246	0.478 ± 0.074

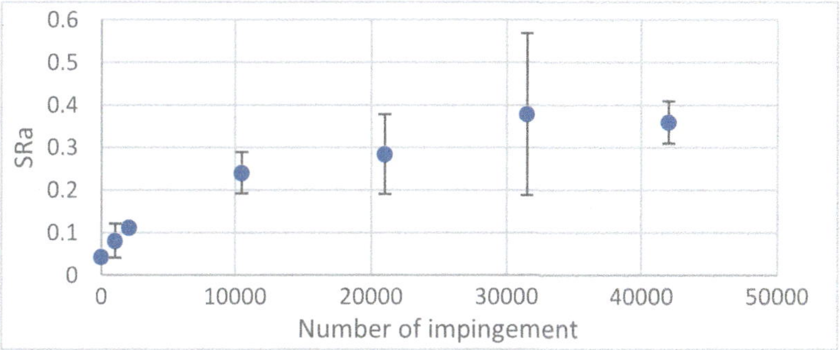

Fig. 4. Area roughness *SRa* based on number of impingement.

3.3 Observation of Interphase Boundaries Affected by PWJ

SEM micrographs of the surface of Ti6Al4V alloy samples in higher magnification show the surface irregularities in relation to the microstructure. Figure 5a shows an example of shallow irregularities developed on the surface after impacted with PWJ for 0.1 and 0.5 s. It responds to the 2100 and 10500 impacts respectively. Figure 5a shows area adjacent to a dimple created on the surface after 2100 impacts of the PWJ. In this area the deformation twins were created as signs of plastic deformation in some of the alpha grains. Increase of effective time to 0.5 s. i.e. after application of 10500 impacts. The deformation twins were formed more uniformly across the impacted area (see Fig. 5b). The highest occurrence of these signs of plastic deformation was observed in the crescent shape area (see OM in Fig. 3). We can see several grains of alpha phase tilted to an angle creating a depressed point with the lowest level being at the interface of the alpha/beta phase. The area around this dimple is heavily occupied with deformation twins in the alpha phase and creates the site of future crack or cavity nucleation.

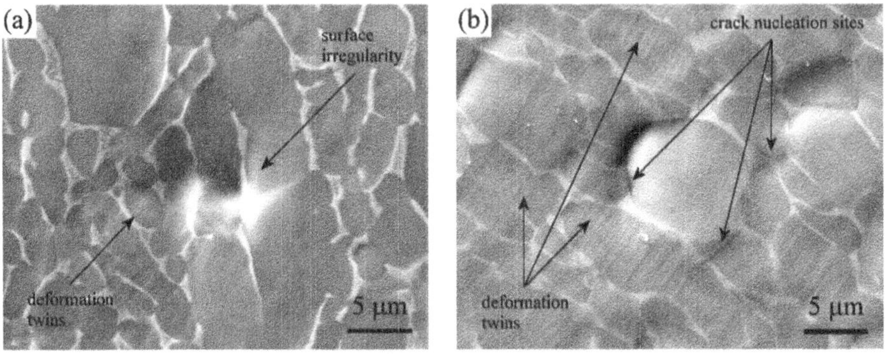

Fig. 5. SEM micrograph of (a) surface irregularity and deformation twins. 0.1 s/2100 impacts. (b) nucleation sites of cracks. 0.5 s/10500 impacts.

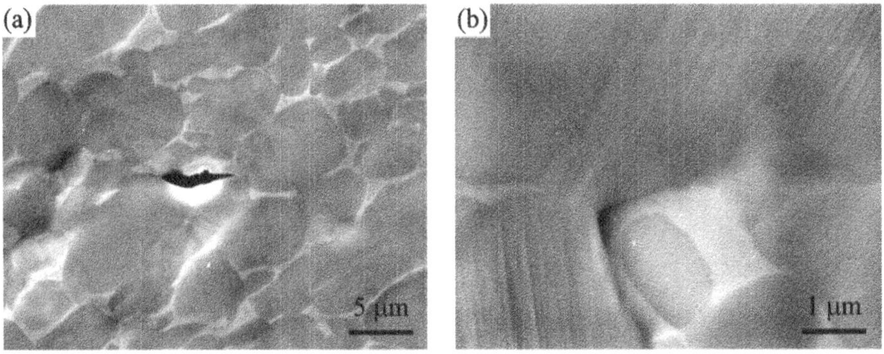

Fig. 6. (a) Surface crack at the interface of phases. 1 s/21000 impacts. (b) SEM micrograph of the surface depression at the intersection of three grains in high magnification. 2.5 s/52500 impacts.

Evolution of surface irregularities with further increase of effective time of PWJ is shown in Fig. 6. The surface of the material after application of PWJ for 1 s. i.e. for the even higher number of impacts (21000). is in Fig. 6a. The crack already nucleated at the interface of the alpha/beta phase. is visible. The crack propagates along the alpha phase boundary and stretches to the length of 5 μm. Figure 6b shows the area of intersection of three grains of the beta phase in very high magnification. The clearly visible depression point was created after subsequent increase in the number of impacts to 52500 (2.5 s).

Figure 7 shows the deformed area after 63000 impacts (3 s). In Fig. 7a several depressions located on the alpha and beta phase interface are visible. Application of

PWJ with this number of impingements resulted in the creation of cavities. The cavity displayed in the micrograph started as the crack at the alpha grain boundary. Nevertheless. Later the removal of material continued through the alpha grains creating the trench of the length of approximately 20 μm.

Except for depressions also elevations were present on the surface of treated material in the area with crescent-like shape. An example is in Fig. 7b. In the middle of the micrograph. There is an elevation of material (pile-up) created at the alpha grains boundaries. The pile up stretches to the length of almost 10 μm. The pile up follows the phase boundaries. The whole area in the micrograph is severely plastically deformed as can be seen from the high amount of deformation twins in all grains of the alpha phase.

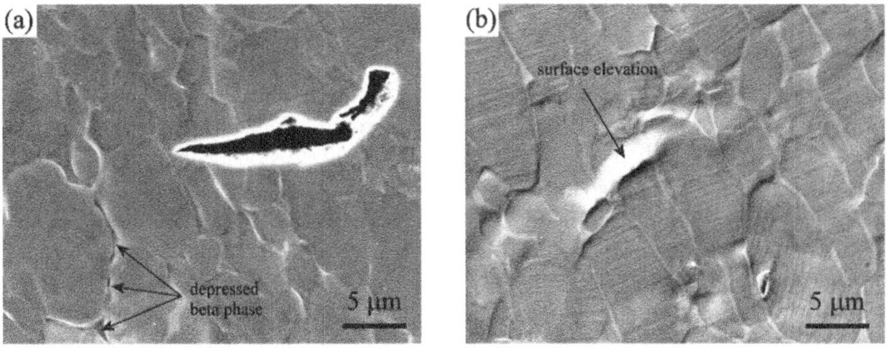

Fig. 7. SEM micrographs of deformed area. 3 s/63000 impacts (a) the depression of material in the location corresponding to the beta phase and the crack/cavity. (b) heavily deformed alpha phase grains with a high number of deformation twins and elevation/pile-up of material.

Figure 8a shows a cavity starting from the phase interface and propagating into the bulk of alpha grain. The cavity reaches the length of about 6 μm. Figure 8b shows surface irregularities stretched across two grains to the length of about 10 μm. Which resemble persistent slip markings.

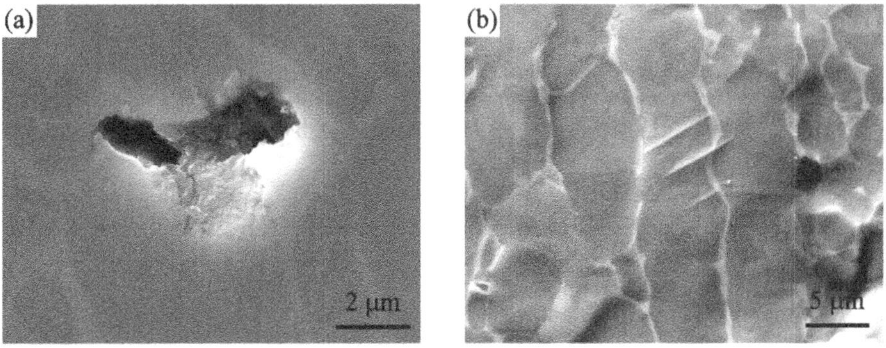

Fig. 8. SEM micrographs (a) detail of cavity created at the beta phase located at the alpha grains boundary after exposition to PWJ for 2.25 s/47250 impacts. (b) persistent slip markings observed on the surface of the sample exposed to the PWJ for 1.5 s/31500 impacts.

4 Discussion

Changes in surface topography in impingent areas caused by microplastic deformation are still not sufficiently understood. It is. However. Necessary to understand these changes in order to optimize erosion damage to use the PWJ technique as a method for material surface strengthening.

From roughness measurement Table 2 it is apparent that surface roughening was present with increasing number of impingements. The surface roughening (Fig. 4) shows some steeper steps. But significance of these is not clear due to the size of the standard deviation. Nevertheless. The roughening of the surface observed was in tenth of microns. Foldyna et al. [2] observed change in roughness parameter Ra. Correlated to significant material loss (second erosion stage). Around 80 μm.

The initial erosion stage exhibiting microplastic deformation is somewhat more challenging to observe (Fig. 3) requiring special techniques. This stage is sensitive to the microstructure of the treated material. Impingement of liquid onto the surface of the Ti6Al4V alloy has already been reported in the literature. Kamkar et al. [12] focused on third erosion stage of rolled Ti6Al4V. They proposed the erosion mechanism which consists of cyclical nucleation of surface and subsurface cracks. Propagation of the cracks and their merging which lead to the removal of large fragments and the final step is the smoothing of the eroded surface [12].

Based on presented research. it can be concluded that before nucleation of crack. Surface depressions are created by tilting of the alpha phase grains. The tilting of the grain due to impingement of liquid was proposed by Thomas et al. [15]. Observations of the rolled Ti6Al4V samples in this paper suggest that the grain tilting is further enhanced by the beta phase surrounding the alpha phase grains (Fig. 6b).

Similar to the research presented in this paper. Huang et al. [13] also observed grain tilting in Ti6Al4V at early deformation after application of high pressure water jet (up to 207 MPa). In the current study. After treating the surface with PWJ (15 MPa) cavities and micro-pits appeared at grain/interface boundaries. Further erosion was not following grain boundaries as significantly as compared to Huang et al. [13]. The results of this paper can be compared only with initial stages of the experiments carried by Huang et al. (after one pass) which may be explained by time of interaction of jet with the surface of treated area.

Kamkar et al. [16] observed initial erosion stage in forged Ti6Al4V. They presented surface upheaval as a sign of initial erosion damage. Protrusions across several grains were documented in their work (width 10–50 μm. Height 2–20 μm) [16]. In samples in this study. no protrusions stretching across entire grains were observed (maximum length of elevation in Fig. 7b is approximately ten microns). The protrusions were always closely related to interphase boundaries. In this research. The signs of plastic deformation are focused around preexisting surface irregularities. Namely depressions. After the low number of impacts (e.g.1050 impacts).

The several cracks and cavities presented in this paper (Fig. 6, Fig. 7 and Fig. 8) are selected examples of cavities observed during conducted research. Generally appearing cavities followed interphase boundaries very closely. Places where concentration of stress may lead to appearance of new cavities is documented at Fig. 5b and Fig. 6b. Cavities presented in this paper ranged from 5 to 20 microns. The cavity captured at

Fig. 8a differentiate from others by shape. Because it propagates into the bulk of alpha grain. The cavities at the phase boundaries are likely caused by interaction of the stress wave and structural discontinuity at the alpha/beta phase boundary [17].

5 Conclusions

The main objective of the present study was to determine the effect of pulsating water jet on the evolution of early stages of damage of Ti6Al4V alloy with the increasing number of impingements. The method used in this study was pulsating water jet. Which allows much higher frequency of impingements than classic method for erosion testing. Conclusions drawn from the results obtained during observations were as follows:

- Changes of the surface morphology at the very early erosion stages were detected using optical microscopy and Differential Interference Contrast. Roughness across the impacted area is non-homogenously distributed probably due to the small angle deviation from the perpendicular direction of the PWJ and the surface of the impacted area.
- The formation of mechanical twins and slip bands due to the application of the PWJ was observed in the alpha grains. These are signs of plastic deformation.
- At the low number of impacts. it was observed that the most signs of deformation occurred in the area adjacent to surface irregularities. These surface irregularities act as stress concentrators. The magnitude of plastic deformation thus depends on the initial surface quality.
- The most observed change at initial stages of erosion of impacted surface was the tilting of the alpha phase.
- The cracks and cavities found on the specimen surface started usually at the phase interface. These sites are (due to the grain tilting) the ideal places for crack initiation creating a good base for material removal and formation of the cavity.

Acknowledgement. The authors are grateful for financial support of the project GACR No 19-000408S.

References

1. Foldyna, J.: Use of acoustic waves for pulsating water jet generation. In: Beghi, M.G. (ed.) Acoustic Waves - from Microdevices to Helioseismology. InTech, Croatia (2011)
2. Foldyna, J., Klich, J., Hlavacek, P., et al.: Erosion of metals by pulsating water jet. Tehn. Vjesnik-Tech. Gazette. **19**(2), 381–386 (2012)
3. Hloch, S., Srivastava, M., Nag, A., et al.: Effect of pressure of pulsating water jet moving along stair trajectory on erosion depth, surface morphology and microhardness. Wear **452–453**, 203278 (2020)
4. Srivastava, M., Hloch, S., Krejci, L., et al.: Residual stress and surface properties of stainless steel welded joints induced by ultrasonic pulsed water jet peening. Measurement **127**, 453–462 (2018)

5. Tönshoff, H.K., Kroos, F., Marzenell, C.: High-pressure water peening-a new mechanical surface-strengthening process. CIRP Ann. **46**(1), 113–116 (1994)
6. Azhari, A., Schindler, C., Li, B.: Effect of waterjet peening on aluminum alloy 5005. Int. J. Adv. Manuf. Technol. **67**, 785–795 (2013)
7. Srivastava, M., Hloch, S., Gubeljak, N., et al.: Surface integrity and residual stress analysis of pulsed water jet peened stainless steel surfaces. Measurement **143**, 81–92 (2019)
8. Hlaváček, P., Kruml, T., Foldyna, J., et al.: Effect of pulsating water jet peening on stainless steel. In: Water Jet 2015. Ústav geoniky AV ČR. v. v. i. (2015). ISBN 978-80-86407-56-2
9. Arola, D., Alade, A.E., Weber, W.: Improving fatigue strength of metals using abrasive waterjet peening. Mach. Sci. Technol. **10**(2), 197–218 (2006)
10. Miao, H.Y., Demers, D., Larose, S., et al.: Experimental study of shot peening and stress peen forming. J. Mater. Process. Technol. **210**(15), 2089–2102 (2010)
11. Srivastava, M., Hloch, S., Tripathi, R., et al.: Ultrasonically generated pulsed water jet peening of austenitic stainless-steel surfaces. J. Manuf. Process. **32**, 455–468 (2018)
12. Kamkar, N., Bridier, F., Bocher, P., et al.: Water droplet erosion mechanisms in rolled Ti–6Al–4V. Wear **301**(1–2), 442–448 (2013)
13. Huang, L., Folkes, J., Kinnell, P., et al.: Mechanisms of damage initiation in a titanium alloy subjected to water droplet impact during ultra-high pressure plain waterjet erosion. J. Mater. Process. Technol. **212**(9), 1906–1915 (2015)
14. Turski, M., Clitheroe, S., Evans, A.D., et al.: Engineering the residual stress state and microstructure of stainless steel with mechanical surface treatments. Appl. Phys. A **99**(3), 549–556 (2010)
15. Thomas, G.P., Brunton, J.H.: Drop impingement erosion of metals. Proc. Roy. Soc. London Ser. A Math. Phys. Sci. **314**(1519), 549–565 (1970)
16. Kamkar, N., Bridier, F., Jedrzejowski, P., et al.: Water droplet impact erosion damage initiation in forged Ti–6Al–4V. Wear **322–323**, 192–202 (2015)
17. Ma, D., Mostafa, A., Kevorkov, D., et al.: Water impingement erosion of deep-rolled Ti64. Metals **5**, 1462–1486 (2015)

Preliminary Testing of Production Thin-Walled Ribs into Aluminum Alloy AW6060 Using the Abrasive Water Jet

Frantisek Botko[1], Dominika Botkova[1,2](✉), Jiri Klich[2], Dagmar Klichova[2], and Vladimir Simkulet[1]

[1] Faculty of Manufacturing Technologies With a Seat in Presov, The Technical University of Kosice, Bayerova 1, 080 01 Presov, Slovakia
dominika.botkova@tuke.sk
[2] Institute of Geonics, The Czech Academy of Sciences, Ostrava-Poruba, Czech Republic

Abstract. Nowadays is placed great emphasis on miniaturization and reduction of weight in many industrial sectors. Application of abrasive water jet to produce thin-walled components hide big potential for future applications, due to small lateral force and absence of heat-affected zone. The presented article is focused on the possibility of creating thin-walled ribs using an abrasive water jet. An experimental procedure was selected to produce the thinnest possible rib in aluminum alloy AW 6060. Ribs were created by two passes of the AWJ with changing of the pitch between two passes. Due to minimal lateral force in the process as possible to create a rib with an average width of 0.4 mm with a height, more than 8 mm. Metallography analysis was performed for a closer look on the subsurface layers of the created rib and root of the rib. Results have the potential to extend knowledge in the field of abrasive water jet machining of thin-walled components.

Keywords: Abrasive water jet · Thin wall · Aluminum alloy

Nomenclature

A	elongation [mm]
AWJ	abrasive water jet
D_F/D_O	focusing tube diameter/ nozzle orifice diameter [mm]
m_a	abrasive mass flow [g.min^{-1}]
MESH	size of abrasive particles
p	pressure of pump [MPa]
R_m	tensile strength [MPa]
$R_{p0.2}$	yield point [MPa]
s	groove pitch [mm]
v	traverse speed [m.min^{-1}]
z	stand-off distance [mm]

S. Hloch et al. (Eds.): ICMEM 2021, LNME, pp. 75–86, 2021.
https://doi.org/10.1007/978-3-030-71956-2_7

1 Introduction

The disintegration effect of water jet technology on different types of materials has been intensively studied in recent decades [1]. Researchers around the world are constantly exploring the possibilities of using water jet modifications in a variety of common as well as specific conditions [2].

In the industrial production area, the serious shortcomings of conventional machining methods are considered to be: HAZ formation, tool wear [3], and the formation of vapors and dust particles [4] in various operations [5].

The use of water as a cutting medium can eliminate the above-mentioned shortcomings. For this reason, intensive efforts are being made to involve or replace certain production operations by water jet [6].

In water jet technology, the tool for cutting and disintegrating materials is a jet of water flowing through the nozzle orifice and impinging on the material [7]. Water jet - the tool can take various forms, which also affects its effectiveness and effects on the surface. The best known is the use of a water jet with or without an abrasive additive [8, 9]. According to the continuity of the flow [10], it is possible to divide the water jet into continuous and discontinuous [11, 12]. From this reason, several modifications of the water jet are known, such as pure continuous [13] and discontinuous (pulsating) [14, 15], abrasive [16, 17], suspension [18, 19], cryogenic [20, 21], etc. Each of the modifications is characterized by specific effects on the target surface [22, 23], therefore the water jet technology is widely applicable in various industries [24–29].

The aim of the described experimental research is focused on the possibilities of using an abrasive water jet in the production of thin walls of metal components [30, 31] as a replacement for conventional milling. Thin-walled components are used in various sophisticated industries [32] and the possibilities of their products have so far been explored by few authors. Application of the thin-walled components is in the heat sink production, in the automotive industry. The possibilities of replacing conventional milling with water jet have been discussed by several authors. Alberdi, et al. [33] focused on the model for prediction of the kerf profile in slot milling using abrasive water jet in aluminum 7075-T651.

The authors [33] found that in AWJ milling, the depth of the machined surface is directly dependent on the feed rate. For this reason, it is difficult to maintain a constant depth of cut profile during milling. The process is also affected by the acceleration and deceleration of the cutting head. Finally, they expressed the need to develop more advanced machining strategies. These machining strategies would be able to adapt the cutting conditions (pressure, the mass flow of the abrasive) so that the influence of the variable feed rate on the final quality of the machined surface is optimized. The authors [33] also expressed the opinion that if the machining strategies were at a more advanced level, AWJ milling could also be used for pocket milling. The idea of pocket milling was also addressed by Botko, et al. [34], who expressed a similar opinion in their study. Other of the few authors dealing with the described issues were Azarsa, et al. [35], who investigated micro-machining of high aspect ratio free-standing structures such as fins used in heat sinks. The authors [35] investigated the effect of changes in the nozzle feed rate, the number of passes, pressure, channel offset, and abrasive mass flow on the final rib shape.

The authors [35] found that at a lower feed rate a wider groove was formed between the walls of the rib (canal), but the formed canal was deeper and less affected by erosion caused by slurry flow. The higher pressure causes the primer of the incident flow to increase, which increases erosion and reduces the required flow shift. At lower feed rates and higher pump pressures, it was possible to machine ribs spaced approximately 500 μm apart machined at aspect ratios >40. Ribs with such dimensions are difficult to achieve with conventional machining methods, so the AWJ milling potential is high in this area.

Due to the fact that the specific researched issue - the creation of ribs using AWJ is described by few authors, the evaluation also took into account the results and procedures of research by other authors [36–39]. In their work, these authors described the issue of division, disintegration, and machining of various materials using AWJ. For the experimental research described in this article, they served as inspiration for the direction of research.

Klich, et al. [36, 37] studied disintegration erosion effects of the pulsating water jet on pre-treated (rough and fine milling, planning, and rolling) aluminum alloy surfaces. Akkurt [38] studied microstructures and variation of the hardness of pure aluminum and aluminum alloy AW 6061 surfaces cut with different processes (sawing, milling, laser, plasma, submerged plasma, wire discharge electric machining, oxy-fuel flame, and abrasive water jet). Research has shown that the AWJ method can be effectively used in industrial applications where no microstructural changes and hardness reduction is essential. Sourd, X., et al. [39] studied the influence of AWJ machining parameters (traverse speed, the pressure of jet, and scan step) on the cutting depth of 3D woven CFRP composite.

Due to the fact that the specific researched issue - the creation of ribs using AWJ is described by few authors, the evaluation also took into account the results and procedures of research by other authors [36–39]. In their work, these authors described the issue of division, disintegration, and machining of various materials using AWJ. For the experimental research described in this article, they served as inspiration for the direction of research.

None of the authors described above dealt with the length of the minimum possible distance at which the rib wall collapsed. For this reason, the "boundary" at which the rib wall is deformed due to previous erosion of the adjacent groove was investigated. This aspect of AWJ milling is important for the future possible use of AWJ in practical applications in order to avoid machining errors and find limits of machining.

2 Experimental Material and Procedure

Presented experimental research was realized on aluminum alloy AW6060 T6 (Table 1), which is a medium strength alloy with good weldability, good corrosion resistance, and good formability. Experimental material was in the state T6, which means the solution heat-treated and artificially aged.

Table 1. Chemical composition and mechanical properties of aluminum alloy AW6060.

Chemical composition									
Fe	Si	Cu	Mg	Mn	Cr	Ti	Zn	Al	
0.2	0.44	0.0071	0.44	0.03	0.004	0.015	0.01	Rest	
Mechanical properties									
R_m [MPa]		$R_{p0.2}$ [MPa]		A [%]		Hardness HB			
218		185		12.56		79			

The experimental investigation was performed in the laboratory at the Institute of Geonics of the Czech Academy of Sciences in Ostrava. Experimental milling was realized on a technological device for the machining of flat materials 2D X-Y cutting table PTV WJ2020-2Z-1xPJ. The abrasive was Australian garnet (MESH 80). The working water pressure was generated by PTV 75–60 pump. Groove measurements were performed with a digital microscope Keyence VHX-5100 optical measuring instrument and evaluated in the VHX-6000 program. Samples for metallographic analysis were cut with a metallographic saw, then ground, polished and etched. Metallographic samples were observed under a light microscope.

The experimental procedure was focused on the creation of thin walls using a single pass of the abrasive water jet. The experiment was performed on 3 separate samples with dimensions 10 × 10 × 100 mm. Walls (ribs) were created by making two grooves with a specific pitch between two passes of cutting head. The starting pitch was selected on 3 mm with lowering to 1.1 mm, which represents a theoretical wall thickness of 0.3 mm. The whole experimental procedure was performed with constant pressure, abrasive mass

Table 2. Technological conditions of the experiment.

Sample	s [mm]	v [m.min^{-1}]	p [MPa]	m_a [g.min^{-1}]	D_F/D_O [mm]	z [mm]	Abrasives
1	3	200	120	100	0.9/0.33	4	Australian garnet MESH 80
2	2.5						
3	2						
4	1.50						
5	1.25						
6	1.10						
7	2	150					
8	1.50						
9	1.25						
10	1.10						

flow, stand-off distance, and DF/DO. The variable parameters were pitch and traverse speed. Experimental technological conditions are shown in Table 2.

3 Results

Figure 1 shows the configuration of the grooves and ribs according to the technologic condition of the experiment listed in Table 1.

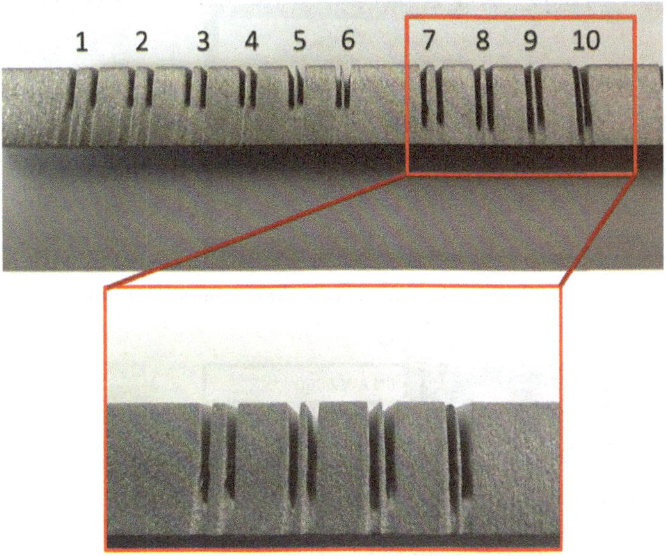

Fig. 1. Configuration of ribs (pairs of grooves).

Figure 2 shows a 3D representation of a pair of grooves with pitch (s) 1.25 mm created with traverse speed (v) 200 mm.min^{-1}. As can be observed rib with the thickness (on the top) approx. 0.2 mm is stable and does not show lateral deformation. The average depth of created groves was 5.5 mm. Figure 2 represents the top view on the pair of grooves (v = 200 mm.min^{-1}, s = 1.25 mm). It should be observed that the width of both grooves is similar from the top view and the width of the rib corresponds with the pitch of the AWJ paths in the material. It can be stated that the created rib is stable, with no measurable lateral deformation.

Figure 3 shows pair of grooves created with parameters (v = 200 mm.min^{-1}, s = 1.10 mm). On the figure can be observed lateral deformation of the rib pointing to the first created groove. The average depth of the grooves was 5.6 mm. The rib created with the pitch 1.1 mm is not stable and is clearly visible deformation towards the first crated groove. Figure represents top view on the pair of grooves (v = 200 mm.min^{-1}, s = 1.10 mm). As is evident from the figure, the rib is inclined laterally towards the first created groove. Inclination on the top level of the rib is approximately 0.3 mm.

Figure 4 shows a 3D representation of a pair of grooves with pitch (s) 1.25 mm created with traverse speed (v) 150 mm.min^{-1}. As can be observed rib with the thickness (on

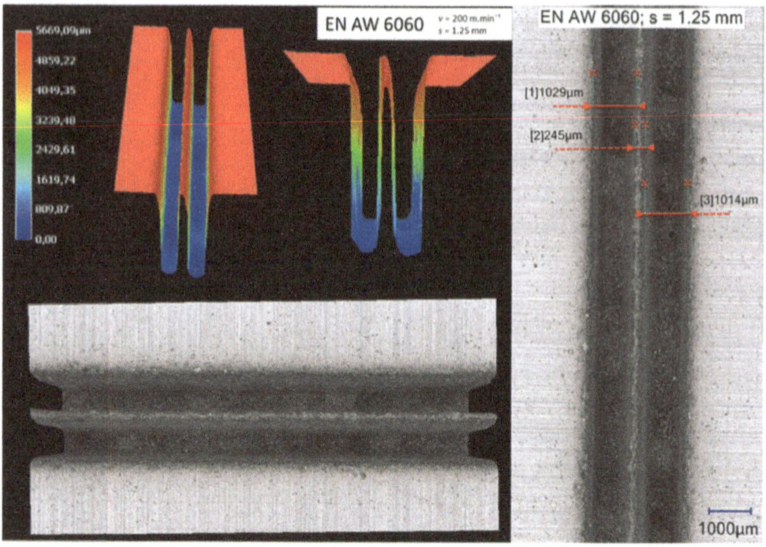

Fig. 2. 3D and top view of the rib ($v = 200$ mm.min^{-1}, $s = 1.25$ mm).

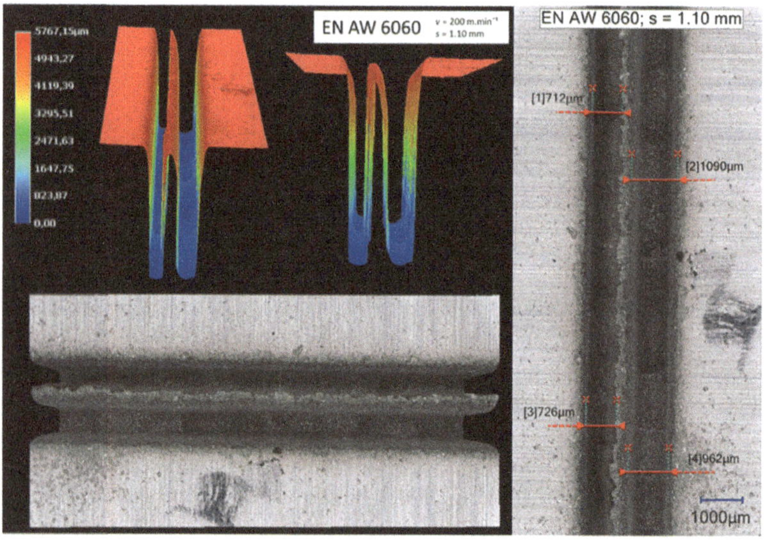

Fig. 3. 3D and top view of the rib ($v = 200$ mm.min^{-1}, $s = 1.10$ mm).

the top) approx. 0.2 mm is stable and does not show lateral deformation. The average depth of created grooves was 8.8 mm. Figure 4 represents the top view on the pair of grooves ($v = 150$ mm.min^{-1}, $s = 1.25$ mm). It should be observed that the width of both grooves is similar from the top view and the width of the rib corresponds with the pitch of the AWJ paths in the material. It can be stated that the created rib is stable, with no measurable lateral deformation.

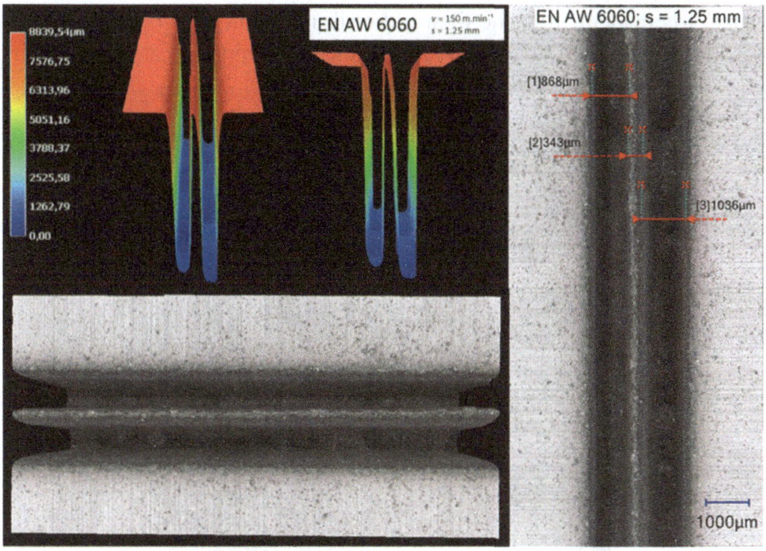

Fig. 4. 3D and top view of the rib ($v = 150$ mm.min^{-1}, $s = 1.25$ mm).

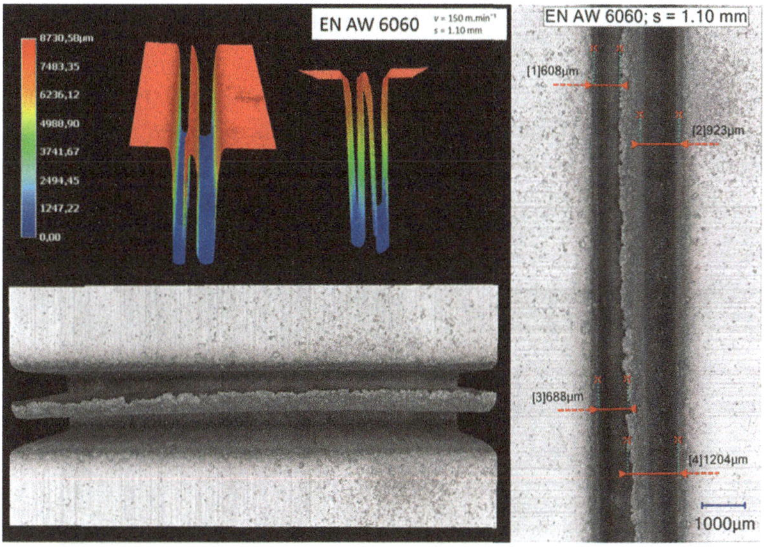

Fig. 5. 3D and top view of the rib ($v = 150$ mm.min^{-1}, $s = 1.10$ mm).

Figure 5 shows pair of grooves created with parameters ($v = 150$ mm.min^{-1}, $s =$ 1.10 mm). On the figure can be observed lateral deformation of the rib pointing to the first created groove. The average depth of the grooves was 8.7 mm. The rib created with the pitch 1.10 mm is not stable and is clearly visible deformation towards the first crated groove. Figure 5 represents top view on the pair of grooves ($v = 200$ mm.min^{-1}, $s =$

1.10 mm). As is evident from the figure, the rib is inclined laterally towards the first created groove. Inclination on the top level of the rib is approximately 0.35 mm.

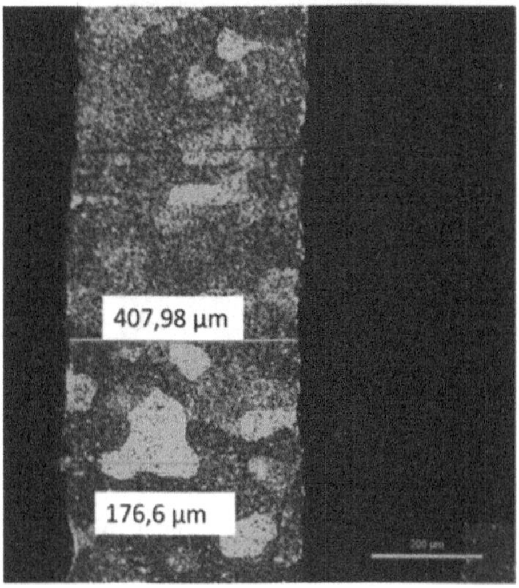

Fig. 6. Metallographic sample of the rib.

Figure 6 shows the metallography of the rib at magnification 100x. The edge of the rib does not exhibit cracks and visible tearing. It can be observed that some grains of alloying compounds reach dimensions 100 µm and more. The biggest grain visible has a critical dimension of 176.6 µm, which is almost half of the total thickness of the created rib.

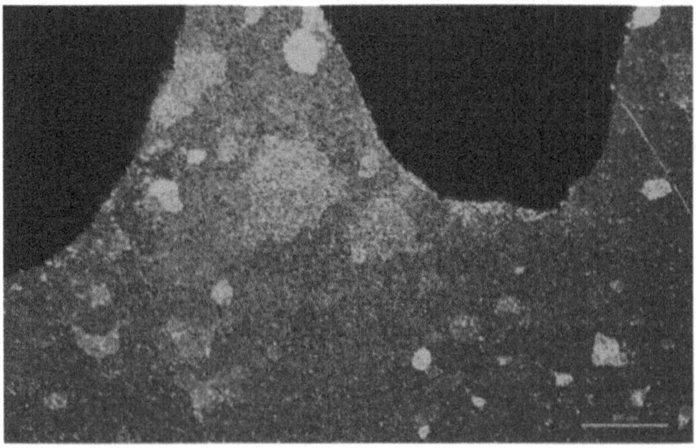

Fig. 7. The root of the rib.

Figure 7 shows the bottom of the groove and root of the rib. The edges are clean, without visible treating and cracks. There was not observed tearing of the grains or cracking on the grain boundaries.

The results for every sample was the same. Ribs created till the pitch 1.25 were stable with no measurable lateral inclination. For the pitch, 1.1 mm was observed inclination of the rib with the direction of the inclination towards the first created groove. Radial forces from the water jet are causing the deformation of the wall. For selected material AW6060 was the thickness of the rib for pitch 1.1 mm on the level of grain size.

4 Conclusions

Production of thin walls in engineering practice meets many difficulties. The creation of thin-walled components by chip machining is difficult and relatively slow due to high lateral forces, which cause deformations. AWJ shows high potential to produce thin-walled components due to the different mechanisms of material removal and small lateral forces. The presented research was focused on the creation of thin-walled ribs into aluminum alloy AW6060 T6.

Obtained findings:

- Stable ribs with height approx. 5 mm and 8 mm with thickness 0.4 mm were created ad once
- No visible cracks and tearing of grains were observed
- The lowest pitch which resulted in stable rib was 1.25 mm
- With pitch, 1.1 mm (theoretical thickness 0.3 mm) created rib for both 5 and 8 mm height was not stable
- Unstable ribs were inclined toward the first crated groove

During the process of cutting and machining with a water jet, not only direct contact with the material occurs. In the case of the deeper holes disintegration occur the radial flow and the flow lag. The disintegration of materials to "depth" is influenced by these factors, therefore the material is not only removed directly along the cut line but there is also a side flow, which removes the material along the sides and thus unevenly weakens the wall of the groove (channel). In the case of AWJ thin wall milling, this factor is decisive when planning the machining process, because the limit at which the yield strength of the material will not be disturbed by the disintegration of the adjacent thin wall must be observed. In case of incorrect determination of the maximum permissible distance, the wall is deformed or collapsed. The determination of the maximum permissible distance between two adjacent grooves depends mainly on the type, composition, and mechanical properties of the material and on the technological condition of cutting. In this regard, new space is opening up to create an advanced way to control AWJ milling operations.

In the future is necessary to perform tests on a wider range of materials and also with shaped paths for obtaining complex information about the production of thin walls in different materials using AWJ.

Acknowledgments. This work was supported by the Slovak Research and Development Agency under the contract no. APVV-17–0490 - Detection of the erosive effect of pulsating water jet on material. This work was further supported by projects: KEGA no. 030TUKE-4/2018 - Popularization and transfer the strategy Industry 4.0 into the technical study programs focusing on secondary schools, VEGA no. 1/0045/18 - Research on selected parameters of structural elements of robots based Al alloys and by the Ministry of Industry and Trade of the Czech Republic, project No. FV 30233. Some aspects associated with the selection of a suitable abrasive for experiments were supported by the project of the Technology Agency of the Czech Republic No. TH04020218. Paper is the result of the Project implementation: University Science Park TECHNICOM for Innovation Applications Supported by Knowledge Technology, ITMS: 26220220182, supported by the Research & Development Operational Programme funded by the ERDF.

References

1. Liu, X.C., Liang, Z.W., Wen, G.L., Yuan, X.F.: Waterjet machining and research developments: a review. Int. J. Adv. Manuf. Technol. **102**(5–8), 1257–1335 (2019). https://doi.org/10.1007/s00170-018-3094-3
2. Kusnerova, M., Foldyna, J., Sitek, L., Valicek, J., Hloch, S., Harnicarova, M., Kadnar, M.: Innovative approach to advanced modulated waterjet technology. Tehn. Vjesnik-Tech. Gazette **19**(3), 475–480 (2012)
3. Hutyrova, Z., Harnicarova, M., Zajac, J., Valicek, J., Mihok, J.: Experimental Study of Surface Roughness of Wood Plastic Composites after Turning. Material Science and Engineering Technology II, pp. 108-+. Trans Tech Publications Ltd., London (2014). https://doi.org/10.4028/www.scientific.net/AMR.856.108
4. Perec, A.: Environmental Aspects of Abrasive Water Jet Cutting. Rocznik Ochrona Srodowiska **20**, 258–274 (2018). Part 1
5. Banon, F., Sambruno, A., Batista, M., Simonet, B., Salguero, J.: Surface quality and free energy evaluation of s275 steel by shot blasting, abrasive water jet texturing and laser surface texturing. Metals **10**(2), article number 290 (2020). https://doi.org/10.3390/met10020290
6. Klichova, D., Klich, J., Zlamal, T.: The use of areal parameters for the analysis of the surface machined using the abrasive waterjet technology. In: Advances in Manufacturing Engineering and Materials, ICMEM 2018, pp. 36–44. Springer, Berlin (2019). https://doi.org/10.1007/978-3-319-99353-9_5
7. Srivastava, M., Hloch, S., Gubeljak, N., Milkovic, M., Chattopadhyaya, S., Klich, J.: Surface integrity and residual stress analysis of pulsed water jet peened stainless steel surfaces. Measurement **143**, 81–92 (2019). https://doi.org/10.1016/j.measurement.2019.04.082
8. Hlavacek, P., Sitek, L., Foldyna, J.: Evaluation of erosion performance of abrasive particles in abrasive water jet cutting process. MM Sci. J. **2020**, 3869–3872 (2020). https://doi.org/10.17973/MMSJ.2020_03_2019140
9. Mital, G., Dobransky, J., Ruzbarsky, J., Olejarova, S.: Application of laser profilometry to evaluation of the surface of the workpiece machined by abrasive waterjet technology. Appl. Sci.-Basel **9**(10), article number 2134 (2019). https://doi.org/10.3390/app9102134
10. Raj, P., Chattopadhyaya, S., Mondal, A.: A review on continuous and pulsed water jet machining. Mater. Today-Proc. **27**, 2596–2604 (2020)
11. Hloch, S., Srivastava, M., Nag, A., Muller, M., Hromasova, M., Svobodova, J., Kruml, T., Chlupova, A.: Effect of pressure of pulsating water jet moving along stair trajectory on erosion depth, surface morphology and microhardness. Wear **452**, article number 203278 (2020). https://doi.org/10.1016/j.wear.2020.203278

12. Lehocka, D., Botko, F., Klich, J., Sitek, L., Hvizdos, P., Fides, M., Cep, R.: Effect of pulsating water jet disintegration on hardness and elasticity modulus of austenitic stainless steel AISI 304L. Int. J. Adv. Manuf. Technol. **107**(5–6), 2719–2730 (2020). https://doi.org/10.1007/s00170-020-05191-3
13. Bodnarova, L., Sitek, L., Hela, R., Foldyna, J.: New potential of high-speed water jet technology for renovating concrete structures. Slovak J. Civ. Eng. **19**(2), 1–7 (2011). https://doi.org/10.2478/v10189-011-0006-z
14. Hloch, S., Srivastava, M., Krolczyk, J.B., Chattopadhyaya, S., Lehocka, D., Simkulet, V., Krolczyk, G.M.: Strengthening effect after disintegration of stainless steel using pulsating water jet. Tehn. Vjesnik-Tech. Gazette **25**(4), 1075–1079 (2018). https://doi.org/10.17559/TV-20170327134630
15. Lehocka, D., Klich, J., Botko, F., Foldyna, J., Hloch, S., Kepic, J., Koval', K., Krejci, L., Storkan, Z.: Pulsating water jet erosion effect on a brass flat solid surface. Int. J. Adv. Manuf. Technol. **97**(1–4), 1099–1112 (2018). https://doi.org/10.1007/s00170-018-1882-4
16. Ganovska, B., Molitoris, M., Hosovsky, A., Pitel, J., Krolczyk, J.B., Ruggierio, A., Krolczyk, G.M., Hloch, S.: Design of the model for the on-line control of the AWJ technology based on neural networks. Indian J. Eng. Mater. Sci. **23**(4), 279–287 (2016)
17. Carach, J., Lehocka, D., Legutko, S., Hloch, S., Chattopadhyaya, S., Dixit, A.R.: Surface roughness of graphite and aluminium alloy after hydro-abrasive machining. In: Advances in Manufacturing (MANUFACTURING 2017), pp. 805–813. Springer, Poznan (2018). https://doi.org/10.1007/978-3-319-68619-6_78
18. Molitoris, M., Pitel, J., Hosovsky, A., Tothova, M., Zidek, K.: A Review of research on water jet with slurry injection. In: International Conference on Manufacturing Engineering and Materials, ICMEM 2016, pp. 333–339. Elsevier Science BV, Novy Smokovec (2016). https://doi.org/10.1016/j.proeng.2016.06.675
19. Perec, A., Radomska-Zalas, A.: Modeling of abrasive water suspension jet cutting process using response surface method. In: Computational Technologies in Engineering (TKI'2018) 2078, article number UNSP 020051. Amer Inst Physics, Jora Wielka (2019). https://doi.org/10.1063/1.5092054
20. Jerman, M., Lebar, A., Sabotin, I., Dresar, P., Valentincic, J.: Ice jet technology. MM Sci. J. 2379–2384 (2018). https://doi.org/10.17973/MMSJ.2018_06_201772
21. Yuvaraj, N., Kumar, M.P.: Cutting of aluminium alloy with abrasive water jet and cryogenic assisted abrasive water jet: a comparative study of the surface integrity approach. Wear **362**, 18–32 (2016). https://doi.org/10.1016/j.wear.2016.05.008
22. Klichova, D., Klich, J.: Study of the effect of material machinability on quality of surface created by abrasive water jet. In: International Conference on Manufacturing Engineering and Materials, ICMEM 2016, pp. 177–182. Elsevier Science BV, Novy Smokovec (2016). https://doi.org/10.1016/j.proeng.2016.06.653
23. Valicek, J., Hloch, S., Fabian, S., Monkova, K., Hatala, M., Luptakova, A., Radvanska, A.: Analysis of signals obtained from surfaces created by abrasive waterjet by means of amplitude-frequency spectra and autocorrelation function. Tehn. Vjesnik-Tech. Gazette **15**(1), 23–31 (2008)
24. Srivastava, A.K., Nag, A., Dixit, A.R., Tiwari, S., Scucka, J., Zelenak, M., Hloch, S., Hlavacek, P.: Surface integrity in tangential turning of hybrid MMC A359/B4C/Al2O3 by abrasive waterjet. J. Manuf. Process. **28**, 11–20 (2017). https://doi.org/10.1016/j.jmapro.2017.05.017
25. Ergene, B., Bolat, C.: A review on the recent investigation trends in abrasive waterjet cutting and turning of hybrid composites. Sigma J. Eng. Nat. Sci.-Sigma Muhendislik Ve Fen Bilimleri Dergisi **37**(3), 989–1016 (2019)
26. Fernandes, R.R., Wilson, D.I.: Modelling the cleaning of viscoplastic layers by impinging coherent turbulent water jets. J. Non-Newtonian Fluid Mech. **282**, article number 104314 (2020). https://doi.org/10.1016/j.jnnfm.2020.104314

27. Carach, J., Hloch, S., Hlavacek, P., Gombar, M., Klichova, D., Botko, F., Mital, D., Lehocka, D.: Hydro-abrasive disintegration of alloy Monel K-500-the influence of technological and abrasive factors on the surface quality. In: International Conference on Manufacturing Engineering and Materials, ICMEM 2016, pp. 17–23. Elsevier Science BV, Novy Smokovec (2016). https://doi.org/10.1016/j.proeng.2016.06.633
28. Hutyrova, Z., Scucka, J., Hloch, S., Hlavacek, P., Zelenak, M.: Turning of wood plastic composites by water jet and abrasive water jet. Int. J. Adv. Manuf. Technol. **84**(5–8), 1615–1623 (2016). https://doi.org/10.1007/s00170-015-7831-6
29. Hreha, P., Hloch, S., Monka, P., Monkova, K., Knapcikova, L., Hlavacek, P., Zelenak, M., Samardzic, I., Kozak, D.: Investigation of sandwich material surface created by abrasive water jet (AWJ) via vibration emission. Metalurgija **53**(1), 29–32 (2014)
30. Michalik, P., Zajac, J., Hatala, M., Mital, D., Fecova, V.: Monitoring surface roughness of thin-walled components from steel C45 machining down and up milling. Measurement **58**, 416–428 (2014). https://doi.org/10.1016/j.measurement.2014.09.008
31. Zajac, J., Hatala, M., Duplak, J., Duplakova, D., Steranka, J.: Experimental high speed milling of the selected thin-walled component. TEM J.-Technol. Educ. Manage. Inf. **6**(4), 678–682 (2017). https://doi.org/10.18421/TEM64-05
32. Michalik, P., Zajac, J., Straka, L., Nowakowski, L., Ambrozy, M.: Draft jig structure design for measuring deformation of thin-walled robotic or conveyor components. Adv. Sci. Technol.-Res. J. **12**(3), 55–60 (2018). https://doi.org/10.12913/22998624/92340
33. Alberdi, A., Rivero, A., de Lacalle, L.N.L., Etxeberria, I., Suarez, A.: Effect of process parameter on the kerf geometry in abrasive water jet milling. Int. J. Adv. Manuf. Technol. **51**(5–8), 467–480 (2010). https://doi.org/10.1007/s00170-010-2662-y
34. Botko, F.,Hlaváček, P., Lehocká, D., Foldyna, V, Hatala, M., Simkulet, V.: Effect of abrasive water jet machining on the geometry of shapes in selected tool steels. In: Lecture Notes in Mechanical Engineering, 6th International Conference on Water Jet - Research, Development, Application, WJ 2019, pp. 49–55. Springer, Celadna (2021). https://doi.org/10.1007/978-3-030-53491-2_5
35. Azarsa, E., Cinco, L., Papini, M.: Fabrication of high aspect ratio free-standing structures using abrasive water jet micro-machining. J. Mater. Process. Technol. **275**, article number 116318 (2020). https://doi.org/10.1016/j.jmatprotec.2019.116318
36. Klich, J., Klichova, D., Foldyna, V., Hlavacek, P., Foldyna, J.: Influence of variously modified surface of aluminium alloy on the effect of pulsating water jet. Strojniski Vest.-J. Mech. Eng. **63**(10), 577–582 (2017). https://doi.org/10.5545/sv-jme.2017.4356
37. Klich, J., Klichova, D., Hlavacek, P.: Effects of pulsating water jet on aluminium alloy with variously modified surface. Tehn. Vjesnik-Tech. Gazette **24**(2), 341–345 (2017). https://doi.org/10.17559/TV-20140219100749
38. Akkurt, A.: The effect of cutting process on surface microstructure and hardness of pure and Al 6061 aluminium alloy. Eng. Sci. Technol.-Int. J.-JESTECH **18**(3), 303–308 (2015). https://doi.org/10.1016/j.jestch.2014.07.004
39. Sourd, X., Zitoune, R., Crouzeix, L., Salem, M., Charlas, M.: New model for the prediction of the machining depth during milling of 3D woven composite using abrasive waterjet process. Compos. Struct. **234**, article number 111760 (2020). https://doi.org/10.1016/j.compstruct.2019.111760

Material Reuse of Waste Abrasive Particles from Abrasive Water Jet Technology in the Field of Polymer Particle Composite Systems

Miroslav Muller[1](✉), Martin Tichý[1], Monika Hromasová[2], Jan Novotný[3], Anna Rudawska[4], and Viktor Kolář[1]

[1] Department of Material Science and Manufacturing Technology, CULS, Kamýcká 129, 165 00 Praha 6 – Suchdol, Czech Republic
muller@tf.czu.cz

[2] Department of Electrical Engineering and Automation, CULS, Kamýcká 129, 165 00 Praha 6 – Suchdol, Czech Republic

[3] Faculty of Mechanical Engineering, J. E. Purkyně University in Ústí nad Labem, Pasteurova 3334/7, 400 96 Ústí nad Labem, Czech Republic

[4] Department of Production Engineering, Lublin University of Technology, Nadbystrzycka 36, 20 618 Lublin, Poland

Abstract. Various natural and synthetic fillers are used in the field of particulate reinforcing composite materials with an aim to increase the wear resistance or to reduce the price. Using a filler that is considered as a waste product can improve sustainable design of these composite materials. The abrasive water jet technology (AWJ) produces a waste product during cutting i.e. a sludge, which contains abrasive particles and residues of the cutting material. This paper researches a possibility of using the residual waste of the AWJ process for the fabrication of polymeric particle composites. Furthermore, the impact of using the waste sludge on the mechanical properties of the composite was determined. To this extent, the tensile strength, maximum elongation, wear resistance and hardness of three polymer systems were compared, two of them used one of the following fillers: (I) Garnet Mesh 80 (standard abrasive; G80N) and (II) used and processed Garnet Mesh 80 (G80R(AWJ)), and the last system was the matrix without the filler. The results proved good wettability between the matrix and both fillers (I) and (II). A significant difference was proved in the size and the shape of (I) and (II) particles. The particles (II) showed rounded edges.

Keywords: Reuse of abrasive residue · Abrasive polymer particle composite filler · Water jet waste · Tensile strength · Elongation at break · Modulus of elasticity · Wear resistance · Hardness · Secondary product · Garnet

S. Hloch et al. (Eds.): ICMEM 2021, LNME, pp. 87–99, 2021.
https://doi.org/10.1007/978-3-030-71956-2_8

1 Introduction

A composite is a material made from two or more constituent materials. These constituent materials have various mechanical and physical properties. Their self-interaction influences resulting mechanical properties [1]. It is not always possible to talk about a synergic effect of mechanical properties. It is usually reached by a reinforcing material (in the case of particles it is filler) in the form of oriented fibres.

The utilization of the particulate reinforcing material (filler) is to reduce cost, improve tribological properties or increase the hardness against a matrix [2–4].

The utilization of the particulate filler usually leads to increase the wear resistance [4–7], as proven by many researchers, i.e. the use of Al_20_3 [1, 8], SiC [9], Si [10], ballotine [11, 12] and biologically based fillers [13].

These composites (Al_20_3, SiC, Si and ballotine) are used as liquid metals. For these composite systems secondary materials are used, i.e. a waste [1, 14]. The filler size and shape influence mechanical properties of a composite [15]. For good functioning of the composite it is necessary to research the interaction between a filler (reinforcing material) and a matrix. A garnet is an example of the waste product, which is used in industries as an abrasive blasting medium, abrasive powder etc. [16]. Some authors have warned for environmental issues and health risk when using waste material due to contamination from cutting material, process liquids [16], oil etc. A waste can be used as filler for a polymeric matrix if it is not classified as dangerous waste, i.e. according to the decree 93/2016 Coll. in the Czech Republic. This waste comes from the machining process, i.e. abrasive particle. This waste production is significant in the Czech Republic and other EU countries [1]. The waste from surface treatment remains a worldwide environmental problem [16]. Authors present the utilization of the garnet [16] as a sand substitute in geopolymer concrete. This reuse of the garnet improves properties of the geopolymer concrete.

The abrasive water jet cutting technology (AWJC) is a method for cutting of various materials for many years. This technology uses different abrasive materials, which are powered by a stream of high-speed water [17].

The water jet technology is an essential unconventional production method at the present. Significant parameters of the technology are: heat unaffected zone, dust particle elimination, dangerous gases are not released etc. [18, 19], which make this technology environmentally friendly.

High operating costs belong to limits of this technology [17]. For example, the garnet consumption creates 56% of operating costs for consumables [17]. This value could be rapidly higher, if we include liquidation costs of used abrasive [17, 20].

The treatment of the waste after the abrasive water jet application was dealt with by many authors who researched further utilisation of waste water and sludge containing abrasive particles and residues of cut materials.

The most frequently discussed issue is the utilisation of water jet waste abrasive in machining.

This is associated with the research on the effect of cutting surface performance and quality [21]. The reason is high financial request on purchase of new abrasive [22]. For utilisation of water jet waste abrasive limits exist [21]. E.g. Kantha Babu and Krishnaiah Chetty state that the reusability of abrasives is determined as 81% [21].

Abrasive particles occur during acceleration and focusing processes and after cutting they are falling apart [21].

An important part of sludge can be reused after treating during the cutting [23]. Babu and Chetty proved effective possibility of abrasive reuse in abrasive water jet machining in their research. The research found that water jet waste abrasive reduced kerf taper and improved cut surface in parallel. The research showed that the choice of abrasive particle size was significant for reaching good results [21].

This research focuses mainly on garnet testing as it is one of the most commonly used abrasive particles in AWJ technology [21]. Habeeb Lateef Musttashar et al. [18] and Castel et al. [24] state that the garnet has high potential for reuse because of the angular fractures and hardness properties.

The research aim is to evaluate mechanical properties of the composite based on waste abrasive particles and primary abrasive particles Garnet MESH80 for a purpose to reach new pieces of knowledge in the area of material engineering. The research also evaluates the application possibilities of waste abrasive particles after the application of AWJ in the field of material reuse.

2 Materials and Methods

The research deals with reuse of the garnet abrasive as a waste from AWJ technology application. The possibility of material reuse is the polymeric particle composite based on waste from AWJ machining process registered in the waste catalogue under number 12 01 15. The material for machining was a polymer composite with a biologic reinforcement. The waste sludge from machining of the composite material was removed from the device CNC AWJ CT 0806. The primary abrasive Garnet MESH 80 used for AWJ technology is presented in Fig. 1A. The waste sludge (Fig. 1B) was dried for 24 h at the temperature 105 ± 5 °C.

The particle analysis was performed by the particle analyser Horiba and by SEM images with Tescan Mira 3. The particle size was established from SEM image analysis by programme Gwyddion.

Epoxide resin LH 288 with hardener H 282 (based on diamine) suitable for the production process using a vacuum infusion was the matrix. The composite materials based on the primary abrasive particle (marked Garnet MESH 80, Fig. 1A) and the waste sludge from the AWJ technology machining process (marked Garnet MESH 80 – AWJ, Fig. 1C) were produced by means of this resin, which mixing ratio was 100:23. The epoxy allows good wettability and saturation of the filler.

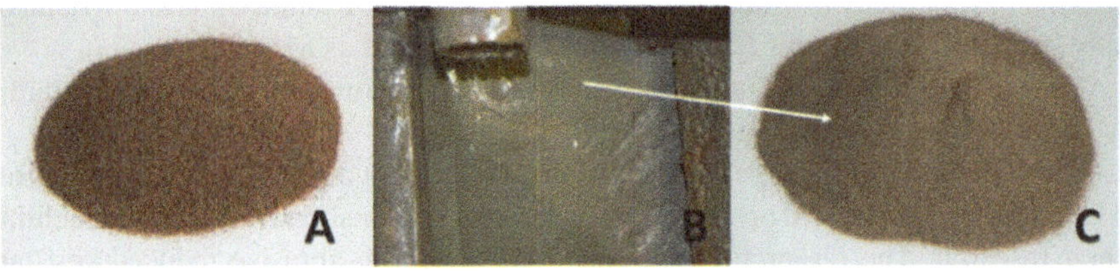

Fig. 1. Filler: A: primary abrasive Garnet MESH 80, B: waste sludge from AWJ process, C: water jet waste abrasive Garnet MESH 80 – AWJ.

The composite board was performed in particle ratio 30 wt.%. The results of maximum filler concentration was used from the previous research [6]. Boards with size 200 × 300 × 4 mm were made by the vacuum infusion. Testing specimens were prepared according to standards. The mechanical properties, i.e. tensile strength, elongation at break and modulus of elasticity were evaluated according to the standard CSN EN ISO 527 on the universal tensile testing machine LABTest 5.50ST at the loading speed 0.6 mm/min^{-1}.

The wear resistance was tested according to the standard Gost 23.208–79 on a device with a rubber wheel and abrasive particles (float sand Otava with fraction size 0.2–0.3 mm) at loading of sample 700 g. The wear was evaluated by the track length after 1000 turns and the mass loss by analytical scales Kern.

The hardness was evaluated according to CSN EN ISO 2039–1 by 5 mm diameter sphere and loading force 961 N on the device DuraJet G5 Rockwell Hardness tester.

Measured values were statistically evaluated by ANOVA F-test method (the hypothesis H_0 means statistically insignificant difference between measured values ($p > 0.05$). The hypothesis H_1 rejects the hypothesis *H0*).

3 Results and Discussion

Histogram of frequencies of used filler in composite materials obtained by the image analysis is shown in Fig. 2. The results from image analysis are in good correlation with results from the particle analyser Horiba. The size of primary abrasive particles Garnet MESH 80 was measured in interval 80–850 μm, where 90% of particles were in interval 100–450 μm. Particles were most often in interval 220–390 μm. The size of water jet waste abrasive particles was measured in interval 30–550 μm, where 90% of particles were in interval 70–400 μm. Particles were most often in interval 110–320 μm.

The hypothesis H1 ($p < 0.05$) was confirmed by the statistical testing, i.e. there is a statistically significant difference between measured size of primary abrasive particles Garnet MESH 80 and water jet waste abrasive particles Garnet MESH 80 - AWJ. Parameter $p = 0.0288$.

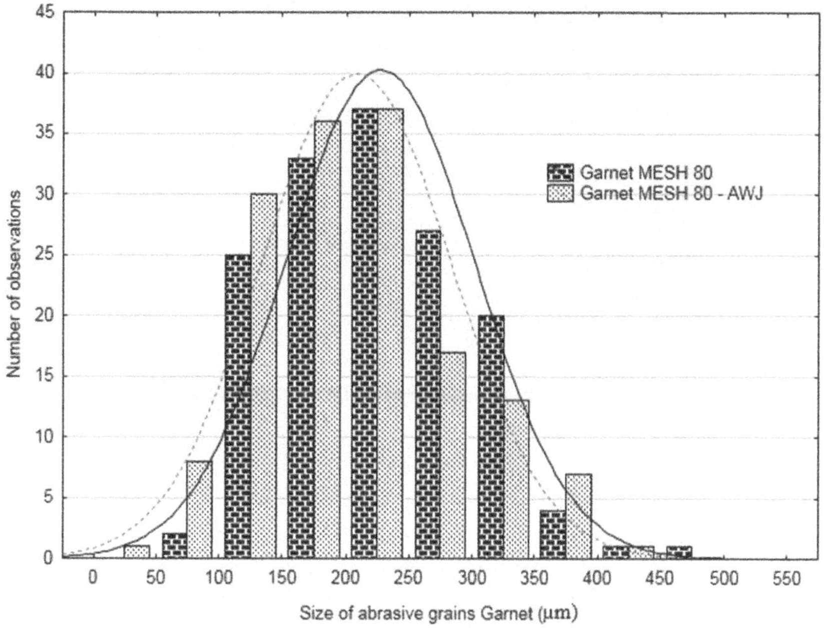

Fig. 2. Histogram of filler frequencies for composite materials, i.e. garnet abrasive particles.

Figure 3 shows SEM images of abrasive particles. Figure 3A, B present the new unused abrasive particles Garnet MESH 80, which are irregularly sharp-edged. Figure 3C, D present the abrasive particles MESH 80 gained from the sludge after AWJ machining process, which are of round edges. Kantha Babu et al. [21] present that larger particles after breakage have sharp edges. Finer particles have no possibility to break, i.e. edges are rounded after machining [21]. This conclusion suggests a complication of AWJ technology. SEM analysis results of used abrasive particles proved that most particles had rounded edges. The difference from Kantaha Babu et al. conclusion is probably due to the machining of polymer composite materials and not metals, which have higher material hardness.

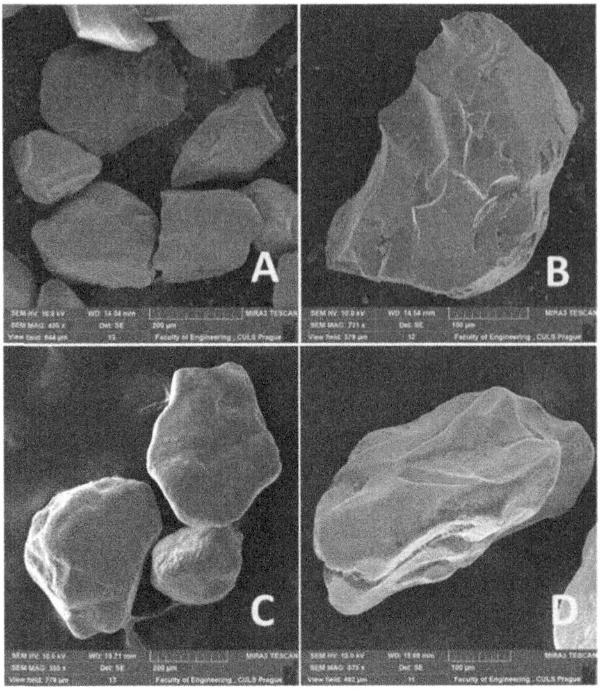

Fig. 3. SEM images of abrasive particles Garnet MESH 80: A: primary abrasive Garnet MESH 80 (MAG 430 x), B: detailed view on primary abrasive (MAG 731 x), C: water jet waste abrasive Garnet MESH 80 – AWJ (MAG 355 x), D: detailed view on water jet waste abrasive Garnet MESH 80 – AWJ (MAG 575 x).

Mechanical properties were compared between etalon (resin – matrix) and the composite system by the tensile strength (Fig. 4), elongation at break (Fig. 5), modulus of elasticity (Fig. 6), wear (Fig. 8) and hardness (Fig. 11). The resin without the filler reached the tensile strength 44.54 $\pm$ 3.51 MPa at the laboratory temperature. Addition of the filler reduced the tensile strength. Significant decrease of the tensile strength up to 32.4% was showed at the composite material based on the primary sharp-edged abrasive particles Garnet MESH 80 (composite G80N). Farraf Noor Ahmed et al. [25] found that irregular shape of the filler decreased mechanical characteristics. The reason is probably by sharp-edged particles, which cause a separation between the matrix and the filler [25].

A slight decrease of the tensile strength up to 9.6% was evaluated at the composite material based on the water jet waste abrasive particles Garnet MESH 80 – AWJ (composite G80R(AWJ)). The decrease of the composite material strength is consistent with other authors' results. The composite materials with the filler also had a negative effect on the elongation at break, where a decrease in interval 79–89% was. The decrease of the elongation at break depends on a type, a fracture size and a wettability of the filler [25, 26]. The modulus of elasticity is visible from Fig. 6. From the results an increase of the modulus of elasticity at composite materials by adding a filler and especially by primary sharp-edged abrasive particles Garnet MESH 80 (composite G80N) is evident.

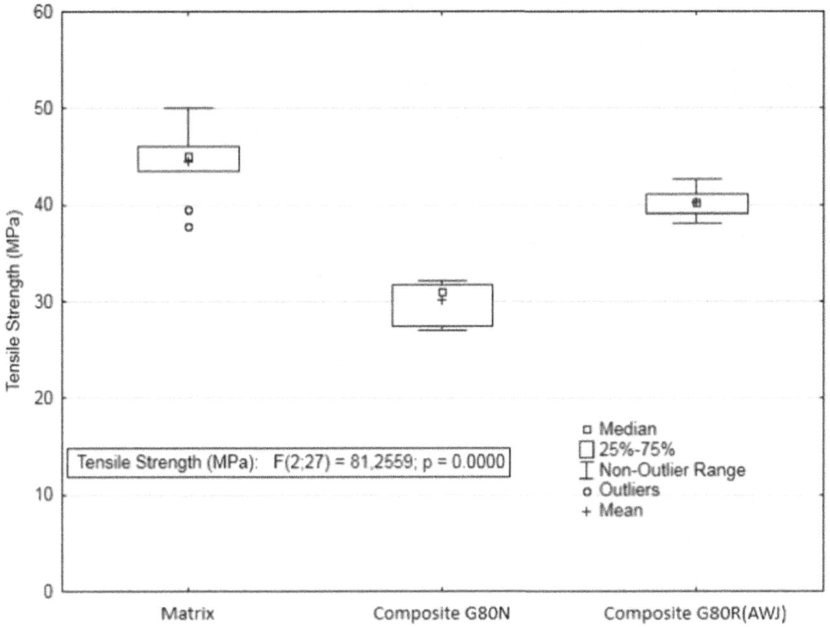

Fig. 4. Tensile strength of matrix and composite materials.

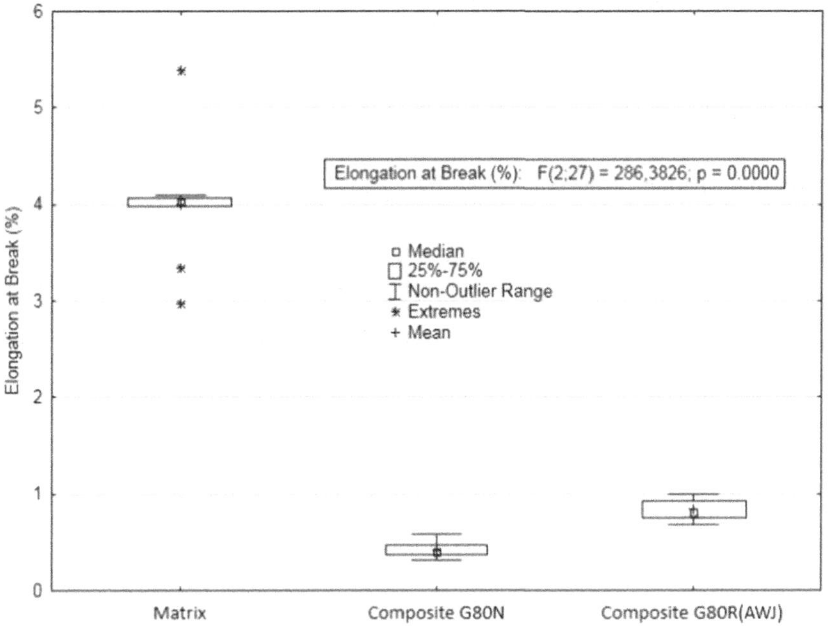

Fig. 5. Elongation at break of matrix and composite materials.

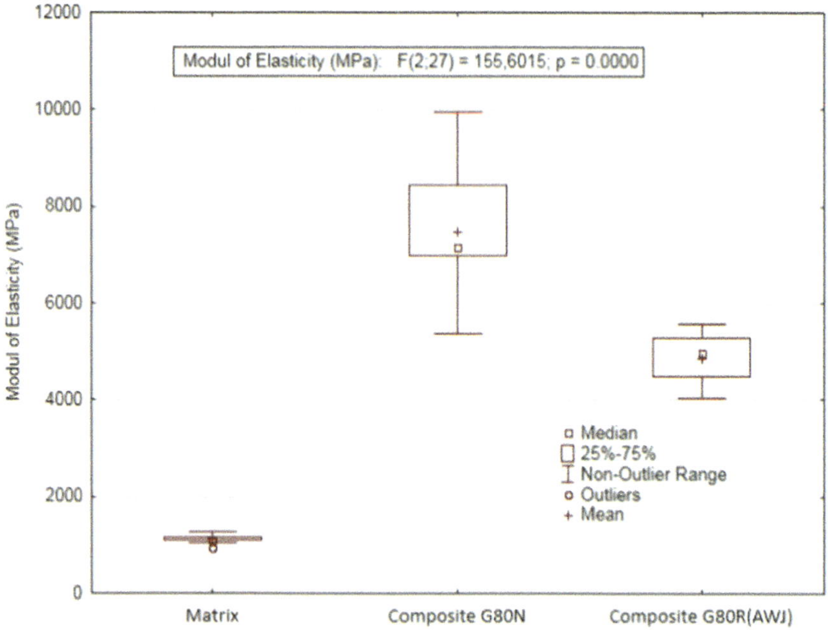

Fig. 6. Modulus of elasticity of matrix and composite materials.

The matrix reinforcing optimization by the fillers depends on the interaction, i.e. a stress transmission between the matrix and the reinforcing filler by adhesion [27–29]. Figure 7 presents fracture surface of tested materials. Figure 7A and B present a good wettability between the matrix and the filler. Conclusions about the utilization of various fillers in composite materials were confirmed, i.e. adding filler into matrix influences mechanical properties of the composite [29–31].

Chang et al. confirmed that the irregular shape of the filler ensured good interaction of materials in the composite [32].

The hypothesis *H1* ($p < 0.05$) was confirmed by the statistical testing, i.e. there is a statistically significant difference between measured matrix, composites G80N and G80R(AWJ) in the tensile strength ($p = 0.000$), the elongation at break ($p = 0.000$), the modulus of elasticity ($p = 0.000$) and the hardness ($p = 0.000$).

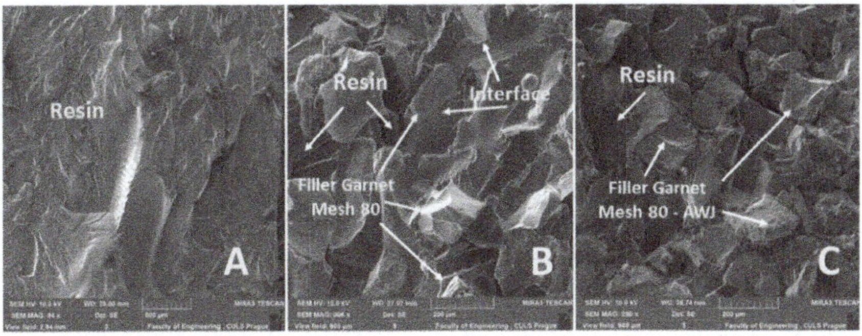

Fig. 7. Static tensile test of fracture surface: A: Pure resin (MAG 94 x), B: polymer composite based on primary abrasive particles Garnet Mesh 80 (MAG 306 x), C: polymer composite based on water jet waste abrasive particles Garnet Mesh 80 – AWJ (MAG 280 x).

Figure 8 shows the wear process of the matrix and composite materials (composite G80N and G80R(AWJ)) at the constant load 700 g to the rubber wheel. The wear process has a linear trend. The results show a linear increase of the mass loss at tested samples according to the wear track. The composite materials showed higher wear resistance, i.e. composite G80N up to 78% and composite G80R(AWJ) up to 68% against the matrix. The dependence tightness, the functional equations of the wear (mass loss) and the wear track are listed in Table 1.

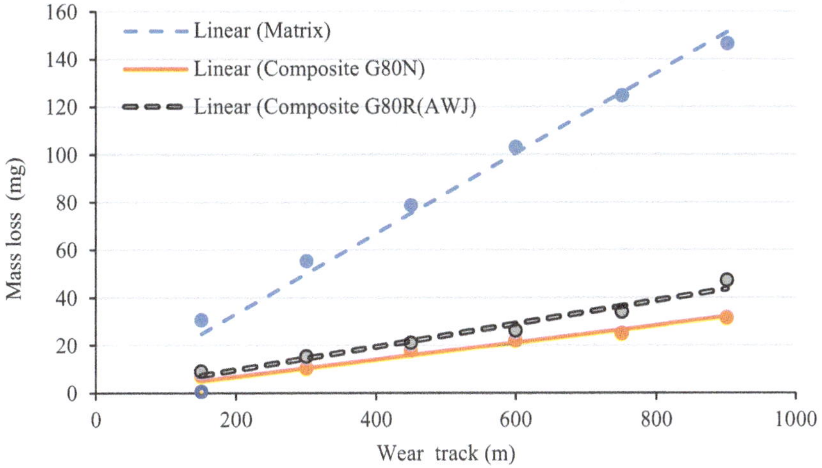

Fig. 8. Wear process of matrix, composite G80N and composite G80R (AWJ).

Table 1. Functional equations of the mass loss dependence on the wear track of composite materials with their determination index.

Designation	Wear test - loading (g)	Functional equation	Determination index
Matrix	700	y = 0.1686x	0.9889
Composite G80N	700	y = 0.0363x	0.9705
Composite G80R (AWJ)	700	y = 0.0363x	0.9671

The SEM analysis of composite surfaces G80N (Fig. 9) and G80R(AWJ) (Fig. 10) did not prove different characteristics. Figures 9 and 10A show a transition between worn out area (left side) and intact area (right side) on the test sample surface. Figures 9 and 10B, C present good interaction in the tested composite. The figures also show that the soft matrix is removed from the test sample surface by the wear process. From the detailed view (Fig. 9 and 10C) of the filler (G80N and G80R (AWJ)) it is clear that the sharp edges of the filler G80N were removed due to the wear, i.e. primary abrasive particles Garnet MESH 80. The hardness of the matrix and the composite material is evident on Fig. 11. The matrix showed a significant difference in the hardness against to the composite materials. The garnet microparticles added to the matrix significantly increased the hardness of the composite materials.

When comparing the wear and hardness results of the composite materials, it is so obvious that there is not a direct correlation between the wear resistance and the hardness.

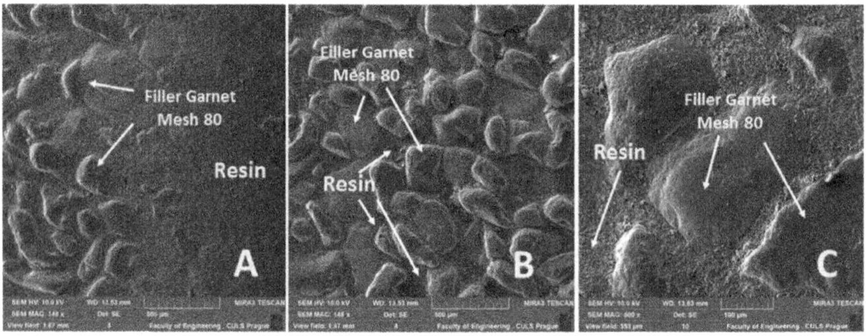

Fig. 9. SEM images of sample surface - composite G80N – loading 700 g: A: transition between worn out area (left side) and intact area (right side) of sample surface (MAG 148x), B: worn out surface (MAG 500 x), C: detailed view on worn out surface and interaction between filler and matrix (MAG 500 kx).

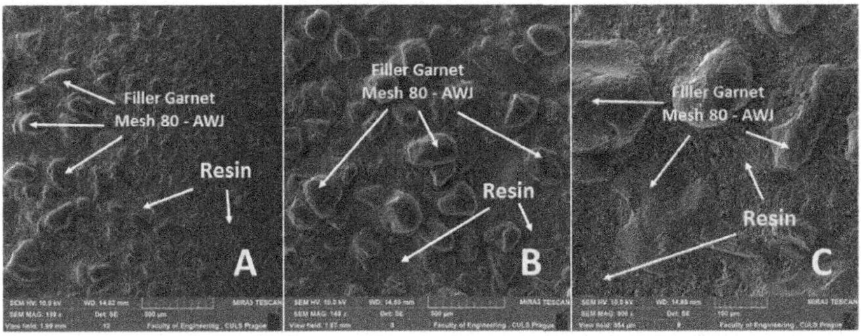

Fig. 10. SEM images of sample surface - composite G80R(AWJ) – loading 700 g: A: transition between worn out area (left side) and intact area (right side) of sample surface (MAG 139x), B: worn out surface (MAG 148 x), C: detailed view on worn out surface and interaction between filler and matrix (MAG 500 kx).

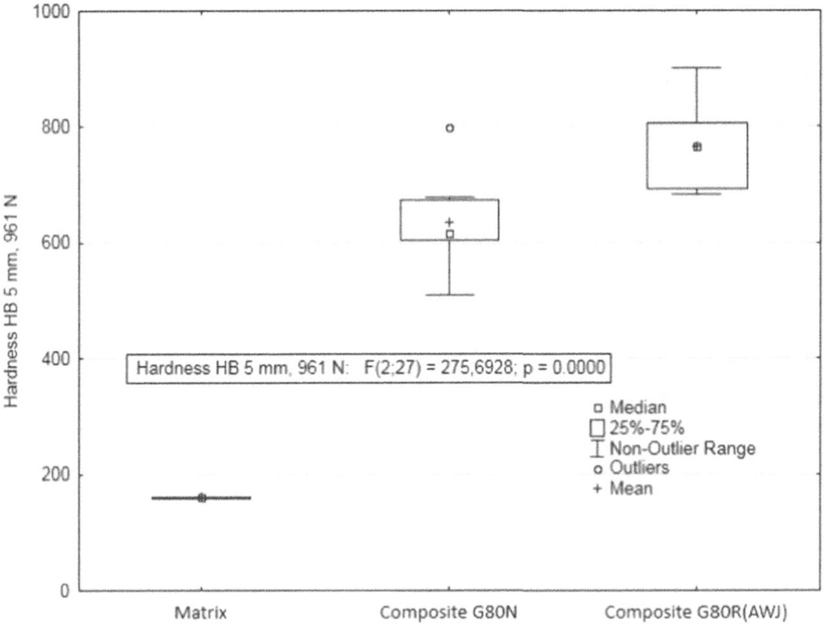

Fig. 11. Hardness of matrix, composite G80N and composite G80R (AWJ).

4 Conclusions

Hard microparticles are used in composite materials for very long time. The research focused on utilization of the waste sludge after cutting of composite materials by the AWJ technology. The filler for composite materials was obtained by treatment of the waste sludge and showed effectiveness in the field of particulate polymer composites. These composites contain hard particles of a garnet, therefore their application is determined for the field of liquid metal intended for a renovation of functional surfaces. The research proved good interaction between the resin and the filler both based on primary abrasive Garnet MESH 80 and water jet waste abrasive from treated sludge after AWJ composite cutting process.

The significant difference was proved in the size and the shape of the primary abrasive particles Garnet MESH 80 and water jet waste abrasive from AWJ process marked as Garnet MESH 80 – AWJ. The water jet waste abrasive particles showed rounded edges. The statistical testing proved a difference between mechanical characteristics of the resin and the composite and between mechanical characteristics of composites with primary abrasive and water jet waste abrasive particles, i.e. results of the tensile strength, elongation at break, modulus of elasticity, wear and hardness.

Acknowledgement. IGA FE CULS in Prague (no. 31140/1312/313108).

References

1. Valášek, P., Müller, M., Hloch, S.: Recycling of corundum particles - two-body abrasive wear of polymeric composites based on waste. Tehn. Vjesnik-Tech. Gazette **22**(3), 567–572 (2015)
2. El-Tayeb, N.S.M., Yousif, B.F., Yap, T.C.: Tribological studies of polyester reinforced with CSM 450-R-glass fiber sliding against smooth stainless steel counter face. Wear **261**, 443–452 (2006)
3. Pihtili, H., Tosun, N.: Effect of load and speed on the wear behaviour of woven glass fabrics and aramid fibre-reinforced composites. Wear **252**, 979–984 (2002)
4. RaVi Kumar, B.N., Suresha, B., Venkataramareddy, M.: Effect of particulate fillers on mechanical and abrasive wear behaviour of polyamide 66/polypropylene nanocomposites. Mater. Des. **30**, 3852–3858 (2009)
5. Kejval, J., Müller, M.: Mechanical properties of multi-component polymeric composite with particles of AL2O3/SiC. Sci. Agric. Bohem. **44**, 237–242 (2013)
6. Müller, M., Kejval, J.: Research on wear resistance of poly-component composite materials. Res. Agric. Eng. (Zemědělská technika) **63**(3), 106–114 (2017)
7. Suresha, B., Ravi Kumar, B.N.: Two-body abrasive wear behavior of particulate filled polyamide66/polypropylene nanocomposites. J. Appl. Polym. Sci. **119**, 2292–2301 (2010)
8. Satapathy, B.K., Bijwe, J.: Analysis of simultaneous influence of operating variables on abrasive wear of phenolic composites. Wear **253**, 787–794 (2002)
9. Basavarajappa, S., Joshi, A.G., Arun, K., Kumar, A.P., Kumar, M.P.: Three-body abrasive wear behaviour of polymer matrix composites filled with SiC particles. Polym.-Plast. Technol. Eng. **14**, 8–12 (2010)
10. Mohan, N., Natarajan, S., Kumaresh Babu, S.P.: The role of synthetic and natural fillers on three-body abrasive wear behaviour of glass fabric-epoxy hybrid composites. J. Appl. Polym. Sci. **124**, 484–494 (2012)
11. Müller, M.: Mechanical properties of resin reinforced with glass beads. Agron. Res. **15**(S1), 1107–1118 (2017)
12. Müller, M., Valášek, P.: Abrasive wear effect on polyethylene, polyamide 6 and polymeric particle composites. Manuf. Technol. **12**(12), 55–59 (2012)
13. Valášek, P., D'Amato, R., Müller, M., Ruggiero, A.: Mechanical properties and abrasive wear of white/brown coir epoxy composites. Compos. Part B-Eng. **146**, 88–97 (2018)
14. Valášek, P., Kejval, J., Müller, M.: Epoxy resin filled with primary and secondary raw material - Useable in agriculture. Res. Agric. Eng. (Zemědělská technika) **60**(4), 165–171 (2014)

15. Fu, S., Feng, X., Luke, B., Mai, Y.W.: Effects of particle size, particle/matrix interface adhesion and particle loading on mechanical properties of particulate-polymer composites. Compos.: Part B **39**, 933–961 (2008)
16. Muttashar, H.L., Ariffin, M.A.M., Hussein, M.N., Hussin, M.W., Ishaq, S.B.: Self-compacting geopolymer concrete with spend garnet as sand replacement. J. Build. Eng. **15**, 85–94 (2008)
17. Gent, M., Menéndez, M., Torno, S., Torano, J., Schenk, A.: Experimental evaluation of the physical properties required of abrasives for optimizing waterjet cutting of ductile materials. Wear **284**, 43–51 (2012)
18. Srivastava, M., Hloch, S., Gubeljak, N., Milkovic, M., Chattopadhyaya, S., Klich, J.: Surface integrity and residual stress analysis of pulsed water jet peened stainless steel surfaces. Measurement **143**, 81–92 (2019)
19. Hloch, S., Ruggiero, A.: Online monitoring and analysis of hydroabrasive cutting by vibration. Adv. Mech. Eng. **5**, 894561 (2013)
20. Kantha Babu, K., Krishnaiah Chetty, I.V.: A study of the use of single mesh size abrasives in abrasive waterjet machining. Int. J. Adv. Manuf. Technol. **29**, 532–540 (2006)
21. Kantha Babu, M., Krishnaiah Chetty, O.V.: A study on recycling of abrasives in abrasive water jet machining. Wear **254**, 763–773 (2003)
22. Dong, Y., Liu, W., Zhang, H., Zhang, H.: On-line recycling of abrasives in abrasive water jet cleaning. Proc. CIRP **15**, 278–282 (2014)
23. Knapp, M., Ohlsen, J.: Recycling of abrasive material in waterjet cutting. In: Proceedings of the 12th International Conference on Jet Cutting Technology, Rouen, France, pp. 511–519 (1994)
24. Castel, A., Vidal, A.T., François, R.: Bond and cracking properties of self-consolidating concrete. Constr. Build. Mater. **24**, 1222–1231 (2010)
25. Ahmad, F.N., Jaafar, M., Palaniandy, S., Azizli, K.A.M.: Effect of particle shape of silica mineral on the properties of epoxy composites. Compos. Sci. Technol. **68**, 346–353 (2008)
26. Nakamura, Y., Yamaguchi, M., Okubo, M., Matsumoto, T.: Effects of particle size on mechanical and impact properties of epoxy resin filled with spherical silica. J. Appl. Polym. Sci. **45**(7), 1281–1289 (1992)
27. Rudawska, A.: Surface free energy and 7075 aluminium bonded joint strength following degreasing only and without any prior treatment. J. Adhes. Sci. Technol. **26**, 1233–1247 (2012)
28. Rudawska, A., Danczak, I., Müller, M., Valášek, P.: The effect of sandblasting on surface properties for adhesion. Int. J. Adhes. Adhes. **70**, 176–190 (2016)
29. Müller, M., Valášek, P., Kolář, V., Šleger, V., Kagan Gürdil, G., Hromasova, M., Hloch, S., Moravec, J., Pexa, M.: Material utilization of cotton post-harvest line residues in polymeric composites. Polymers **11**(7), 1106 (2019)
30. Kim, H.S., Khamis, M.A.: Fracture and impact behaviours of hollow micro-sphere/epoxy resin composites. Compos. A Appl. Sci. Manuf. **32**(9), 1311–1317 (2001)
31. Agoudjil, B., Ibos, L., Majesté, J.C., Candau, Y., Mamunya, Ye.P.: Correlation between transport properties of Ethylene Vinyl Acetate/glass, silver-coated glass spheres composites. Compos. Part A: Appl. Sci. Manuf. **39**(2), 342–351 (2008)
32. Chang, J.J., Yeih, W.C.: The effect of particle shape on bond strength improvement of epoxy particle coating composites. J. Mar. Sci. Technol. **9**, 153–160 (2001)

On-Line Monitoring of In-Vitro Application of PWJ for Bone Cement Disintegration

Akash Nag[1](✉), Sergej Hloch[2,3], and Amit Rai Dixit[1]

[1] Department of Mechanical Engineering, Indian Institute of Technology (Indian School of Mines), Dhanbad 826004, India
akashnag1992@gmail.com

[2] Institute of Geonics, Czech Academy of Sciences, Ostrava-Poruba, Czech Republic

[3] Faculty of Manufacturing Technologies, TUKE with a Seat in Presov, 08001 Presov, Slovakia

Abstract. The use of pulsating water jet as a minimal invasive method for disintegration of bone cement requires optimal determination of its machine settings which effects the erosion depth. The volume of fluid required and the distance of the nozzle from the bone cement surface during its disintegration are one of the important machine settings for its in-vivo applications. Moreover, controlling of the technology during its action without effecting the phenomenon responsible for erosion is a challenging task. Therefore, in this study influence of variation of the nozzle diameter and standoff distance on the disintegration depth have been studied. Acoustic emission signals in form of acceleration values, recorded during the disintegration process are analyzed and correlated with the achieved groove depth trends. The results showed similar trend of acceleration values and disintegration depth when varying the nozzle diameter or standoff distance. Both the acceleration and disintegration depth, increases with an increase in the standoff distance till an optimal limit and decreases after it. Also, with the increase in the nozzle diameter, disintegration depth and acceleration value increase due to increase in the water flow rate. The analogy of recorded acoustic emission signals with the depth values achieved during the process can be used in further studies for controlling of the water jet process.

Keywords: Pulsating water jet · Bone cement · Disintegration depth · On-line monitoring

1 Introduction

Bone cement or Polymethylmethacrylate (PMMA) is an important constituent for many joint arthroplasty surgeries or operations. It is majorly used in cemented arthroplasty, where it acts as a transfer medium of the body weight and loads during movement from implants to the bone [1]. The load-carrying capacity and the stability of the whole prothesis-bone cement-bone system increases. Bone cement is used to create a compact space between the implant and the affected bone [2]. It is commercially available as two separate components. One being the polymer powder, and the other is the liquid monomer. Powder component contains initiators of the polymerization process, some

S. Hloch et al. (Eds.): ICMEM 2021, LNME, pp. 100–110, 2021.
https://doi.org/10.1007/978-3-030-71956-2_9

agents which are opaque to X-rays and antibiotics if required. The liquid monomer contains accelerators, stabilizers, and coloring agents for easier visual access. These two components are usually mixed in a ratio of 2:1 or else as prescribed by the manufacturer [3]. Proper techniques for mixing of the cement is required, which is important for the longevity of the service life and load-carrying capacity of the bone cement. The primary joint arthroplasty surgeries are majorly failed due to infection [4], aseptic loosening [5], and cement bone interface failure of the prothesis-bone cement-bone system [6]. This medical condition requires a revision arthroplasty surgery in which the damaged component in which case of hip arthroplasty can be either the acetabular or the femoral component [7]. In both cases, the components have to be taken out of their position and replaced with a new one. The removal of the fixed components from its position is the major bottleneck of the whole revision joint arthroplasty procedure. It involves disintegration and extraction of the well-fixed or hardened bone cement without destroying or damaging the adjacent bone [8]. The bone present is compromised, and the properties of the bone change depending upon the age of the person undergoing the revision surgery. Various methods for extraction of bone cement have been discovered and also being used in which mechanical tools and means are most commonly used [9]. Currently, mechanical tools like chisels, high-speed burrs, hammers, and oscillating saws are most commonly used in practice. However, chances of damage to the femoral shaft, compromises its strength and post-surgery fractures increases [10]. Another method in which mechanical tools are employed is the segmental extraction of bone cement [9]. This method is applied when the defect is in the femoral component and is tightly fixed to the surrounding cement. Therefore, the entire bone cement is extracted in segments to allow direct access to the femoral channel. The drawback of this approach is that it can only be done through the posterior approach during the revision surgery [11]. Cortical windowing and extended trochanteric osteotomy approaches are also used in some cases, but more blood and time loss is observed [12]. Cement in cement technique is used [13] to overcome these limitations. In this technique, entire old bone cement is not extracted. New cement is inserted in the femoral canal, where it mixes with the old intact cement. It shows good results in terms of stability and strength of the whole bone-bone cement-implant system. However, the presence of blood and debris in the interface of the new and old cement can reduce the strength of the bond [14]. The next area of exploration for the extraction of bone cement is using ultrasonic or high energy waves. Extracorporeal shock wave lithotripter [15] and ultrasonically driven chiseling system [16] are the examples of technique which uses ultrasonic energy to disintegrate the cement mantle by melting it. The use of ultrasound waves to melt the bone cement has also been researched [17]. The main disadvantage of this method is the generation of high temperatures for the melting of the bone cement, which causes the induction of heat into the adjacent bone and tissues causing thermal necrosis [18]. The use of laser is also explored for extraction and removal of bone cement from the femoral canal. Solid-state lasers such as excimer, Nd:YaG lasers have been experimented to disintegrate the bone cement [19]. The demerits of these lasers are a slow disintegration rate and not deeper disintegration grooves. CO_2 laser showed a higher disintegration rate [20] but was accompanied by a wider thermal necrosis region and generation of toxic fumes at the operation site during the process [21]. The next form of energy that was explored was the use of hydrodynamic energy generated by the high-speed flow of liquid and impacting the material ahead of it [22]. The first use of pure water jet for the disintegration of bone cement was carried out by Honl et al. [23]. The experiments showed promising results by the use of water jet for extraction of bone cement. The motivation for using water jet for the disintegration of bone cement is because it is a cold cutting process, which eliminates the negative effect of thermal necrosis of neighboring bone and tissues [24]. Due to the property of liquid to travel in a less resistive path, selective cutting can be performed based on the mechanical properties of the materials [25]. Furthermore, the use of flexible tubing allows regions to be reached that cannot be reached with other rigid tools cannot reach [23]. As it is a continuous flow of liquid, so always a clean and new cutting edge is achievable [16]. Pure waterjet and abrasive waterjet were used in past studies to predict the feasibility of this technology to be used in vivo. Honl M. et al. [23] performed a study for disintegration of bone cement and bone with pure water jet and abrasive water jet. Using pure water jet, visible disintegration grooves were formed with a minimum pressure of 40 MPa. Whereas, while using abrasive water jet, water pressure of 20 MPa was sufficient for disintegration of bone cement. Kraaij G. et al. [26] used pure water jet for cutting of periprosthetic tissue. Minimum water pressure of 10 MPa with 0.2 mm nozzle diameter and 5 MPa with 0.6 mm nozzle diameter was sufficient to disintegrate the tissue. Honl M. et al. [25] further compared pure water jet and abrasive water jet erosion ability to disintegrate bone cement with different pressure levels. Rough cutting traces were observed while disintegrating with pure water jet as compared to deeper and uniform traces when abrasive water jet was used. However, with these advantages of disintegration of bone cement or bone with pure water jet and abrasive water jet, some limitations of the processes are also associated with it. High flow rate of water is required for appropriate disintegration rate [26]. Also, non-biocompatible abrasives are also not advisable to use during cutting of bone cement because of chances of infections [27]. Moreover, estimation and controlling of the disintegration depth during the water jet process has been rarely studied in past. Overcoming these limitations, a hybrid version of water jet technology with ultrasonic technology known as pulsating water jet has been developed at Institute of Geonics Czech Republic [28]. The modified liquid is generated due to induction of pressure fluctuation in the high velocity stream after it enters into the high-pressure acoustic chamber. These pressure fluctuations are produced by the sonotrode which is ultrasonically excited with help of piezoelectric crystal transducer [29]. These pressure fluctuations are tuned in the high-pressure chamber by varying the length of the chamber [30]. Standing waves are produced in the chamber whose amplitude increase towards the nozzle exit due to the shape of the chamber. When it exits the nozzle, these pressure fluctuations are transformed in variable axial velocity fluctuation which causes the continuous jet to bunch up alternatively with regions of higher and lower density of water cluster at some specific distances from the nozzle [31]. These water bunches or slugs impact the surface of the material repeatedly causing damage to the material by water hammer effect. These repetitive impingement of the water droplet initiates erosion in the material with low technological input in terms of flow rate or water pressure required [32]. The erosion of the material depends on different machine settings such as pressure, traverse speed of the nozzle, nozzle diameter, standoff distance, acoustic chamber length, frequency and power of the sonotrode. The influence of these input parameters can be observed in form of erosion depth created by the action of PWJ at different input parameters. However, to control or determine the erosion caused during the process can be measured using on-line monitoring system such as recording of acoustics emission signals [33]. Acoustic emission monitoring [34] gives information about the dynamic events appearing during the deformation or disintegration of the material without affecting the phenomenon responsible for erosion. The acoustic sensors measure the surface energy generating due to formation of surface cracks and slips inside the material due to the action of the impact of the jet in form of acceleration of the material. These signals and values of the acceleration can be used for controlling the technology during the disintegration process by the operator to achieve desired erosion results.

In past literature, the influence of the input parameters on the disintegration of both engineering and biomedical material has been studied, which confirms the hypothesis that this technological modification of the water jet can be used for various applications and needs further in-depth investigation. This paper deals with the study of the influence of different nozzle diameter ($d = 0.25$ mm, 0.3 mm, 0.35 mm, and 0.4 mm) and the standoff distance ($z = 1$–19 mm) on groove depth formed during the disintegration of bone cement. The depth measured was further correlated with the acceleration value of the acoustic emission signals detected on the bone cement during the experiments. This correlation between the actual depth achieved and the acceleration values can lead to on-line control of the technology while disintegrating bone cement.

2 Experimental Setup

C-ment 1 bone cement manufactured by Leader Biomedical was used during the experiments [35]. It has a standard viscosity. Both components of the bone cement i.e. the monomer liquid and polymer powder, are mixed manually. Heat is produced during the polymerization process with a maximum of 70 °C. Dough formation starts after approximately 1 min of mixing, and it ultimately sets within 8 min in the desired shape or pattern. This bone cement was used for its better compressive strength (93 MPa), bending strength (65 MPa), and intrusion depth for long-lasting results.

The experiments were conducted using Hammelmann HDP 253 high-pressure pump with maximum operating conditions as 160 MPa at 67 l/min. A robot manipulator (IRB 6640-180) arm was used for the movement of the nozzle head. A circular nozzle of varying diameters $d = 0.25$ mm, 0.3 mm, 0.35 mm, 0.4 mm was used to study the influence of the groove depth created during the process. The trajectory of the robotic arm with the nozzle head was programmed to follow a stair trajectory starting with 1 mm standoff distance with a consecutive step height distance 2 mm and a horizontal distance of 15 mm in between the steps till 19 mm standoff distance (Fig. 1). The other machine settings were kept constant and have been mentioned in Table 1. Ecoson WJ-UG 630-40 ultrasonic generator was used to actuate the piezoelectric crystals, which further transfers the oscillation to the sonotrode for the formation of pressure fluctuations. The mean oscillating frequency of the vibrating sonotrode was kept at 20.20 kHz (Fig. 2).

For the on-line measurement of the acceleration values during the experiments, NI PXI-1031 measurement system was used. This system could work in the AC/DC module.

Table 1. Experimental condition.

Frequency f [kHz]	Pressure p [MPa]	Nozzle diameter d [mm]	Standoff distance z [mm]	Traverse speed v [mm/s]	Chamber length lc [mm]	Material
20.20	10	0.25, 0.3, 0.35, 0.4	1–19	1	12	C-ment 1

Data collected were processed using LabView, sound and vibration measurement suite, spectral measurement toolkit, and advanced signal processing toolkit. Accelerometers used for detecting the vibrations were PCB-352A60 with frequency ranging from 5–60,000 Hz. 10 acceleration values for each standoff distance and experimental conditions were used for statistical correctness and plotted against standoff distance (Fig. 3).

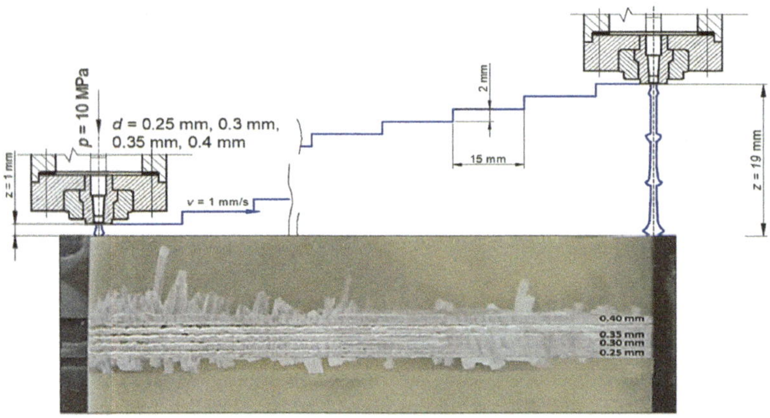

Fig. 1. Schematic representation of experimental setup with disintegrated bone cement material at specific conditions.

For depth measurement, the disintegrated material was scanned using MicroProf FRT, and the scanned files were processed for depth measurement by SPIP software. Three depth measurements (located at 5 mm distant from each other) for each standoff distance were considered for calculating the mean depth and its deviation from the mean value for statistical correctness (Fig. 4).

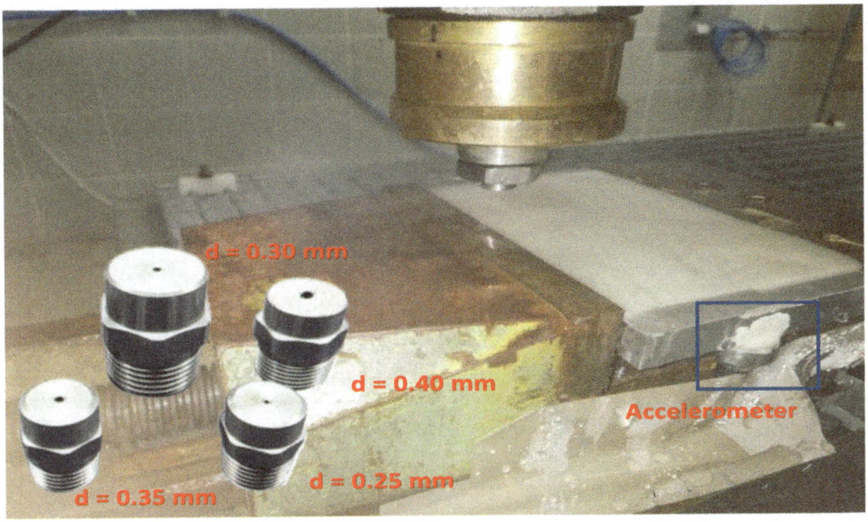

Fig. 2. Actual experimental setup with the different nozzles of varying diameters and accelerometer used.

3 Results and Discussion

Figure 3 and Fig. 4 shows variation in the acceleration (*ACC*) and disintegration depth (*h*) values with variation in the machine settings (d = 0.25 mm–0.4 mm and z = 1 to 19 mm), respectively. Acceleration values recorded during the process shows an increasing nature till certain standoff distance depending upon the nozzle diameters used during the experimental run. After reaching the maximal value, it decreases until the maximum standoff distance. With an increase in the d from 0.25 mm to 0.4 mm, the overall acceleration values increase for the entire range of standoff distance z – 1 mm to 19 mm. This increase in acceleration (ACC = 3.55 m/s^2, 5.75 m/s^2, 10.03 m/s^2 and 14.01 m/s^2 for d = 0.25 mm, 0.3 mm, 0.35 mm and 0.4 mm respectively at z = 3 mm) values depicts faster movement of the disintegrated cement particles during the experimental run. Higher acceleration values depict higher erosion phenomenon when the material particles undergo stresses induced in it higher than its ultimate strength (93 MPa), resulting in disintegration and tearing out of small particles from the bulk material creating grooves. This is attributed to the increase in the volume of water interacting with the bone cement surface such as while using a nozzle with d = 0.25 mm, 0.38 l/min volume of water is used as compared to d = 0.4 mm where 0.96 l/min volume of water impacts the surface. For each nozzle diameter, the acceleration trend is similar to increasing standoff distance (z = 1 mm to 19 mm). The increasing phase of the acceleration values is due to the gradual formation of increasing amplitude of axial velocity of the jet with an increase in the standoff distance [36]. After reaching a certain standoff distance, the acceleration values begin to decrease till the entire range of standoff distance selected for the study. The decrease is due to the aerodynamic resistance caused by the surrounding air to the pulsed water jet. This resistance decreases the effective energy of the jet with which it impacts the surface of the bone cement resulting in lesser movement of the material particles and lesser values of acceleration. For $d = 0.4$ mm, $ACC = 10.37$ m/s^2, 14.01 m/s^2 and 8.18 m/s^2 at $z = 1$ mm, 3 mm and 19 mm respectively.

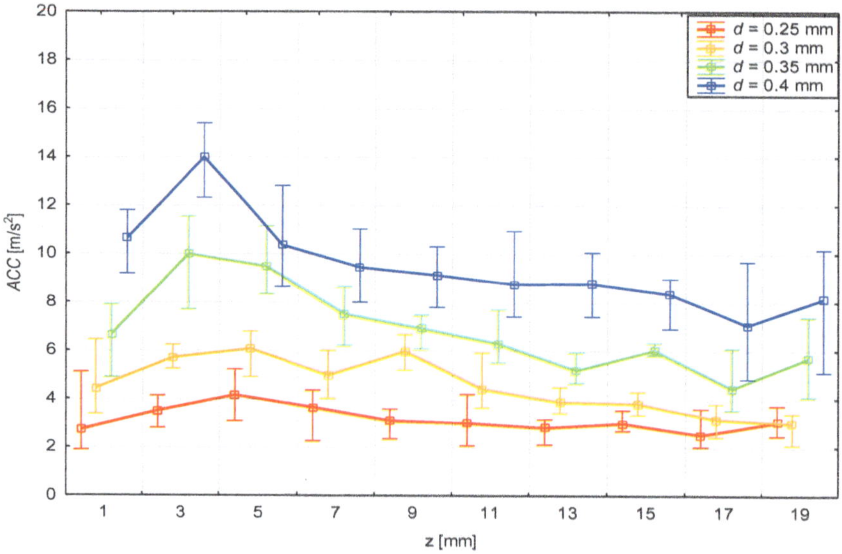

Fig. 3. Variation of acoustic emission acceleration (ACC) values with variation of nozzle diameter (d) and standoff distance (z).

Figure 4 shows the depth of the grooves created during the experimental runs using selected parameters. The depth curve shows typical groove depth variation generated by PWJ with increasing standoff distance keeping all other parameters same. The depth values increase up to a certain standoff distance then decreases gradually. Groove depth was measured from the starting value of standoff distance $z = 1$ mm, for each experimental run. The values of the disintegrated depth increase with an increase in the nozzle diameter from $d = 0.25$ mm to 0.4 mm for the entire selected standoff distance range z = mm to 19 mm. This is attributed to the increase in the volume of water (0.38 l/min to 0.96 l/min) impacting the surface of the bone cement leading to the initiation and propagation of cracks into the material and finally detaching from the parent material ($h = 577$ µm, 625 µm, 699 µm and 772 µm for $d = 0.25$ mm, 0.3 mm, 0.35 mm and 0.4 mm respectively at $z = 5$ mm). For all nozzle diameter, the deepest groove depth is recorded for $z = 5$mm, where the amplitude of the variable axial velocity fluctuation is maximum, leading to the higher energy of the jet impacting the material. Beyond this standoff distance, the negative aerodynamic drag comes into a role and lowers the resultant energy of the jet interacting with bone cement. Therefore, the depth achieved at $z = 1$ mm, 5 mm, and 19 mm are 651 µm, 772 µm, and 534 µm, respectively, for $d = 0.4$ mm. The disintegration grooves formed by each nozzle diameter shows a uniform width throughout its trajectory path.

By comparing Fig. 3 and Fig. 4, it can be inferred that on-line monitoring of the process with the help of accelerometers and recording acceleration value can be used

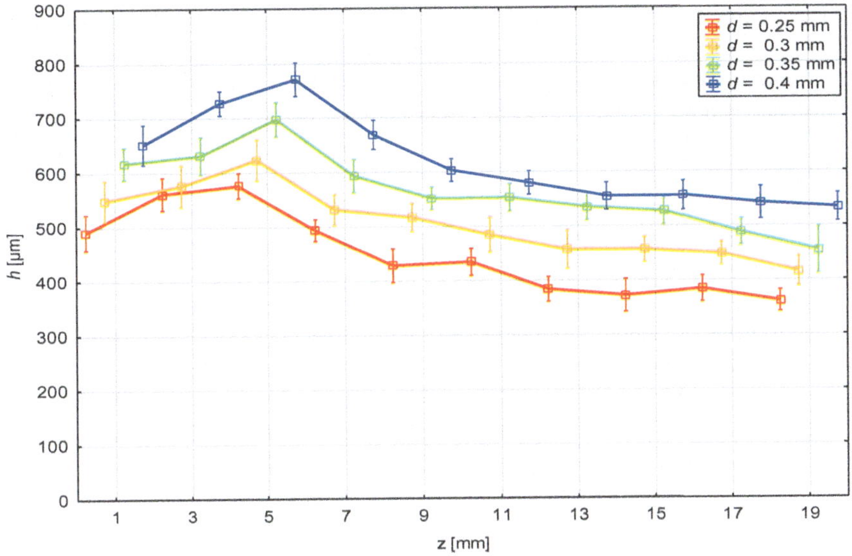

Fig. 4. Variation of disintegrated depth (h) with changing nozzle diameter (d) and standoff distance (z).

to predict the relative depth or other information about the process. It can be observed that the trend of both the curves is quite similar, with maximum values occurring at approximately the same standoff distance. For instance, depth curve and acceleration curve plotted for d = 0.25 mm and 0.3 mm has their respective maximum values at z = 5 mm (For d = 0.25 mm and 0.35 mm, ACC = 4.19 m/s^2 and 6.11 m/s^2 and h = 577.33 μm and 625.67 μm respectively). Also, the nature of the curve, which is increasing till z = 5 mm and the gradual decreasing is similar. However, for d = 0.35 mm and 0.4 mm, the maximal value of acceleration and depth occurs for different standoff distances (z = 5 mm for maximal depth and z = 3 mm for maximal acceleration value). This measurement difference may be due to some error in the acquiring sensors or while measurement of the depth profile. However, acquiring an acoustic signal during the process can be used to predict and control the groove depth. It can be also be used by the surgeons to adjust the machine settings under which they can achieve desirable results. Moreover, if some changes in the signals occur during the process influenced by the difference in the erosion nature, the operator or surgeon can control the process easily.

4 Conclusions

The main aim of the experiment was to determine similarity in the curves between the acceleration values recorded by acoustic emission sensors during the experimental run and the depth of the grove generated during the same experimental run. The result of the study concludes:

1) The nature of both the acceleration curve (Fig. 3) and the groove depth curve (Fig. 4) shows a similar trend and can be correlated to each other.
2) Acceleration values for a constant nozzle diameter, increases to certain standoff distance and then decreases with further increase in the standoff distance (For d = 0.35 mm, ACC = 6.70 m/s^2, 10.03 m/s^2 and 5.735 m/s^2 at $z = 1$ mm, 3 mm and 19 mm respectively).
3) Overall acceleration values of the bone cement subjected to water jet increase by increasing the nozzle diameter, resulting in larger volume of water interacting with the surface (For d = 0.25 mm and 0.3 mm, ACC = 4.19 m/s^2 and 6.11 m/s^2 respectively at $z = 5$ mm).
4) Depth curves also show a similar nature when increasing to a certain standoff distance and then decreasing for higher standoff distances (For $d = 0.3$ mm, $h = 618.67$ μm, 699.33 μm and 456.33 μm at $z = 1$ mm, 5 mm and 19 mm respectively). Moreover, the groove depth increases with the use of larger nozzle diameter (For $d = 0.25$ mm and 0.3 mm, $h = 577.33$ μm and 625.67 μm respectively at $z = 5$ mm).

Acknowledgment. This study was supported by the Slovak Research and Development Agency under Contract No. APVV-17-0490.

References

1. Hloch, S., Foldyna, J., Sitek, L., Zeleňák, M., Hlaváček, P., Hvizdoš, P., Kľoc, J., Monka, P., Monková, K., Kozak, D., Magurová, D.: Disintegration of bone cement by continuous and pulsating water jet. Tech. Gaz. **20**, 593–598 (2013)
2. Lee, C.: The mechanical properties of PMMA bone cement. In: The Well-Cemented Total Hip Arthroplasty: Theory and Practice, pp. 60–66. Springer, Heidelberg (2005). https://doi.org/10.1007/3-540-28924-0_6
3. Deb, S., Koller, G.: Acrylic bone cement: genesis and evolution. In: Orthopaedic Bone Cements, pp. 167–182 (2008)
4. Reina, N., Delaunay, C., Chiron, P., Ramdane, N., Hamadouche, M.: Infection as a cause of primary total hip arthroplasty revision and its predictive factors. Orthop. Traumatol. Surg. Res. **99**, 555–561 (2013)
5. Marcos, L., Buttaro, M., Comba, F., Piccaluga, F.: Femoral cement within cement technique in carefully selected aseptic revision arthroplasties. Int. Orthop. **33**, 633–637 (2009). https://doi.org/10.1007/s00264-008-0516-0
6. Ulrich, S.D., Seyler, T.M., Bennett, D., Delanois, R.E., Saleh, K.J., Thongtrangan, I., Kuskowski, M., Cheng, E.Y., Sharkey, P.F., Parvizi, J., Stiehl, J.B., Mont, M.A.: Total hip arthroplasties: what are the reasons for revision? Int. Orthop. **32**, 597–604 (2008). https://doi.org/10.1007/s00264-007-0364-3
7. Kavanagh, B.F., Ilstrup, D.M., Fitzgerald, J.R.H.: Revision total hip arthroplasty. J. Bone Joint Surg. Am. **67**, 517–526 (1985)
8. Masri, B.A., Mitchell, P.A., Duncan, C.P.: Removal of solidly fixed implants during revision hip and knee arthroplasty. JAAOS-J. Am. Acad. Orthop. Surg. **13**, 18–27 (2005)
9. Lombardi, A.V: Cement removal in revision total hip arthroplasty (1992). https://www.ncbi.nlm.nih.gov/pubmed/19997818. https://doi.org/10.3109/17453679209154735

10. Zweymüller, K., Steindl, M., Melmer, T.: Anterior windowing of the femur diaphysis for cement removal in revision surgery. Clin. Orthop. Relat. Res. **441**, 227–236 (2005). https://doi.org/10.1097/01.blo.0000192042.05584.9c
11. Megas, P., Georgiou, C.S., Panagopoulos, A., Kouzelis, A.: Removal of well-fixed components in femoral revision arthroplasty with controlled segmentation of the proximal femur. J. Orthop. Surg. Res. **9**, 137 (2014). https://doi.org/10.1186/s13018-014-0137-9
12. Chen, W.-M., McAuley, J.P., C Anderson Engh, J., Robert H Hopper, J., Engh, C.A.: Extended slide trochanteric osteotomy for revision total hip arthroplasty. J. Bone Jt. Surg. **82-A**, 5 (2000)
13. Keeling, P., Prendergast, P.J., Lennon, A.B., Kenny, P.J.: Cement-in-cement revision hip arthroplasty: an analysis of clinical and biomechanical literature. Arch. Orthop. Trauma Surg. **128**, 1193–1199 (2008)
14. Li, P.L.S., Ingle, P.J., Dowell, J.K.: Cement-within-cement revision hip arthroplasty; should it be done? J. Bone Jt. Surg. Br. **78**, 809–811 (1996)
15. Weinstein, J.N., Oster, D.M., Park, J.B., Park, S.H., Loening, S.: The effect of the extracorporeal shock wave lithotriptor on the bone-cement interface in dogs. Clin. Orthop. Relat. Res. **235**, 261–267 (1988)
16. Porsch, M., Schmidt, J.: Cement removal with an endoscopically controlled ballistically driven chiselling system. Arch. Orthop. Trauma Surg. **121**(5), 274–277 (2001). https://doi.org/10.1007/s004020000233
17. Goldberg, S.H., Cohen, M.S., Young, M., Bradnock, B.: Thermal tissue damage caused by ultrasonic cement removal from the humerus. J. Bone Jt. Surg. - Ser. A. **87**, 583–591 (2005). https://doi.org/10.2106/JBJS.D.01966
18. Gardiner, R., Hozack, W.J., Nelson, C., Keating, E.M.: Revision total hip arthroplasty using ultrasonically driven tools. J. Arthroplasty. **8**, 517–521 (1993)
19. Zimmer, M., Klöbl, R., De Toma, G., Jansson, V., Refior, H.J., Heimkes, B., Kühne, J.-H.: Bone-cement removal with the excimer laser in revision arthroplasty. Arch. Orthop. Trauma Surg. **112**, 15–17 (1992)
20. Scholz, C., Matthes, M., Kar, H., Boenick, U.: Die Knochenzemententfernung mit dem laser-bone cement removal with the laser. Biomed. Tech. Eng. **36**, 120–128 (1991)
21. Sherk, H.H., Lane, G., Rhodes, A., Black, J.: Carbon dioxide laser removal of polymethylmethacrylate. Clin. Orthop. Relat. Res. 67–71 (1995)
22. Hreha, P., Hloch, S., Magurovd, D., Valicek, J., Kozak, D., Harnicdrovd, M., Rakin, M.: Water jet technology used in medicine. Teh. Vjesn. **17**, 237–240 (2010)
23. Honl, M., Rentzsch, R., Müller, G., Brandt, C., Bluhm, A., Hille, E., Louis, H., Morlock, M.: The use of water-jetting technology in prostheses revision surgery - First results of parameter studies on bone and bone cement. J. Biomed. Mater. Res. **53**, 781–790 (2000). https://doi.org/10.1002/1097-4636(2000)53:6%3c781::AID-JBM20%3e3.0.CO;2-G
24. Schmolke, S., Pude, F., Kirsch, L., Honl, M., Schwieger, K., Krömer, S.: Wärmeentwicklung bei der Wasser-Abrasivstrahl-Osteotomie/Temperature Measurements During Abrasive Water Jet Osteotomy. Biomed. Tech. Eng. **49**, 18–21 (2004)
25. Honl, M., Dierk, O., Küster, J.R., Müller, G., Müller, V., Hille, E., Morlock, M.: Die Wasserstrahldiskotomie im mikroinvasiven Zugang-In-vitro-Testung und erste klinische Aspekte eines neuen Verfahrens. Z. Orthop. Ihre Grenzgeb. **139**, 45–51 (2001)
26. Kraaij, G., Tuijthof, G.J.M., Dankelman, J., Nelissen, R.G.H.H., Valstar, E.R.: Waterjet cutting of periprosthetic interface tissue in loosened hip prostheses: an in vitro feasibility study. Med. Eng. Phys. **37**, 245–250 (2015). https://doi.org/10.1016/j.medengphy.2014.12.009
27. Schwieger, K., Carrero, V., Rentzsch, R., Becker, A., Bishop, N., Hille, E., Louis, H., Morlock, M., Honl, M.: Abrasive water jet cutting as a new procedure for cutting cancellous bone—in vitro testing in comparison with the oscillating saw. J. Biomed. Mater. Res. Part B Appl. Biomater. **71**, 223–228 (2004)

28. Foldyna, J., Svehla, B.: Method of generation of pressure pulsations and apparatus for implementation of this method (2010)
29. Nag, A., Hloch, S., Dixit, A.R., Cuha, D.: Investigation on pulsating liquid jet with physiological saline on aluminium surface. In: Advances in Manufacturing Engineering and Materials, pp. 63–71. Springer (2019)
30. Nag, A., Hloch, S., Čuha, D., Dixit, A.R., Tozan, H., Petrů, J., Hromasová, M., Müller, M.: Acoustic chamber length performance analysis in ultrasonic pulsating water jet erosion of ductile material. J. Manuf. Process. **47**, 347–356 (2019)
31. Srivastava, M., Hloch, S., Tripathi, R., Kozak, D., Chattopadhyaya, S., Dixit, A.R., Foldyna, J., Hvizdos, P., Fides, M., Adamcik, P.: Ultrasonically generated pulsed water jet peening of austenitic stainless-steel surfaces. J. Manuf. Process. **32**, 455–468 (2018). https://doi.org/10.1016/j.jmapro.2018.03.016
32. Hloch, S., Adamčík, P., Nag, A., Srivastava, M., Čuha, D., Müller, M., Hromasová, M., Klich, J.: Hydrodynamic ductile erosion of aluminium by a pulsed water jet moving in an inclined trajectory. Wear **428–429**, 178–192 (2019). https://doi.org/10.1016/j.wear.2019.03.015
33. Hloch, S., Nag, A., Pude, F., Foldyna, J., Zeleňák, M.: On-line measurement and monitoring of pulsating saline and water jet disintegration of bone cement with frequency 20 kHz. Measurement. **147**, 106828 (2019)
34. Hloch, S., Ruggiero, A.: Online monitoring and analysis of hydroabrasive cutting by vibration. Adv. Mech. Eng. **5**, 894561 (2013)
35. Product Brochure | C-ment | Leader Biomedical. https://www.leaderbiomedical.com/brochure/c-ment-pmma-bone-cement/. Accessed 01 Feb 2020
36. Hloch, S., Srivastava, M., Nag, A., Muller, M., Hromasová, M., Svobodová, J., Kruml, T., Chlupová, A.: Effect of pressure of pulsating water jet moving along stair trajectory on erosion depth, surface morphology and microhardness. Wear. **452**, 203278 (2020)

Effect of Surface Modification and Standoff Distance on Hydrodynamic Ductile Erosion

Ashokkumar Ravichandiran[1], Peter Pavol Monka[1], Katarína Monková[1](✉), and Monika Hromasova[2]

[1] Faculty of Manufacturing Technologies (TUKE) with a seat in Presov, 08001 Presov, Slovakia
katarina.monkova@tuke.sk

[2] Czech University of Life Sciences, Prague, Czech Republic

Abstract. The disintegration of material surface exposed to pulsating water jet depends upon the technological parameters as well as interacting surface properties. Surface modified by various machining operations results in the generation of different surface roughness. These differences in surface asperities cause variation in the shear stress-induced into the material during the lateral jetting phase of the water hammer phenomenon. Therefore, in the present study influence of different modified surfaces (0A, 1A, 2A, 3A and 4A) and variation in standoff distance ($z = 1$ to 57 mm) on material erosion in terms of disintegration depth is studied. The surfaces are modified using milling operation with variation in tool diameter, cutting speed and number of tool inserts. The result shows a bell curve pattern of disintegration depth with increasing standoff distance. The maximal disintegration depth of the samples varied from $h_{max} = 523$ to 773 µm depending upon the surface finish with minimum maximal depth ($h_{max} = 523$ µm) observed for the polished sample.

Keywords: PWJ · Surface roughness · Disintegration depth · Face milling

1 Introduction

Hydrodynamic erosion [1] is a phenomenon in which material disintegration occurs when impacted with the high-speed liquid stream. Water droplets, when impinges material surface, cause damages to aircraft wings [2], turbine blades [3] and pump components [4]. This erosion is based upon the water hammer effect, which induces compressive stresses into the material when impacted by water droplets [5]. The onset of impact is characterized by radial outward flow of water. The calculated velocity of outward flow is calculated in range of 5 - 10 times the impact velocity of the jet [6]. During this lateral jetting, it induces shear stress on the material surface, depending upon its interaction with the surface asperities [7]. This initiates fatigue cracks and assists the surface cracks to deepen into the material. These cracks act as stress concentration sites and then interconnect with neighboring cracks to cause material failure. More recently, investigations on the effect of surface roughness on water droplet erosion has been conducted by many researchers. Hancox and Brunton [7] experimentally studied it by

S. Hloch et al. (Eds.): ICMEM 2021, LNME, pp. 111–119, 2021.
https://doi.org/10.1007/978-3-030-71956-2_10

using specially designed rotating arm apparatus on stainless steel. It interrupted the continuous flow and caused water droplets to impinge the roughened surfaces. Results showed that surface polished with finer abrasive grain required 3.3 times more impacts to produce pits than polished with rougher grains. Heymann [5] analyzed the induction of the impact force into the material, which was subjected to the various machining process. It was concluded that due to the presence of surface asperities generated by various machining processes, shear force induced into the material increased. This helped in the initiation of fatigue crack and its propagation into the material. Another study by Hattori and Takinami [8], showed that surface asperities in the form of grain boundaries also affect the droplet erosion process and the erosion initiates from the grain boundaries itself. Kirols et al. [9] using surface morphology images showed that the incubation period of erosion caused during water droplet erosion decreases with an increase in surface roughness. Fujjisawa et al. [10] also studied numerically and experimentally, the erosion mechanism during impingement of liquid droplet on single groove roughness. The results showed that for large groove roughness, erosion starts from the bottom of the groove as compared to groove edges for small groove roughness.

Currently, research on utilizing water hammer erosion phenomenon for controlled disintegration of materials is carried out in the form of the pulsating water jet [11, 12]. Continuous water jet utilizes initial impact pressure followed by dominant stagnation pressure when interacting with the material. Whereas, pulsating water jet works on the principle of repetitive impingement of material with impact pressure [13]. Foldyna J. [14] invented a prototype for a generation of pulsating water jet utilizing ultrasonic energy. Piezoelectric ceramics attached to ultrasonic sonotrode are excited by the ultrasonic generator, which causes disturbances in the high-pressure water entering into the acoustic chamber [15]. These disturbances inside the acoustic chamber generate pressure fluctuations in the water, which intensify near the nozzle exit due to its converging shape [16]. After the nozzle exit, the jet starts separating into discrete water clusters classified as impact waves depending upon the flow rate and distance before it impinges the material surface [17]. These repetitive impingements of impact waves on the material at a certain standoff distance induce fatigue cracks, which propagates in the materials causing material failure and dislodgement [18].

Determination of an effective range of standoff distance is required for the efficient disintegration of the materials. Also, the past literature showed a dependence of the surface finish of the material on erosion rates. Therefore, in the present study, the erosion effect of PWJ in terms of disintegration depth (h) impacting different modified material surfaces (0A, 1A, 2A, 3A and 4A) with variable standoff distance ($z = 1$ to 57 mm) is studied.

2 Experimental Setup

Aluminium alloy AW-6060 was used during the experiments [19]. It is commonly used for various industrial applications in the automotive and aviation industries. The physical properties of the material used during the experiments are shown in Table 1. Its ductility property provides better visibility of the grooves created by the PWJ. The aluminium workpiece was subjected to face milling process to generate different surfaces with varying surface roughness. The operating parameters used are listed in Table 2.

Table 1. Physical properties of AW - 6060.

Physical properties	Values
Density	2.70 g/cm^3
Thermal conductivity	210 W/m-K
Tensile strength	215 MPa
Young modulus	696.5 GPa
Shear modulus	26.1 GPa
Brinell hardness	70 HB
Elongation min.	8%

Table 2. Technological parameters for modifying the aluminium surface.

Sample	Tool diameter (mm)	Number of tool inserts	Cutting speed (m/min)
0A	Polished surface		
1A	63	5	400
2A	50	4	400
3A	42	3	500
4A	20	4	200

Hammelmann high-pressure pump was used for delivering the water with the required flow rate ($\overset{\circ}{Q} = 4.32$ l/min) for the experiments. A robotic manipulator (ABB IRB 6640) was used for providing the desired motion to the nozzle head. The ultrasonic generator was used for supplying ultrasonic excitation to the piezoelectric ceramics to oscillate at a specific frequency ($f = 20.2$ kHz). Sonotrode attached to piezoelectric ceramics amplify these oscillations and creates pressure pulsation in the continuous water jet to transform it into pulsating water jet. The circular nozzle of $d = 0.6$ mm was used with a supply pressure of 40 MPa. Standoff distance was varied from $z = 1$ to 57 mm from above the material surface in the step of 2 mm. Due to sample length constraint, the whole range of z is achieved 4 runs, i.e. $z = 1$ to 15 mm, 15 to 29 mm, 29 to 43 mm and 43 to 57 mm. The disintegration length for each z was kept 20 mm (Fig. 1). The technological parameters used for the experiments are given in Table 3. Samples after disintegration were scanned using MicroProf FRT. These scanned filed were exported to SPIP software for measurement of disintegration depth. 5 depth readings for each z was noted for each sample. The maximum, minimum and average values of h at each z are plotted and shown in Fig. 2.

Table 3. Technological parameters of PWJ used for the experiments.

Technological parameter	Level
Supply pressure	40 MPa
Operating frequency	20.2 kHz
Acoustic chamber length	12 mm
Water flow rate	4.32 l/min
Traverse speed	2 mm/s
Standoff distance	1–57 mm
Nozzle diameter	0.6 mm

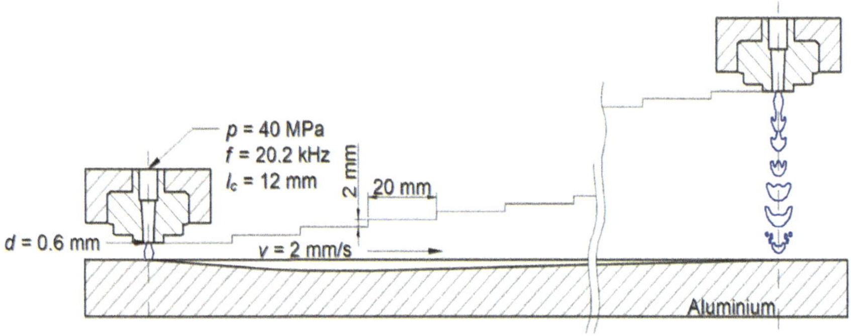

Fig. 1. Experimental setup.

3 Result and Discussion

All the modified surface samples were exposed to the PWJ according to the technological conditions mentioned in Table 3. Fig. 2 shows the images of the samples after disintegration. The variations of h are plotted against increasing z of all the samples (0A, 1A, 2A, 3A, 4A) and shown in Fig. 3.

The disintegration curve (Fig. 3) for all the samples followed a similar trend of h. No measurable erosion depth was measured until certain standoff distance ($z = 13$ and 15 mm for 0A and 1A, 2A, 3A, 4A, respectively) for all the samples depending upon the roughness of the modified surfaces. This range of z corresponds to the incubation regime of erosion in which subsurface properties changes, but no measurable erosion depth is obtained [20]. With a further increase in standoff distance from $z = 13$ or 15 mm till 27 mm, h shows an increasing trend. This is attributed to surface impact with growing clusters of water waves which are formed at acceleration regime of erosion [21].

The impact pressure induced by these impact waves exceeds the ultimate strength of the material causing material disintegration. At $z = 29$ mm, maximum disintegration depth is obtained for every sample (Table 4). This is attributed to the development of discrete clusters of impact waves leading to water hammer phenomenon and corresponds to culmination regime of erosion [22]. With a further increase in z from 31 mm to 57 mm,

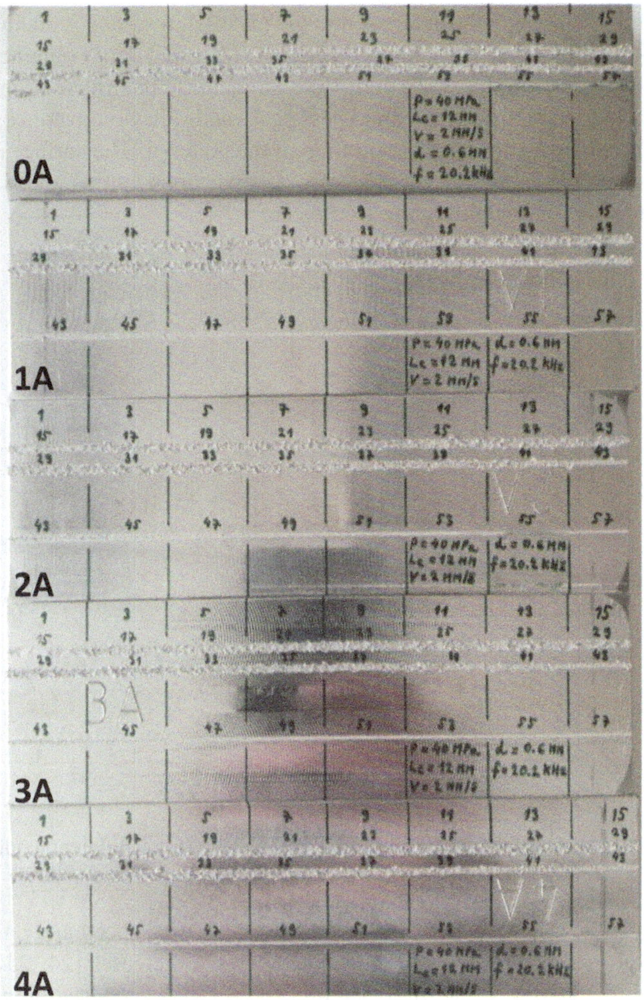

Fig. 2. Disintegrated sample after exposure to PWJ.

h shows a decreasing trend. It is due to the de-concentration of the water clusters by aerodynamic drag resulting in the reduction of subsequent impact pressure. This range of z corresponds to the depletion regime of erosion. 0A and 4A samples showed no measurable disintegration above $z = 53$ mm and 55 mm, respectively. This regime of erosion is known as a termination, which is due to the dispersion of the water cluster into further smaller water droplets, showing no material loss [23].

The material disintegration depth on different sample surface depends on the machining parameters used during milling of the surface. 0A sample was polished, which removed any surface asperities, resulting in mirror finish surface. While impacting the polished surface with PWJ, the lateral flow of the jet generated onset of the water hammer phenomenon does not interact with these surface irregularities. This affects the generation and induction of shear stress into the material, resulting reduction of surface disintegration as compared to roughened surfaces. Maximal depth obtained on sample 0A was $h_{max} = 523$ µm at $z = 29$ mm. With an increase in surface roughness of the surface, lateral jet interacts with the roughness peaks and valleys and induces shear stress into the material causing to the disintegration of the material. The surface roughness of the machined surface using milling operation depends upon cutting speed and number of tool inserts used for machining the workpiece.

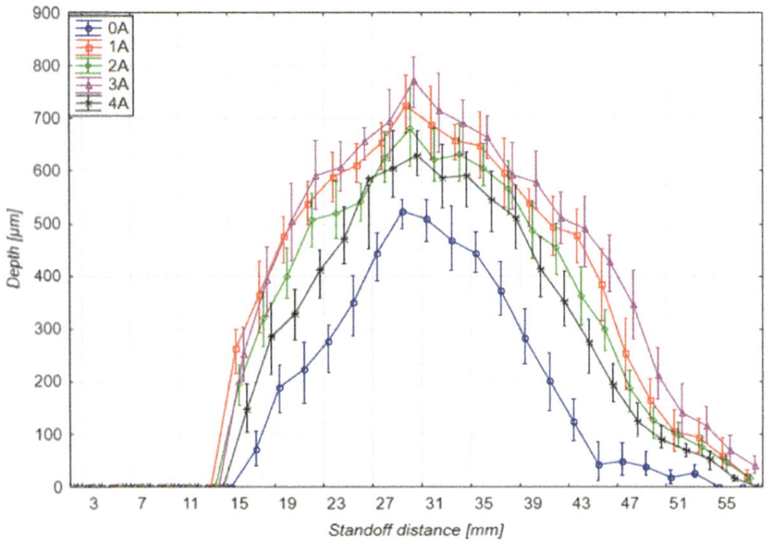

Fig. 3. Effect of standoff distance (z) on disintegration depth (h) on different modified surfaces.

Previous studies showed that with an increase in cutting speed and number of tool inserts, the surface roughness of the machined surface increases [24, 25]. Therefore, sample 4A with a cutting speed of 200 m/min and 4 tool inserts has minimum surface roughness corresponding overall lower h values as compared to other machined samples. The $h_{max} = 630$ µm was obtained at $z = 29$ mm. With the increase of cutting speed to 400 m/min, the surface of sample 2A was generated. The h values obtained for this modified surface were deeper than 4A. This attributed to the increase in the surface roughness due to the increase in cutting speed from 200 m/min to 400 m/min. Maximum depth obtained was $h_{max} = 680$ µm at $z = 29$ mm. Sample 1A surface was prepared using 5 tool inserts at 400 m/min, which deteriorated the surface quality more as compared to sample 2A. The maximum h obtained for sample 1A was $h_{max} = 726$ µm at $z =$ 29 mm. Sample 3A was machined using the cutting speed of 500 m/min with 3 tool inserts. This resulted in a rougher surface among all the machined samples. Maximum $h_{max} = 772$ µm was obtained at $z = 29$ mm. The h_{max} and z at which it was obtained for all the samples are summarized in Table 4.

Table 4. Maximum disintegration depth and corresponding standoff distance for all samples.

Sample	0A	1A	2A	3A	4A
h_{max} [µm]	523	726	680	772	630
z [mm]	29	29	29	29	29

4 Conclusions

The study showed that surface asperities created by various machining setting have a proportional effect on disintegration depth of the grooves produced by PWJ. Major conclusions of the experimental work are enlisted hereafter:

1) Disintegration depth corresponding to the erosion rate showed five regimes such as incubation, acceleration, culmination, depletion and termination regime of erosion sequentially with increasing $z = 1$ to 47 mm.
2) Disintegration depths measured for the polished sample was lower than other samples for the entire standoff distance range ($h_{max} = 523$ µm) due to absence of surface asperities and irregularities.
3) Variations in face milling machining parameters change the surface finish of the workpiece generating surface roughness and defects.
4) Variation in surface quality affects the interaction of the liquid with the material in turn affecting the disintegration depths of the material exposed to PWJ ($h_{max} =$ 772 µm as compared to 630 µm for 3A and 4A, respectively).

In future work, the effect of initial residual stresses of the material induced by the machining operations on disintegration depth when exposed to PWJ should be analyzed.

Acknowledgement. This study was supported by the Slovak Research and Development Agency under Contract No. APVV-17–0490.

References

1. Hloch, S., Adamčík, P., Nag, A., Srivastava, M., Čuha, D., Müller, M., Hromasová, M., Klich, J.: Hydrodynamic ductile erosion of aluminium by a pulsed water jet moving in an inclined trajectory. Wear **428–429**, 178–192 (2019). https://doi.org/10.1016/j.wear.2019.03.015
2. Kamkar, N., Bridier, F., Jedrzejowski, P., Bocher, P.: Water droplet impact erosion damage initiation in forged Ti-6Al-4V. Wear (2015). https://doi.org/10.1016/j.wear.2014.10.020
3. Mahdipoor, M.S., Kirols, H.S., Kevorkov, D., Jedrzejowski, P., Medraj, M.: Influence of impact speed on water droplet erosion of TiAl compared with Ti6Al4V. Sci. Rep. (2015). https://doi.org/10.1038/srep14182
4. Fujisawa, N., Wada, K., Yamagata, T.: Numerical analysis on the wall-thinning rate of a bent pipe by liquid droplet impingement erosion. Eng. Fail. Anal. **62**, 306–315 (2016). https://doi.org/10.1016/j.engfailanal.2016.01.005
5. Heymann, F.J.: Erosion by liquids. Mach. Des. **10**, 118–124 (1970)

6. Thomas, G.P., Brunton, J.H.: Drop impingement erosion of metals. Proc. R. Soc. A Math. Phys. Eng. Sci. **314**, 549–565 (1970). https://doi.org/10.1098/rspa.1970.0022
7. Hancox, N.L., Brunton, J.H.: The erosion of solids by the repeated impact of liquid drops. Philos. Trans. R. Soc. London. Ser. A, Math. Phys. Sci. (1966). https://doi.org/10.1098/rsta.1966.0036
8. Hattori, S., Takinami, M.: Comparison of cavitation erosion rate with liquid impingement erosion rate. Wear. **269**, 310–316 (2010). https://doi.org/10.1016/j.wear.2010.04.020
9. Kirols, H.S., Kevorkov, D., Uihlein, A., Medraj, M.: The effect of initial surface roughness on water droplet erosion behaviour. Wear **342–343**, 198–209 (2015). https://doi.org/10.1016/j.wear.2015.08.019
10. Fujisawa, K., Yamagata, T., Fujisawa, N.: Liquid droplet impingement erosion on groove roughness. Nucl. Eng. Des. **330**, 368–376 (2018)
11. Hloch, S., Foldyna, J., Sitek, L., Zeleňák, M., Hlaváček, P., Hvizdoš, P., Kľoc, J., Monka, P., Monková, K., Kozak, D., Magurová, D.: Disintegration of bone cement by continuous and pulsating water jet. Tech. Gaz. **20**, 593–598 (2013)
12. Hloch, S., Nag, A., Pude, F., Foldyna, J., Zeleňák, M.: On-line measurement and monitoring of pulsating saline and water jet disintegration of bone cement with frequency 20 kHz. Measurement **147**, 106828 (2019)
13. Foldyna, J., Sitek, L., Ščučka, J., Martinec, P., Valíček, J., Páleníková, K.: Effects of pulsating water jet impact on aluminium surface. J. Mater. Process. Technol. **209**, 6174–6180 (2009)
14. Foldyna, J., Klich, J., Hlavacek, P., Zelenak, M., Scucka, J.: Erosion of metals by pulsating water jet. Teh. Vjesn. Gaz. **19**, 381–386 (2012)
15. Hloch, S., Foldyna, J., Pude, F., Kľoc, J., Zeleňák, M., Hvizdoš, P., Monka, P., Smolko, I., Ščučka, J., Kozak, D.: Experimental in-vitro bone cements disintegration with ultrasonic pulsating water jet for revision arthroplasty. Teh. Vjesn. Znan. časopis Teh. Fak. Sveučilišta u Osijeku. **22**, 1609–1615 (2015)
16. Nag, A., Hloch, S., Čuha, D., Dixit, A.R., Tozan, H., Petrů, J., Hromasová, M., Müller, M.: Acoustic chamber length performance analysis in ultrasonic pulsating water jet erosion of ductile material. J. Manuf. Process. **47**, 347–356 (2019)
17. Nag, A., Hloch, S., Dixit, A.R., Pude, F.: Utilization of ultrasonically forced pulsating water jet decaying for bone cement removal. Int. J. Adv. Manuf. Technol. **110**, 829–840 (2020)
18. Srivastava, M., Hloch, S., Tripathi, R., Kozak, D., Chattopadhyaya, S., Dixit, A.R., Foldyna, J., Hvizdos, P., Fides, M., Adamcik, P.: Ultrasonically generated pulsed water jet peening of austenitic stainless-steel surfaces. J. Manuf. Process. **32**, 455–468 (2018). https://doi.org/10.1016/j.jmapro.2018.03.016
19. Čuha, D., Nag, A., Chlupová, A., Hloch, S.: Erosion of titanium and aluminium alloys using pulsating water jet: effect of standoff distance. In: International Conference on Water Jet-Research, Development, Applications, pp. 56–66. Springer (2019)
20. Srivastava, M., Hloch, S., Gubeljak, N., Milkovic, M., Chattopadhyaya, S., Klich, J.: Surface integrity and residual stress analysis of pulsed water jet peened stainless steel surfaces. Measurement **143**, 81–92 (2019)
21. Hloch, S., Srivastava, M., Nag, A., Muller, M., Hromasová, M., Svobodová, J., Kruml, T., Chlupová, A.: Effect of pressure of pulsating water jet moving along stair trajectory on erosion depth, surface morphology and microhardness. Wear **452–453**, 203278 (2020)
22. Nag, A., Hloch, S., Dixit, A.R., Cuha, D.: Investigation on pulsating liquid jet with physiological saline on aluminium surface. In: Advances in Manufacturing Engineering and Materials, pp. 63–71. Springer (2019)
23. Raj, P., Hloch, S., Tripathi, R., Srivastava, M., Nag, A., Klichová, D., Klich, J., Hromasová, M., Muller, M., Miloslav, L., Chattopadhyaya, S., Adamcik, P.: Investigation of sandstone erosion by continuous and pulsed water jets. J. Manuf. Process. **42**, 121–130 (2019). https://doi.org/10.1016/j.jmapro.2019.04.035

24. Yalcin, U., Karaoglan, A.D., Korkut, I.: Optimization of cutting parameters in face milling with neural networks and Taguchi based on cutting force, surface roughness and temperatures. Int. J. Prod. Res. **51**, 3404–3414 (2013)
25. Nas, E., Demir, H.: The influence of number of inserts and cutting parameters on surface roughness in face milling. Technology **13**, 1–7 (2010)

Effect of Acoustic Chamber Length on Disintegration of Ductile Material with Pulsating Water Jet

Akash Nag[1](✉), Sergej Hloch[2], Avinash Babu Ramesh Babu[3], Monika Hromasova[4], and Amit Rai Dixit[1]

[1] Department of Mechanical Engineering, Indian Institute of Technology (Indian School of Mines), Dhanbad 826004, India
akashnag1992@gmail.com
[2] Institute of Geonics, Czech Academy of Sciences, Ostrava-Poruba, Czech Republic
[3] Faculty of Manufacturing Technologies TUKE With a Seat in Presov, 08001 Presov, Slovakia
[4] Czech University of Life Science, 16500 Prague, Czech Republic

Abstract. Pulsating water jet is a hybrid process of ultrasonic machining and waterjet machining. The continuous jet of the water is induced with ultrasonic disturbances resulting in forced breakup of the jet. The morphology of the jet is stimulated by ultrasonic transducer depend upon the frequency and power transmission. The transmission of the fluctuations is influenced by variation in acoustic chamber length which in turn leads to variation in disintegration depth. Acoustic chamber is a space where the inlet water from the pump is induced with the ultrasonic fluctuations generated by the ultrasonic sonotrode. Acoustic chamber length is varied for tuning the system to generate maximum output power transmission with specific input conditions. In the present study, the acoustic chamber length is changed from 5 mm to 22 mm with inlet water pressure kept constant at 30 MPa. An inclined trajectory starting with standoff distance 5 mm till 101 mm at an angle of 16° with traverse speed of 1 mm/s is used. 18 experimental runs are conducted, and all the grooves created by the impact of pulsating water jet were measured and plotted for observing the influence of acoustic chamber length with varying standoff distance. Surface morphology of the eroded grooves were studied using SEM images which showed surface features such as voids, cavities, micro-channels and upheaved surfaces along the periphery of the grooves.

Keywords: Pulsating water jet · Acoustic chamber length · SEM analysis · Disintegration depth

1 Introduction

Hydrodynamic erosion [1] caused by flowing water is one of the oldest naturally occurring erosion phenomenon which leads to the creation of valleys, canyons etc. The erosion caused by high-speed liquid (>300 m/s) impacting the material surface has been always a problem for various mechanical components such as leading edges aircraft wings

S. Hloch et al. (Eds.): ICMEM 2021, LNME, pp. 120–131, 2021.
https://doi.org/10.1007/978-3-030-71956-2_11

[2], turbine blades [3], windmill blades [4] and water pipelines [5]. In recent decades, researchers are developing new methods to use this erosive property of fluid into an advantage in the form of controlled disintegration of the material. One of the methods is in the form of water jet (WJ) technology [6], which utilizes the kinetic energy of the high-speed water to disintegrate or erode the material. The disintegration process does not depend upon the physical properties of the materials and due to the cold cutting process, no alteration in the material thermal properties occurs. Continuous WJ is a conventional form of the technology which utilizes large flow rate such as 6 l/min for the disintegration of the material. Modification of continuous WJ is abrasive WJ (AWJ) [7]. AWJ utilizes abrasives in the water stream to disintegrate materials. This variant of water jet utilizes pumps generating 600 MPa pressure, leading to consistently machining of the entire range of material. AWJ can be used for cutting [8], turning [9] and milling [10] operations. With the advancement in the material science field [11], advanced materials with specific physical properties are developed and used are being used production process. To machine these advance materials into the required form, WJ and AWJ technology are not suitable for achieving the desired outcome (material removal rate and surface roughness) with permissible input conditions (inlet pressure and flow rate). The major shortcomings are 1) Use of high-pressure pumps (>200 MPa). 2)Use of large volume of flow rate. 3) Recurring cost of abrasives and machinery parts 4) Wearing out of nozzles after certain work hours. These shortcomings make this technology as non-eco-friendly technology. Therefore, the development of technology to achieve same output in terms of disintegration with minimal possible technological input is being focused on which would be also eco-friendly. Pulsating water jet [12] is one of the recent developments in the water jet technology field. Continuous water jet stream is broken into bunches of droplets and water slugs downstream to nozzle exit. These water droplets when interacting with the surface induces impact pressure in the material which is more than the stagnation pressure induced by continuous jet. Repetitive pressure transient waves are formed which induces compressive stress wave at the site of contact between the droplet and material. The lateral flow of the jet starts when the deformation compressibility limit of the water droplet is exceeded. The magnitude of pressure-induced during lateral jetting is same as stagnation pressure. This flow induces shear stress in the material which exceeds the ultimate strength of the material resulting in its failure and disintegration. Therefore, with pressure as low as $p = 8$ MPa, erosion of the material is possible as compared to continuous jet. Different mechanisms for the generation of pulsating water jet are used to in past studies such as internal mechanical modulators, self-resonating nozzles, vibrating transformer near the nozzle exit. A new method of generation of PWJ [13] has been developed in the Institute of Geonics which overcomes the shortcomings of previous methods of pulsating the continuous jet. In this method, ultrasonic sonotrode is used to generate pressure fluctuation in the water entering the acoustic chamber. These pressure fluctuations are caused by the movement of the sonotrode attached with piezo-electric crystals which are excited by the ultrasonic generator. The pressure fluctuation in the acoustic chamber generates pressure standing waves whose amplitude increases towards the nozzle exit due to the geometry of the chamber. Acoustic chamber is a cylindrical space between the tip of the oscillating sonotrode and converging waveguide

connected to the nozzle exit in which standing waves produced by the pressure fluctuations are formed [14]. The length of the acoustic chamber can be adjusted to tune the system to maximize the impact effect of the jet. At the nozzle, these pressure fluctuations are transformed into velocity fluctuation causing the variable axial velocity of the jet. Due to the variable axial velocity, the jet downstream the nozzle begins to bunch into consecutive clusters of water droplets and slugs. These clusters when interacting with the surface induces cyclic stresses exceeding the ultimate strength of the material leading to failure and erosion of the materials. The output of the process depends on various technological parameters such as inlet pressure (p), traverse speed (v), acoustic chamber length (l_c), nozzle diameter (d)and standoff distance (z). Many studies related to their influence on the disintegration depth and course of erosion has been already available in the open literature. However, influence of l_c and its interactional effect with varying z is not much explored. Some of the previous studies related to the influence of l_c and shape of the acoustic chamber has been discussed here. Pochyly F. et al. [15] designed different geometrical configuration of the pressured chamber for propagation of pressure waves based on transfer matrix model. The results showed that circular trapezoidal shaped tube was optimal for wave transmission in the high-pressure system. Foldyna J. et al. [14] studied effect of different configuration of acoustic wave propagation in high-pressure system by using simulation. It concluded that different configuration length resulted different standing wave amplitude at the nozzle end. Riha Z. & Foldyna J. [16] performed comparative study on different types of transitional space for efficient transmission of pressure pulsations through high-pressure system using CFD modelling. The results showed that higher amplitude of velocity was obtained using tool with radius transition with the given conditions.

Although many research works dealing with influence of input parameters of PWJ has been carried out. The influence of the acoustic chamber configuration on the disintegration of material with varying standoff distance has not still much researched. Therefore, in the present study influence of acoustic chamber length (l_c) with varying standoff distance (z) during disintegration of aluminium surface with pulsating water jet on disintegration depth (h) was studied.

2 Experimental Setup

In the experimental study, the AW-6060 aluminium alloy was used as the workpiece. The mechanical properties, like lower density (2700 kg/m^3), corrosion resistance, better strength (170 MPa) and non-toxicity makes it widely acceptable non-ferrous metal. Moreover, compared to other metals with comparable properties, it is cheaper from an economic point of view. Its ductility, better visibility of grooves is there when exposed to PWJ.

The experimental setup for the generation of pulsating water available in Institute of Geonics, Czech Republic was used (Fig. 1). The ultrasonic generator Ecoson WJ – UG 630-40 was used for exciting the piezoelectric transducer to oscillate the sonotrode with a spatial frequency of 20 kHz and maximum power output of 800W. Pressure pump Hammelmann HDP 253 for delivering supply input pressure water during the experiments. It can operate with a maximum pressure of 160 MPa with a maximum flow rate of 67 l/min. Robot manipulator ABB IRB 6640-180 was used for handling the nozzle head over the material. Circular nozzle with diameter $d = 0.6$ mm of StoneAge company was used during the experiments.

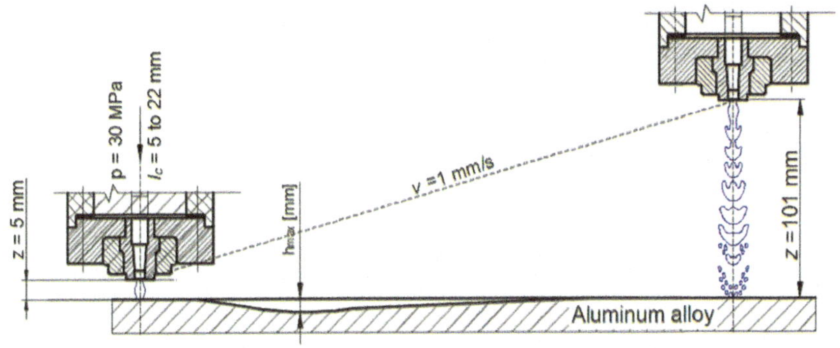

Fig. 1. Schematic diagram of the experimental runs.

An inclined trajectory movement at an angle of 16° was used during the test. The initial standoff distance was 5 mm over the material and was varied to 101 mm at the end. Traverse speed $v = 1$ mm/s and supply water pressure $p = 30$ MPa was kept constant for the movement of the nozzle head throughout the experiments. The acoustic chamber length was varied by rotating a mechanical nut fitted outside the pulsating head. The pitch of the nut is 2 mm, so for every experimental run the nut was rotated by half a rotation to vary the chamber length by 1 mm (Fig. 1). Initial chamber length was set to 5 mm and was changed to 22 mm with a variation of 1 mm in every experimental run. These values of chamber length were decided on the basis of pilot test which when performed with values beyond this range resulted into instability of the frequency of the oscillating sonotrode ($f = 20.4$ kHz). The experimental conditions are summarized in Table 1.

Table 1. Experimental conditions.

Acoustic chamber length (l_c)	Pressure (p)	Traverse speed (v)	Standoff distance (z)	Nozzle diameter (d)	Material
5–22 mm	30 MPa	1 mm/s	5–101 mm	0.6 mm	Aluminium alloy

Depth measurements of the grooves were carried out by scanning the sample with MicroProf FRT optical profilometer and then importing the data into SPIP software to measure the depth values. For each standoff distance 3 depth measurements were taken for statistical validity of the results. The mean and their deviation were plotted to observed the influence of the standoff distance for each chamber length. Surface morphology of the generated grooves were studied using SEM analysis (TESCAN MIRA 3 GME).

The surface was coated with gold layer in the presence of argon environment using QUARUM rotating coater pump. The SEM analysis was done in nitrogen environment using secondary electron (SE) detector.

3 Results and Discussion

All 18 experimental runs were carried out according to Table 1. The resultant disintegrated sample is shown in Fig. 2.

Fig. 2. Image of disintegrated material with specified experimental conditions.

The groove depth for each experimental run were calculated by importing the scanned data obtained from MicroProf to SPIP 6.7. Three measurement points for each standoff distance were taken in consideration and their mean and standard deviation were plotted for observing the influence of variation of standoff distance on erosion capability of the PWJ (Fig. 3, 4, 5, 6 and 7). It was measured in terms of disintegration depth for every trace created at different acoustic chamber length.

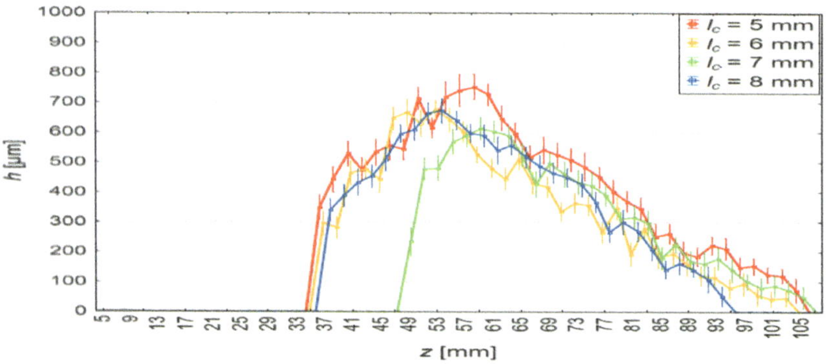

Fig. 3. Effect of variation of standoff distance with changing acoustic chamber length $l_c = 5$–8 mm on disintegrated depth.

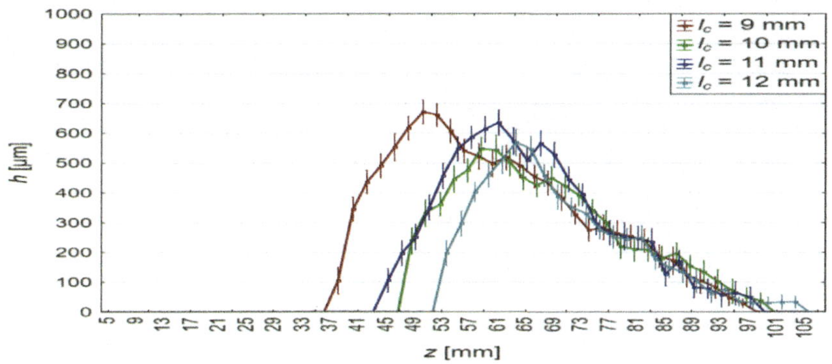

Fig. 4. Effect of variation of standoff distance with changing acoustic chamber length $l_c = 9$–12 mm on disintegrated depth.

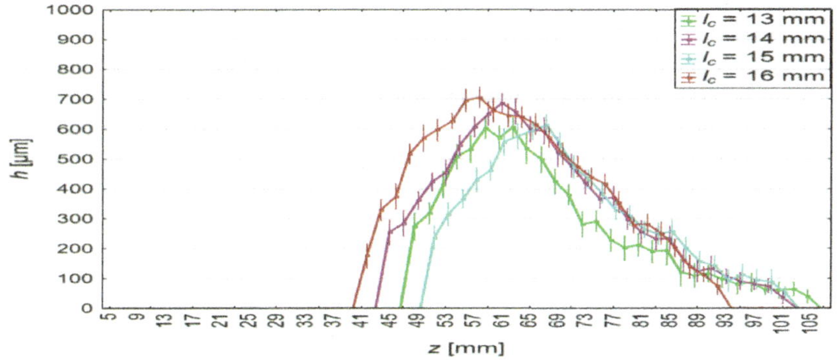

Fig. 5. Effect of variation of standoff distance with changing acoustic chamber length $l_c = 13$–16 mm on disintegrated depth.

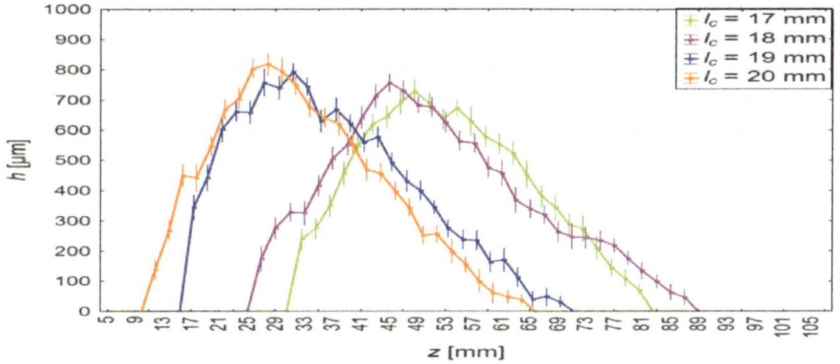

Fig. 6. Effect of variation of standoff distance with changing acoustic chamber length $l_c = 17$–20 mm on disintegrated depth.

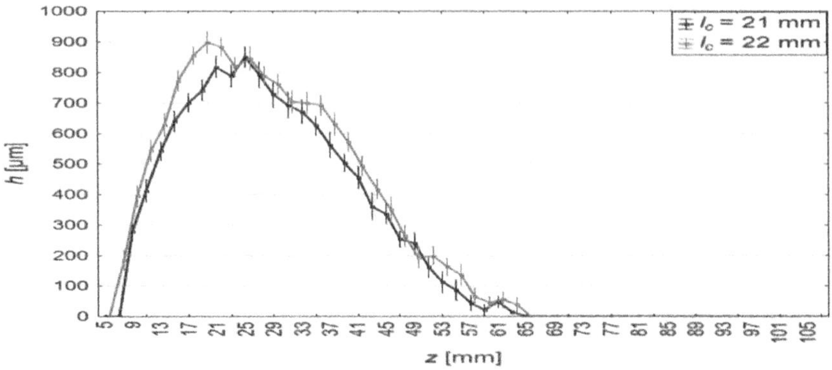

Fig. 7. Effect of variation of standoff distance with changing acoustic chamber length $l_c = 21$–22 mm on disintegrated depth.

The overall trend of the disintegration depth (h) with increase in standoff distance (z) follows a bell shaped curved with maximal depth achieved at mid values of the standoff distance and 2 ends at minimum and maximum standoff distance respectively (Fig. 3, 4, 5, 6 and 7). The overall course of the disintegration depth can be divided into 5 major regimes namely incubation phase, accelerating phase, maximal phase, declining phase and termination phase [17]. Every point of the material impacted by the PWJ undergoes through some of these phases and depending upon the energy of the jet and the position of the nozzle over the material surface i.e. standoff distance, it belongs to any one of the five phases. In incubation phase, the standoff distance of the nozzle is less and the jet do not reach breakup length limit before impacting the surface. Therefore, no detectable or visible erosion of the material surface is observed [18]. With increase in standoff distance, breakup length of PWJ is achieved and the jet starts to form well developed discrete bunches of water slugs and impact the surface. The energy of PWJ impacting the surface, overcomes the incubation phase and shows detectable erosion of the surface. This disintegration depth increases with increase in the standoff distance and this section is termed as accelerating phase of erosion. This is due to increase in the amplitude of the axial velocity fluctuation with increase in standoff distance resulting into increase in impact energy [19]. With further increase in the standoff distance to a certain limit, a stagnation stage with maximal disintegration depth is reached. This phase is marked with insignificant increase in disintegration depth with increase in standoff distance [20]. With more increase of the nozzle distance, aerodynamic resistance comes into consideration leading to decrease in the energy of the jet. Therefore, disintegration depth recorded in between this range of standoff distance decreases as compared to the previous section and named as declining phase of erosion. With further increase in standoff distance, the jet loses its entity and splits in scattered bunch of smaller water droplets and shows no surface erosion. Material exposed above this standoff distance belongs to termination stage of erosion. For instance, disintegration depth curve formed corresponding to $l_c = 5$ mm (Fig. 3), incubation stage continues till standoff distance $z = 35$ mm, from $z = 37$ mm to 51 mm accelerating phase of erosion takes place. From $z = 51$ mm to 61 mm, maximal erosion phase is observed. With further increase in

$z = 62$ mm till 105 mm, declining phase of disintegration is recorded and no visible erosion belong to termination phase is observed after $z = 107$ mm. Variation in ranges of standoff distance belong to the different phases depend upon the input technological parameters which in this study is acoustic chamber length. With increase in chamber length, the amplitude of the standing waves formed by the pressure fluctuation at the exit of the nozzle changes. These changes in the amplitude of the pressure fluctuation also determines the axial velocity fluctuation amplitude with which the PWJ impacts the material surface. Therefore, with variation of l_c from 5 to 22, the amplitude of the pressure fluctuation first decreases till 12 mm and then again increase till 22 mm. It can be attributed to the formation of node and antinode of the pressure waves at the exit of the nozzle which then propagate through the standoff distance to achieve required amplitude corresponding to the energy of the PWJ before impacting the surface. With node at the exit of the nozzle ($l_c = 12$ mm), higher standoff distance is required to obtained the desired amplitude of velocity fluctuation, thereby experiencing more aerodynamic resistance along the path. These negative energy counters the energy of the PWJ for generating deeper grooves as compared to condition were antinode is formed at the nozzle exit ($l_c = 22$ mm). For example, deepest disintegration depth recorded in experimental run with $l_c = 12$ mm (Fig. 4) is $h = 568.67$ μm as compared to $h = 898.33$ μm for $l_c = 22$ mm.

Figure 8 depicts the vertical lower limit z_{vll}, maximal disintegration z_{md} and vertical upper limit z_{vul} of erosion trace created by varying the acoustic chamber length from $l_c = 5$ to 22 mm in terms of standoff distance z, keeping all other technological parameters same. Z_{vll} represents the standoff distance at which visible and measurable material erosion is observed i.e. interface of incubation phase and accelerating phase of erosion curve whereas z_{md} represent the maximal disintegration depth which lies in the maximal phase of erosion curve achieved at particular technological setting. Z_{vul} represents the standoff distance after which no disintegration is observed i.e. end of declining phase and starting of termination phase of erosion curve.

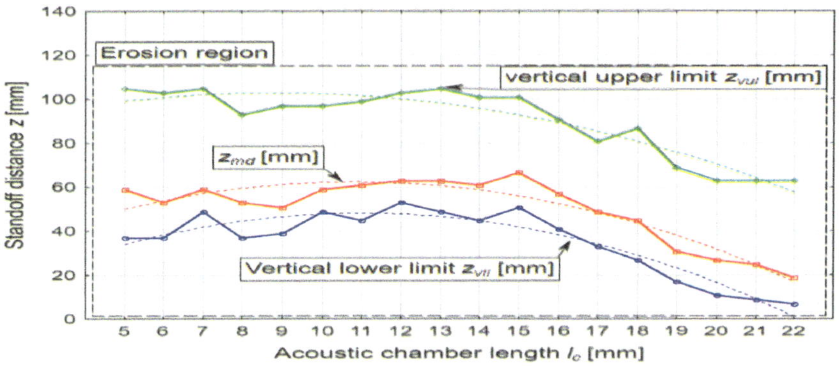

Fig. 8. Variation of Z_{vul}, Z_{md} and Z_{vll} with varying acoustic chamber length ($l_c = 5$ to 22 mm).

z_{vll} gradually increases to higher standoff distance with increase in acoustic chamber length to a certain level $l_c = 12$ mm and then starts to decrease with further increase in l_c ($z_{vll} = 37$, 53 and 7 mm for $l_c = 5$, 12 and 22 mm, respectively). The variation in

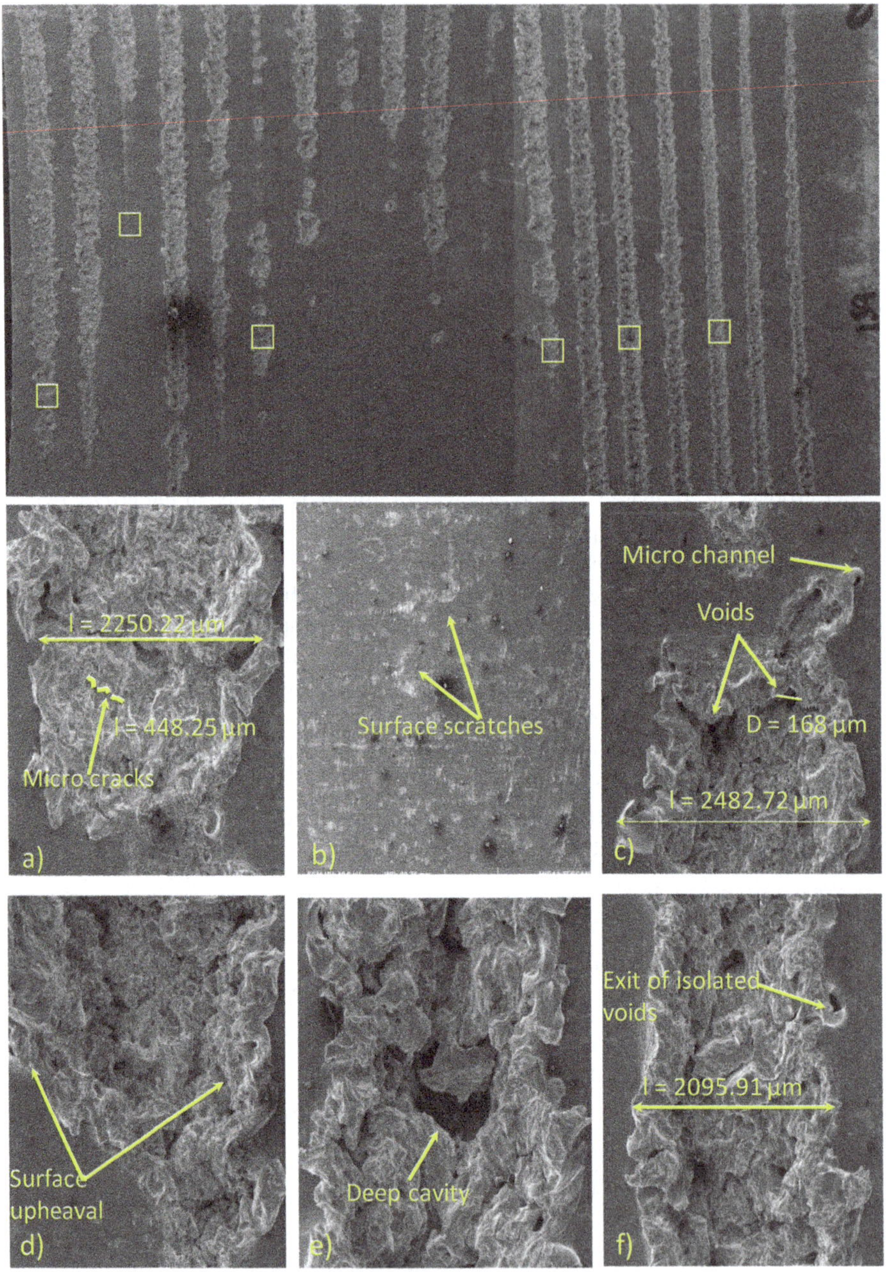

Fig. 9. SEM images of the disintegrated grooves for $l_c = 5$ to 22 mm (left to right).

z_{vll} is attributed to the amplitude of the standing waves at nozzle exit formed due to the pressure fluctuation in the acoustic chamber. Z_{md} also follows similar trend as z_{vll} for every l_c level. This standoff distance represents the distance at which maximal erosion for the current parametric setting occurs. It also depicts the standoff distance at which maximum amplitude of the axial velocity fluctuation representing energy of the PWJ impact with the material surface occurs. Z_{vul} represents the termination or the upper limit at which disintegration of the material surface ends. This is due to the aerodynamic resistance induced by the surrounding air acting against the direction of flow and PWJ energy. The uneven spacing between the z_{vll} and z_{md} curve and z_{md} and z_{vul} curve reveals the fact that accelerating phase of erosion requires smaller range of standoff distance and then the declining phase which takes larger range of standoff distance for each setting of l_c. Fig. 8 also can be helpful to identify region where PWJ can be used for peening [21, 22] disintegration [23, 24] and surface treatment purposes. For larger peening region, l_c = 10 to 15 mm can be used whereas, for longer disintegration region, l_c = 5 to 9 mm can be implemented. For surface roughening, l_c = 17 to 22 mm can be set keeping all other parameters constant. Tuning of the ultrasonic system can also be used to improve output power transmission by setting it into its resonance frequency regime by adjusting the l_c.

Figure 9 shows the morphology of the disintegrated material surface exposed to PWJ. The images were captured using SEM to observe in depth eroded surface features and characteristics. Width of the grooves formed are approximately in range of 2482 μm–2095 μm (Fig. 9). Micro cracks, voids and cavities are observed in the surface which are a common feature of eroded material surface. These are formed due to the compressive forces taking on the surface when impacted by the PWJ with appropriate standoff distance. The compressive waves are induced in the material followed by the shear stress travelling parallel to the surface which leads to the upheaved surface at the boundaries of the disintegrated grooves (Fig. 9d). Presence of surface asperities enhances the strength of the PWJ to cause erosion. When the ultimate strength of the material (σ = 220 MPa) is exceeded by the induced stresses, it fails and breakaway to form disintegration grooves. Surface scratches (Fig. 9b) are observed in the incubation phase of the trace made with l_c = 7 mm which depicts that z_{vll} (49 mm) is not reached attributed to insufficient energy of PWJ to cause measurable disintegration. Micro channels (Fig. 9c) and exit of isolated voids (Fig. 9e) are also observed near by the periphery of the grooves which may be due to the constructive interference of the pressure waves induced in the material, resulting in sudden increase in stresses in some locations near by the footprint of the PWJ [26].

4 Conclusions

The main aim of the study was to analyse the effect of variation in acoustic chamber length (l_c = 5 mm to 22 mm) with increasing standoff distance (z = 5 mm to 110 mm) on disintegration depth of aluminium workpiece. The following observations can be concluded from the study:

1) Maximum disintegration depth h, decreases from l_c = 5 mm (h = 753 μm) to 12 mm (h = 568 μm) and then increases to deepest for 22 mm (h = 898 μm).
2) z_{vll}, z_{md} and z_{vul} depends upon the amplitude of the pressure fluctuation at the nozzle tip which can be modified by tuning the system by changing the acoustic chamber length.

3) With increase in acoustic chamber length from $l_c = 5$ to 12 mm, z_{vll} shifts gradually to higher values whereas with further increase in l_c from 13 to 22 mm, z_{vll} shifts to lower values. ($z_{vll} = 37, 53$ and 7 mm for $l_c = 5$, 12 and 22 mm).
4) SEM analysis showed eroded surface characteristics like voids, cavities and cracks along with upheaved surface along the periphery of the eroded grooves. Some isolated voids created due to constructive interference of the surface pressure waves were also observed.

Acknowledgment. This study was supported by the GAČR 19-00408S.

References

1. Hloch, S., Adamčík, P., Nag, A., Srivastava, M., Čuha, D., Müller, M., Hromasová, M., Klich, J.: Hydrodynamic ductile erosion of aluminium by a pulsed water jet moving in an inclined trajectory. Wear **428–429**, 178–192 (2019). https://doi.org/10.1016/j.wear.2019.03.015
2. Tobin, E.F., Young, T.M., Raps, D., Rohr, O.: Comparison of liquid impingement results from whirling arm and water-jet rain erosion test facilities. Wear **271**, 2625–2631 (2011)
3. Lee, B.-E., Riu, K.-J., Shin, S.-H., Kwon, S.-B.: Development of a water droplet erosion model for large steam turbine blades. KSME Int. J. **17**, 114–121 (2003)
4. Keegan, M.H., Nash, D.H., Stack, M.M.: On erosion issues associated with the leading edge of wind turbine blades. J. Phys. D. Appl. Phys. **46**, 383001 (2013)
5. Crockett, H.M., Horowitz, J.S.: Erosion in nuclear piping systems. J. Press. Vessel Technol. **132**, 024501(1–3) (2010)
6. Farmer, I.W., Attewell, P.B.: Rock penetration by high velocity water jet: a review of the general problem and an experimental study. In: International Journal of Rock Mechanics and Mining Sciences & Geomechanics Abstracts, pp. 135–153. Elsevier (1965)
7. Hloch, S., Valíček, J.: Topographical anomaly on surfaces created by abrasive waterjet. Int. J. Adv. Manuf. Technol. **59**, 593–604 (2012)
8. Hreha, P., Radvanská, A., Hloch, S., Peržel, V., Królczyk, G., Monková, K.: Determination of vibration frequency depending on abrasive mass flow rate during abrasive water jet cutting. Int. J. Adv. Manuf. Technol. **77**, 763–774 (2015). https://doi.org/10.1007/s00170-014-6497-9
9. Nag, A., Ščučka, J., Hlavacek, P., Klichová, D., Srivastava, A.K., Hloch, S., Dixit, A.R., Foldyna, J., Zelenak, M.: Hybrid aluminium matrix composite AWJ turning using olivine and Barton garnet. Int. J. Adv. Manuf. Technol. **94**, 2293–2300 (2018)
10. Fowler, G., Pashby, I.R., Shipway, P.H.: The effect of particle hardness and shape when abrasive water jet milling titanium alloy Ti6Al4V. Wear **266**, 613–620 (2009)
11. Hreha, P., Hloch, S., Magurovd, D., Valicek, J., Kozak, D., Harnicdrovd, M., Rakin, M.: Water jet technology used in medicine. Teh. Vjesn. **17**, 237–240 (2010)
12. Foldyna, J., Svehla, B.: Method of generation of pressure pulsations and apparatus for implementation of this method (2011)
13. Foldyna, J., Svehla, B.: Method of generation of pressure pulsations and apparatus for implementation of this method (2010)
14. Foldyna, J., Říha, Z., Sitek, L., Švehla, B.: Numerical simulation of transmission of acoustic waves in high-pressure system. In: Proceedings of the International Congress on Ultrasonics, Vienna (2007)

15. Pochylý, F., Habán, V., Foldyna, J., Sitek, L.: 3D problem of pressure wave propagation in the tube with inconstant cross section. In: Proceedings of the International Congress on Ultrasonics, Vienna (2007)
16. Říha, Z., Foldyna, J.: Ultrasonic pulsations of pressure in a water jet cutting tool. Teh. Vjesn. Gaz. **19**, 487–491 (2012)
17. Hloch, S., Srivastava, M., Nag, A., Müller, M., Hromasová, M., Svobodová, J., Kruml, T., Chlupová, A.: Effect of pressure of pulsating water jet moving along stair trajectory on erosion depth, surface morphology and microhardness. Wear **452–453**, 203278 (2020)
18. Srivastava, M., Hloch, S., Tripathi, R., Kozak, D., Chattopadhyaya, S., Dixit, A.R., Foldyna, J., Hvizdos, P., Fides, M., Adamcik, P.: Ultrasonically generated pulsed water jet peening of austenitic stainless-steel surfaces. J. Manuf. Process. **32**, 455–468 (2018). https://doi.org/10.1016/j.jmapro.2018.03.016
19. Nag, A., Hloch, S., Čuha, D., Dixit, A.R., Tozan, H., Petrů, J., Hromasová, M., Müller, M.: Acoustic chamber length performance analysis in ultrasonic pulsating water jet erosion of ductile material. J. Manuf. Process. **47**, 347–356 (2019)
20. Raj, P., Hloch, S., Tripathi, R., Srivastava, M., Nag, A., Klichová, D., Klich, J., Hromasová, M., Muller, M., Miloslav, L., Chattopadhyaya, S., Adamcik, P.: Investigation of sandstone erosion by continuous and pulsed water jets. J. Manuf. Process. **42**, 121–130 (2019). https://doi.org/10.1016/j.jmapro.2019.04.035
21. Srivastava, M., Hloch, S., Gubeljak, N., Milkovic, M., Chattopadhyaya, S., Klich, J.: Surface integrity and residual stress analysis of pulsed water jet peened stainless steel surfaces. Measurement **143**, 81–92 (2019)
22. Srivastava, M., Hloch, S., Krejci, L., Chattopadhyaya, S., Dixit, A.R., Foldyna, J.: Residual stress and surface properties of stainless steel welded joints induced by ultrasonic pulsed water jet peening. Meas. J. Int. Meas. Confed. **127**, 453–462 (2018). https://doi.org/10.1016/j.measurement.2018.06.012
23. Nag, A., Hloch, S., Dixit, A.R., Cuha, D.: Investigation on pulsating liquid jet with physiological saline on aluminium surface. In: Advances in Manufacturing Engineering and Materials, pp. 63–71. Springer (2019)
24. Hloch, S., Nag, A., Pude, F., Foldyna, J., Zeleňák, M.: On-line measurement and monitoring of pulsating saline and water jet disintegration of bone cement with frequency 20 kHz. Measurement **147**, 106828 (2019)

Investigation of Different Abrasives Mixtures Ratio in Abrasive Water Jet Drilling of SS304

Sounder Rajan Rajendran, Balaji Kuppusamy, Yuvaraj Natarajan(✉), and Sivakumar Mahalingam

Department of Mechanical Engineering, Vel Tech Rangarajan Dr. Sagunthala R&D Institute of Science and Technology, Chennai, India
yuvaceg09@gmail.com

Abstract. The present work deals with effect of different abrasives mixtures ratio on Abrasive Water Jet (AWJ) drilling of SS304. In this study, the drilling operations were conducted with a combination of different abrasives with different mixture ratios such as 100% garnet, 60% garnet & 40% aluminum oxide, 60% garnet & 40% silicon carbide, 40% garnet & 60% aluminum oxide, 40% garnet & 60% silicon carbide, 50% silicon carbide & 50% aluminum oxide. Machining parameters namely, abrasive types, stand-off distance and feed rate were also considered for this study. Based on these factors, Taguchi L_{18} orthogonal array was designed for the drilling operations. The hole geometrical features such as cylindricity and surface roughness were investigated. The experimental results indicate that 40% garnet & 60% aluminum oxide and 60% garnet & 40% aluminum oxide has produced a better cylindricity and surface finish respectively. In addition, the application of taguchi with grey analysis has produced optimal parameter settings to better hole features in AWJ drilling process. The optimal parameter settings found to be 40% of garnet & 60% of aluminum oxide, 5 mm of stand-off distance and 170 mm/min of feed rate.

Keywords: Abrasive water jet · SS304 · Abrasive mixture ratio · Hole geometry

1 Introduction

Abrasive Water Jet (AWJ) machining is one of the nontraditional machining processes for machining different materials with better quality surfaces. In this technique, the material removal takes place through an erosive action. During machining, the cut-ting and deformation wear modes occur in the materials [1]. It is happening through an impingement of the high velocity AWJ stream and is formed while high pressure water passing through a tiny orifice in the mixing chamber increases the kinetic energy of water jet and thus, accelerates the abrasive particles during mixing. The parameters considered in AWJ machining are water pressure, feed rate, stand-off distance (SOD), type of abrasives, nozzle diameter, etc. In AWJ machining process, the abrasive feature has contributed more in cutting action rather than other parameters. AWJ technique has also proved its versatile in different operations such as cutting, turning, milling, drilling,

S. Hloch et al. (Eds.): ICMEM 2021, LNME, pp. 132–141, 2021.
https://doi.org/10.1007/978-3-030-71956-2_12

polishing and peening [2]. However, it is observed that the AWJ drilling operations are not explored well. In these regards, few reports of them presented below.

Nair and Kumanan [3] have investigated the AWJ drilling of Inconel 617 alloy by considering garnet as an abrasive with a mesh size of #80. They studied about different hole geometrical features such as circularity, cylindricity, material removal rate, perpendicularity and parallelism. The results indicate that the SOD and abrasive flow rate were identified as the influencing factors for better hole geometry. Kumar et al. [4] have analyzed the AWJ parameters for drilling Al/WC composites using garnet abrasive with a mesh size of #80. Among other operational parameters, the SoD was influencing parameter in material removal rate and surface roughness. Karatas et al. [5] examined the surface roughness in the drilled holes of CFRP composites by AWJ. They noted that the water jet pressure and stand-off distance were the significant factors on surface roughness of the drilled holes. Aydin et.al. [6] analysed the performances of different abrasives in AWJ cutting of marbles. Various abrasives used in the study were garnet, white fused alumina, brown fused alumina, glass, silicon carbide and emery powder. The results reported that the cutting performance of the AWJ was varied with abrasive density and hardness. Khan and Haque [7] performed the AWJ cutting of glass with garnet, silicon carbide and aluminium oxide abrasives. They found that the silicon carbide and aluminium oxide had more cutting energy than garnet abrasive. Shibin et al. [8] have analyzed the performance of silicon carbide abrasives in AWJ cutting of aluminium alloy.

Perec [9] investigated the AWJ machining of titanium alloy with different abrasive particles namely garnet, olivine, and crushed glass abrasive. His reported that the garnet abrasive was produced a deepest cut in titanium alloy than other abrasives. Yu et al. [10] have examined the mixing of garnet and alumina abrasives in AWJ cut-ting of aluminium alloy. They found that 75% of alumina and 25% of garnet produced a higher depth of cut and a lower surface roughness. And, this is the only article reported on the abrasive mixture in AWJ machining process till now. However, the research on different types of abrasive mixtures with various compositions further needs to be explored in different AWJ operations. Articles reported that different abrasives were employed in the AWJ process for the examination of alternate abrasives in process features, and also confirmed that better results were attained with each type of abrasive. Researchers have attempted or adopted various techniques for the improvement in performance of the AWJ process. However, the past reports have concluded that the most cost effective technique is use of abrasive features in an effective mode. Most of the researchers often used the single type of abrasives and abrasive mesh sizes in AWJ operations [2]. Therefore, in this work, the abrasive mixtures are prepared with different ratios and its effect on AWJ drilling SS304 is studied. The following parameters like abrasive mixture ratio, SoD and feed rate are considered as the inputs and cylindricity, roughness are considered as the output parameters.

2 Experimental Plan

In this study, the work material namely stainless steel 304 was considered for drilling operations. This process was done through a trepanning mode. The dimensions of the SS 304 are 200 mm × 100 mm × 10 mm. Drilling experiments were done on the AWJ machine tool (model: WSI) with a maximum water jet pump capacity of 400 MPa (E60 pump) was employed. Fig. 1 shows the experimental setup for drilling operations. The hole was drilled with a diameter of 10 mm. The AWJ drilling parameters and its conditions are shown in Table 1. In this work, abrasive type, stand-off distance (SOD) and feed rate have been considered for the variable parameters. Regarding the selection of abrasives, alluvial Indian garnet (purity 98%), black silicon carbide (purity 98%) and black aluminium oxide (electrocorundum – 99% purity) were used. With respect to these operational parameters, L_{18} orthogonal array was designed for the drilling experiments using MiniTab 17 and the experimental design is shown in Table 2. Different types of abrasives were used in this study with different mixtures ratio for drilling operations. Also, the machining time of each experiment was varied with an input of feed rate of the nozzle. The hole geometrical features considered were cylindricity and surface roughness. Hole cylindricity was measured by using 3D Co-ordinate measuring machine (CMM – Accurate make). The average surface roughness parameter (Ra) was measured in the drilled holes. It was measured using a contact type roughness tester by considering an evaluation length of 5 times of sampling length (5 × 0.8 = 4 mm).

Fig. 1. Experimental setup of AWJ drilling process.

Table 1. AWJ drilling parameters and its conditions.

Parameters	Units	Condition
Water jet pressure	MPa	380
Feed rate	mm/min	100, 170 and 230
Standoff distance	mm	1, 3, 5
Abrasive type	–	Garnet, silicon carbide (SiC) and aluminum oxide (Al_2O_3)
Abrasive mixture ratio	% Vol	100% garnet, 60% garnet + 40% SiC, 60% garnet + 40% Al_2O_3, 40% garnet + 60% SiC, 40% garnet + 60% Al_2O_3 and 50% garnet + 50% Al_2O_3
Abrasive mass flow rate	g/min	360
Abrasive mesh size	#	80
Impact angle	Degree	90
Water orifice diameter	mm	0.35
Focusing nozzle diameter	mm	1.1
Length of the focusing tube	mm	76

Table 2. Experimental design for AWJ drilling process.

Ex. No	Abrasive mixture ratio	SOD (mm)	Feed rate (mm/min)
1	100% garnet	1	100
2	100% garnet	3	170
3	100% garnet	5	230
4	60% garnet & 40% silicon carbide	1	100
5	60% garnet & 40% silicon carbide	3	170
6	60% garnet & 40% silicon carbide	5	230
7	60% garnet & 40% Aluminum oxide	1	170
8	60% garnet & 40% Aluminum oxide	5	100
9	60% garnet & 40% Aluminum oxide	3	230
10	40% garnet & 60% Aluminum oxide	1	230
11	40% garnet & 60% Aluminum oxide	3	100
12	40% garnet & 60% Aluminum oxide	5	170
13	40% garnet & 60% silicon carbide	1	170
14	40% garnet & 60% silicon carbide	3	230
15	40% garnet & 60% silicon carbide	5	100
16	50% Aluminum oxide & 50% silicon carbide	1	230
17	50% Aluminum oxide & 50% silicon carbide	3	100
18	50% Aluminum oxide & 50% silicon carbide	5	170

3 Results and Discussion

Figure 2 shows the cylindricity and surface roughness values of 18 drilled holes under various AWJ conditions. Cylindricity and roughness are important geometrical features in the drilled holes. Cylindricity feature ensures the roundness and straightness of the hole along its axis. From the results, it is noticed that a lowest cylindricity was obtained at the 10th drilled hole. It is attributed to the employment of 40% of garnet & 60% of Aluminum oxide with 1 mm of SoD and 230 mm/min of feed rate. As result of this, least cylindrical variation was formed in the hole profile and its variation is found to be 0.171 mm. A higher cylindricity variation was also noticed at 40% of garnet & 60% of Aluminum oxide with 3 mm SoD and 100 mm/min feed rate, and the result is found to be 0.298mm. These results indicate that feed rate and SoD influences the cylindrical shape of the drilled holes even though a same type of abrasive mixture is used.

Similarly, a lowest surface roughness was observed in the 11th hole and is result found to be 2.3853 μm. This result was obtained due to the involvement of 40% of garnet & 60% of Aluminum oxide with a SoD 3 mm and feed rate 100 mm/min. Among various experimental conditions, 40% of garnet & 60% of silicon carbide with a 3mm SoD and 230 mm/min feed rate produced a higher surface roughness and the result is observed to be 4.28 μm. It is noticed that the abrasive mixture was the most influencing factor than the pure garnet. It is attributed to that the effect of total mass of abrasive mixture produced a higher cutting energy which resulted better hole features in SS 304. The mass of abrasive mixture and cutting energy of the jet was varied with mixing of different abrasive particles as the density and hardness of each abrasive type is varied. Among the different abrasives, the silicon carbide and aluminium oxide abrasives are higher density and hardness than garnet. From the results, it is noted that abrasive mixture ratio was found suitable for the improvement in performance of the AWJ process.

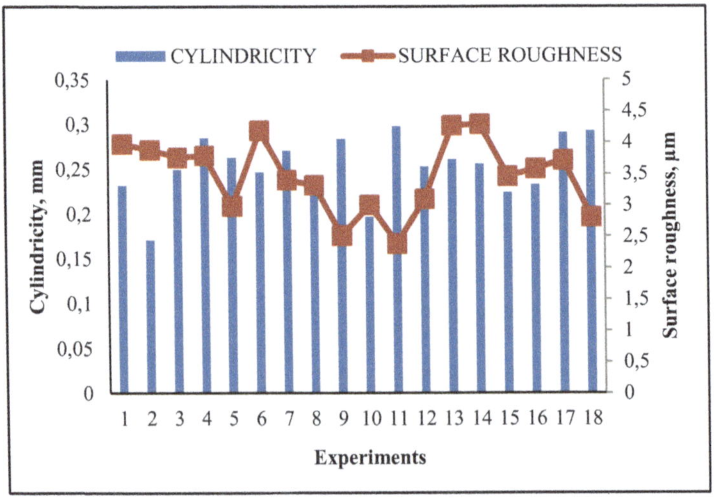

Fig. 2. Cylindricity and suface roughness of the AWJ drilled holes.

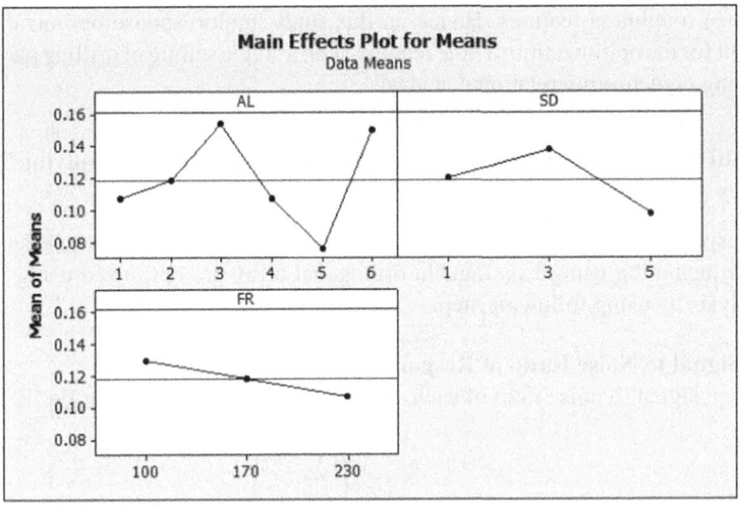

(a)

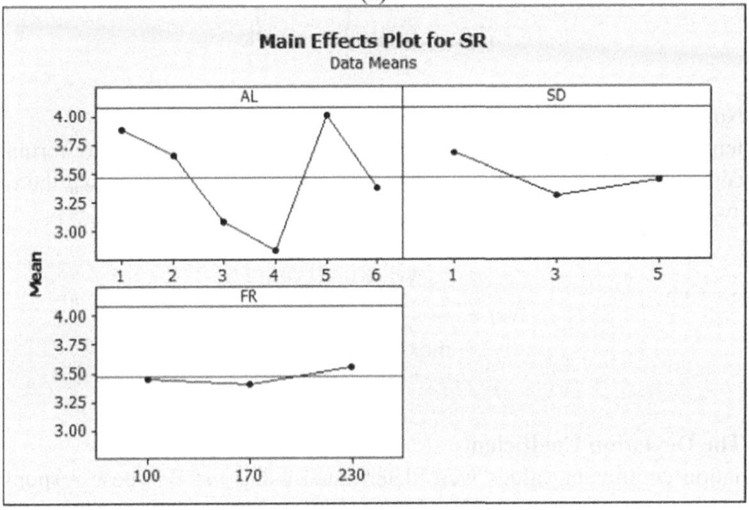

(b)

Fig. 3. Main effects plot for (a) Cylindricity error and (b) Surface roughness.

Figure 3 shows the main effects plot for the hole features such as cylindricity and roughness. This plot indicates that third level of abrasive mixture type (40% garnet & 60% Aluminum oxide), second level of SoD (3 mm) and first level of feed rate (100 mm/min) were produced an optimal circularity error in the drilled hole. Similarly, fifth level of abrasive mixture type (40% of garnet & 60% of silicon carbide), first level of SoD (1 mm) and second level of feed rate (170 mm/min) were produced an optimal surface finish. It is inferred that the selection of proper abrasive mixture ratio improves the hole geometrical features of the AWJ drilled holes rather than the use of single type abrasive. However, the different process parameters settings were found for cylindricity and surface roughness features. Hence, in this study, multiresponse optimization was carried out for the optimization of hole features with a single setting of drilling parameters by applying taguchi-grey relational analysis.

3.1 Multi Performance Optimization of Process Parameters by Applying Taguchi Grey Analysis

The corresponding response values of cylindricity and surface roughness measured during experimentation using L_{18} Taguchi orthogonal array are optimized using Taguchi grey analysis by using following steps.

Step 1: Signal to Noise Ratio of Response
In this step, signal to noise ratio of each response was determined using Eq. 1.

$$Sx_i = -10\log\left(\frac{\sum_{j=1}^{nr} x_{ij}^2}{nr}\right) \tag{1}$$

Step 2: Normalization of S/N Values
In this step, the S/N ratio values of different responses are converted to normalization of S/N values. Normalizing S/N ratio values is carried out to bring down the response ratios between 0 and 1 using Eq. 2.

$$Nx_i = \frac{Sx_i - \min_{i=1}^{ne}(Sx_i)}{\max_{i=1}^{ne}(Sx_i) - \min_{i=1}^{ne}(Sx_i)} \tag{2}$$

Step 3: The Deviation Coefficient
The deviation coefficient values were determined using Eq. 3. These responses also calculated between values 0 to1. Maximum value is 1 and minimum value is 0.

$$Dx_i = 1 - Nx_i \tag{3}$$

Step 4: The Grey Relational Coefficient (GRC)

$$Gx_i = \frac{MDx_i + \phi XDx_i}{Dx_i + \phi XDx_i} \tag{4}$$

Step 5: Grey Relational Grade (GRG)
In this step, the grey relational grades were obtained using Eq. 5 and the results are shown in Table 3.

$$GRG_i = \frac{GCE_i + GSR_i}{2} \tag{5}$$

Table 3. GRC and GRG/MRPI values.

Hole No	GCE	GSR	GRG/MRPI	Rank
1	0.7422	0.5339	0.6380	12
2	0.5134	0.5459	0.5296	18
3	1	0.5626	0.7813	2
4	0.5162	0.5588	0.5375	17
5	0.7365	0.7227	0.7296	5
6	0.7010	0.5104	0.6057	13
7	0.5177	0.6226	0.5702	15
8	0.6382	0.6383	0.6383	11
9	0.5	0.9176	0.7088	7
10	0.7155	0.7185	0.7170	6
11	0.5300	1	0.7650	3
12	0.9324	0.6910	0.8117	1
13	0.9852	0.5020	0.7436	4
14	0.8775	0.5	0.6887	8
15	0.7205	0.6114	0.6659	9
16	0.5690	0.5902	0.5796	14
17	0.5383	0.5692	0.5537	16
18	0.5383	0.7811	0.6597	10

In Table 3, the GRG values were arranged in decreasing order and the higher value of GRG represents the first rank which gives the optimum responses with the drilling parameters.

Table 3 shows the ranking of various AWJ drilling conditions on SS 304. From this table, a higher grey relational grade indicates the Rank 1 which was obtained at hole no.12 and the corresponding parameter setting is 40% of garnet & 60% of Aluminium oxide with a 5mm SoD and 170 mm/min feed rate. These combined parameters setting was produced the uniform material removal rate during AWJ drilling process. The 40% of garnet & 60% of Aluminium oxide was found suitable abrasive mixture ratio for drilling SS 304 along with other appropriate parameter conditions. This type of mixture ratio has generated a higher cutting energy, in which individual abrasive is uniformly accelerated by high kinetic energy of water jet rather than the employment of single type abrasive and other mixtures. Due to the proper mixing ratio of different mass of abrasives in the mixing chamber, distribution of abrasives and their impingement over the target surface become uniform. As result of this, a better cylindrical shaped hole with surface finish was attained.

The optimum cylindricity error and surface roughness are obtained at 40% garnet & 60% aluminum oxide, 5 mm SoD and 170 mm/min feed rate. This optimized parameter setting can further minimize the cylindricity error and surface roughness in the drilled holes.

4 Conclusion

In this experimental work, abrasive water jet (AWJ) drilling experiments were conducted on SS 304 with different types of abrasives mixtures ratio. In SS 304, 18 holes are drilled by varying the abrasives mixture ratios, feed rate and SoD. The effect of abrasive mixture ratios and other variable parameters on hole geometrical features namely, cylindricity and surface roughness were examined. The major results are concluded below.

1. The different abrasive mixtures and its ratio were improved the AWJ drilling performance features than the pure garnet. It is happened due to the attainment of higher cutting energy through the employment of critical mass of abrasive mixtures in the AWJ. These effects were produced based on the density and hardness of the abrasive types.
2. A lowest cylindricity error was obtained by the employment of 40% garnet & 60% aluminum oxide with 1 mm SoD and 230 mm/min feed rate in the AWJ process.
3. Among various abrasive mixture types, 60% of garnet & 40% of aluminum oxide was produced a quality finish with a SoD of 3 mm and feed rate of 100 mm/min were used.
4. A Taguchi with grey analysis exhibited the optimal parameters settings for better hole features in AWJ drilling operations. The 40% of garnet & 60% of aluminum oxide, 5 mm of SoD and 170 mm/min of feed rate were found as optimal parameter settings for the production of minimum cylindricity error and surface roughness in the AWJ drilled holes.

From the above conclusions, it is noticed that the proper selection of abrasive mixture ratio plays a vital role for obtaining better hole features in SS 304 using the AWJ.

References

1. Momber, A.W., Kovacevic, R.: Principles of Abrasive Water Jet Machining. Springer, London (1998)
2. Natarajan, Y., Murugesan, P.K., Mohan, M., Khan, S.A.L.A.: Abrasive water jet machining process: a state of art of review. J. Manuf. Process. **49**, 271–322 (2020)
3. Nair, A., Kumanan, S.: Optimization of size and form characteristics using multi-objective grey analysis in abrasive water jet drilling of Inconel 617. J. Braz. Soc. Mech. Sci. Eng. **40**, 121 (2018)
4. Kumar, K.R., Sreebalaji, V.S., Pridhar, T.: Characterization and optimization of abrasive water jet machining parameters of aluminium/tungsten carbide composites. Measurement **117**, 57–66 (2018)

5. Karatas, M.A., Gokkaya, H., Nalbant, M.: Optimization of machining parameters for abrasive water jet drilling of carbon fiber-reinforced polymer composite material using Taguchi method. Aircr. Eng. Aerosp. Technol. **92**(2), 128–138 (2019)
6. Aydin, G., Kaya, S., Karakurt, I.: Effect of abrasive type on marble cutting performance of abrasive waterjet. Arab. J. Geosci. **12**(11), 357 (2019)
7. Khan, A.A., Haque, M.M.: Performance of different abrasive materials during abrasive water jet machining of glass. J. Mater. Process. Technol. **191**(1–3), 404–407 (2007)
8. Shibin, R., Anandakrishnan, V., Sathish, S., Sujana, V.M.: Investigation on the abrasive water jet machinability of AA2014 using SiC as abrasive. Mater. Today: Proc. **21**(1), 519–522 (2019)
9. Perec, A.: Experimental research into alternative abrasive material for the abrasive water-jet cutting of titanium. Int. J. Adv. Manuf. Technol. **97**(1–4), 1529–1540 (2018)
10. Yu, Y., Sun, T., Yuan, Y., Gao, H., Wang, X.: Experimental investigation into the effect of abrasive process parameters on the cutting performance for abrasive waterjet technology: a case study. Int. J. Adv. Manuf. Technol. **107**, 2757–2765 (2020)

Effect of Water Flow Rate on Operating Frequency and Power During Acoustic Chamber Tuning

Akash Nag[1](✉), Gabriel Stolárik[2], Brano Svehla[3], and Sergej Hloch[2,4]

[1] Department of Mechanical Engineering, Indian Institute of Technology (Indian School of Mines), Dhanbad 826004, India
akashnag1992@gmail.com

[2] Faculty of Manufacturing Technologies (TUKE) with a seat in Presov, 08001 Prešov, Slovakia

[3] Ecoson Ltd., 91501 Nové Mesto nad Váhom, Slovakia

[4] Institute of Geonics, Czech Academy of Sciences, Ostrava-Poruba, Czech Republic

Abstract. Tuning of an ultrasonic system for pulsating water jet technology for efficient power transmission is discussed in this paper. The efficiency of the ultrasonic transducer depends upon the stiffness of the water interacting with it. The stiffness of water depends upon the inlet volume flow rate inside the acoustic chamber controlled by supply pump pressure and diameter of the nozzle. The interactional effect of varying water stiffness and the sonotrode efficiency can be adjusted by varying the acoustic chamber length. It changes the space required by the standing waves inside the chamber to achieve optimal amplitude near the nozzle exit. Therefore, supply pressure of 20, 30 and 40 MPa, nozzle diameter of 0.3–1 mm and chamber length from 0 to 22 mm were used in this study to observe the change in the operating frequency and output power of the sonotrode. The results showed that at lower flow rate (<3.0 l/min), the optimal chamber length depends upon both supply pressure and nozzle diameter. However, for flow rate above 3.06 l/min, the achieved resonance frequency depends solely on volume flow rate. The optimal chamber lengths observed are 20 and 21 mm for flow rate ranging from 3.06 to 5.43 l/min and 5.88 to 12.01 l/min, respectively. These results can be used as quick settings of the system in the resonance regime for future experiments.

Keywords: Acoustic chamber length · Resonance frequency · Water impedance · Ultrasonic power

1 Introduction

Ultrasonic technology is a non-conventional technology which is used for machining of conductive and non-metallic materials [1]. The machining process does not depend upon the physical and electrical characteristic of the workpiece. It is preferably used for machining brittle material with hardness more than 40 HRC [2]. Ultrasonic technology is being used for various applications such as cutting, drilling and grinding operations

S. Hloch et al. (Eds.): ICMEM 2021, LNME, pp. 142–154, 2021.
https://doi.org/10.1007/978-3-030-71956-2_13

[3]. High-frequency (>20 kHz) electrical energy is transformed into mechanical oscillations using transducers and then focused using a horn or sonotrode to disintegrate the materials [4]. The typical frequency of oscillation is more than 20 kHz with amplitude of 5–50 μm. However, the efficiency of the ultrasonic transducer for conversion of electrical energy into mechanical energy depends upon various factors. It depends upon electrical factors such as impedance of the circuit [5], resonance frequency [6], type of piezoceramic material and power transmission efficiency [6]. Mechanical factors, namely sonotrode design [7], material of the transducer and external loading of the sonotrode, along with environmental factors, such as ambient temperature, humidity or medium of application, affect the working efficiency of the ultrasonic system [8]. Considering electrical factors, the ultrasonic systems are designed to operate efficiently either in the resonance frequency regime or the anti-resonance frequency regime [9]. Applications such as ultrasonic welding and die polishing are performed at the anti-resonance frequency where speed and amplitude of vibration are maximum [10]. Applications like ultrasonic cleaning and cavitating of materials can done at the resonance frequency of the ultrasonic system [11]. In resonance frequency, the magnitude of vibration is maximum. Resonance and anti-resonance frequency also depend on the impendence of the electrical circuit. Electrical impedance is the ratio of voltage to current in an AC circuit [12]. In an ultrasonic system, resonance frequency is achieved when impendence of the electrical circuit is minimum, and anti-resonance frequency corresponds to highest electrical impendence of the circuit [13]. Piezoceramics have their own resonance frequency depending upon the material properties and dimension of the piezoceramics. However, this frequency changes when piezoceramic stacks are coupled with the horn and booster of the ultrasonic sonotrode to enhance its vibrational amplitude [14]. This variation in the resonance frequency depends upon the material and dimension of the sonotrode and the acoustic speed of the medium.

Ultrasonic technology in underwater applications generates cavitation bubbles which can be used for controlled disintegration of materials [15]. It creates disturbances in the water to form water bubbles which explode to do cavitation of the material surface. This phenomenon was implemented for forced decaying of continuous water jet into pulsating water jet (PWJ). PWJ is a hybrid technology which combines both ultrasonic technology and water jet technology [16]. The setup designed by Foldyna et al. [17] uses an ultrasonic oscillating sonotrode to enhance pressurized liquid volume entering the acoustic chamber with specific geometry. Ultrasonic vibrations are induced in the liquid volume to generate local pressure fluctuations [18]. These pressure fluctuations are responsible for forced decaying of the jet outside the nozzle head. Decay of the jet causes the formation of impact waves which interact with the surface with impact pressure [19]. Therefore, induction of ultrasonic disturbances in the pressurized fluid in the acoustic chamber is important. According to previous studies, the frequency–impedance characteristic of piezoelectric ceramics changes when used in air medium as compared to liquid medium [20]. The water adds liquid impedance to the original impedance of the ultrasonic transducer when operated in air. The variation in the overall impedance of the circuit depends upon the volume and resistance of the water toward the oscillating sonotrode [21]. It can be also considered as a water sonotrode coupled with a metallic sonotrode.

Variation in the impedance of the circuit also affects the power transmission efficiency of the ultrasonic generator. Generally, at higher impedance values, the output power lowers due to lowering of the current flowing in the electronic circuit keeping the operating voltage constant [13]. Therefore, for effective cavitation results, the whole ultrasonic system including metallic sonotrode and water sonotrode should have lower impedance value for efficient transmission of ultrasonic power. It also corresponds to lower actual frequency of the oscillating sonotrode with higher power output.

Wave propagation through water medium depends on the wave generating source and also on the stiffness of the medium. In PWJ, water acts as a medium for the pressure fluctuations inside the acoustic chamber whose stiffness is determined by the volume flow rate of the water [22]. The flow rate is determined by the supply pressure and the nozzle diameter [23]. Therefore, with changing these parameters, the stiffness of the water changes, which in turn changes the water impedance. Efficient transmission of ultrasonic energy requires the system to be in the least impedance region. For adjusting the system, provision in form of variable acoustic chamber length is provided in the design. It can be lengthened or shortened with the help of a mechanical nut, to tune the system in the resonance regime of the ultrasonic transducer [24]. Therefore, in the present study, the optimal operating frequency (f_o) and acoustic chamber length (l_c) for different nozzle diameter varying from $d = 0.3$ to 1 mm with supply pressure $p = 20$, 30 and 40 MPa are determined. These values can be used for quick tuning of the system in future experiments.

2 Experimental Setup

The experiments were conducted using a Hammelmann high-pressure plunger pump with maximum pressure of $p = 160$ MPa with flow rate of $\dot{Q} = 67$ l/min. A robot manipulator (ABB IRB 6640) is used to hold the pulsating head in a desired position. Ecoson WJ-UG 630–40 ultrasonic generator was used to supply the electrical signals to the piezoelectric ceramics for its actuation [1] (Fig. 1). The maximum operation power (P) of the generator is 800 W, and the operating frequency (f_o) range is 19–22 kHz. The experiments were conducted by changing the nozzle insert diameters from $d = 0.3$ to 1 mm. For each d, supply p was changed to 20, 30 and 40 MPa. The acoustic chamber length (l_c) was varied from 0 to the value at which there was a sudden change in f_o and P. Experimental conditions are as shown in Table 1.

Table 1. Experimental conditions.

Nozzle diameter [mm]	Supply pressure [MPa]	Acoustic chamber length [mm]	Initial frequency [kHz]
0.3, 0.4, 0.5, 0.6, 0.7, 0.8, 0.9, 1	20, 30, 40	0–22	21.00

The piezoceramics were put into vibration by using the hand-held control unit (Fig. 2). Also, the operating frequency and power output for every experimental run are displayed in the unit. The generator is kept functioning for 1 min for each technological setting. It was done to keep uniformity in all the runs. The ultrasonic generator was switched on with an initial frequency (f_i) of 21 kHz for each parametric setting. The f_o and P are noted from the display. The l_c was changed by rotating a mechanical nut which extends the length of the chamber. The pitch of the rotating nut is 2 mm. Therefore, for each half rotation, l_c increases by 1 mm. Again, the same procedure for determining the f_o and P was repeated. The upper limit of the l_c was determined by change in the f_o and P value higher than 21 kHz (f_s) and 200 W. The values of f_o and P were plotted graphically to observe the influence of l_c on them (Figs. 3, 4, 5, 6, 7, 8, 9 and 10).

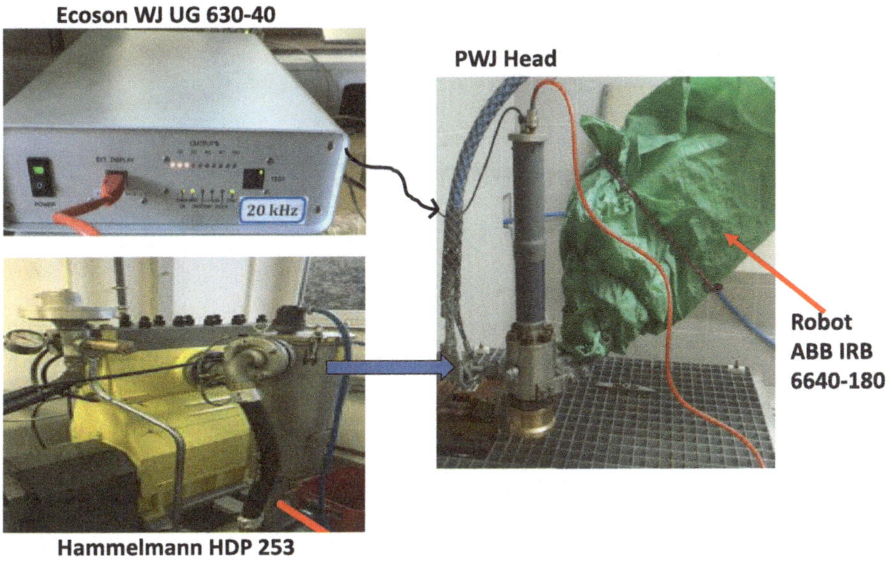

Fig. 1. Experimental setup.

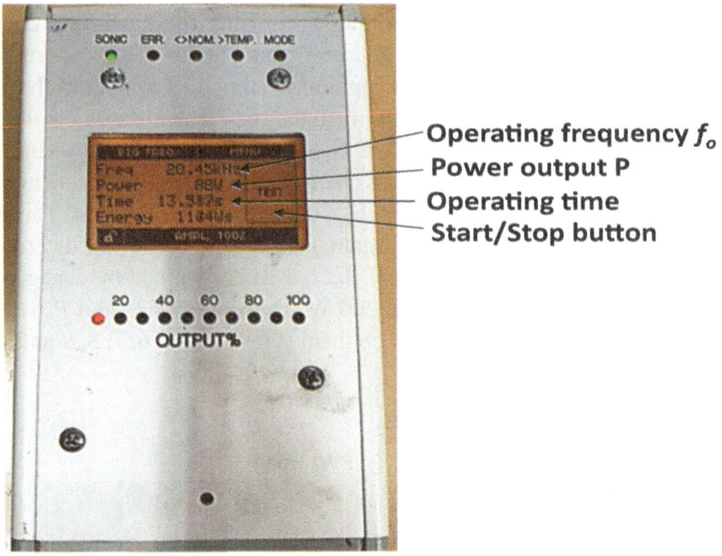

Fig. 2. Hand-held control panel.

3 Result and Discussion

The following graphs shows the variation in the actual frequency recorded after powering the ultrasonic generator and the instantaneous power output for d varying from 0.3 to 1 mm and p of 20, 30 and 40 MPa.

Figure 3 shows a graphical representation of the P and f_0 generated as a function of the l_c at $d = 0.3$ mm. Optimal l_c values representing the resonance regime in terms of frequency and power were noted at 18 mm for 20 MPa, 20 mm for 30 MPa, and 21 mm for 40 MPa. With increase in the pressure from $p = 20$ to 40 MPa, the l_c setting for obtaining the minimum frequency increases to $l_c = 18$, 20 and 21 mm, respectively. This is attributed to the increased water volume entering the acoustic chamber per unit time (0.76, 0.98 and 1.08 l/min for 20, 30 and 40 MPa, respectively), increasing the water impedance [2]. This in turn increases the space required for achieving the optimal amplitude of the pressure fluctuation near the nozzle exit. At $l_c = 19$, 21 and 22 mm for $p = 20$, 30 and 40 MPa, respectively, the frequency suddenly rises and the power increases abruptly, interrupting normal operating conditions.

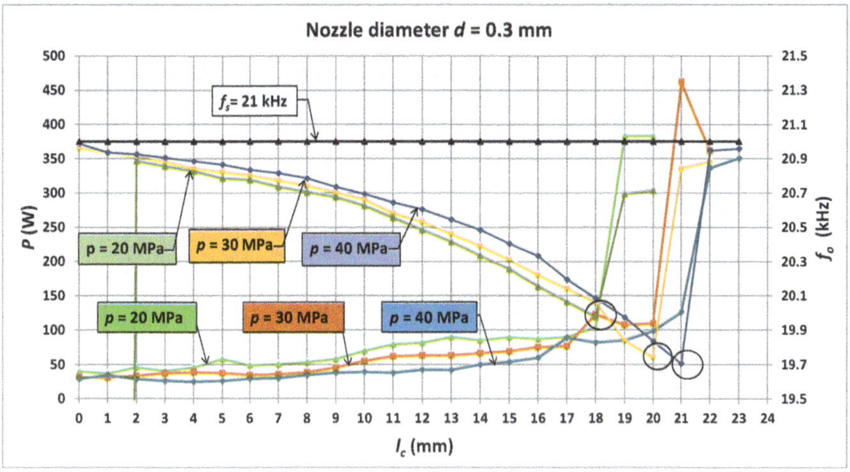

Fig. 3. Variation in f_o and P with l_c at $p = 20$, 30 and 40 MPa with nozzle diameter $d = 0.3$ mm.

Figure 4 shows the variation of output power and operating frequency with varying pressure p = 20 MPa, 30 MPa and 40 MPa with changing l_c at nozzle diameter d = 0.4 mm. A Similar trend as d = 0.3 mm was obtained with optimal l_c values changing for each p value. The optimal l_c points were marked in the graph.

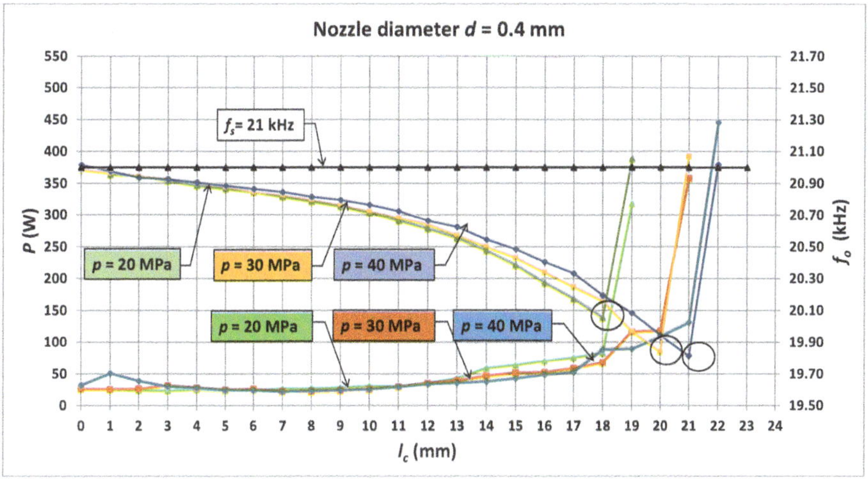

Fig. 4. Variation in f_o and P with l_c at $p = 20$, 30 and 40 MPa with nozzle diameter $d = 0.4$ mm.

Figure 5 shows the variation of f_o and P with l_c at $d = 0.5$ mm. The graph shows a similar trend with $d = 0.3$ and 0.4 mm for $p = 20$ and 40 MPa, but for $p = 30$ MPa the f_o occurs at $l_c = 18$ mm, as well as for $p = 20$ MPa. This can be attributed to the combined effect of fluctuations in the electrical signals from the ultrasonic generator and variation in the water flow during the run.

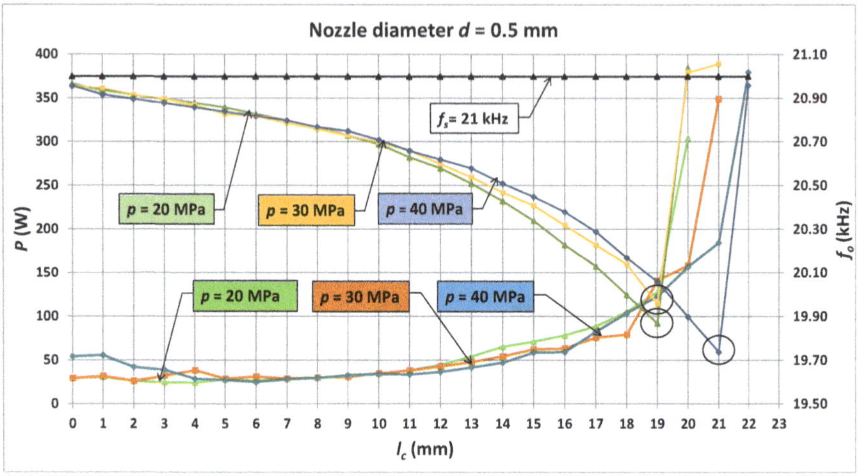

Fig. 5. Variation in f_o and P with l_c at $p = 20$, 30 and 40 MPa with nozzle diameter $d = 0.5$ mm.

Figure 6 shows the variation of f_o and P with l_c at $d = 0.6$ mm. The graph shows coincidence of the l_c value at 20 mm for all p values. It is attributed to the higher values of flow rate (3.06–4.32 l/min), and the l_c value at which the minimum f_o occurs are not significantly affected by the flow rates. The formation of the optimal amplitude of the pressure fluctuation near the nozzle exit is achieved at $l_c = 20$ mm.

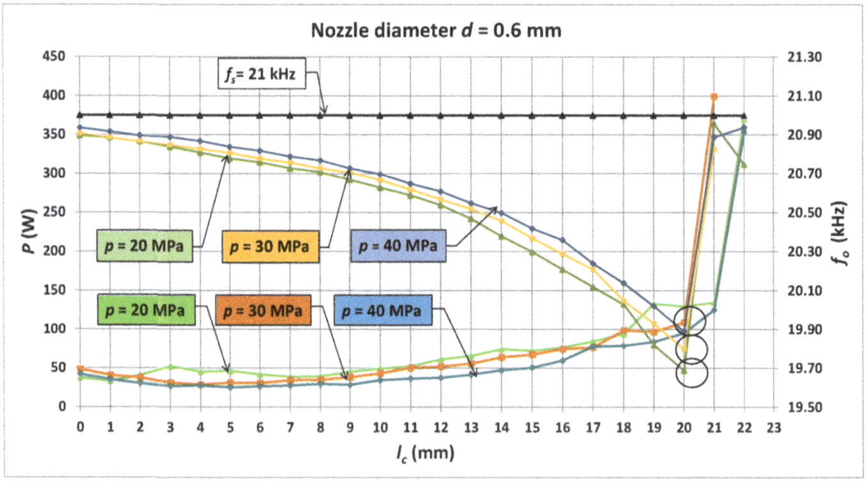

Fig. 6. Variation in f_o and P with l_c at $p = 20$, 30 and 40 MPa with nozzle diameter $d = 0.6$ mm.

Figure 7 shows the variation of f_o and P with l_c at $d = 0.7$ mm. The graph shows coincidence of the l_c value at 20 mm for all $p = 20$ and 30 MPa values, but at $p = 40$ MPa the minimum f_o was obtained at $l_c = 21$ mm. It is in agreement with the l_c values obtained for $d = 0.7$ mm. Therefore, it can be concluded that for volume rate values (4.16 and 5.10 l/min at $p = 20$ and 30 MPa, respectively), minimum f_o is achieved at $l_c = 20$ mm. For 40 MPa, $l_c = 21$ mm was obtained, which is attributed to increased volume flow rate (5.88 l/min) at the specified conditions. With increase in flow rate, a larger volume of the acoustic chamber is required to obtain the optimal amplitude of the standing wave near the nozzle exit.

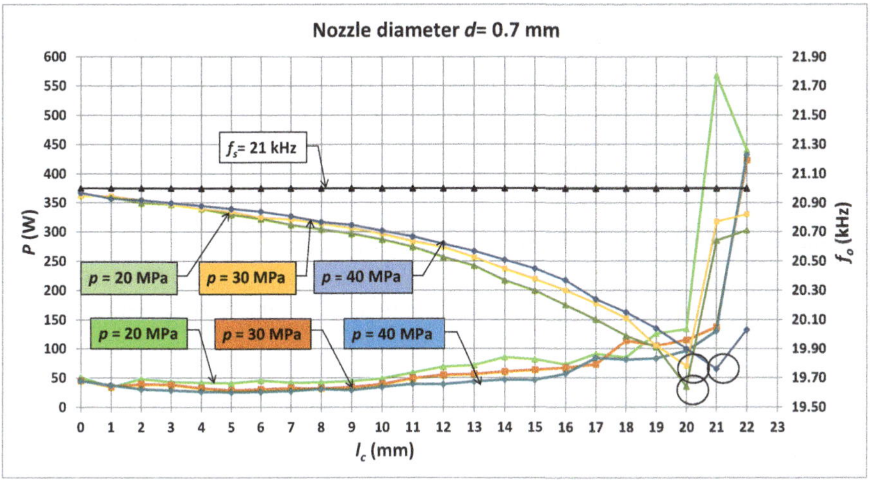

Fig. 7. Variation in f_o and P with l_c at $p = 20$, 30 and 40 MPa with nozzle diameter $d = 0.7$ mm.

Figure 8 shows the variation of f_o and P with l_c at $d = 0.8$ mm. The graph shows the optimal l_c value at 20 mm for $p = 20$ MPa, whereas $l_c = 21$ mm for $p = 30$ and 40 MPa. This variation is attributed to the volume flow rate at the specified flow conditions such as 5.43, 6.66 and 7.68 l/min for $p = 20$, 30 and 40 MPa, respectively.

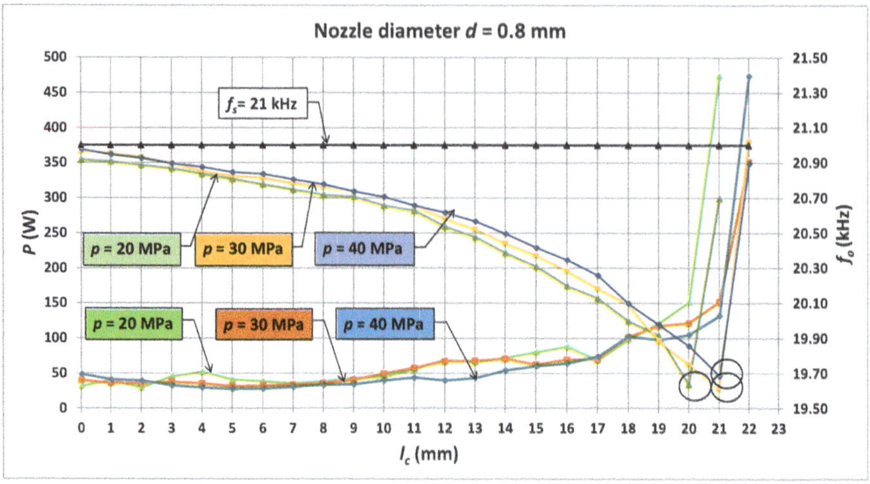

Fig. 8. Variation in f_o and P with l_c at $p = 20$, 30 and 40 MPa with nozzle diameter $d = 0.8$ mm.

Figures 9 and 10 represent the variation of P and f_o with change in l_c for $d = 0.9$ and 1 mm at $p = 20$, 30 and 40 MPa. Due to the higher volume flow rate (6.88 to 12.01 l/min), the optimal l_c value was obtained at $l_c = 21$ mm at which the f_a was minimum for all the technological settings.

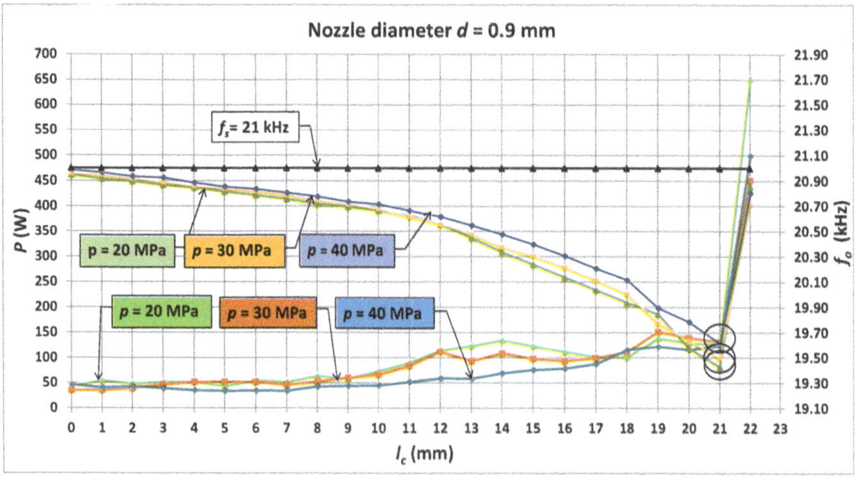

Fig. 9. Variation in f_o and P with l_c at $p = 20$, 30 and 40 MPa with nozzle diameter $d = 0.9$ mm.

Figure 11 shows the variation of the resonance frequency (f_r), which is the minimum f_o and l_c for which f_r is achieved for each volume flow rate. The graph shows that at lower flow rates (<3 l/min), the l_c setting for achieving f_r fluctuates depending upon the nozzle diameter and supply pressure. For flow rates between 3.05 and 5.43 l/min, l_c is constant at 20 mm unaltered by the nozzle diameter or the supply pressure. For flow

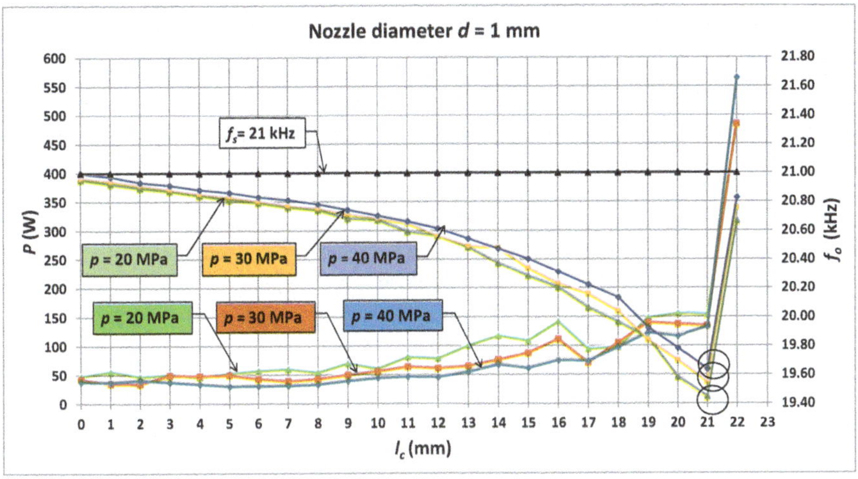

Fig. 10. Variation in f_o and P with l_c at $p = 20$, 30 and 40 MPa with nozzle diameter $d = 1$ mm.

rate > 5.88 l/min, the l_c at which f_r is obtained is fixed at $l_c = 21$ mm. This is attributed to the increase in the water impedance generated by higher flow rates (>5.88 l/min) which requires larger space ($l_c > 21$ mm) to obtain the desired amplitude of the pressure fluctuation inside the acoustic chamber. The f_r also tends to decrease with increasing volume flow rate attributing to the increase in the resistance for the movement of the sonotrode at increased flow rates. Therefore, oscillation of the sonotrode achieved per second decreases. However, the exact trend of the f_r for each volume flow rate is still not clear.

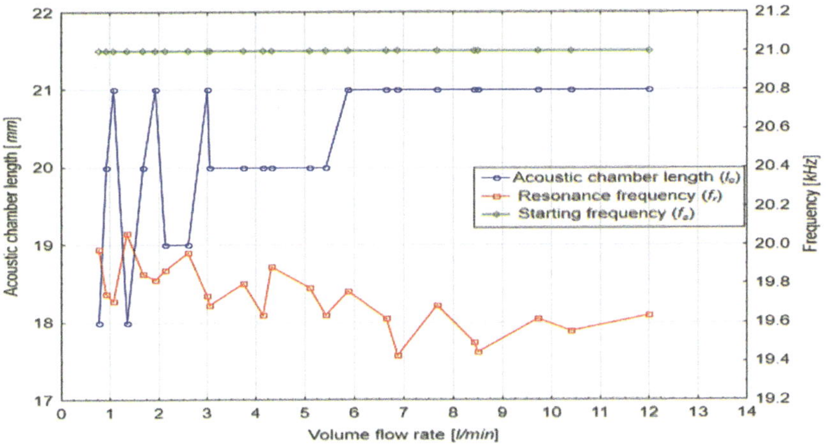

Fig. 11. Variation in f_r and corresponding l_c with increasing volume flow rate.

The optimal values of l_c, f_r and P obtained at various technological settings are listed in Table 2. These results can be used in future studies for a quick adjustment of PWJ technology, for optimal l_c values, which would ensure that the f_o is in the impedance range and the power transmission is efficient.

Table 2. Optimal setting of acoustic chamber length and corresponding values of resonance frequency and output power.

Nozzle diameter d [mm]	Supply pressure p [MPa]	Volume flow rate $\dot{Q}$ [l/min]	Optimal acoustic chamber length l_c [mm]	Resonance frequency f_r [kHz	Output power P [W]
0.3	20	0.76	18	19.98	106
	30	0.94	20	19.75	112
	40	1.08	21	19.71	127
0.4	20	1.36	18	20.06	86
	30	1.66	20	19.85	120
	40	1.92	21	19.82	132
0.5	20	2.12	19	19.87	129
	30	2.60	19	19.96	142
	40	3.00	21	19.74	185
0.6	20	3.06	20	19.69	130
	30	3.74	20	19.80	110
	40	4.32	20	19.89	125
0.7	20	4.16	20	19.64	134
	30	5.10	20	19.78	115
	40	5.88	21	19.76	131
0.8	20	5.43	20	19.64	153
	30	6.66	21	19.62	153
	40	7.68	21	19.69	133
0.9	20	6.88	21	19.43	128
	30	8.42	21	19.50	134
	40	9.73	21	19.62	125
1	20	8.49	21	19.45	155
	30	10.40	21	19.56	136
	40	12.01	21	19.64	133

4 Conclusion

The study shows the influence of water volume flow rate determined together with supply pressure and nozzle diameter on sonotrode operating frequency and power output. The main conclusions of the study are summarized as follows:

1. Power transmission and operating efficiency of the oscillating sonotrode depend upon the external forces acting on it.

2. Increase in the volume flow rate of the water inside the acoustic chamber varies the water impedance, leading to variation of optimal impedance of the system.
3. With increase in the flow rate ($\dot{Q} = 0.76$ to 12.01 l/min), the space required to overcome the stiffness of the medium increases ($l_c = 18$ to 21 mm).
4. The resonance frequency for the sonotrode lowers from $f_r = 19.89$ to 19.64 kHz with increase in the flow rate from $\dot{Q} = 0.76$ to 12.01 l/min.

This study shows the optimal l_c setting for tuning the ultrasonic system in the resonance regime to obtain desired results. In further investigation of the performance of PWJ, researchers can use this study or the methodology as a standard method for determining and setting the optimal chamber length on basis of the fluid flow rate. A similar study with a 40 kHz sonotrode and all other flow rates should be performed for quick tuning of the PWJ. Also, direct measurement of electric impedance of the circuit can be performed for more accurate and exact results.

Acknowledgement. This study was supported by the Slovak Research and Development Agency under Contract No. APVV-17-0490 and VEGA 1/0096/18.

References

1. Kumar, J.: Ultrasonic machining—A comprehensive review. Mach. Sci. Technol. **17**, 325–379 (2013)
2. Thoe, T.B., Aspinwall, D.K., Wise, M.L.H.: Review on ultrasonic machining. Int. J. Mach. Tools Manuf. **38**, 239–255 (1998)
3. Singh, R.P., Singhal, S.: Rotary ultrasonic machining: a review. Mater. Manuf. Process. **31**, 1795–1824 (2016)
4. Singh, R., Khamba, J.S.: Ultrasonic machining of titanium and its alloys: a review. J. Mater. Process. Technol. **173**, 125–135 (2006)
5. Balamuth, L.: Mechanical impedance transformers in relation to ultrasonic machining. J. Acoust. Soc. Am. **26**, 934 (1954)
6. Yao, Z., Guo, Z.N., Zhang, Y.J., Deng, Y., Zhang, W.T.: Research on the frequency tracking in rotary ultrasonic machining. Proc. CIRP **6**, 556–560 (2013)
7. Sherrit, S., Badescu, M., Bao, X., Bar-Cohen, Y., Chang, Z.: Novel horn designs for power ultrasonics. In: IEEE Ultrasonics Symposium, pp. 2263–2266. IEEE (2004). https://doi.org/10.1109/ULTSYM.2004.1418291
8. Upadhye, V., Agashe, S.: Effect of temperature and pressure variations on the resonant frequency of piezoelectric material. Meas. Control **49**, 286–292 (2016)
9. Shuyu, L.: Optimization of the performance of the sandwich piezoelectric ultrasonic transducer. J. Acoust. Soc. Am. **115**, 182–186 (2004)
10. Hirose, S., Aoyagi, M., Tomikawa, Y., Takahashi, S., Uchino, K.: High power characteristics at antiresonance frequency of piezoelectric transducers. Ultrasonics **34**, 213–217 (1996)
11. Arnold, F.J., Mühlen, S.S.: The resonance frequencies on mechanically pre-stressed ultrasonic piezotransducers. Ultrasonics **39**, 1–5 (2001). https://doi.org/10.1016/S0041-624X(00)00047-0
12. Cheney, M., Isaacson, D., Newell, J.C.: Electrical impedance tomography. SIAM Rev. **41**, 85–101 (1999)

13. Davari, P., Ghasemi, N., Zare, F.: Power converters design and analysis for high power piezoelectric ultrasonic transducers. In: 2014 16th European Conference on Power Electronics and Applications. IEEE (2014). https://doi.org/10.1109/EPE.2014.6910986
14. Hocheng, H., Kuo, K.L.: On-line tool wear monitoring during ultrasonic machining using tool resonance frequency. J. Mater. Process. Technol. **123**, 80–84 (2002)
15. Shchukin, D.G., Skorb, E., Belova, V., Moehwald, H.: Ultrasonic cavitation at solid surfaces. Adv. Mater. **23**, 1922–1934 (2011)
16. Hloch, S., Adamčík, P., Nag, A., Srivastava, M., Čuha, D., Müller, M., Hromasová, M., Klich, J.: Hydrodynamic ductile erosion of aluminium by a pulsed water jet moving in an inclined trajectory. Wear **428–429**, 178–192 (2019). https://doi.org/10.1016/j.wear.2019.03.015
17. Foldyna, J., Klich, J., Hlavacek, P., Zelenak, M., Scucka, J.: Erosion of metals by pulsating water jet. Teh. Vjesn. Gaz. **19**, 381–386 (2012)
18. Raj, P., Hloch, S., Tripathi, R., Srivastava, M., Nag, A., Klichová, D., Klich, J., Hromasová, M., Muller, M., Miloslav, L., Chattopadhyaya, S., Adamcik, P.: Investigation of sandstone erosion by continuous and pulsed water jets. J. Manuf. Process. **42**, 121–130 (2019). https://doi.org/10.1016/j.jmapro.2019.04.035
19. Hloch, S., Srivastava, M., Nag, A., Muller, M., Hromasová, M., Svobodová, J., Kruml, T., Chlupová, A.: Effect of pressure of pulsating water jet moving along stair trajectory on erosion depth, surface morphology and microhardness. Wear **452–453**, 203278 (2020)
20. Chilibon, I.: Underwater flextensional piezoceramic sandwich transducer. Sens. Actuators A Phys. **100**, 287–292 (2002)
21. Wevers, M., Lafaut, J.-P., Baert, L., Chilibon, I.: Low-frequency ultrasonic piezoceramic sandwich transducer. Sens. Actuators A Phys. **122**, 284–289 (2005)
22. Hloch, S., Nag, A., Pude, F., Foldyna, J., Zeleňák, M.: On-line measurement and monitoring of pulsating saline and water jet disintegration of bone cement with frequency 20 kHz. Measurement **147**, 106828 (2019)
23. Nag, A., Hloch, S., Dixit, A.R., Cuha, D.: Investigation on pulsating liquid jet with physiological saline on aluminium surface. In: Advances in Manufacturing Engineering and Materials, pp. 63–71. Springer (2019). https://doi.org/10.1007/978-3-319-99353-9_8
24. Nag, A., Hloch, S., Čuha, D., Dixit, A.R., Tozan, H., Petrů, J., Hromasová, M., Müller, M.: Acoustic chamber length performance analysis in ultrasonic pulsating water jet erosion of ductile material. J. Manuf. Process. **47**, 347–356 (2019)
25. Srivastava, M., Hloch, S., Gubeljak, N., Milkovic, M., Chattopadhyaya, S., Klich, J.: Surface integrity and residual stress analysis of pulsed water jet peened stainless steel surfaces. Measurement **143**, 81–92 (2019)

The Study of Method of Acoustic Generation of Pulses for Creating Pulsating Water Jet

Piush Raj(✉), Somnath Chattopadhyaya, and Amitava Mondala

Department of Mechanical Engineering, IIT (ISM), Dhanbad, India
piush.raj@gmail.com

Abstract. In discussions a modern technology Pulsating water jet (PWJ), has created a space because of its characteristic of cutting ability with appreciable surface finish that are used in industrial application, mining industry and hard to cut materials. As many as five techniques have been reported so far, for creation of pulses employed in generation of pulsating water jet. Sequential progress and development of each method has its own importance, instantly the focus is to provide mathematical approach in understanding of one of the method of creation of pulsating water jet that is based on acoustic generation of pulses and transmission of generated waves to pressurized water right to the nozzle exit. Geometrically and physically the explanation of this system is very complex. However, few important parameters that play vital role in this process are second kinematic viscosity of water, speed of sound in water, force of jet, energy of jet, breakup length of pulses, resonating fundamental frequency, resonating amplitude, and water hammer expressions in pipe flow as a function of pressure. The purpose of the current review work to present the reader with different approaches adopted by authorities of the field about describing the phenomenon mathematically of acoustically generation of pulses for PWJ. Therefore, discussions among all methods have been done based on above described parameters. This will provide researches with better understanding of instant generation of pulsating water jet with overarching goal of mathematically placement of all related equations at a platform as a standard source of reference.

Keywords: Acoustic generation · Equations · Literature survey · Review

List of Symbols

U	- Mean exit velocity
t	- Time
$u\ (0, t)$	- Uniform velocity
ΔU	- Amplitude of modulation
f	- Frequency
$r_0(x,\ t)$	- Surface configuration of jet at any instant
x	- Distance below the nozzle
p	- Mass density
p	- Pressure
μ	- Viscosity

S. Hloch et al. (Eds.): ICMEM 2021, LNME, pp. 155–182, 2021.
https://doi.org/10.1007/978-3-030-71956-2_14

u	- Axial X velocity vector
V	- Radial r component of velocity
$\acute{u}$	- Variation in un-modulated jet velocity
R	- Nozzle radius of un-modulated jet
λ	- Wavelength of modulation
Re	- Reynolds number
We	- Weber Number
St	- Strouhal Number
ζ	- A positive constant
S_d	- Strouhal Number
f	- Resonating frequency
V	- Velocity
d	- Diameter of jet
ρ	- Density of water
ρ_{ref}	- Reference density of water (998.2 kg.m^{-3})
p	- Real pressure
p_{op}	- Operating pressure
K	- Bulk modulus of water (2.2 $\times$ 10^9 Pa)
a	- Velocity of sound
Π_{ij}	- Non- reversible stress tensor
Θ	- Dynamic second viscosity
C_{ij}	- Tensor of deformation speed
ω	- Frequency domain
δ_{ij}	- Kronecker delta
η	- Dynamic viscosity
ξ	- Kinematic second viscosity
γ	- Kinematic viscosity
s	- Lap lace operator
$\xi(s)$	- Lap lace function of second kinematic viscosity
P, Q	- Special Legendre Polynomial
c_r	- Mean velocity of liquid along r
P	- Transfer matrix representing pressure and discharge pulsations
m_a	- Mass acoustic
c_a	- Capacitance acoustic
w_0	- Angular frequency
f_0	- Basic frequency
k_{rcc}	- Accurate resultant force
f_{zm}	- Calculated frequency
$\rho_{\Delta f}$	- Relative difference between measured frequency and predicted fundamental frequency
x_j	- Constituent of coordinate
u_i	- Constituent of velocity
μ	- Viscosity
μ_t	- Turbulent viscosity
k	- Turbulent kinetic energy
ε	- Turbulent dissipation energy

C_μ	- Constant (in two – equation models)
σ_K	- Constant (in modeled kinetic energy transport equation)
σ_ε	- Constant (in modeled dissipation rate transport equation)
$C_{1\varepsilon}$, $C_{3\varepsilon}$, S_ε	- Constants (in modeled dissipation rate transport equation)
G_k, G_b	- Constants
a_q	- Volume fraction at qth phase
ρ_q	- Density at qth phase
K	- Bulk modulus of water
a	- Speed of sound
b, d	- Constants
p_i	- Impact pressure
c	- Shock velocity
p_s	- Stagnation pressure
$\Delta\tau$	- Impact time
E_k	- Kinetic energy
v_l	- Liquid discharge velocity
v_s	- Sound velocity
m_d	- Approximate weight of pulse
m_0	- Weight of sample before test
m_l	- Weight of sample after test
ς	- Second kinematic viscosity
Θ	- Second viscosity coefficient
L_0	- Breakup length
M	- Velocity modulation ratio
f	- Frequency of velocity modulation
P_h	- Hydraulic power
d	- Diameter of nozzle orifice
Δm	- Mass loss
v_{trs}	- Traverse velocity
s	- Length of cut
p	- Ambient pressure upstream of nozzle exit
Q	- Actual flow rate of water
P_u	- Ultrasonic power
Δm_s	- Rate of mass loss
c_0	- Shock velocity
v_0	- Velocity of jet
ρ_1	- Density of liquid
ρ_2	- Density of solid
c_1	- Shock velocity in liquid
c_2	- Shock velocity in solid
a	- Acoustic (water hammer) wave speed
P	- Piezo metric pressure
Z	- Elevation of pipe centre line from a given datum
H	- Piezometric head
V	- Cross sectional mean velocity
u	- Local longitudinal velocity

A	- Cross sectional area of pipe
g	- Gravitational acceleration
τ_w	- Shear stress at pipe wall
D	- Pipe diameter
x	- Spatial coordinate along the pipeline
t	- Temporal coordinate
l_0	- Turbulent length scale
τ_0	- Turbulent time scale

Abbreviations

PWJ	- Pulsating water jet
CFD	- Computational fluid dynamics
DFT	- Discrete fourier transform
UNESCO	-United nation's education scientific and cultural organization
N-S	- Navier stokes
RANS	- Reynolds averaged Navier Stocks equation
DoE	- Design of experiment
UDF	- User defined function
CS	- Control surface
CV	- Control volume
UST	- Ultrasonic tool

1 Introduction

The potential outcomes of pulsating water jet on interaction accompanied with solid plane have been interest of research community to understand the mechanism and validate it mathematically since last decade [1]. Pulsating water jet (PWJ) has created a space in manufacturing arena wherein growing demand of high strength materials can be disintegrated. PWJ is a clean technology where water itself acts as cutting tool with cleaning agent of burrs thus obtained due to cutting. It is gaining popularity as a sustainable non-traditional machining process being used today in the industry. PWJ is commonly used in rock disintegration [2–4], rock mining [5], metal cutting [6–8], concrete cutting [9, 10], in medical applications [11–13] and cleaning and cooling [14, 15] of microchips using PWJ in the field of electronics.

In this paper selective research articles are reviewed which have tried to explain the process of acoustically generation of pulses for PWJ with some mathematical expressions and compare based on the parameters that are second kinematic viscosity of water, speed of sound in water, force of jet, energy of jet, breakup length of pulses, resonating frequency, resonating amplitude, and water hammer expressions in pipe flow. The aim of the paper is to focus the mathematical process and principles that are used to explain the acoustically generated PWJ system so far by eminent researchers in the field. The initial section of the paper describe the overview with brief history of development of pulsating water jets and utilization of water hammer phenomenon to achieve betterment in cutting and finally a comprehensive discussion will be provided based on deduced parameters by each study.

2 Brief Historical Background

Studies have been reporting evolution of modulating jet. Chronological arrangement of development of modulation of jet started with internal mechanical flow modulators device which was integrated within nozzle [16]. The basic idea for unpopularity was short life span of moving components inside nozzle. Second method was based on the principle of self-resonating nozzles. A PULSER-FED configuration which is a union of PULSER and "ORGAN PIPE" configuration consists of tandem-orifice Helmholtz resonating chamber followed by an organ pipe section. A shock pressure are created when liquid flow to exit of resonating chamber. These pressures are carried back to inlet of the tube creating standing wave by adding with pressure pulsations. While the frequencies are paired with Helmholtz chamber and the organ pipe wave are pressure resonance takes place and jet starts to create distinct annular eddy ending up with creation of pulses [17] refer Fig. 1(A). The device failed due to low modulation of liquid jet. Third technique for modulating jet was external use of rotating disc containing slots, holes, may be sprockets reported by [3, 4, 18] refer Fig. 1(B). However, the technique has limited its application to rock cutting only. In fourth method ultrasonic vibrating transformers were placed upstream the vicinity of the nozzle exit in a way that fluid in pressure flow past through annulus between the transformer and nozzle wall. The transformer generates high intensity ultrasound field upstream that modulates the water jet [19] refer Fig. 1(C). Major drawbacks were bulkiness of cutting tool with high wear of tip of vibrating transformer because of excessive cavitational erosion.

Another approach adopted for modulating the jet is established by acoustic creation of pulses [20] refer Fig. 1(D). The acoustic actuator generates pressure pulsations with pressurized liquid available in the sonic chamber. Mechanically operated amplifying device amplifies these pressure pulsations and onward forwarded to liquid waveguide equipped with pressurized liquid feed to the nozzle. Acoustic actuator, acoustic chamber, pulsations mechanically amplified and fluid waveguide are basic components which are used for productive transmission of pulsating energy from creation point that is generator to the exit point that is nozzle. This effective system has overcome the excessive wear of acoustic originator reported in the work of Vijay et al. [19] of pulsations as the setting of acoustic actuator in acoustic chamber is not subject to the immense wear due to cavitations erosion. Another physical situation for creating pulsating water jet is by use of fluidic nozzle device. Stouffer et al. [21] explains the device build up for basically apparatus for spraying water not directly on target material but cyclically deflecting off the target with help of this fluidic nozzle, jet start to breakup into small droplets refer Fig. 1(E). The use of this system fluidic nozzle was applied in cleaning of windshield of cars.

In all of the above methods water jet emanates from the nozzle and interacts with air; process is described as flow of free jets. Many mathematical models with various equations are put to explain the phenomenon discretely. However, the collective literature review to explain mathematically in this area is unsaid. Our intent is to bring all the work in one platform and analyze the governing equations that effectively explain the flow of free jet mathematically considering factors like density, pressure, viscosity, shock velocity, nozzle size and nozzle configuration, and surface tension, frequency of the water jet. In the process of reviewing for this purpose all the available research literature are considered with specific aim to filter the existing research base on acoustic generation of pulses method to produce pulsating water jet.

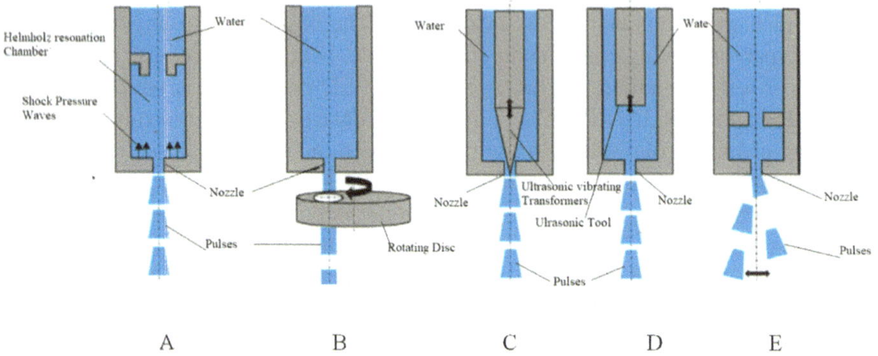

Fig. 1. Different methods of PWJ production A) Helmholtz Method B) Rotating Disc Method C) Vibrating transformer method D) Acoustic actuator Method E) Fluidic Nozzle Method [22].

3 Research Methodology

The intention behind this review work to fabricate all related mathematical model in one place, jotting down serially development year wise and to serve the reader with better grip of equation of motion and dynamics under lying the phenomenon of acoustic pulse generation. The major problem related to this review addressed "Mathematical model that are responsible for pulse generation of pulsed/modulated water jet".

4 Literature Review

Sami et al. [23] has tried predicting the breakup length of a modulated jet emanating a small orifice into air and to develop useful parameters. The flow of free jets are ruled by factors of viscosity, pressure, nozzle size and nozzle configuration, density, and surface tension of jet fluid and completely described by local momentum conditions so controlled by simple mathematical models. Author has considered jet with nozzle radius R at time t and at velocity u (0, t) consistent along the departure point but repeat periodically about mean value U and can be written as

$$\mathrm{u}\,(\mathrm{o}, \mathrm{t}) = \mathrm{U} + \Delta\mathrm{U}\ \sin\, 2\pi \tag{1}$$

U represents the mean exit velocity and ΔU the amplitude and f frequency of modulation. Continuity equation considering the control volume (C.V) that is bounded by control surface (C.S) and the integral form using Reynolds transport theorem the conservation of mass applied to control volume expressed [24, 25] by

$$\frac{\partial}{\partial t}\int_{cv}^{\cdot} \rho d\forall + \int_{cs}^{\cdot} \rho(v.n)dA = 0 \tag{2}$$

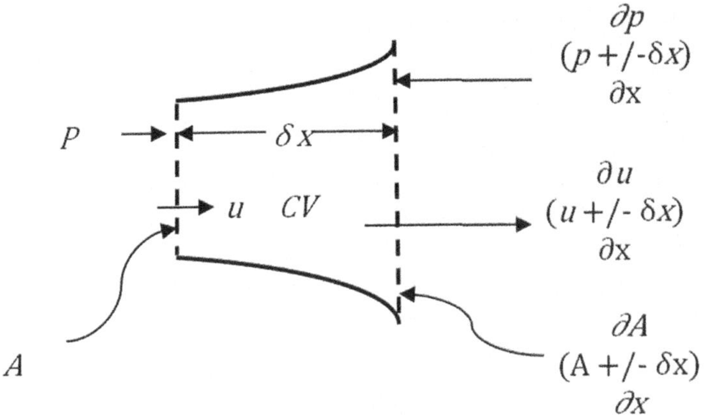

Fig. 2. Control Volume figure for continuity equation [23].

n = unit outward normal vector to control surface, v = velocity vector.

The control volume can be written as (A. dx) as per Fig. 2 and applying the fact that control volume moves to zero, Eq. 2 yields

$$\frac{\partial r_0^2}{\partial t} + \frac{\partial}{\partial x}\left(r_0^2\right) = 0 \tag{3}$$

With aim to establish surface configuration of jet (r_0 (x, t)) adopting polar cartesian coordinate system with x express the distance below from nozzle and r denotes the radial coordinate. Establishing the continuity equation and equation of motion for asymmetric flow of jet without swirl and rest parameter set constant.

Consider for circular jet with horizontal discharge velocity form the equation for continuity written in terms of r_0 as

$$\frac{\partial r_0}{\partial t} + \mathrm{u}\frac{\partial r_0}{\partial x} = -\frac{r_0}{2}\left(\frac{\partial u}{\partial x}\right) \tag{4}$$

Writing the equation of motion, the N-S equation for axisymmetrical jet flow applied to obtain equation of motion of modulated fluid where in viscous force and gravitational forces are not considered while pressure and velocity along length outline are constant along fluid flow and expressed as

$$\frac{\partial u}{\partial t} + \mathrm{u}\frac{\partial u}{\partial x} = -\frac{1}{p}\frac{\partial p}{\partial x} + \frac{\mu}{p}\left(\frac{\partial_u^2}{x^2}\right) \tag{5}$$

$$\frac{\partial v}{\partial t} + \mathrm{u}\frac{\partial v}{\partial t} + \mathrm{v}\frac{\partial v}{\partial r} = \frac{u}{p}\left(\frac{\partial^2 \mathrm{v}}{x^2} + \frac{\partial^2 \mathrm{v}}{r^2} + \frac{1\partial v}{r\partial r}\frac{v}{r^2}\right) \tag{6}$$

With p representing the density of mass, p the pressure, μ the viscosity, and u represents axial velocity along x and v represents axial velocity along y and radial r components of the velocity vector. Consider the pressure term expression can be written considering r_N and r_T f curvatures I the combination of two perpendicular planes assuming small amplitude modulation as

$$\frac{\partial u}{\partial t} + \mathrm{u}\frac{\partial u}{\partial x} = \frac{\sigma}{p}\left[\frac{1\partial r_0}{r_0^2\partial x} + \frac{\partial^3 r_0}{\partial x^3}\right] + \frac{\mu}{p}\left[\frac{\partial^2 u}{\partial x^2}\right] \tag{7}$$

The continuity Eq. (4) and equation of motion Eq. (7) are simultaneously solved to find fluids bunching eventually breakup length are calculated along the jet axis. Further velocity u can be written in respect of uniform velocity U (un-modulated jet velocity) and a changing u' which intern modifies the two Eqs. (4) and (7), Also, consider transforming the frame work at un-modulated jet velocity U, the investigation of a wavelength of the stationary stream expression as

$$\frac{\partial r_0}{\partial t} + u'\frac{\partial r_0}{\partial z} = -\frac{r_0}{2}\left(\frac{\partial u'}{\partial x}\right) \tag{8}$$

$$\frac{\partial \mathrm{u}'}{\partial \mathrm{t}} + \mathrm{u}'\frac{\partial \mathrm{u}}{\partial \mathrm{z}} = \frac{\sigma}{\rho}\left[\frac{\partial \mathrm{r}_0}{\mathrm{r}_0^2\partial \mathrm{z}} + \frac{\partial^3 \mathrm{r}_0}{\partial \mathrm{z}^3}\right] + \frac{\mu\partial^2 \mathrm{u}'}{\rho\partial \mathrm{z}^2} \tag{9}$$

Equation (8) and (9) can be written with help of new non-dimensional parameters as

$$\zeta = r_0/R;\ \mathrm{u} = u'/U;\ \eta = z/\lambda;\ \tau = tU/\lambda;\ R_e = \rho UR/\mu;\ We^2 = \frac{\sigma}{\rho RU^2};\ \mathrm{St} = R/\lambda \tag{10}$$

Where R is radius of the un-modulated jet, λ the wavelength and f the frequency of modulation, *Re* Reynolds numbers of flow, *We* Weber numbers of flow, and *St* Strouhal numbers of flow.

$$\partial\xi/\partial\tau + \mathrm{u}\frac{\partial\xi}{\partial\eta} = -\frac{\xi}{2}\left[\frac{\partial \mathrm{u}}{\partial\eta}\right] \tag{11}$$

$$\frac{\partial u}{\partial \tau} + \mathrm{u}\frac{\partial u}{\partial \eta} = We^2\left(\frac{\partial\zeta}{\zeta^2\partial\eta} + St^2\frac{\partial^3\zeta}{\partial\eta^3}\right) + \frac{st\partial^2 u}{Re\partial\eta^2} \tag{12}$$

For high Reynolds number flows the last term of Eq. (10) may be neglected and written as

$$\frac{\partial u}{\partial \tau} + u\frac{\partial u}{\partial \eta} = We^2\left(\frac{\partial\zeta}{\zeta^2\partial\eta} + St^2\frac{\partial^3\zeta}{\partial\eta^3}\right) \tag{13}$$

Equations (11) and (13) are simultaneously solved, and author suggested three methods for solutions tabulated below with corresponding dimensionless break-up length (X_R or a_b St) for each of three solutions.

Out of three methods, method II and method III have identical solution for breakup lengths which is defined with respect to the finite ratio between decreased and the undisturbed smooth jet radii.

Chahine et al. [17] has proposed a new type pulsating device based on the principle of self-resonance to create pressure fluctuations. The author designed "PULSER-FED"

Table 1. A comparison on Solutions of equations [23].

Method	Z	x_b/R
I	$Exp\,[-\pi M\tau\ cos\ 2\pi\eta]$	$ln\zeta/\pi M\ St$
II	$1-\pi M\tau\ cos\ 2\pi\eta$	$1/\pi M\ St$
III	$1+\pi M\alpha\ cos\ 2\pi\tau$	$1/\pi M\ St$

self-resonating jet 'SERVOJET' for experimentations which produce extremely high frequency intervals of water driven in air and forecasted its application of high frequency jet pulsations on improving action of cleaning by air only or water jet lead dissection. The non-dimensional number suggested for defining periodic characteristic of any periodic oscillating system that is Strouhal number S_d.

$$S_d = \frac{fd}{V} \tag{14}$$

Where;
f represents the resonating frequency;
d represents the jet diameter and V represents velocity.

For optimum performance of jet S_d should be in range $0.3 < Sd \leq 1.2$ provided that i) recreation of the impetuous stress formed by every slug ii) buffering of the stress by a liquid surface from the last slug and iii) slug creation due to aerodynamic drag. The author confirmed that the range proposed for S_d has not been achieved by any other researcher. It was further ascertained that increase in erosive sharpness depends upon frequency variations.

Foldyna et al. [26] worked on numerical simulation of transmission of 3D, unsteady, turbulent fluid flow of compressible water in peak pressure frame of reference fitted with acoustic originator. The CFD model was utilized using Fluent code of CFD to reproduce mathematically the effect of operation and layout factors of acoustic originator and those in transmission line on creation and behaviour of pulsating jet. The results were verified using dynamic pressure sensors at nozzle exit in upstream movement of high pressure pulsations. The geometry is shown in Fig. 3. At the inlet fluid 30 MPa pressure was set and mass flow rate was determined.

Further, geometry was configured with reference of change in length "L" as shown in Fig. 3, which the biggest diameter of pipe with a unit of 8 mm (1 unit equal to 8 mm) and accordingly corresponding lengths were set as 64, 72, 80, and 88 respectively based on these configuration results have been plotted.

At the outlet zero friction wall was defined by frequency of 20 kHz and amplitude of 6μm using goniometric functions. The model was tested first for steady flow of water with above boundary conditions and then with unsteady reciprocal motion of vibrating wall. The turbulent part was taken care by k-ε turbulence model. Water compressibility was taken care by UDF equation for speed of sound in water and water density by

$$\rho = \frac{\rho_{ref}}{1-\frac{\Delta p}{K}} = \frac{\rho_{ref}}{1-\frac{p-p_{op}}{K}} \tag{15}$$

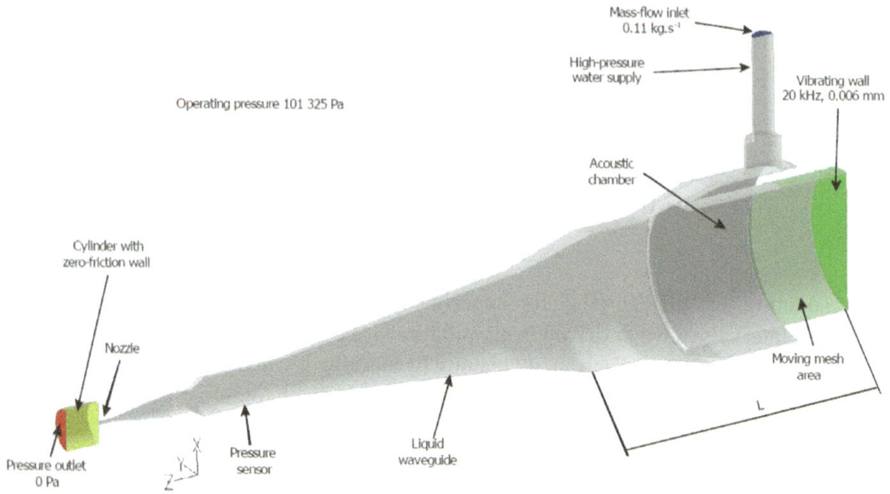

Fig. 3. Model geometry and boundary conditions [26].

$$a = 2.10^{-6}p + 1432\,[31] \tag{16}$$

where ρ is water density, ρ_{ref} is reference water density, p is real pressure, p_{op} represents operating pressure in MPa, K express bulk modulus of water and a velocity of sound. Graphical representations of simulated numerical process of acoustic generation on time behaviour of pressure were in good agreement with experimental measurements. Further the change in geometry influence on simulation those to numerical model on amplitude of pressure are accurate. The acoustic actuator generates standing waves in the system and the amplitude of wave increase towards nozzle exit due to shape of waveguide placement as seen in Fig. 4. As the cavitations effects are not taken into consideration the numerical model may not be accurate.

Foldyna et al. [27] performed experiments for determining velocity of sound and water's second kinematic viscosity for frequency equals to 20 kHz based on calculation of forced oscillation of water by transfer matrixes method. The transfer matrix P with circular cross section and the area contraction through the linear axis (conffuser) are obtained by [28]. Foldyna et al. [29] had Also, performed similar experiments to evaluate velocity of sound and water's second kinematic viscosity adopting modal analysis where a pipe filled with water were impact was applied on the face further response were collected after impact. Pressure sensors were deployed at both ends of tube to collect to collect the response. In this study the results were compared with respect to under water acoustics provided by Wong et al. [30] and were in close agreement with "UNESCO" equation" Also, known as Chen and Millero equation. The set of equation developed in 1977 came to known as United Nations Education Scientific and Cultural Organization Equation which provide the speed of sound in sea water at high pressure and one set was revised by Del Grosso (1974) with temperature scale ITS-90.

The experiment was conducted with big capacity pipe with a round cross section containing water at high pressure. A frequency of 20 kHz was produced by acoustic

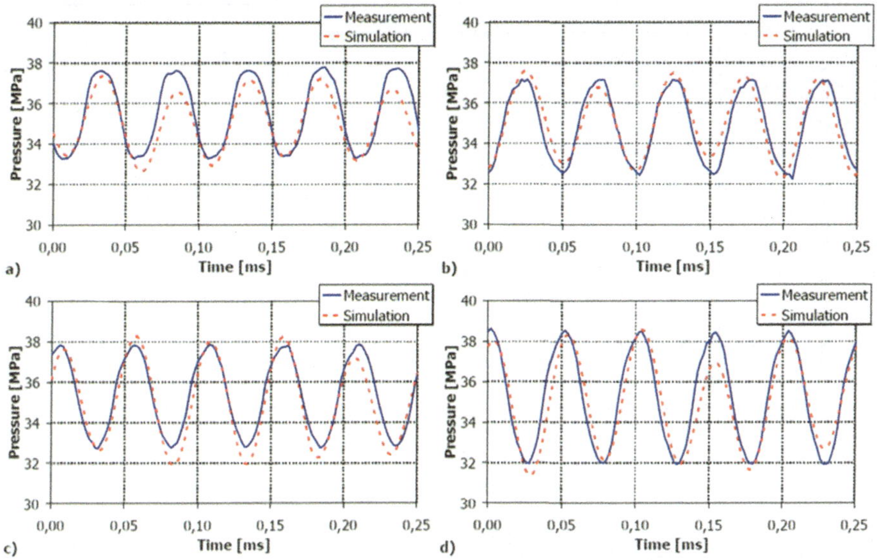

Fig. 4. Comparison of time behaviour of relative static pressure obtained from numerical simulation and laboratory measurement: a) configuration 64, b) configuration 72, c) configuration 80, d) configuration 88 [26].

originator of pressure pulsations. The boundary conditions were enforced using acoustic originator at one edge of pipe while other edge of pipe kept under curtain. The stagnant pressure was maintained from 10 to 100 MPa. The measured waveforms recorded with the help of sensors KISTLER 6211 and 6229 via Discrete Fourier Transform (DFT) to calculate existing amplitude, frequency and the stages for total recorded pressure. Eventually velocity of sound and water's second viscosity had been found with help of genetic algorithm. It can be observed refer Fig. 5 that determining speed of sound in liquid has dependency upon static pressure. The experimental output is in close agreement with "UNESCO equation" [30].

On comparison with details of [29] a variation are observed on account of non-assumption of testing pipe like an oscillatory elastic shape. Figure 6 represents that stagnation pressure is key factor for water's second viscosity. Some minor deviations are observed most probably on account of non-assumption of testing pipe like an oscillatory elastic shape.

Foldyna et al. [31] worked out transformation of peak frequency pulsation energy to farthest distance that fulfils basic need of the system refer the following Fig. 7. For achieving this target amplification of pressure pulsations are done through proper liquid waveguide to transfer pulsations to nozzle. Optimal results are obtained when the whole peak pressure system from acoustic originator to nozzle are at resonance. The author has developed analytical and numerical modelling with inbuilt acoustic originator.

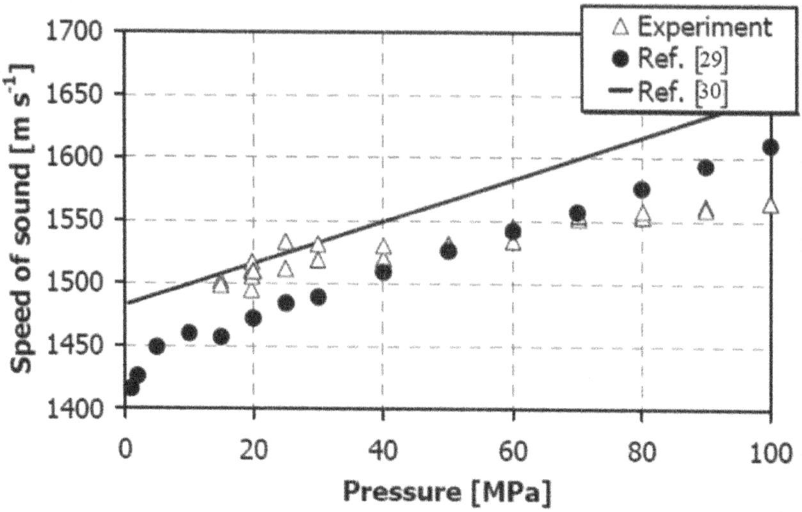

Fig. 5. The velocity of sound in water like a function of operating pressure. Experiments of [29] and study [27, 30].

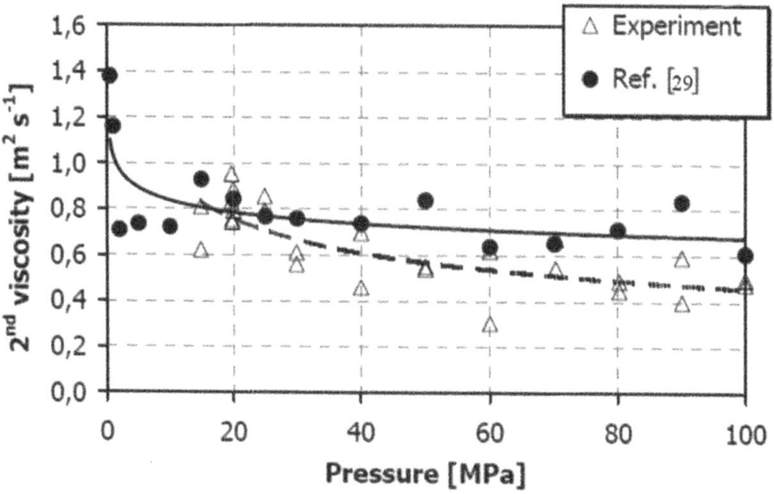

Fig. 6. The water's second (voluminous) viscosity as a function of working pressure. Laboratory data compared with experiment [27, 29].

4.1 Analytical Solution

Consider the conffuser-shaped tube refer Fig. 8 with circular cross-section where in two waveforms pressure and flow oscillations are solved hinged upon Navier-Stokes equation and mathematical expression for wave for propagation of pressure wave. The standard kinematic viscosity and kinematic second viscosity are taken care in wave equations which are related to liquid compressibility.

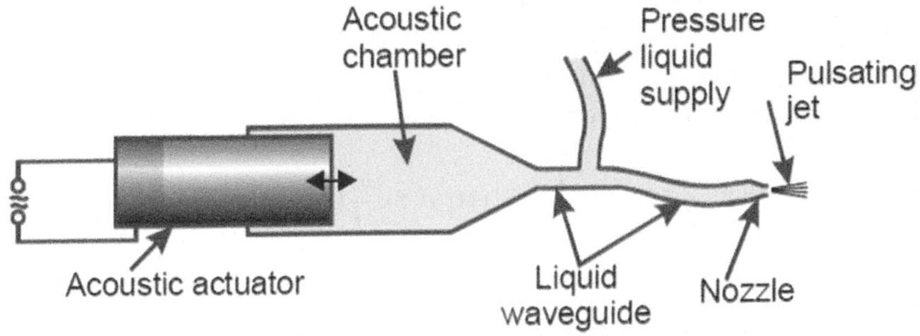

Fig. 7. Schematic sketch of the high-pressure setup with unified acoustic originator of pressure pulsations [31].

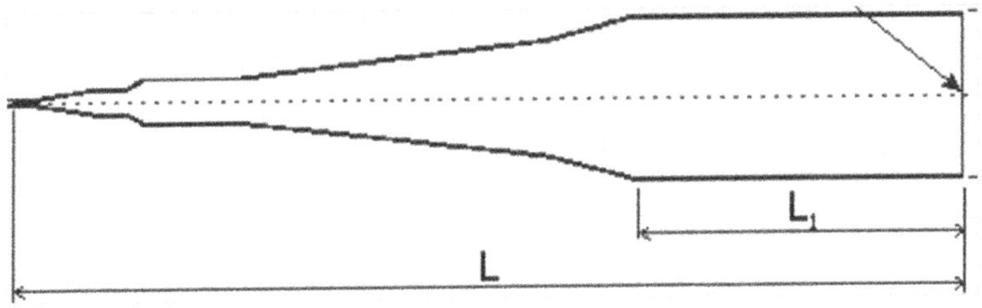

Fig. 8. Schematic sketch of the conffuser-shaped pipe with round cross-section [31].

The non-reversible stress tensor Π_{ij} upon viewpoint of wave equation can be expressed by

$$\Pi_{ij} = 2\eta c_{ij} + \delta_{ij} \int_0^t \Theta(t-\tau) C_{kk}(\tau) \mathrm{d}\tau \tag{17}$$

here the function ⊙ represents dynamic second viscosity which has been associated to voluminous recollection and C_{ij} is tensor of deformating speed. Experimentally verifyed above Eq. (17) and can be rewritten in frequency domain (ω)

$$\Pi_{ij\omega} = 2\eta c_{ij\omega} + \delta_{ij} \frac{k}{\omega} C_{kk\omega} \tag{18}$$

Where by δ_{ij} acts as Kronecker delta, η dynamic viscosity. The Kinematic second viscosity can be expressed by following equation and reflects frequency dependence.

$$\xi = \frac{k}{\rho\omega} \tag{19}$$

Where ρ represents liquid density.

4.2 Wave Equation

The pressure function of wave equation can be expressed considering N-S equation expressed with help of Laplace operator Δ in representation as

$$\frac{\partial^2 p}{\partial t^2} - 2\gamma\frac{\partial}{\partial t}(\Delta p) - \int_0^t \rho^{-1}\Theta(t-\tau)\frac{\partial}{\partial \tau}(\Delta p)d\tau - v^2\Delta p = 0 \tag{20}$$

Where γ represents kinematic viscosity, t represents time, p represents pressure, and v represents velocity of sound in medium water. Apply initial condition of zero to Eq. (20) it transforms to

$$s^2\sigma - s[2\gamma + \xi(\mathrm{s})]\Delta\sigma - v^2\Delta\sigma = 0 \tag{21}$$

Where s is Laplace operator and ξ Laplace function of second kinematic viscosity. Now wave equation represented in spherical coordinate system (r, ø, v) as

$$\frac{\partial^2\sigma}{\partial r^2} + \frac{2\partial\sigma}{r\partial r} + \frac{1}{r^2}\frac{\partial^2\sigma}{\partial v^2} + \frac{1}{r^2}\mathrm{Cotg}\frac{\partial\sigma}{\partial v} + \frac{1}{r^2}\frac{1}{\sin v^2}\frac{\partial^2\sigma}{\partial\varphi^2} + k^2\sigma = 0 \tag{22}$$

Assuming the solution of Eq. (20) as a product of functions

$$\sigma = \mathrm{Z}(r)\mathrm{W}(\cos v)\Phi(\varphi) \tag{23}$$

Then independent particular integrals written by

$$\Phi_p = A_n \cos \mathrm{n}\varphi + B_n \sin \mathrm{n}\varphi \tag{24}$$

$$W_p(\cos v) = M_{nm}P_m^n(\cos v) + N_{nm}Q_m^n(\cos v) \tag{25}$$

P, Q represents special Legendre polynomials.

4.3 Transfer Matrix

In conffuser-type tubes (Fig. 8) flow pulsations and pressure pulsations in hydraulic arrangement can be determined by help of transfer matrix P. In such situation let us establish the average speed of liquid C_r along r with help of term

$$\tilde{c}_r = \frac{1}{2v_0}\int_0^{v_0} C_r(r,\varphi,v)dv \tag{26}$$

Assuming flow is rotary symmetrical and derived assumption P_m^n, Q_m^n will be streamlined to the following form P_m^0, Q_m^0. On the basis of NS equations mean velocity C_r can be rewritten considering pressure component p changes according to angle v,

$$\frac{\partial\,\widetilde{C}_r}{\partial t} = -\frac{2v_0}{\rho}\frac{\partial p}{\partial r} - \frac{2v_0\xi}{\rho^2 v^2}\frac{\partial^2 p}{\partial r\partial t} \tag{27}$$

On assuming the solution subsequent Laplace pictures of the pressure component and the speed along C_r

$$\sigma = \frac{1}{\sqrt{r}}[\mathrm{F}J_{0.5}(kr) + \mathrm{G}Y_{0.5}(kr)] \tag{28}$$

At m = 0; D = 1/2

$$W_r = \frac{\alpha}{s\sqrt{r}}\left[\mathrm{F}\frac{\partial J_{0.5}(kr)}{\partial \gamma} + \mathrm{G}\frac{\partial Y_{0.5}}{\partial r}\right] \tag{29}$$

Suppose we suggest r = r_0 as aphysical state vector

$$u^T = [W_r(r_o, \mathrm{s}), \sigma(r_o, \mathrm{s})] \tag{30}$$

And for r the state vector

$$u^T = [W_r(\mathrm{r}, \mathrm{s}), \sigma(\mathrm{r}, \mathrm{s})] \tag{31}$$

The dependency in neighbourhoods of r and r_0 could again be written by help of a transfer matrix

$$\mathrm{u}(\mathrm{r}, \mathrm{s}) = \mathrm{Pu}(r_0, \mathrm{s}) \tag{32}$$

The matrix P obtained from Eq. (29) and (30) eliminating integration constant F, G.

$$\mathrm{P} = \frac{1}{\delta\sqrt{rr_0}}\begin{Vmatrix} \alpha\frac{J_{0.5}}{\partial r}(kr) & \alpha\frac{Y_{0.5}}{\partial r}(kr) \\ J_{0.5}(kr) & Y_{0.5}(kr) \end{Vmatrix}\begin{Vmatrix} Y_0(kr_o) & -\alpha\frac{\partial Y_{0.5}(kr_o)}{\partial r} \\ -J_{0.5}(kr_0) & \alpha\frac{\partial J_{0.5}}{\partial r}(kr_0) \end{Vmatrix} \tag{33}$$

The matrix P in Eq. (33) represents pressure as well as discharge pulsations of hydro mechanical arrangement with conffuser-shaped pipes. However, individual elements are supported by the second viscosity and velocity of sound, which are in turn dependent upon static pressure and frequency respectively.

Kusnerova et al. [32] has proposed analytical method to predict the frequency of liquid oscillations. The generation of high frequency pressure pulsations as the output of the PWJ system has geometric dimension of elements and flow velocities as major input parameter. The construction of modulated water jet technical arrangement with the dissecting head established on hydrodynamic oscillating setup has great importance as tubes represents acoustic mass and chambers an acoustic capacitance [33]. The guiding principle behind this is fine molecules oscillate within the pipe as a compact body without any compression or dilutions Therefore, carrying medium show least elasticity. At the same time chamber witness particle oscillations as rigid body and experience compression with dilution at all positions of the control volume. Acoustic theory suggests mass m_a and acoustic capacitance C_a [34–40] solving equation of motion (34) which describes self-oscillations in the system where ω_0 as the most basic angular frequency and the most basic frequency f_0 is written by.

$$f_0 = \frac{1}{2\pi}\sqrt{\frac{1}{C_a m_a.k_{rcc}}}\mathrm{Hz.} \tag{34}$$

The correct acoustic matter content of total tubes vibrates within the hydrodynamic setup and expressed by equation

$$m_{as} = k_{rcc}.m_a \tag{35}$$

The term k_{rcc} supply accurate resultant forces that are proved by experiments and in match with theoretical forecasting of strong turbulent water movement. The total acoustic mass m_a and C_a capacitance acoustic for i = 3 and j = 2 by summing individual impedance and given by

$$m_a = \sum\nolimits_{i=1}^{3} m_{ai},\ kg/m^4;\ \frac{1}{c_a} = \sum\nolimits_{j=1}^{2} \frac{1}{C_{aj}};\ C_a,\ m^4.s^2/\text{kg} \tag{36}$$

After direct measurement the data obtained were tabulated, calculated and regressions were drawn. The relative dissimilarity among calculated frequencies f_{ZM} and forecasted basic frequencies f_0 according to the proposed relation

$$\rho_{\Delta f}\cdot = \frac{|f_0 - f_{ZM}|}{f_{ZM}}.100\% \tag{37}$$

The variations on theoretically evaluated and calculated outcomes of the fundamental frequency of oscillations of the setup are limited to 6.4%. However, the proposed model results are satisfactorily permissible for the technical practice for predicting the fundamental oscillation frequency.

Riha et al. [41] has focused studied on influential parameter of changing geometries in area between big and narrow diameters of pipe and subsequent impact on amplification of pressure pulsations. The investigation revolved around calculating output of heavy pressure pulsation amplification according to geometry in transition area. Three fundamental changes with respect to geometry and production were defined refer Fig. 9 for all important parts, a) step diametric change b) change of diameters of cone and c) radius variations of diameters.

The water jet tool is a complicated physical problem. Author has created mathematical model using commercial CFD software ANSYS for water jet tooling. The flow consists of 3D two phase turbulent compressible and flow which is time dependent of fluid with walls in motion. The turbulent part is defined using N-S equation given by RANS (Reynolds averaged Navier-stokes equation) method. For calculation purpose k-ε turbulence model is under use. For accurate result of model with regions of low and high velocities following equation are solved;

$$\frac{\partial \rho}{\partial t} + \frac{\partial}{\partial x_j}(\rho u_i) = 0 \tag{38}$$

Where the above equation represents continuity equation, ρ represents density; t represents time, x_j constituent of coordinate and u_i constituent of velocity.

$$\frac{\partial}{\partial t}(\rho u_i) + \frac{\partial}{\partial x_j}(\rho u_i u_j) = -\frac{\partial p}{\partial x_i} + \frac{\partial}{\partial x_j}\left[\mu\left(\frac{\partial u_i}{\partial x_j} + \frac{\partial u_j}{\partial x_i} - \frac{2}{3}\delta_{ij}\frac{\partial u_l}{\partial x_l}\right)\right] + \frac{\partial}{\partial x_j}\left(\mu_t\left(\frac{\partial u_i}{\partial x_j} + \frac{\partial u_j}{\partial x_i}\right) - \frac{2}{3}\left(\rho k + \mu_t\frac{\partial u_l}{\partial x_l}\right)\delta_{ij}\right) \tag{39}$$

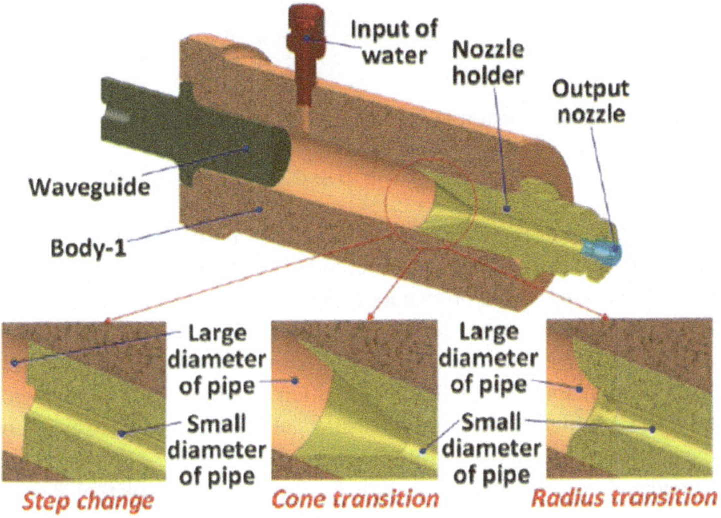

Fig. 9. Tool assembly for the water jet cutting and provided transition locality [41].

Equation (39) represents the momentum equation and depicts special geometry of N-S equation for turbulent fluid flow, p represents pressure, μ represents viscosity, μ_t turbulent viscosity and k represents turbulent kinetic energy. Both of above Eq. (38) and (39) is termed as Reynolds - averaged equation.

$$\mu_t = \rho C_\mu \frac{k^2}{\varepsilon} \tag{40}$$

$$\frac{\partial}{\partial t}(\rho k) + \frac{\partial}{\partial x_i}(\rho k u_i) = \frac{\partial}{\partial x_j}\left(\left(\mu + \frac{\mu_t}{\sigma_k}\right)\frac{\partial k}{\partial x_j}\right) + G_k + G_b - \rho\varepsilon - Y_M + S_k \tag{41}$$

$$\frac{\partial}{\partial t}(\rho \varepsilon) + \frac{\partial}{\partial x_i}(\rho \varepsilon u_i) = \frac{\partial}{\partial x_j}\left(\left(\mu + \frac{\mu_t}{\sigma_\varepsilon}\right)\frac{\partial \varepsilon}{\partial x_j}\right) + \rho C_1 S_\varepsilon + \rho C_2 \frac{\varepsilon^2}{k + \sqrt{\varepsilon v}} + C_{1\varepsilon}\frac{\varepsilon}{k}C_{3\varepsilon}G_b + S_\varepsilon \tag{42}$$

Equation from (38) to (42) forms closed system for one phase compressible turbulent isothermal fluid flow. Further the turbulent flows at wall are different than flow at middle. Therefore, need for describing turbulence with assistance of non-equilibrium wall function. The approach for solution of fluid flow in sub-layers of boundary layer with help of empirical equation given by following two Eqs. (43) and (44);

$$\frac{1}{\rho_q}\left[\frac{\partial}{\partial t}(\alpha_q \rho_q) + \nabla.(\alpha_q \rho_q \vec{v_q}) = S_{\alpha_q} + \sum\nolimits_{p=1}^{n}(\dot{m}_{pq} - \dot{m}_{qp})\right] \tag{43}$$

$$\sum\nolimits_{q=1}^{n} \alpha_q = 1 \tag{44}$$

For the water flow to be compressible let us write the relation for speed of sound and density likewise a function of pressure as follows;

$$\mathrm{K} = -\rho \frac{dp}{d\rho} \tag{45}$$

$$\rho = \frac{\rho_{ref}}{1 - \frac{K}{p - p_{ref}}} \tag{46}$$

For the speed of sound, a simple linear Eq. (47) would be used [29].

$$a = b.p + d \tag{47}$$

As above equations forms closed system and the partial differential equation are solved using numerical method known as method of control volumes. The suitable model for studying high pressure pulse propagation in given geometry was formed. Hence, the work performed evaluated the appropriate parameter that is velocity at the output nozzle in the smallest cylinder. In the direction of output nozzle normalized velocity are produced with maximum amplitude while in transition of radius. This normalized velocity amplitude is proves efficiency of high pressure pulsations transfer in the geometry of tool and the output of the velocity from nozzle would produce pulsating waterjet with well-developed pulses. The study concluded that CFD model of tool discloses three geometrical variants of tool for water jet cutting. It is evident that transition of radius produces best pulsating of water jet in the given boundary condition whereas the other two variants produce low amplitude of velocity at nozzle exit giving poor pulsations of jets.

In one of the recent studies carried by Lehocka et al. [8] proposed and validated a mathematical model (DoE)3^3. The work signifies the effect of hydraulic power on pulsating water jet disintegration of brass CW614N. The work was carried out on full factorial design of experiment (DoE)3^3, with validation of experiment worked out with changing the traverse velocity from 0.2 to 1.4 mm/s. The variable parameters were traverse velocity, nozzle diameter and hydraulic power. The experiment observed significant mass removal, which proves efficient metallic erosion by PWJ. The author suggested impact pressure p_i (MPa) can be written by equation

$$p_i = v_l.\rho.c \tag{48}$$

Where c is shock velocity and approximated using following equation

$$c = v_s + k.\, v_l \tag{49}$$

The duration for which these waves are active can be given by

$$\Delta\tau = \frac{3.r.v_l}{2.c2} \tag{50}$$

During the process impact pressure are relived the second stage of liquid impact known as stagnation pressure which is written by

$$p_s = 1/2\,\rho\, v_l^2 \tag{51}$$

And the kinetic energy of a pulse can be written as

$$E_k = 1/2\, m_d.v_l^2\ (\mathrm{J}) \tag{52}$$

The experiment was performed with eighty-one experiment samples and each sample was weighted before and after cutting by pulsating water jet and mass removal Δ m (mg /s) was calculated by

$$\Delta \mathrm{m} = \frac{(\mathrm{m}_0 - \mathrm{m}_1).v}{L} \times 1000 \tag{53}$$

The significant factors which turn up important are hydraulic power of pump P_h(kW), nozzle orifice diameter d (mm), and traverse speed v (mm/s). Based on data probability plot were drawn shows a normal distribution. The test namely Kolmogorov-Smirnov showed significant level of 5% of each sign of file arise from normal distribution. Another test Durbin-Watson approves the criterion of interval of the critical values and prove no correlation, However, the results were verified of residuals according to hypothesis H0: $\mathrm{p} = p_0$ and H1: $\mathrm{p} \neq p_0$, one of the sample at t test in agreement with hypothesis H0 about residual normality. The mathematical model were formulated in scale generated from the estimated regression coefficients by following Eq. (54),

$$y = 15.331037 + 7.8551667x_1 + 4.0745x_3 - 4.656333x_2 + 3.4281667\, x_1x_3 - 1.820056x_3^2 - 1.847417x_1x_2 - 1.318833x_2x_3 + 3.9287778\, x_2^2 + 1.7805833\, x_2x_1^2 - 1.025\, x_1x_2x_3 + 1.9670833\, x_1x_2^2 + 1.9005\, x_3x_2^2 \tag{54}$$

The above equation were transformed into natural scale as (55), the natural scale formulated from the estimation of values in selected intervals considering effects of hydraulic power P_h (kW), nozzle orifice diameter d (mm), and traverse speed v (mm/sec) variation respectively.

$$f(x) = 0.2208706805\,\mathrm{P}_h^2.\mathrm{d} - 0.2945310525.\,\mathrm{P}_h^2 + 5.035542319\,d^2\mathrm{P}_h - 2.796834385.P_h.\mathrm{d.v} - 18.55608193.P_h.\mathrm{d} + 6.222715743\,P_h.\mathrm{v} + 15.34860576.P_h + 107.0383324\,d^2.\mathrm{v} - 92.94679292.d^2 - 267.5099307.\mathrm{d.v} + 274.2151671.\mathrm{d} - 29.121264.v^2 + 192.7083708.\mathrm{v} - 207.6966906 \tag{55}$$

The above compiled regression equation predicts the values of mass material removal for provided intermission of process factors. The generated equations are practically applicable for pulsating water jet.

5 Discussion on Different Parameters

As discussed in details many methods and approaches to explain the acoustic generation of pulses in previous section. An overview of the outcome of the studies by authorities of the subject is presented in this section specifically under heads of parameters such as below mentioned, then followed by conclusion.

i) Second kinematic viscosity and velocity of sound
ii) Breakup length of Jet after nozzle exit

iii) Force effect analysis
iv) Energy equation of jet
v) Water hammer pressure of jet
vi) Study of fundamental frequency approach
vii) k-ε turbulence model

5.1 Second Kinematic Viscosity and Velocity of Sound

In high pressure system with acoustic wave propagation sound speed and second kinematic viscosity as a function of pressure are vital factor as damping is inversely related to the natural frequency. Foldyna et al. [29] used modal analysis approach to reach to fact experimentally calculating with a tube filled with water and sufficient impact of pressure was given. The result achieved was plotted graphically in below Fig. 10. a) and 10 b).

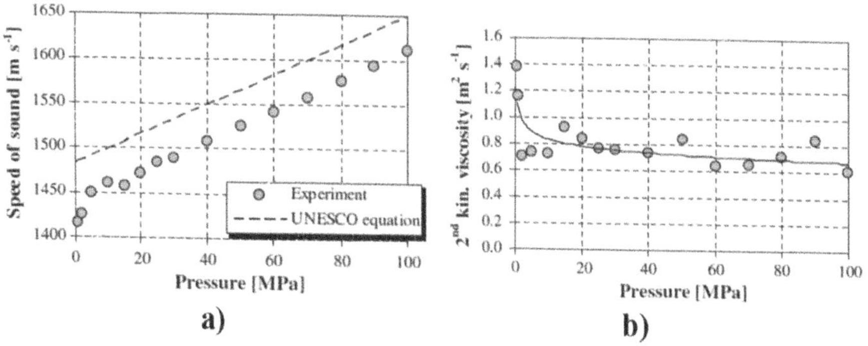

Fig. 10. a) Speed of sound in water compared with UNESCO equation and b) Second kinematic viscosity of water as function of pressure.

The comparison was made with speed of sound in sea water discussed by Wong et al. [30] and with 'UNESCO Equation" ('UNESCO Equation" already discussed in above section). The results were in good agreement to above equation with minor impurities of air bubbles trapped in water initially. In another study of Foldyan et al. [27] has again verified the two important parameters on the basis of deduction of forced oscillation of liquid by transfer matrix method. In this method genetic algorithm were placed in work and results obtained were plotted graphically refer Fig. 5 and Fig. 6 which is also, in close agreement with 'UNESCO Equation". In work of Pochyly et al. [42] performed mathematical analysis considering compressible nature of fluid at high frequency of pressure pulsations with steady flow and applying Einstein summation convention concluding that seconds kinematic viscosity with expression

$$\varsigma = \frac{\lambda}{\rho} = \frac{\odot}{m_r};\ f_r \leq 1000\ \text{Hz} \tag{56}$$

For water holds the second viscosity coefficient $\Theta = 9800\,\text{m}^2\text{s}^2$.

5.2 Breakup Length of Jet After Nozzle Exit

Pressure pulsations after passing by nozzle exit need to be suitable breaking up for bunching phenomenon, which is one of the main characteristics of pulsating water jet method. Shami et al. [23] has come up with three possible results refer Table 1 above for possible breakup length. However, method III was cited better describing breakup length by author which is

$$\zeta = 1 + \pi \mathrm{M}\alpha \cos 2\pi\tau \tag{57}$$

where it is known from [23] the diemntionless parameters for solving

$$\zeta = \frac{r_0}{R}, \ \alpha = \frac{xf}{U} = \frac{x}{\lambda}, \quad \tau = \frac{t}{T} = \mathrm{ft} = \frac{tU}{\lambda}, \quad \mathrm{M} = \frac{\Delta U}{U}, \quad \mathrm{St} = \frac{R}{\lambda} \tag{58}$$

$$\frac{r_0}{R} = 1 + \pi \mathrm{M}\alpha \ \cos \ 2\pi\tau \tag{59}$$

Which gives breakup length

$$L_o = \frac{U}{\pi Mf} = \frac{UT}{\pi M} \tag{60}$$

M is velocity modulation ratio, U average velocity of jet, f is frequency of the velocity modulation.

Whereas Chahine and Conn et al. [43] argued that distance L_o after which bunching phenomenon starts dependent on time of crest of wave taking over trough of wave after a time interval T given by

$$\mathrm{T} = \frac{\lambda}{4\Delta U} \tag{61}$$

and breakup length by

$$L_o = TU = \frac{\lambda U}{4\Delta U} = \frac{U}{4Mf} \tag{62}$$

However, both expression are nearly same and establish the fact that breakup length is inversely proportional to frequency and modulation ratio of velocity and directly to average velocity of jet.

5.3 Force Effect Analysis

Foldyna et al. [44] has studied specifically force effect with recording dynamic pressure inside the nozzle. Lot of data were accumulated based on experiments on force effect of PWJ with two sets of nozzle adjustments of UST (ultrasonic tool) having cone vibrating tip of 2 mm in diameter (I) and 10 mm of diameter (II). A variation in UST was Also, made in experiment s with (I) position changed from 6 to 12 mm and for (II) 10 to 25 mm which was the distance from exit of nozzle to end of nozzle tip. However, no mathematical calculations were presented but graphically strong evidence were proved that position of UST inside the nozzle plays a vital role as nozzle (I) exhibits lower force effect when compared to nozzle (II) which produce higher effects. Results were plotted graphically shown below Fig. 11 a) and b). However, the study endorsed performance of PWJ is free from position of UST.

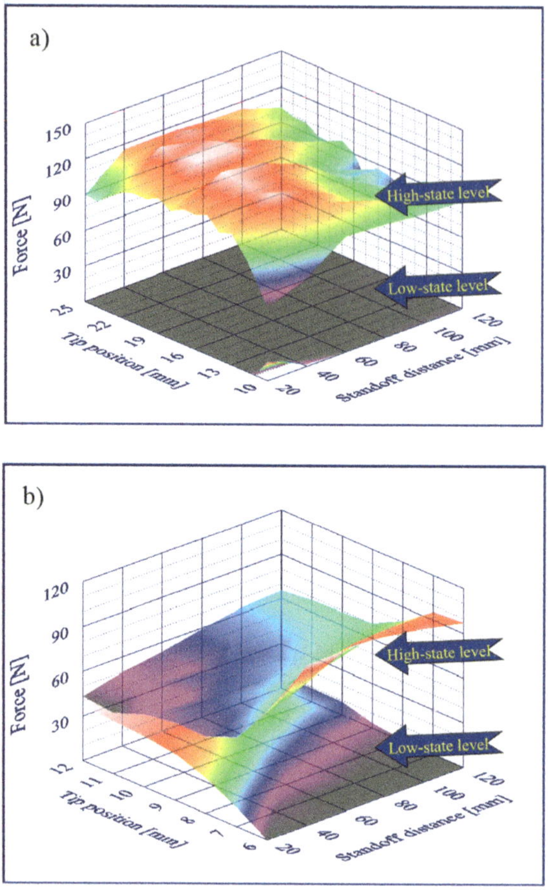

Fig. 11. a) Effect of tip position and standoff distance on high and low state level of the pulse at p = 20 MPa, Nozzle diameter = 1.19 mm a) pulse nozzle (I) b) pulse nozzle (II) [44].

5.4 Energy Equation of Jet

The one of the important factors when studying a system is energy calculation. On analysis of present research Lehocka et al. [8] has come up with kinetic energy of pulse with equation expression at (52). However, in the expression $E_k = 1/2m_d . v_l^2, m_d$ that is weight of pulse calculated experimentally. In a similar situation Vijay et al. [45] argued about specific energy of jet and come up with expression E_v as

$$E_v = \frac{(pQ + P_U)60\,p_S}{\Delta m_s} \tag{63}$$

Where Δm_s is rate of mass loss given by Foldyna et al. [44]

$$\Delta m_s = \frac{\Delta m \cdot v_{trs}}{s} \tag{64}$$

Where s represents length of cut and Δm mass loss for other symbols refer nomenclature of symbol.

5.5 Water Hammer Pressure of Jet

Foldyna et al. [44] explained impact pressure of jet so called water hammer pressure with the expression

$$p_i = 1477000\ v_0 \tag{65}$$

Where v_0 is velocity of jet and the stagnation pressure of continuous water jet which is the maximum pressure at the point of collision of solid and liquid expressed as

$$p_s = 500\ v_0^2 \tag{66}$$

In another research of Foldyna et al. [46] described water hammer pressure developed by impact can be derived by 1 D Rankine-Hugoniot analysis with

$$p_i = \rho\ v_0 c_0 \tag{67}$$

Where c_0 shock velocity and the value of the shock velocity can be addressed from Lehocka et al. [8] study by the expression (49), whereas Bowden and Brunton et al. [47] has defined the similar pressure that would be generated during cylindrical jet of fluid impact the surface, having compressible nature persists until release waves reaches the central axis as

$$p_i = \frac{v\rho_1 C_1 \rho_2 C_2}{\rho_1 c_1 + \rho_2 c_2} \tag{68}$$

Where v represents impact velocity and ρ_1, ρ_2 are the densities and $c_1 c_2$ shock velocities in liquid and the solid respectively.It is worth mentioning that Foldyna et al. [31] has come up with pressure matrix Eq. (33) which describes pressure as well as discharge pulsations of the instant study in focus for creation of wave acoustic with confuser shaped pipes. Pochyly et al. [28] presented similar expression for amplification of pressure pulsations through confuser shaped tube tuned in resonance. Further, Joukowsky et al. [48] deduced equation to explain transient flow theory and describe as the "fundamental equation of water hammer" that can be given by

$$\Delta \mathrm{P} = \pm \rho a \Delta V \quad \text{or} \quad \Delta \mathrm{H} = \pm \frac{a \Delta V}{g} \tag{69}$$

Here a = acoustic wave speed (water hammer), ρ = fluid density, $P = \rho\ g$ (H-Z) piezometric pressure $V = \int u\ dA$ cross sectional mean velocity. The positive sign implies water hammer wave moving downstream and negative sign implies moving upwards. Moreover, in above Eq. (69) the dynamics of gas are provided with momentum shock condition where in flow velocity is negligible to wave speed in special case. Many researchers have worked to refine the fundamental equation of water hammer in pipe flow. Around 1960s the equation were established fully with induction of two dimensionless parameters and combined study resulted in classical mass and momentum equation for 1-D (one dimensional) water hammer flow [49–52] below

$$\frac{a^2}{g}\frac{\partial V}{\partial x} + \frac{\partial H}{\partial t} = 0 \tag{70}$$

$$\frac{\partial V}{\partial t} + g\frac{\partial H}{\partial x} + \frac{4}{\rho D}\tau_w = 0 \tag{71}$$

The above equation constitutes the one dimensional wave propagation fundamental equation and consists of all physics involved in a complex pipe system.

5.6 Approach to Predict Fundamental Frequency

In a wide range of unsteady flow frequency analysis are applied successfully. This analysis is used in flow measurement [53] and in investigation of visco-elastic behaviour of fluid [54]. However, Kusnerova et al. [32] has applied this approach in setting up model of PWJ for which measurement were taken for fundamental oscillation frequency against prediction and confirmation. However, the results obtained in comparison to theoretical fundamental frequency were in limit of 6.4%. In study by Lee et al. [55] endorsed the fact that frequency approach can be handful in guessing the unsteady system response. Pipe transmission signals of fluid are hyperbolic in nature and well versed to change in transient due to pipe wall visco-elasticity and unsteady friction. The most basic feature is signal repeat periodically.

5.7 k-ε Turbulent Model

Riha et al. [41] described the 3 dimensional compressible turbulent time dependent fluid flow taking place inside conffuser tube and outflow in incompressible environment of air verified with CFD model using k-ε turbulent method. According to Giancarlo et al. [56] k-ε turbulent model is based on two-equation models where in two different transport equations are solved for two independent turbulent quantities, in this case a) turbulent length b) time scale and expressed as

$$l_o \alpha \frac{K^{3/2}}{\varepsilon}, \quad \tau_0 \alpha \frac{K}{\varepsilon} \tag{72}$$

The rate of decomposition of turbulent kinetic energy has been in close agreement with Eqs. (38) to (42) which constitute a close system and explain well the transition of area the fluid experience while moving between big diameter to small diameter of confusser tube and subsequent amplification of pressure pulsations. Further the short comings of k- ε turbulent model as reported by [56] lies on the fact that model cannot provide clear description on flow structures developed by normal Reynolds-stress anisotropies also model is based on dissipative energy basis which has no justification for relaxation effect of Reynolds stress.

6 Conclusion

The instant holistic study of mathematical models prescribed by various researchers on pulsating water jet is handful source of information while studying the complex phenomenon of pulse generation by acoustic actuator. As mentioned in previous section, many methods have been adopted to iterate the governing parameters of PWJ. The conclusive remarks drawn based on these studies are listed below.

I. Calculation of second kinematic viscosity and velocity of sound by modal analysis and by transfer matrix method are in close agreement with international standard set for speed of sound in sea water "UNESCO" equation and has been proved it dependency on pressure. Also, second kinematic viscosity is frequency dependent Eq. (56).
II. The breakup length of PWJ after nozzle exit has been wisely constructed considering energy conservation and control volume concept. Further the relation established for breakup length with respect to frequency and velocity of jet are in agreement with physical verification as seen in practical experiments.
III. Force analysis has been done without governing force equations but the effect is drawn graphically which establish that placement of UST inside the nozzle plays a vital role exhibiting higher and lower force effect as placement of UST at a larger distance and smaller distance respectively. To account for the unavailability of literature phenomenon will be investigated in detail with governing equations. However the energy equation produced are in good agreement.
IV. The water hammer equations are the most basic parameter on which effectiveness of PWJ depends. Water hammer of jet is broader area where a number of equations are dealt. However, the physical experimentation and verification can be further investigated.
V. As k-ε model is based on dissipative energy and has reported short comings better model such as anisotropic generalization of eddy-viscosity can be applied to iterate the process again.

The above approach explains water jet generation to transmission in conical tube since all the equations are tested in real condition. However, more studies can be performed to accurately understand the values of each parameter with practical verification to attain further improvement on erosion front of pulsating water jet and attain overarching goals.

Acknowledgments. The authors wish to thank the Department of Mechanical Engineering IIT (ISM) Dhanbad and valuable contribution of Faculty of Manufacturing Technologies TUKE with a seat in Prešov, Slovakia and Institute of Geonics of the Czech Academy of Sciences, Ostrava, Czech Republic for constructive analysis and support.

References

1. De Haller, P.: Untersuchungen über die durch Kavitation hervorgerufenen Korrosionen. Schweizerische Bauzeitung **101**(21&22), 243–246 (1933)

2. Foldyna, J., Sitek, L., Martinec, P., Scucka, J., Jekl, P., Mlynarczuk, M.: Rock cutting by pulsing water jets. International Society for Rock Mechanics and Rock Engineering (2005)
3. Dehkhoda, S., Hood, M.: An experimental study of surface and sub-surface damage in pulsed water-jet breakage of rocks. Int. J. Rock Mech. Min. Sci. **63**(November), 138–147 (2013). https://doi.org/10.1016/j.ijrmms.2013.08.013
4. Polyakov, A., Zhabin, A., Averin, E., Polyakov, A.: Generalized equation for calculating rock cutting efficiency by pulsed water jets. J. Rock Mech. Geotech. Eng. **11**(4), 867–873 (2019). https://doi.org/10.1016/j.jrmge.2018.11.009
5. A review of rock cutting for underground mining: past, present, and future. J. South. Afr. Inst. Min. Metall., 1–27 (2017). WITS Special Edition Volume I
6. Hloch, S., Adamčík, P., Nag, A., Srivastava, M., Čuha, D., Müller, M., Klich, J.: Hydrodynamic ductile erosion of aluminium by a pulsed water jet moving in an inclined trajectory. Wear **428**, 178–192 (2019)
7. Foldyna, J., Sitek, L., Ščučka, J., Martinec, P., Valíček, J., Páleníková, K.: Effects of pulsating water jet impact on aluminium surface. J. Mater. Process. Technol. **209**(20), 6174–6180 (2009)
8. Lehocká, D., Klich, J., Botko, F., Foldyna, J., Hloch, S., Kepič, J., Storkan, Z.: Pulsating water jet erosion effect on a brass flat solid surface. Int. J. Adv. Manuf. Technol. **97**(1–4), 1099–1112 (2018)
9. Sitek, L., Foldyna, J., Martinec, P., Scucka, J., Bodnarova, L., Hela, R.: Use of pulsating water jet technology for removal of concrete in repair of concrete structures/Pulsuojancios vandens sroves technologijos naudojimas betonui pasalinti remontuojant betoninius statinius/Pulsejosas augstspiediena udens struklas tehnologijas izmantosana betona slana nokalsanai remontejamam betona konstrukcijam/Pulseeriva veejoa tehnoloogia kasutamine betooni eemaldamiseks betoonkonstruktsioonide remondil. Baltic J. Road Bridge Eng. **6**(4), 235 (2011)
10. Foldyna, V., Foldyna, J., Klichova, D., Klich, J., Hlaváček, P., Bodnarova, L., Jarolim, T., Mamulová Kutláková, K.: Effects of continuous and pulsating water jet on CNT/concrete composite (2017)
11. Hloch, S., Foldyna, J., Sitek, L., Zeleňák, M., Hlaváček, P., Hvizdoš, P., Kloc, J., Monka, P., Monkova, K., Kozak, D., Magurová, D.: Disintegration of bone cement by continuous and pulsating water jet. Technical Gazette **20**(4), 593–598 (2013)
12. Sharma, N.C., Lyle, D.M., Qaqish, J.G., Galustians, J., Schuller, R.: Effect of a dental water jet with orthodontic tip on plaque and bleeding in adolescent patients with fixed orthodontic appliances. Am. J. Orthod. Dentofac. Orthop. **133**(4), 565–571 (2008)
13. Kural, C., Kaya, I., Yilmaz, M., Demirbaş, E., Yücel, B., Korkmaz, M., Cetinus, M.E.: A comparison between three irrigation methods in the debridement of contaminated bovine cancellous bone and the effect of duration of irrigation on the efficiency of debridement. Acta Orthop. Traumatol. Turc. **43**(4), 359–365 (2009)
14. Kondo, T., Ando, K.: Simulation of high-speed droplet impact against a dry/wet rigid wall for understanding the mechanism of liquid jet cleaning. Phys. Fluids **31**(1), 013303 (2019)
15. Hew, F.L., Timchenko, V., Reizes, J.A., Leonardi, E.: Numerical evaluation of the effectiveness of micro pulsating water jets for cooling of microchips. In: International Conference on Micro/Nanoscale Heat Transfer, vol. 43901, pp. 625–633 (2009)
16. Nebeker, E.B.: Percussive jets-state-of-the-art. In: Proceedings of the 4th US Water Jet Symposium, pp. 28–41 (1987)
17. Chahine, G.L., Conn, A.F., Johnson Jr., V.E.: Cleaning and cutting with self-resonating pulsed water jets. In: 2nd US Water Jet Conference (1983)
18. Summer, D.A.: Presentation during Session 3(untitled). In: Workshop on the Application of High Pressure Water Jet Cutting Technology, Rolla, Missouri, pp. 107–133 (1975)
19. Vijay, M.M.: U.S. Patent No. 5,154,347. U.S. Patent and Trademark Office, Washington, DC (1992)

20. Foldyna, J.: U. S Pat No. US 7, 934, 666 B2 Method of generation of pressure pulsations and apparatus for implementation of this method (2011)
21. Stouffer, R.D., Bray Jr, H.C.: U.S. Patent No. 4,052,002. Washington, DC: U.S. Patent and Trademark Office (1977)
22. Zeleňák, M., Říha, Z., Jandačka, P.: Visualization and velocity analysis of a high-speed modulated water jet generated by a hydrodynamic nozzle. Measurement **159**, 107753 (2020)
23. Sami, S., Ansari, H.: Governing equation in a modulated liquid jet. In: Proceedings of the 1st US Water Jet Symposium Water Jet Technology Association (WJTA), St. Louis, USA, pp. I-2.1–I-2.9 (1981)
24. Chaudhry, M.H.: Applied hydraulic transients (2014)
25. Watters, G.Z.: Analysis and control of unsteady flow in pipe systems, vol. 336 (1984)
26. Foldyna, J., Riha, Z., Sitek, L., Svehla, B.: Numerical simulation of transmission of acoustic waves in high-pressure system. In: Proceedings of the International Congress on Ultrasonics, Vienna, 9–13 April 2007 (2007). https://doi.org/10.3728/ICUltrasonics.2007.Vienna.1456_foldyna
27. Foldyna, J., Habán, V., Pochylý, F., Sitek, L.: Transmission of acoustic waves. In: Proceedings of the International Congress on Ultrasonics, Vienna (2007)
28. Pochylý, F., Habán, V., Foldyna, J., Sitek, L.: 3D problem of pressure wave propagation in the tube with inconstant cross section. In: Proceedings of the International Congress on Ultrasonics, Vienna (2007)
29. Foldyna, J., Sitek, L., Haban, V.: Acoustic wave propagation in high–pressure system. Ultrasonics **44**(1), e-1457–e1460 (2006)
30. Wong, G.S., Zhu, S.M.: Speed of sound in seawater as a function of salinity, temperature, and pressure. J. Acoust. Soc. Am. **97**(3), 1732–1736 (1995)
31. Foldyna, J.: Use of acoustic waves for pulsating water jet generation. In: Acoustic Waves—From Microdevices to Helioseisomology, pp. 323–342. Intech (2011)
32. Kušnerová, M., Foldyna, J., Sitek, L., Valíček, J., Hloch, S., Harničárová, M., Kadnár, M.: Innovative approach to advanced modulated waterjet technology (2012)
33. Kuttruff, H.: Room Acoustic, p. 457. Applied Science, Publishers Ltd., London (1973)
34. Kušnerová, M., Hlaváč, L.: Self-vibrating chambers with continuous passage of liquid (2006)
35. Kušnerová, M., Valíček, J., Müller, M., Hrabě, P., Pella, A., Radvanská, A.: Analysis of acoustic parameters of hydrodynamic system and its application. In: 8th International Scientific Conference on Technology Systems Operation, Prešov, pp. 214–216 (2007)
36. Kušnerová, M., Valíček, J., Müller, M., Hrabě, P., Radvanská, A., Pella, A.: Derivation of hydrodynamic system acoustic parameters. In: 8th International Scientific Conference on Technology Systems Operation, Prešov, pp. 217–219 (2007)
37. Kušnerová, M., Valíček, J., Hloch, S., Radvanská, A.: Convert coefficients for the analytical model of the frequency of the water dynamic system. In: Conference on Applications and Design in Mechanical Engineering Universiti Malaysia Perlis (UniMAP) (2007)
38. Kušnerová, M., Valíček, J., Hloch, S., Ergić, T., Ivandić, Ž.: Derivation and measurement of the velocity parameters of hydrodynamics oscillating system. Strojarstvo: časopis za teoriju i praksu u strojarstvu **50**(6), 375–379 (2008)
39. Sitek, L.: Investigation of high pressure water jet behaviour using jet/target interaction. In: 12th International Conference on Jet Cutting Technology, pp. 59–66 (1994)
40. Vala, M.: The measurement of the non-setting parameters of the high pressure water jets. In: Geomechanics 93-Strata Mechanics/Numerical Methods/Water Jet Cutting, pp. 333–336. Routledge (2018)
41. Říha, Z., Foldyna, J.: Ultrasonic pulsations of pressure in a water jet cutting tool. Tehnicki vjesnik/Technical Gazette **19**(3), 487–491 (2012)
42. Pochylý, F., Habán, V., Fialová, S.: Bulk viscosity-constitutive equations. Int. Rev. Mech. Eng. (IREME) **5**(6), 20 (2011)

43. Chahine, G.L., Conn, A.F., Johnson, V.E., Frederick, G.S.: Passively interrupted impulsive water jets. In: 6th International Conference on Erosion by Liquid and Solid Impact (1983)
44. Foldyna, J., Sitek, L., Švehla, B., Švehla, Š: Utilization of ultrasound to enhance high-speed water jet effects. Ultrason. Sonochem. **11**(3–4), 131–137 (2004)
45. Vijay, M.M.: Ultrasonically modulated pulsed jets: basic study. Jet Cut. Technol. 15–35 (1994)
46. Foldyna, J., Jekl, P., Sitek, L.: Possibilities of utilization of modulated jets in rock cutting. In: Proceedings of the 1st International Conference Mining Techniques, pp. 85–96 (2001)
47. Bowden, F.P., Brunton, J.H.: Proc. R Soc. London Ser. A **263**, 331 (1961)
48. Joukowski, N.E.: Memoirs of the imperial academy society of St. Petersburg. In: Proceedings of the American Water Works Association, vol. 24, pp. 341–424 (1898)
49. Chaudhry, M.H.: Applied Hydraulic Transients. Inc. (1987)
50. Watters, G.Z.: Analysis and control of unsteady flow in pipelines (1984)
51. Wylie, B.E., Streeter, V.L.: Fluid Transients. Feb Press, Ann Arbor (1983)
52. Wylie, E.B., Streeter, V.L., Suo, L.: Fluid Transients in Systems, vol. 1, p. 464. Prentice Hall, Englewood Cliffs (1993)
53. Kashima, A., Lee, P.J., Nokes, R.: Numerical errors in discharge measurements using the KDP method. J. Hydraul. Res. **50**(1), 98–104 (2012)
54. Suo, L., Wylie, E.B.: Hydraulic transients in rock-bored tunnels. J. Hydraul. Eng. **116**(2), 196–210 (1990)
55. Lee, P.J., Duan, H.F., Ghidaoui, M., Karney, B.: Frequency domain analysis of pipe fluid transient behaviour. J. Hydraul. Res. **51**(6), 609–622 (2013)
56. Alfonsi, G.: Reynolds-averaged Navier-Stokes equations for turbulence modeling. Appl. Mech. Rev. **62**(4), 1–20 (2009)

Abrasive Water Jet Piercing of Superalloys: A Study of Small Diameter Deep Holes

Natarajan Yuvaraj[1(✉)], Chinnakulanthai Sampath Shamli[1], Mohan Mughilvalavan[2], and Radhakrishnan Muruganandhan[2]

[1] Vel Tech Rangarajan Dr. Sagunthala R&D Institute of Science and Technology, Chennai, Tamil Nadu, India
yuvaceg09@gmail.com

[2] College of Engineering Guindy, Anna University, Chennai, Tamil Nadu, India

Abstract. This paper reports on the abrasive water jet (AWJ) piercing of small holes with high aspect ratios on superalloys. Different superalloys such as C22, C276, Inconel 625 and 718 and Hastelloy X, each with a depth of 10 mm, 20 mm, 30 mm, 40 mm and 50 mm were considered in the present work. The hole geometrical features, namely diameter, circularity, cylindricity and perpendicularity were studied in this study. Experimental findings were discussed by analyzing various depths of the superalloys. The results of superalloys are dissimilar in hole geometrical features except for the hole diameter. In piercing, an average size of the hole diameter was found as 2.288 mm by AWJ. The other hole features namely, circularity and cylindricity were examined, and the variation of these features were higher at the exit side of the holes. Besides, a barrel shape was observed in the deep holes of superalloys, and is confirmed through the hole profile morphology. It is suggested that the use of a proper work holding device (fixture) would be recommended for the restriction of target material deflection by the impingement of high velocity AWJ. Unlike other machine tools, the results found in the present study confirmed that AWJ brings an indication to the manufacturing Industries for the production of deep holes in superalloys for the needful applications. However, still a lot of deep hole studies can be required for the betterment of hole features.

Keywords: Abrasive water jet · Deep holes · Piercing · Hole geometry

1 Introduction

In the current scenario, the production of holes with high aspect ratio structures is a major concern in the modern manufacturing Industries especially tiny deep holes are more demand in modern applications such as aerospace [1], petro chemical [2], automotive [3] etc. The specific components of these applications are made of different superalloys as which contains high corrosion resistance, strength and temperature resistance properties. Usually, the deep hole refers to a depth to diameter ratio of the hole which is more than 10 and the hole diameter is less than 3 mm which is needed for the automobile, aerospace and petro chemical sectors [1–4]. The detailed description of deep holes is

S. Hloch et al. (Eds.): ICMEM 2021, LNME, pp. 183–196, 2021.
https://doi.org/10.1007/978-3-030-71956-2_15

described by Zabel [4]. Besides, the drilling of micro and macro size deep holes on superalloys is quite difficult. Due to this, different machine tools, namely conventional drilling, Electrical Discharge Machining (EDM) and Laser Beam Machining (LBM) are used for processing different materials. However, these machining techniques often significant limitations while processing superalloys because these alloys exhibit the extreme strength at higher cutting zone temperature, formation of residual stress by high cutting force and temperature and tool wear due to the presence of carbide particles which causes the poor formation of hole features. In existing techniques, hole geometry and surface integrity are major concerns due to the presence of thermal effect in the cutting zone [5]. Among the advanced machining techniques, abrasive water jet (AWJ) is a cold process which offers the significant benefits such as less heat affected zone, lower cutting forces and better surface integrity to the drilling operations, especially in low thermal conductivity materials including super alloys. In AWJ, material removal takes place through an erosion process by a continuous impact of abrasives on the target material. A majority of the material removal takes place through a cutting wear mechanism by shearing action of abrasives with the target material [6]. Despite AWJ machining is well established in cutting, turning and milling operations, potential in drilling operations are not explored well, particularly in production of small diameter deep holes [7].

Orbanic and Junkar [8] have produced the small and deep blind holes on 40 mm depth of aluminium alloy and tool steels using piercing mode of AWJ operation. In the piercing operation, the hole depth and tapered hole were increased with increase in piercing time. Li and Wang [9] have performed the AWJ piercing experiments on 30 mm depth of titanium alloy Their results indicated that hole depth and diameter were increased by increasing water jet pressure. Bhavani et al. [10] studied the piercing performance on 50 mm depth of brass 353. They reported that water jet pressure has found as an influencing factor on material removal rate, kerf taper and surface roughness. Li [11] has monitored the jet penetration on 30 mm depth of Inconel 718 and carbon steel under piercing mode of AWJ operation with varying water jet pressures and piercing times. Drilling performance was varied based on the material properties. And, the results found that water jet pressure was the influencing factor in jet penetration. Besides, they reported that higher water jet pressure increased the hole diameter. Leninraj and Rajadurai [12] have investigated the deep hole capability of AWJ on Titanium alloy. In this study, they have adopted the hole correction method in the drilling process. Holes were made with 3 mm diameter and 100 mm depth. The results indicated that the uniform hole diameter and lower surface roughness was attained with a hole correction method in AWJ drilling process.

Akkurt [13] has drilled holes on different materials such as brass, pure aluminum, aluminum alloy, stainless steel and tool steel materials. He has observed that the drilled surface quality was reduced at higher depth of the work materials by losing cutting efficiency of the jet. ShivajiRao and Satyanarayana [14] have studied the drilling performance in glass using different mesh sizes of garnet, and the results found that stand-off distance (SOD) found as a significant factor for hole quality. Arun Raj et al. [15] investigated the AWJ drilling of 10 mm thickness plate of Inconel 625. Their experimental results portrayed that SOD is the most influencing parameter among the chosen machining parameters. Zhang et al. [16] have investigated the viability of AWJ drilling of titanium with 13 mm thickness. Various operations were performed on this material such as static piercing, dynamic piercing and hole cutting. Also, drilling was carried out by variable hole diameter. Fredin et al. [17] examined the linear piercing on stainless steel of varying thickness from 10 to 70 mm. Among the three approaches of piercing investigated, linear piercing is more effectual. Investigational outcomes revealed that for thicker metal plate higher feed rates did not pierce the material whereas, lower feed rates produced a great impact on it. Deepak et al. [18] have studied the AWJ drilling of D2 steel. To carry out this drilling process, machining parameters such as feed rate, SOD and pressure were taken and their impact on surface roughness was evaluated. Investigation revealed that pressure is the most significant parameter and surface roughness lessened with increase in pressure. Balaji et al. [19] have carried out the AWJ drilling on 6.35 mm depth of Al 6061. These operations were conducted by differing nozzle traverse speed, abrasive mass flow rate and SOD, and its effect on surface roughness was studied.

Based on the literatures, it is found that no investigations were made on deep hole drilling of different nickel alloys by using AWJ machining process. Also, lack of studies on small and high deep holes with small diameters (<3 mm). So far, few articles were reported in the production of deep holes on aluminium alloy, tool steel and titanium alloy with a maximum depth of 40 mm [7]. However, the past researchers reported on deep hole geometry and surface features are scarce. Hence, the studies taken in the present work has not been matter to any technical articles till date. Hence, this machinabiltiy study aims to pierce small size of deep holes on different nickel alloys with various depths, and to measure the hole geometry and surface features of the drilled holes.

2 Materials and Methods

In this study, the deep hole drilling operations were carried out on the AWJ machining center (Make: Waterjet Germany, Model: S3015) is shown in Fig. 1. This machine tool has following operational features, such as cutting table of 3 m × 1.5 m, maximum pressure of 400 MPa, traverse rate of X/Y direction of 10 m/min, Z axis travel of 200 mm and water flow rate of 3.2 l/min. A simple fixture was employed for AWJ drilling, and the high aspect ratio holes were produced in the range of 8:1 to 40:1. In this study, the cylindrical shaped nickel alloys such as C22, C276, Inconels 625, 718 and Hastelloy X were considered for hole making operations. The hardness of superalloys are given in Table 1. The different depths of superalloys such as 5 mm, 10 mm, 20 mm, 30 mm, 40 mm and 50 mm were chosen and are shown in Fig. 2. In this study, the diameter of superalloys was taken as 10 mm. In AWJ, drilling was usually done by using trepanning and piercing techniques. In the present study, deep holes were made by using piercing process (static method) in which the feed rate of the nozzle was zero. The piercing experiments were conducted with different thickness of superalloys with fixed level setting of AWJ parameters and is shown in Table 2. The input setting of the piercing process is shown in Table 3. The processing time of 7 to 15 min was taken for piercing different thicknesses of superalloys. In this study, one minute was also allowed after piercing the exit side of each material. In this study, the hole diameter, circularity, cylindricity and perpendicularity were measured by using the Co-ordinate measuring machine (CMM TESA MICROHITE 3D). The width of the hole cross sectional profiles was measured by using a video measuring system. Besides, the hole profiles were taken using an Image-J software.

Fig. 1. AWJ deep hole drilling.

Table 1. Hardness of superalloys.

Materials	C 22	C276	Inconel 625	Inconel 718	HAX
Hardness, HB	240	230	220	352	92

Table 2. Deep hole piercing parameters and its levels.

Parameters	Levels
Water jet pressure, MPa	380
Orifice diameter, mm	0.25
Focusing nozzle diameter, mm	1.8 mm (Used Condition)
Abrasive type	Garnet
Abrasive mesh size, #	80 (Coarse particles)
Impingement angle, deg	90
SOD, mm	3
Abrasive flow rate, g/min	450
Piercing depth, mm	10, 20, 30, 40 & 50

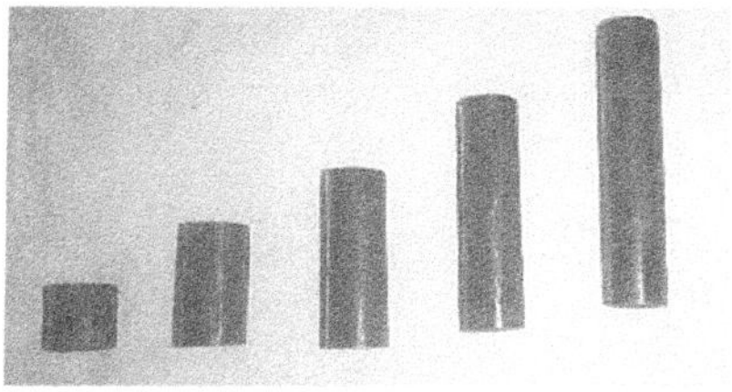

(a) C 22

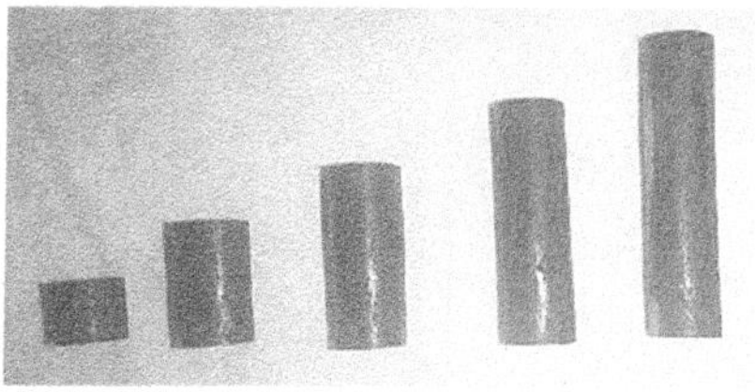

(b) C 276

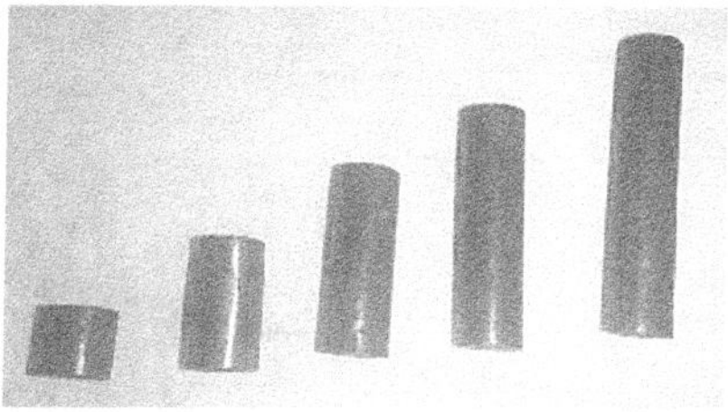

(c) Inconel 625

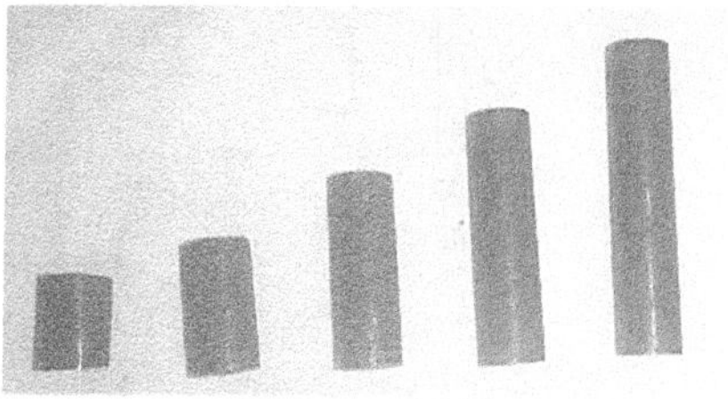

(d) Inconel 718

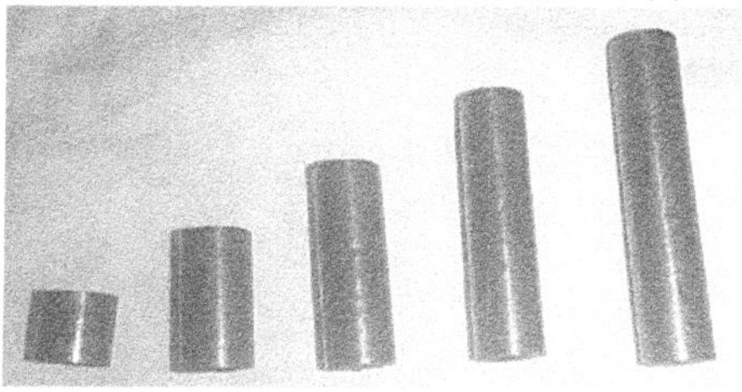

(e) HAX

Fig. 2. Various depth of superalloys.

Table 3. Experimental design for AWJ piercing superalloys.

Ex. No	Superalloys	Water jet pressure, MPa	Abrasive flow rate, g/min	SOD, mm	Piercing depth, mm
1	C22, C276, Inconel 625, Inconel 718, HAX	380	450	3	10
2					20
3					30
4					40
5					50

3 Results and Discussion

3.1 Effect of Various Superalloys with Different Depths on Hole Diameter

The entry and exit diameters of the deep holes were measured by using the Co-ordinate measuring machine and the results are presented in Fig. 3. It indicates that hole diameter was increased by increasing the depth of the superalloys. And, the results found that

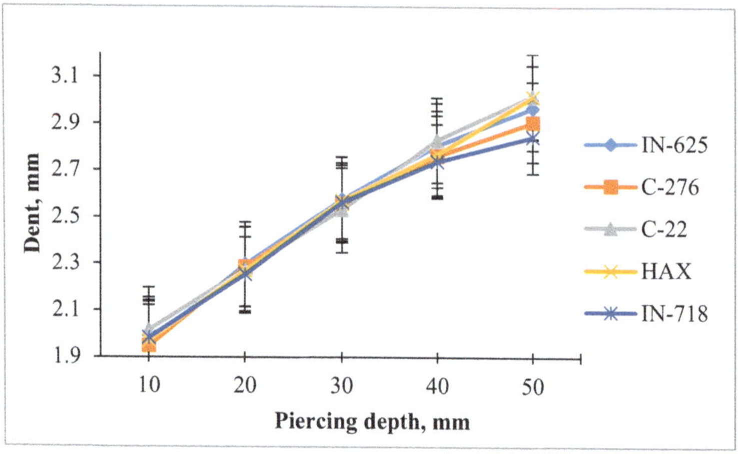

(a)

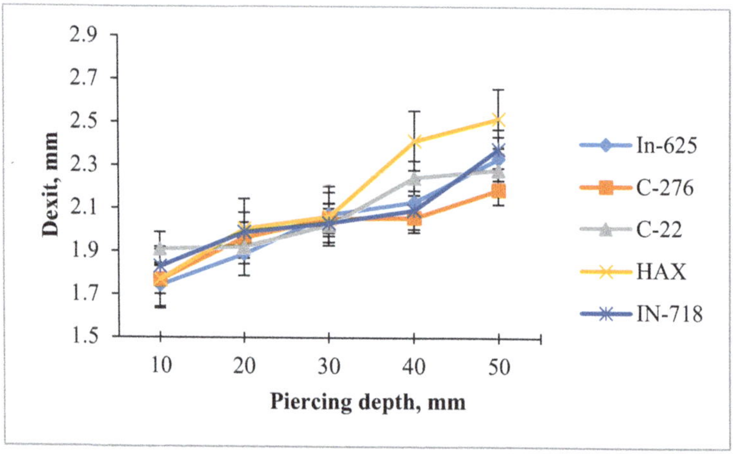

(b)

Fig. 3. Diameter of the drilled holes a) entry and b) exit.

the entry and the exit hole diameter of different super alloys is remain same. It is also observed that the HAX alloy exhibited a higher hole diameter at exit region. This result found due to its lower hardness than other super alloys which results in more material removal takes place at lower cutting region. However, a few variations were observed in the entry and exit diameter of the holes with irrespective of the materials. The entry of the hole is larger than the exit of the hole is found. Water jet has a higher potential to enlarge the hole dimensions in the entry section of the various materials. As the production of miniature hole and pressure variations, the turbulence flow is produced in the water jet. The turbulent flow causes the secondary erosion in the entry side of the material by the deflection of abrasives. This action increases the number of abrasives hitting on the existed hole side wall surface. Therefore, the hole diameter was increased and overcut occurs at the entry side of the deep hole, in comparison to the exit hole. Despite, hardness of the HAX alloy has greater than C-22 and C-276 alloys, the hole diameter was increased at the exit side. This hardness nature of HAX alloy produced a strong flow of the turbulence jet while exceeding a 30 mm depth. Due to the continuous impingement of AWJ with strong turbulent flow causes increase the diameter of the exit hole through a large destructive energy of the jet. Furthermore, it is claimed that an average hole diameter is found to be 2.288 mm in the superalloys. From the results, it is observed that the deep hole diameter was increased by 21% compared to the nozzle diameter. To prevent those variations in the hole diameter, water jet flow can be controlled in the inside of the hole during the drilling operations.

3.2 Effect of Various Superalloys with Different Depths on Perpendicularity

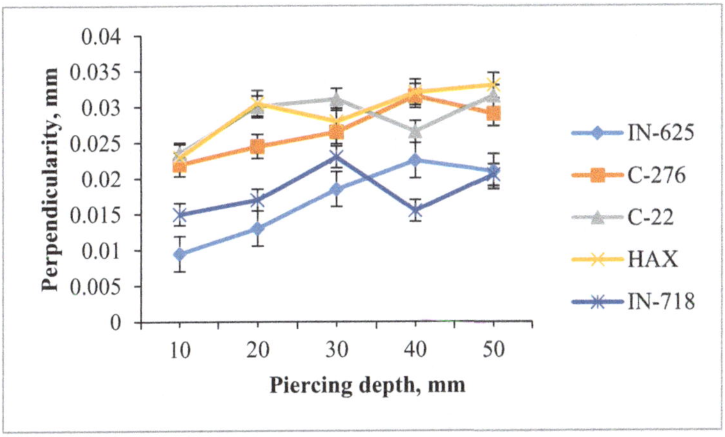

Fig. 4. Perpendicularity of drilled holes.

In this study, axis perpendicularity was studied and presented in Fig. 4. Axis perpendicularity is measured between the drilled hole axis with a specified datum. Entry and exit side of the hole perpendicularity was measured and their average values are plotted against the various depth of the work materials. An axis of the hole deviation from the datum plane is considered as perpendicularity error and these variations are described. It reveals that a slight perpendicularity variation was observed in the super alloys used in the present study. The result indicates that the perpendicularity error increases with increase in thickness of the super alloys. This is because of jet penetration capability was reduced while increasing the thickness of the alloys. As a result, overall drilling time was increased and also leads to produce the formation of taper holes. However, the perpendicularity was decreased at different depth of the holes with various super alloys. This result happened as threshold energy of the jet could be varied while interaction with various depth of the super alloys. It is also noticed that, Inconels 625 and 718 have similar trends in hole perpendicularity upto 30 mm depth. After that, the perpendicularity was varied due to the difference in hardness of the work materials.

In addition, a few perpendicularity variations were noticed in different depths of the holes. At a thickness of 40 mm, the perpendicularity variation was decreased in C-22 and Inconel 718 alloys. This result happened due to the action of the secondary erosion process through the occurrence of backflow of the water jet. This additional erosive action further removes the material in the side walls from bottom to top by returning a severe deflection of the jet [8, 12, 17]. This process owing to happened a higher hardness of C-22 and Inconel 718 alloys produced a strong jet deflection while crossing the penetration depth of 30 mm. Further, while jet moving upwards, the residue of the jet energy with scattered abrasive particles removes the small volume of material (hole wall sides) through the deformation wear process rather than the cutting wear mode. As a result, a lower perpendicularity of the hole is found. It is confirmed that the threshold region is observed at 40 mm on those materials which may turn to produce the hole with a less taper formation. Threshold region for the other alloys were observed at depth of 50 mm. The variations in the threshold region were observed as the backflow of the jet make strong erosive action in the hole side wall with different level of penetration rate. As a result, material removal of the side wall was further altered by a backflow of the jet which may synchronize with the reference axis of the hole.

3.3 Effect of Various Superalloys with Different Depths on Circularity

Figure 5 shows the circularity of the holes in different superalloys. Circularity measures the closeness of the obtained hole from the true shape. Different results were obtained by varying the depth of the superalloys at entry and exit sides. Of all superalloys, Inconel 718 exhibited a lower circularity in the entry and exit side of the holes. This result produced due to the higher hardness of the material. This harder material limits the unwanted material removal in the side wall region as well as the entry region even though a deflection of the jet is severe. As a result, circularity is controlled in the entry region of the Inconel 718. Similarly, circularity is also controlled in exit side of the deep hole. It may be produced since the energy of the jet backflow is increased while increasing the penetration depth; at threshold depth level, the accumulated jet (unused energy of the AWJ and backflow of the jet) with higher cutting energy could pierce the material with least circularity variation in the exit side. In addition, a lower circularity was also noticed on the entry side of the C-276. In this case, hardness of the material is lower than the other alloys which in turn to regenerate a hole with lower circularity variation through an occurrence of uniform material removal action from a depth of 50 mm to top cutting region. Despite, Inconel 625 has second hardest material among the superalloys, a poor circularity was observed on both the sides. This variation is produced as the energy of jet including backflow of the jet or deflection did not the compensate with the properties and structure of the Inconel 625. Furthermore, a high circularity variation might be found by the mixing of unused jet and deflection of the jet is not uniform; as a result, higher circularity variations were produced in the exit of deep hole even though an axis of the hole is coincident with the true shape. From these results, it is inferred that the non-linear results were found in superalloys due to the occurrence of secondary erosive wear at different depths of the holes. These results were formed due to the severe alterations in the jet profile's backflow when the jet deflection occurs. In this region, the material removal takes place through the deformation erosion process. Therefore, the piercing mode often produces secondary erosion process at various depths of the target material.

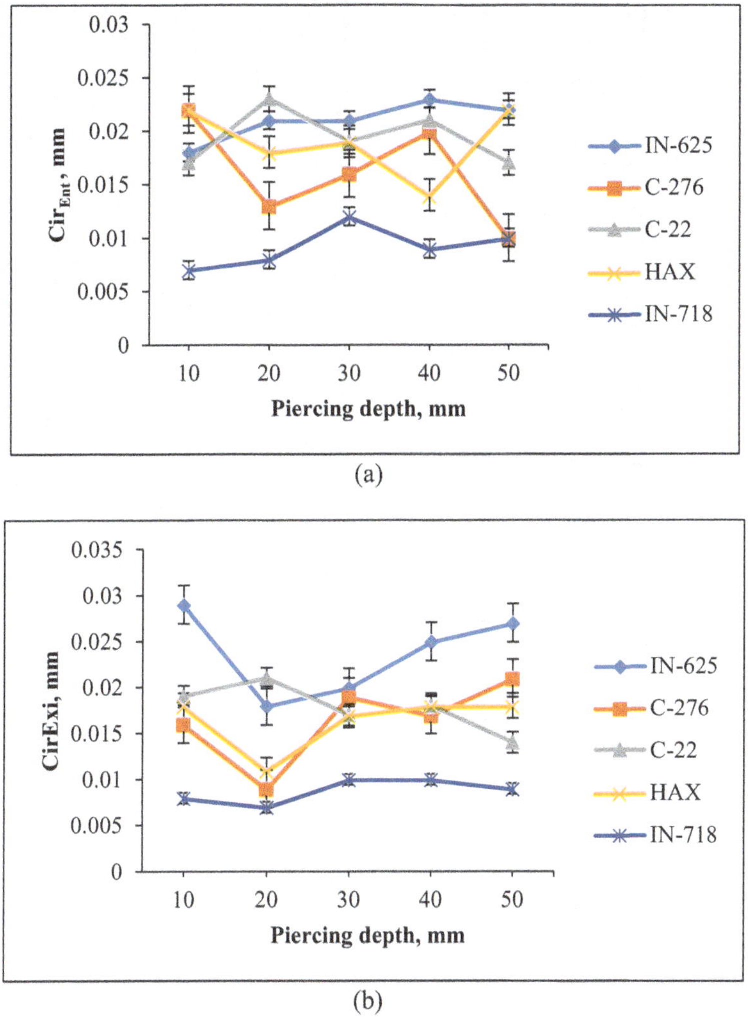

Fig. 5. Circularity of drilled holes a) entry b) exit.

3.4 Effect of Various Thickness on Hole Profile Morphology

The various depth cross sectional hole profiles of different superalloys are shown in Fig. 6. It is remarked that few profiles were bent due to the vibration of target materials through the impingement of AWJ with higher cutting force.

It is noticed that taper formation occurred at more than 10 mm thickness of pierced superalloys. This formation produced was as the abrasive particles were scattered from

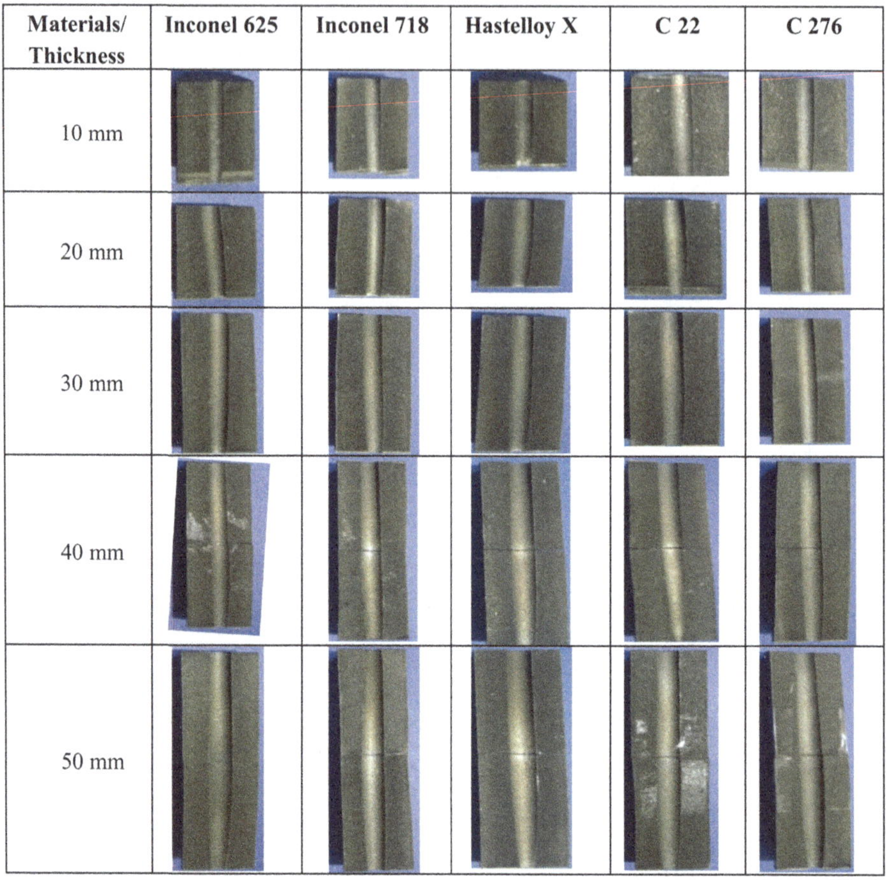

Fig. 6. Various cross sectional depth profiles of superalloys.

the periphery of the water jet through a poor structure of the coherent jet stream that results in taper formed. During piercing operations, the width of the AWJ impact region was also varied while increasing the depth of the material. It is confirmed through piercing 50 mm depth of superalloys, are shown in Fig. 7. The changes in hole profile was occurred due to the secondary erosion in the side walls which further increases the material removal rate. It is attributed to the effect of the back flow through the remaining erosive action AWJ. After hitting the nickel alloys with certain limit of the depth, the used AWJ possesses a considerable kinetic energy which tends to alter the hole profile by the secondary erosion process, which is limited to the certain extent of the hole depth. After that, the size of the hole was decreased by decreasing the cutting energy of AWJ. This leads to form the taper formation at the exit of the hole. This could reflect the inherent feature of the machining process. It is confirmed by formation of different top kerf profile in the nickel alloys. It is also observed that the enlarged profiles i.e. barrel shape were observed at the center of the hole. This shape was clearly seen at 50 mm thickness of superalloys, as shown in Fig. 7. It is understood that the barrel shapes were produced due to the lack of consistency in the jet profile as the accumulation of more volume of air phase in the AWJ leads to produce the poor form of cohesive jet. It is noticed that, barrel shape was clearly observed on 50 mm depth of the materials. In addition, taper formation was found by decreasing the width of the cross sectional hole at the top and bottom of the holes. At the top, secondary erosion and poor formation of the cohesive jet influences the formation of wider kerf (middle) whereas at the bottom, the AWJ loses its kinetic energy while increasing the penetration depth. Further, these changes were confirmed by measuring hole cross sectional width of superalloys.

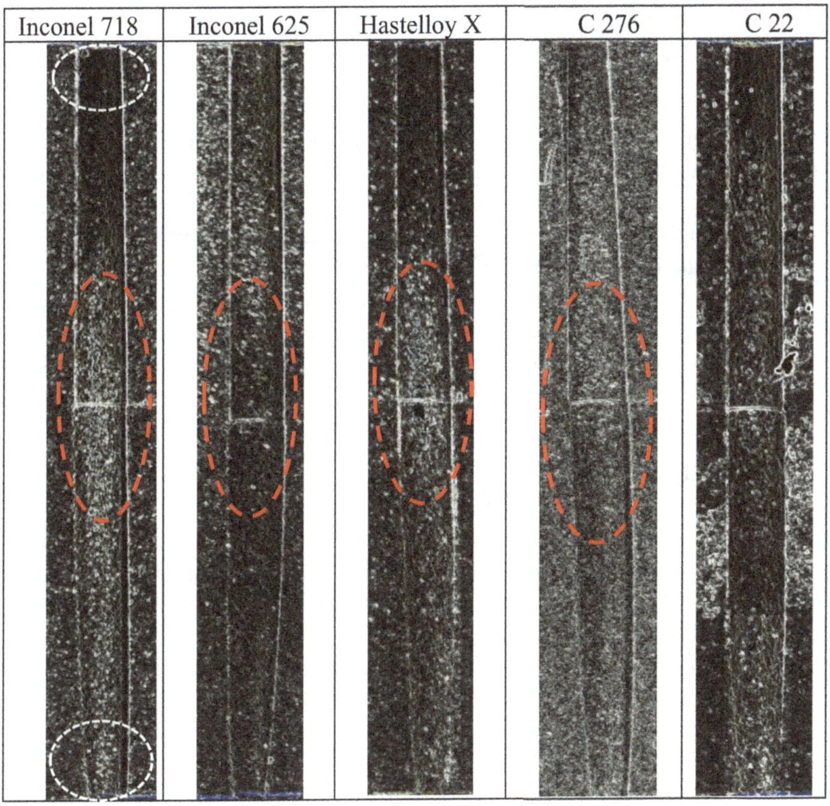

Fig. 7. Enlarged hole cross sectional profiles for 50 mm depth pierced super alloys.

Table 4 presents the hole cross sectional width for superalloys. It was measured by using video measuring system. The hole width measurements were varied from 1.961 mm to 3.539 mm. From the results, it is observed that hole inward (0 mm) consists of a lower hole cross sectional width compared to the other measurements made at different distance levels. It is also noticed that AWJ piercing increases the hole cross sectional profile at the depth of 20 mm and 30 mm from a hole inward. By claiming of this, barrel shapes were found in all materials, and are shown in Fig. 7. A double tapered profile is observed in the drilled holes at entry and exit regions, are encircled by

Table 4. Hole cross sectional width of superalloys.

Nominal, mm	Hastelloy-X, mm	Inconel 718, mm	Inconel 625, mm	C-276, mm	C-22, mm
Inward	3.037	2.264	2.116	1.961	2.352
10	3.119	3.006	3.347	2.892	3.092
20	3.389	3.176	3.528	3.144	3.226
30	3.539	3.177	3.345	3.152	3.144
40	3.372	3.041	3.118	3.041	2.950
50	3.155	3.025	3.222	2.912	2.811

white color. This type of hole profile was formed as the strong influence of secondary erosive action takes place at hole depth of 20 mm to 30 mm for various superalloys by AWJ piercing. After exceeding the jet penetration depth of 30 mm, the jet piercing energy and secondary erosion tends to weak which results in convergent shaped holes were formed in the lower cutting regions. This result was confirmed by the smaller size of hole width in the lower cutting region. It is also confirmed that drilled hole geometry was significantly affected by piercing energy of the AWJ. Among various superalloys, Hastelloy X was produced a higher hole width since the hardness of its material was lower than other target materials.

4 Conclusions

In the present study, the deep holes were made on nickel super alloys by using abrasive water jet (AWJ) drilling process as the piercing mode. The major findings are given below.

i. A piercing mode of AWJ produced a severe jet deflection that modifies the geometry of the deep hole. It may happen due to the occurrence of loss of jet energy and its structure, and severe secondary erosive wear while piercing hard materials.
ii. In AWJ piercing process, an average hole diameter was found to be 2.288 mm. This result was due to the formation of divergence of jet stream with the nickel super alloys. As a result of this, circularity error was observed in all materials. This outcome arises owing to the production of undesirable shape of the jet by the influence of scattered abrasive particles.
iii. Axis perpendicularity values were varied for all work materials as the stability of the jet was decreased by increasing the jet penetration depth of super alloys. This jet stability was affected by the influence of major contribution of air phase in the mixing chamber as well as exit of the nozzle before the cutting process.
iv. A barrel kind of the shape was observed while increasing the penetration depth in all nickel super alloys. It is happened due to the occurrence of secondary erosive wear by the deflection of AWJ. This may be attributed to that the reduction in kinetic energy of AWJ.

v. Hardness of superalloys influences the geometrical features of the AWJ drilled holes. However, at some cases, the process attributes could not be predicted with the hardness conditions that needs to be enhanced with optimum parameters for the generation of linear relationship between parameters and attributes.
vi. During AWJ piercing process, the employed parameters modified the jet deflection after 20 to 30 mm depth of the superalloys, and resulted in a double tapered hole.

The results of this study can bring the confidential level to the researchers to overcome the constraints involved in the production of deep holes by AWJ for useful applications. However, the results obtained in this study would require further investigations with optimal parameter conditions in order to improve the hole profile with free of hole geometry defects. In addition, the surface texture of the hole profiles will be explored in future work.

Acknowledgement. Dr. N. Yuvaraj has acknowledge the Director (R&D), Vel Tech Rangarajan Dr. Sagunthala R&D Institute of Science and Technology, Avadi, Chennai-62, Tamil Nadu for providing financial assistance to this research work under the scheme of VEL TECH SEED FUND (2018–19).

References

1. Machno, M.: Impact of process parameters on the quality of deep holes drilled in Inconel 718 using EDD. Materials **12**, 1–8 (2019)
2. Mannan, S., Patel, S.: A new Ni-base superalloy for oil and gas applications. In: Superalloys-The Minerals, Metals & Materials Society (TMS), pp. 31–39 (2008)
3. Bhole, P.H., Chaudhari, G.E.: Optimization of Crankshaft oil hole drilling process-a review. IOSR J. Eng. **10**(1), 29–32 (2020)
4. Zabel, A.: Deep hole drilling with small diameters. In: Laperrière, L., Reinhart, G. (eds.) The International Academy for Production Engineering, CIRP Encyclopedia of Production Engineering. Springer, Heidelberg (2014)
5. Biermann, D., Kirschner, M.: Experimental investigations on single-lip deep hole drilling of superalloy Inconel 718 with small diameters. J. Manuf. Process. **20**, 332–339 (2015)
6. Momber, A.W., Kovacevic, R.: Principles of Abrasive Water Jet Machining, vol. XXV, p. 394. Springer-Verlag, London (1998)
7. Natarajan, Y., Murugesan, P.K., Mohan, M., Ahmed, S.: Abrasive water jet machining process: a state of art of review. J. Manuf. Process. **49**, 271–322 (2020)
8. Orbanic, H., Junkar, M.: An experimental study of drilling small and deep blind holes with an abrasive water jet. Proc. Inst. Mech. Eng. Part B: J. Eng. Manuf. **218**(5), 503–508 (2004)
9. Li, H., Wang, J.: An experimental study of abrasive waterjet machining of Ti-6Al-4V. Int. J. Adv. Manuf. Technol. **81**(1–4), 361–369 (2015)
10. Bhavani, B.G., Muthukrishanan, G., Ramasubramani, M.S., Balaji, A.K., Mariappan, S.K., Leninraj, S., Varatharajan, R.: An experimental study on deep hole machining in brass using abrasive water jet machine. Int. J. Mod. Manuf. Technol. X **1**, 27–36 (2018)
11. Li, H.: Monitoring the abrasive waterjet drilling of Inconel 718 and steel: a comparative study. Int. J. Adv. Manuf. Technol. **107**, 3401–3414 (2020)
12. Lenin Raj, S., Rajadurai, A.: Experimental study on deep-hole making in Ti-6Al-4V by abrasive water jet machining. Mater. Res. Express **6**, 066532 (2019)

13. Akkurt, A.: The effect of material type and plate thickness on drilling time of abrasive water jet drilling process. Mater. Des. **30**(3), 810–815 (2009)
14. ShivajiRao, M., Satyanarayana, S.: Abrasive water jet drilling of float glass and characterization of hole profile. Glass Struct. Eng. (2019). https://doi.org/10.1007/s40940-019-00112-7
15. Arun Raj, A.C., Senkathir, S., Abhijit, J.: Experimental investigation of abrasive waterjet machining of Nickel based superalloys (Inconel 625). IOP Conf. Ser.: Mater. Sci. Eng. **402**, 012181 (2018)
16. Zhang, S., Wu, Y., Chen, D.: Hole-drilling using abrasive water jet in titanium. Int. J. Mach. Mach. Mater. **9**(1/2), 47–65 (2011)
17. Fredin, J., Jönsson, A.: Experimentation on piercing with abrasive waterjet world academy of science, engineering and technology. Int. J. Ind. Manuf. Eng. **5**(11), 2393–2399 (2011)
18. Deepak, D., Anjaiah, D., Shetty, S.: Optimization of process parameters in abrasive water jet drilling of D2 steel to produce minimum surface roughness using Taguchi approach. In: 6th International Conference on Electronics, Computer and Manufacturing Engineering (ICECME 2017), Singapore, pp. 236–239 (2017)
19. Balaji, D.S., Prabhakaran, S., Umanath, K.: Multi-objective optimization in abrasive water jet drilling of AA6061 alloy. J. Adv. Res. Dyn. Control Syst. **9**(2), 791–798 (2017)

Manufacturing Engineering Session

Sustainability of the Production Process by Applying Lean Manufacturing Through the PDCA Cycle – A Case Study in the Machinery Industry

Marko Milosevic[1], Marko Djapan[2], Roberto D'Amato[3], Nicolae Ungureanu[4], and Alessandro Ruggiero[1](✉)

[1] Department of Industrial Engineering, University of Salerno, Via Giovanni Paolo II, 132, 84084 Fisciano, SA, Italy
ruggiero@unisa.it
[2] Faculty of Engineering, University of Kragujevac, Sestre Janjic 6, 34000 Kragujevac, Serbia
[3] Departamento de Ingeniería Mecánica, Química y Diseño Industrial, Universidad Politecnica de Madrid (UPM), 28040 Madrid, Spain
[4] Faculty of Engineering, Department of Engineering and Technology Management, Technical University of Cluj-Napoca, 400114 Baia Mare, Romania

Abstract. This paper presents a case study of the practical application of the PDCA cycle as a problem-solving methodology in the company for the production of welded excavator frames, occured due to the increasing of the customer demand resulting from the variation of the global market. The objective of this research is to implement the Lean tool to ensure the sustainability of the production process while improving efficiency by at least 8–10% on a monthly basis. The Value Stream Map was applied to visually see the entire production process, as well as to identify the root causes of losses. Thereafter, Heijunka was employed to the leveling production operations, ensuring the rational use of all available resources. By applying the Lean and PDCA approach based on the developed action plan the performance of efficiency was significantly increased by 10.67% compared to the state before process improvement. The management team determined to implement further improvements relying on the methodology presented in this paper.

Keywords: PDCA cycle · Value stream mapping · Production improvement · Lean manufacturing

1 Introduction

The characteristics of the business environment on a global level experienced significant changes in the late 20th and early 21st centuries. Current world economic and political trends, as well as the effects of changes in the markets, have led most companies to a situation where their future depend on the possibility and speed of adapting to the new

S. Hloch et al. (Eds.): ICMEM 2021, LNME, pp. 199–211, 2021.
https://doi.org/10.1007/978-3-030-71956-2_16

situation and conditions. Flexibility and adaptability of business strategy is a characteristic that separates successful and unsuccessful companies. Those who achieve growth in production and profits, conquer new markets and those companies that stagnate or are often forced to completely cease their business activities, regardless of the capacities and potentials they possess [1]. Such a requirement is an imperative of any production and business system, regardless of its size and impact. Recent economic history has shown that it is companies that are global leaders and industrial giants that first identified the need to adapt their business strategy to the new conditions dictated by the market and the global business environment.

Having a strong relationship between the amount of work-in-process (WIP) and manufacturing lead time (LT) and operational variables [2], it is possible to identify and then eliminate losses. Fighting losses is a constant problem of companies struggling to survive in a dynamic market. There are many different approaches. One of them is the lean manufacturing approach. Given the typical production problems reflected in redundancies, machine availability, line efficiency, high level of the WIP, and a large inventory, companies strive for low production costs and a quick response to market demands. In this sense, it is essential to use methods/tools to identify losses and improve the overall performance of companies. In addition to those applied in the first step such as 5S [3, 4], one of the widely accepted tools is Value Stream Mapping - VSM [5, 6]. VSM can reflect production problems [7] and lay the groundwork for solving them. The efficiency of VSM implementation is shown in numerous examples [8]. Besides, there is research that improves VSM by combining it with some other available methods/tools such as Value Stream Analysis (VSA) [9]. One of the essential characteristics of VSM is the necessity to apply it correctly. Otherwise, there is a complication of the identification of waste, lead to misinterpretations and assessment errors, and undermine the implementation of future improvements [10].

As a result of the VSM analysis, facility layout problems are the number one priority. Amine Drira et al. [11] gave extensive and in that period literature review in detail about the facility layout problems based on the survey. Nowadays, there are research papers that improve this problem using the different types of optimization tools, a multi-objective particle swarm optimization algorithm [12], integrated slicing tree approach [13], multi-objective ant colony optimization algorithm [14], different mathematical optimization approaches [15]. In this paper, the authors applied VSM using Deming's PDCA (Plan-Do-Check-Act) approach [16] on the example of the machine industry. The authors focused on PDCA because it represents a four-step tool that systematically provides planning, testing, and developing changes through the problem solving prism as well as the VSM application. In [17] PDCA was mentioned as an effective methodology for improving quality in production. Also, the employment of PDCA methodology in health care is becoming more frequent [18–20]. The increase in the company's performance has resulted in the application of PDCA on the example of continuous improvement of production in the textile industry [21]. In [22], relying on the PDCA approach to problem solving, the electronic company reduced production defects by up to 77%. The objective of this research is to apply the PDCA approach to improve process performance by mapping the value stream. This paper will cover the application of VSM through each step of the PDCA, as well as monitoring the operational performance of the production process.

2 Material and Methods

The case study has been carried out in the machinery manufacturing company located in Kragujevac, Serbia. The analyzed product is excavator frame. The Lean manufacturing tools are implemented at the production line of the frame unites. Due to increased customer demand, the company had to reorganize the manufacturing process with the focus to reduce and eliminate Non-Value-Added activities (NVA). Based on the PDCA cycle the project was conducted into the four main steps, Plan - Establish objectives and processes required to deliver the desired results; Do - Carry out the plan from previous step; Check - Evaluate gathered data and results from the "Do" phase; Act - Establish the sustainability of achieved results and implement them to the next problem station, following A3 problem-solving tool.

2.1 Process Background

The only customer is the parent company, which deals with the assembly of excavators and placement on the global market. The first contact with the customer is made by the sales department. The weekly delivery plan is submitted by the sales department to the production planning department, which applies the Kanban system to define the material flow and analyze the necessary resources for the uninterrupted production mix. A schematic view of the production process is shown in Fig. 1.

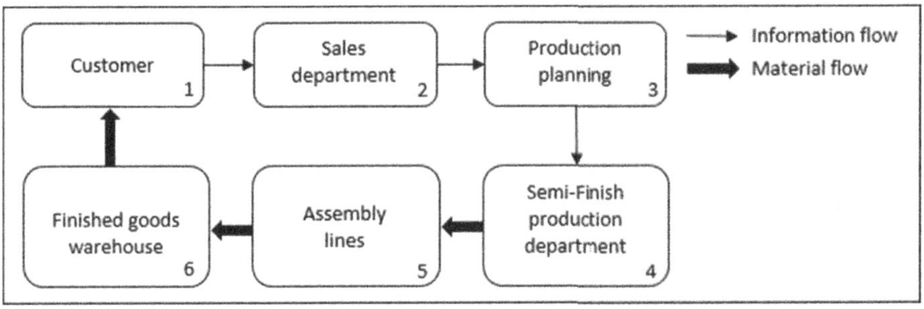

Fig. 1. Schematic view of the process background.

The analyzed production line in this paper through the 2-bin Kanban system pulls the semi-finished products necessary for the further production process of the metal construction of the frames. After consuming one bin, the production line pulls the material from the other, while the other empty bin and its associated Kanban card are transported to the semi-finished product department to reloading. The reloading process begins with laser or gas plasma cutting of steel plates according to defined technical documentation. The further flow of cut semi-finish material was established by the FIFO principle

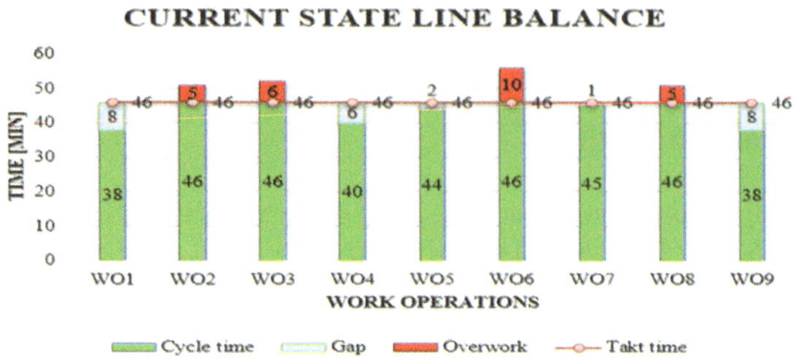

Fig. 2. Current state line balance.

(First in-First out). Depending on the characteristics of the required material (shape and dimensions), the next production steps are rolling, drilling, milling, or bending. A saw and a CNC lathe are employed for cylindrical semi-finish materials.

The bin ready for transport is returned to the production line with the associated Kanban card for further consumption according to the plan defined by the production planning department. The first production operation on the production line consists of the production of subassemblies which, after machining, with the side panels (third production operation), are assembled and pre-welded into one unit at the fourth workplace. The welding process is supported by three workstations in a row, after which the process continues through two cleaning parts of unwanted welding products. The last tenth step consists of the final processing and control of the finished product. All the ten workstations operate respecting the One-Piece-Flow principle. Further, logistics department ships finished products from the production line to the internal warehouse for customer delivery preparation.

2.2 Plan Phase

The first step of the implementation of the PDCA cycle are the activities on defining the current state of the process, the problems within it, analyzing the root causes of the problem, setting goals, and a plan for achieving them. The driver of the process reorganization is the increased customer demand for the analyzed product from 48 to 54 frames per week. The previously projected capacity of the production line supported 48 pieces of products per week. In such a defined production process, serving only in one shift, the takt time (TT) required 46 min/pc.

Operations Time Measurement and Analysis. Even distribution of work activities during the production cycle and a balanced production line ensure an efficient production process. To achieve the necessary process efficiency, the measurement and analysis of the production cycle time (CT) was approached. Cycle time is defined as the total time spent from the beginning to the end of the process, including the delay time, as well as the time required for manipulation between the two processes. The measurement was performed on each working operation of the analyzed production line. The stopwatch is used for time measurement. The authors, in collaboration with the company's expert team, analyzed the measured time, considering the recorded activities on the value-added (VAA) and the non-value-added (NNVA and NVA), where NNVA is the necessary non-value-added activities, while NVA is the non-value-added activities or waste. Figure 2 shows a graphical presentation of the analyzed cycle times, the takt time of the line based on customer demand of 48 pieces per week, as well as imbalances due to inadequate balance of the production line.

The analysis of the production process by observing the time of all work operations of one cycle gives a clear picture of the production line balance. Shown in Fig. 2, an inequality between work operations was observed. It was noticed an operators WO1, WO4, WO5, WO7 and WO9 will work less, while an operators WO2, WO3, WO6 and WO8 will not be able to achieve the daily set norm. As a result of such an imbalance, the production lines will at the end of the shift produce only as many products as the slowest operator produced, in this case, work station n.6 will operate 8 instead of 10 pieces.

Loss Analysis. To define the current state of the process and find the bottleneck and its root cause, the analysis of the measured times was approached, with the observed activities divided into three categories VAA, NNVA, and NVA (see Fig. 3).

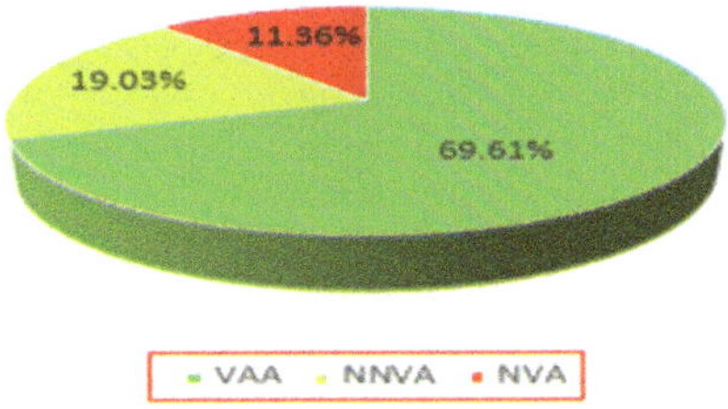

Fig. 3. Categorization of operator activities during the cycle time.

Current State Value Stream Map. To create a current state map, the authors and the expert team of the company collected the necessary data. The data were collected by walking through the production plant, interviewing the operators who carry out individual tasks within the production process. The advantage of walking through the production plant is that during the process observation can be observed production wastes. Also, the conversations with operators can give the answers to key questions necessary to draw a value stream map (VSM). Following this methodology of creating the value stream map [9], Fig. 4 shows a current state map of the production process of the analyzed product.

As one of the VSM outputs, the overall performance of the process is presented through the Manufacturing Cycle Efficiency (MCE) parameter. The summative indicator is presented at the bottom of the current state map. It represents production waiting time (minutes and days), obtained by summing the lead times of each process parts taking into account all inventory between each operation.

Key Performance Indicators. Measuring performance in production organizations is important not only for managers but also for stakeholders. It shows where the organization is currently located and where it is moving on the path to achieving its goals. On the

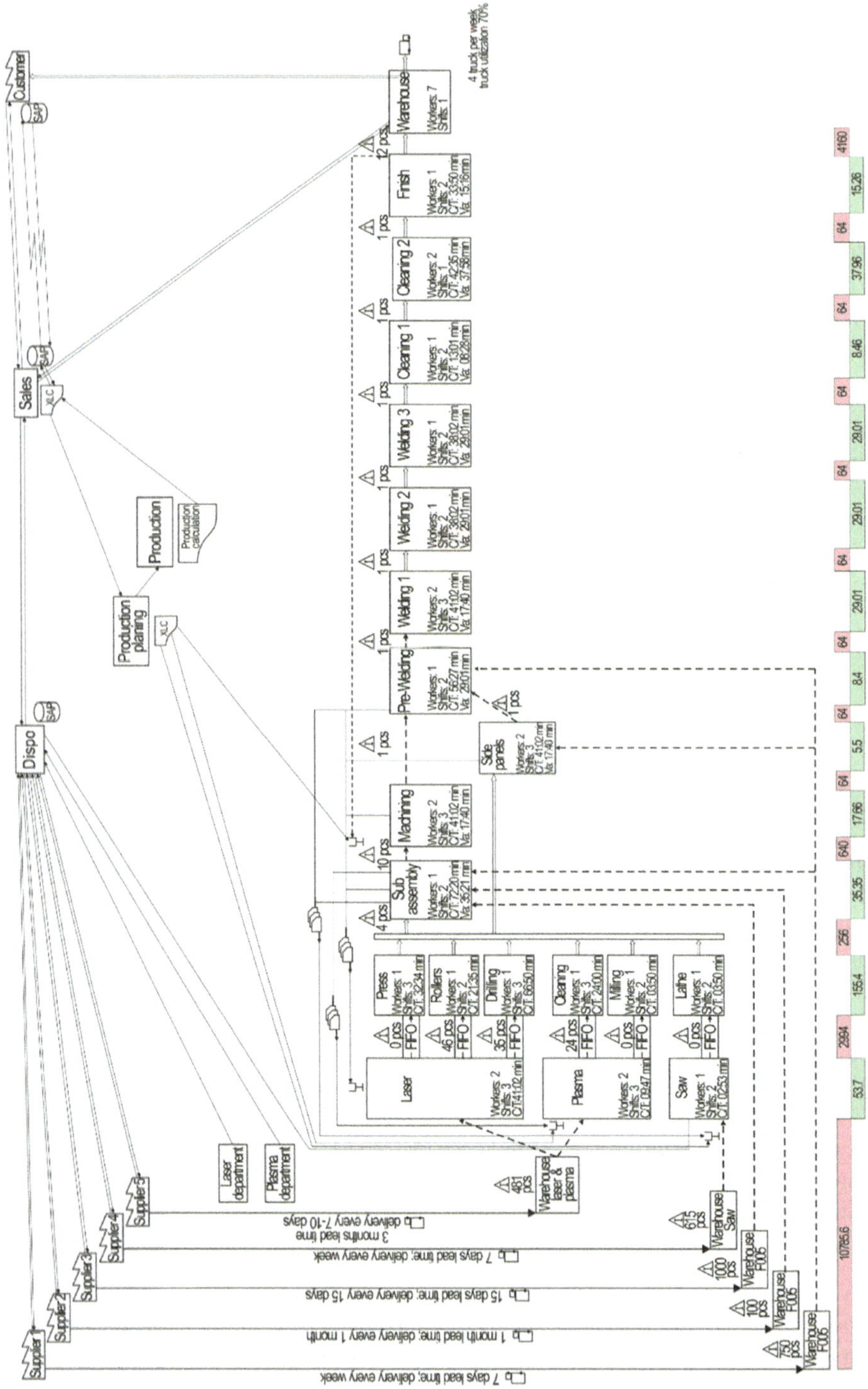

Fig. 4. Current State Map of the analysed production process.

one hand, it is possible to see the effects of past decision-making, ie whether operational and profitable goals have been achieved (through the dimensions of time, quality, and flexibility), while on the other hand, performance evaluation and measurement show financial results, the way consumers perceive the organization and cultural aspects of the work environment. In this way, insight is gained into whether existing resources are used efficiently [23].

Attempts to improve the quality of the production process, in a continuous manner, start with the assessment and evaluation of the quality of activities, subprocesses and production processes [24]. To achieve production goals that are in line with business goals, it is necessary to measure the performance of the production process. As with other production functions, performance measurement is an important tool for managing the production process. An effective method for measuring and evaluating the performance of the production process leads to improved performance and reduced production costs [25–27].

In this case, for the analyzed product as an indicator of the stability of the production process, the daily efficiency of the production line was monitored, taking into account all the delays during the cycle time, which are not a direct consequence of the operator's NVA on the line. The efficiency of the production line was measured by the following formula (1):

$$\text{Efficiency} = \frac{\text{Produced quantity} \times \text{Total production time per unit}}{(\text{Number of available workers} \times 450\text{ min}) - \text{Production downtime}} \times 100, [\%] \tag{1}$$

The efficiency was measured and tracked daily in the period of the 21 working days (see Fig. 5).

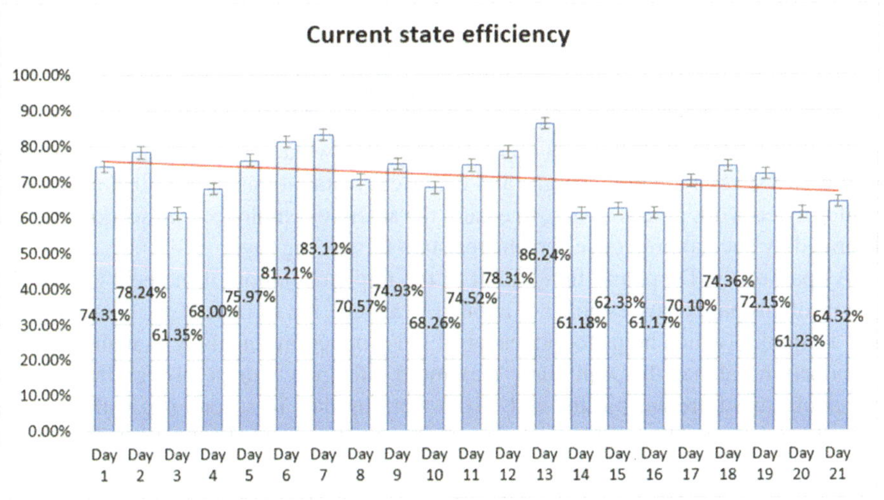

Fig. 5. The current state efficiency of a 21 days.

The obtained data gives to the authors a picture of the unstable production process. The consequence of this situation is reflected in the accumulation of time lost due to NNVA and NVA presented during work. It is noticed over motion of workers and materials between processes.

Future State Value Stream Map. The purpose of a value stream map is to find sources of losses and to eliminate them by implementing a future value stream map. When performing an analysis of the current state of the production process, it is necessary to project the future state by adhering to the rules of the Lean philosophy, but in places where this is possible. One of the biggest mistakes is trying to subordinate all processes to the concept of Lean production [1].

When defining the goal of improvement, it is necessary to follow the SMART criterion (S-Specific; M-Measurable; A-Achievable; R-Relevant; T-Timely).

New circumstances, ie increased demand for the observed product by the customer by 12.5%, as a solution to improve KPI's process, imposed a reorganization of the production line and rebalancing in the next 5 working weeks. The set goal was to increase monthly efficiency of the production line by 8–10% average in the next 10 weeks. To achieve this goal, shown in Fig. 6, a future value stream map is drawn to present a comprehensive picture of the new process and to indicate potential errors that occurred during its planning. Shown on the map, instead of the previous 9 workstations, the new projection of the production line requires 10 workstations and employing a new operator or pulling him from the production line where there are no needs for his activities.

In cooperation with the company's expert team, but also through discussions with the operators on the production line, based on the future value stream map, the authors drafted a new layout of the production line (see Fig. 7), establishing the "U shape" concept, thus reducing transport losses and distance of the sharing work.

2.3 Do Phase

The second "Do" phase of the PDCA cycle is the implementation phase of the solution planned through the first "Plan" phase. Within this step, it is necessary to carry out all planned activities so that each of them contains its own PDCA cycle.

Since the production process consists mostly of welding activities, it was necessary to pay special attention to gas and electrical installations to establish a safe and uninterrupted work process. Also, to ensure the safety of employees, all workplaces are separated by special protective screens resistant to welding rays.

As part of the Do phase, in addition to the implementation of the new layout, the authors of this paper have established another Lean tool, the so-called Line balance board (jap. Heijunka) which helps to the leader of the production line daily to maintain such a production balance that will allow maximum efficiency shown through the number of manufactured products at the end of the working day, in accordance with the time defined by customer demand.

3 Results and Discussion (Check and Act Phase)

3.1 Check Phase

The third (Check) phase of the PDCA cycle is defined by monitoring and analysis of the obtained results after the implementation of the first two phases. It should show the

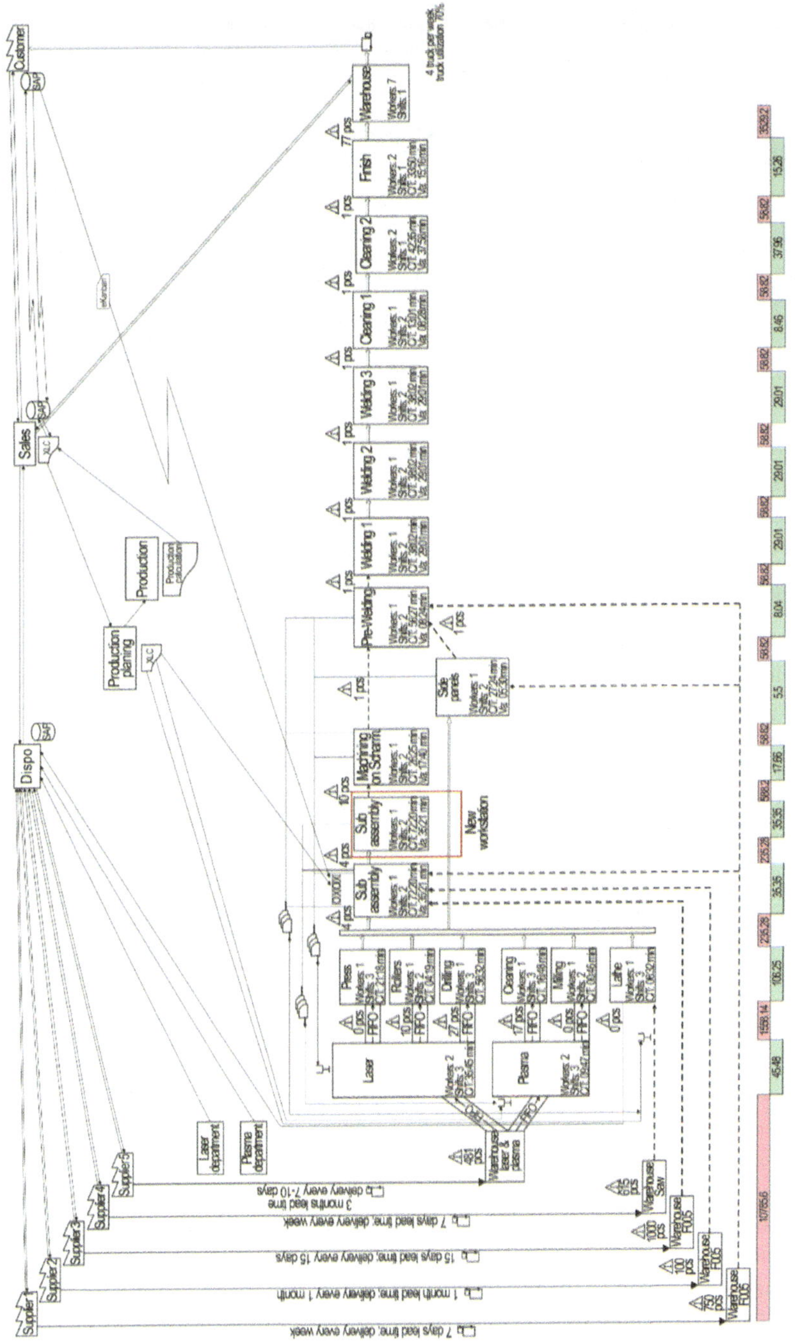

Fig. 6. Future State Map of the new production process.

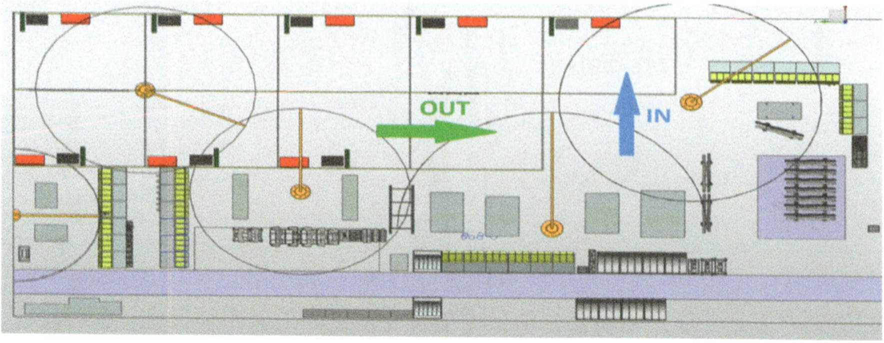

Fig. 7. New layout of the production line.

progress of the KPI's after the improvement and answer whether the goal set in the Plan phase has been achieved or additional activities need to be carried out for the achieved goal to be in line with the management plans. However, if the goal is not fully achieved, it is necessary to return the process to the Plan phase and reconsider the root cause of the problem because it is not uncommon for inadequate analysis of the root cause to lead to negative results of the implemented solution. It is also very important that the set goal is defined under the previously described SMART criterion, because otherwise if the goal is not achievable or relevant, it results in additional losses for the company to redefine and re-achieve it.

Based on the improvements implemented in phase 2, the results showed a sustainable flow of production and delivery of finished products to the customer. The following Table 1 gives a comparative view of the KPI's before and after improvements. The efficiency is presented as a monthly (21 days) average.

Table 1. Measured KPI's before and after improvements.

KPI's	Before	After
Efficiency	71.52%	82.19%
MCE	9.76	12.92

Variations in daily efficiency are now a consequence of hiring new employees on the production line who have undergone initial training but still do not have enough experience to reach the production rhythm defined by a takt time of 41 min per piece.

Table 2 shows the percentage ratio of the VAA-NNVA-NVA analyzed before and after the improvement of the production process. The increase in VAA after the implementation of the solution is a consequence of the reduced movement of people and materials between the sub-processes, as well as the implemented 5S systems when setting up a new layout. Also, new standard work instructions have been developed, set up at each workplace, which describes in detail but sufficiently clearly the procedure of each work operation so that the possibility of an error during work is minimized.

Table 2. Classification of activities before and after applied improvement.

Activity	Before	After
VAA	69.61%	72.17%
NNVA	19.03%	14.38%
NVA	11.36%	13.45%

Following the monitored and analyzed parameters, the authors and the management of the company concluded that the required increase in production line efficiency of 8–10% average on a monthly basis was achieved within a defined time frame. Specifically, after implemented activities, the efficiency monitored 21 working days was increased in the average by 10.67% in comparison to the same period before the improvement. The goal achieved in this way ensured a sustainable production process with stable production capacity and regular supplying to the customer with finished products in the quantities he requires and when he requires.

3.2 Act Phase

The fourth phase of the cycle is the process of standardizing a successfully implemented solution and establishing its sustainability. In this case study, the company's management implemented the Act phase by applying VSM and other Lean tools implemented in this case on three new production lines where due to variations in demand it was necessary to change the production process to ensure its sustainability.

The key to the success of the PDCA cycle is that it never ends. With each Act phase, a new Plan phase should begin and a new improvement process should start.

4 Conclusions

Answers to the questions why Lean production is considered such a unique concept and why such a concept is necessary to establish successful production, but almost all other questions related to the success of Lean production, are obtained by improving the technical and organizational view of production.

The case study presented in the paper regarding the company from the machinery industry provides proof that the PDCA cycle is a powerful and reliable Lean tool that can be applied in any company striving for continuous improvement, regardless of whether it is the manufacturing sector or services. The company has reached the set goal of 10.67% and stabilized the flow of production and delivery, but this must not be the end of improvements. As already mentioned, the end of this project should be the initial signal to launch a new improvement that will raise production standards by relying on Lean tools applied in this case study such as VSM, 5S, Heijunka, but also other tools depending on the problem which the company is facing.

References

1. Macuzic, I., Djapan, M.: Lean koncept u upravljanju proizvodnjom, 1st edn. Faculty of Engineering, University of Kragujevac, Kragujevac (2016)
2. Cuatrecasas-Arbós, L., Fortuny-Santos, J., Ruiz-de-Arbulo-López, P., Vintró-Sanchez, C.: Monitoring processes through inventory and manufacturing lead time. Ind. Manag. Data Syst. **115**(5), 951–970 (2015)
3. Tonic, N., Arsovski, S., Djapan, M., Macuzic, I.: Manufactury effectiveness improving using LEAN and 5S method. In: 8th Quality Festival, pp. 417–426. Faculty of Engineering, Kragujevac, Serbia (2014)
4. Milosevic, M., Macuzic, I., Todorovic, P., Djapan, M., Giagloglou, E., Vuckovic, D.: Implementation of the 5S system as a factor for improving the quality management. In: 7th International Quality Conference, pp. 585–589. Faculty of Engineering, Kragujevac, Serbia (2013)
5. Das, B., Venkatadri, U., Pandey, P.: Applying lean manufacturing system to improving productivity of airconditioning coil manufacturing. Int. J. Adv. Manuf. Technol. **71**(1–4), 307–323 (2014)
6. Garza-Reyes, J.A., Kumar, V., Chaikittisilp, S., Tan, K.H.: The effect of lean methods and tools on the environmental performance of manufacturing organisations. Int. J. Prod. Econ. **200**, 170–180 (2018)
7. Shou, W., Wang, J., Wu, P., Wang, X., Chong, H.Y.: A cross-sector review on the use of value stream mapping. Int. J. Prod. Res. **55**(13), 3906–3928 (2017)
8. Singh, H., Bahl, A., Kumar, A., Singh Mann, G.: Materials and information flow analysis and optimization of manufacturing processes in MSMEs by the application of value stream mapping (VSM) technique. Mater. Today: Proc. **5**(14, Part 2), 28420–28426 (2018)
9. Stadnicka, D., Litwin, P.: Value stream mapping and system dynamics integration for manufacturing line modelling and analysis. Int. J. Prod. Econ. **208**, 400–411 (2019)
10. Dal Forno, A.J., Pereira, F.A., Forcellini, F.A., Kipper, M.L.: Value Stream Mapping: a study about the problems and challenges found in the literature from the past 15 years about application of Lean tools. Int. J. Adv. Manuf. Technol. **72**, 779–790 (2014)
11. Drira, A., Pierreval, H., Hajri-Gabouj, S.: Facility layout problems: a survey. Ann. Rev. Control **31**(2), 255–267 (2007)
12. Guan, C., Zhang, Z., Liu, S., Gong, J.: Multi-objective particle swarm optimization for multi-workshop facility layout problem. J. Manuf. Syst. **53**, 32–48 (2019)
13. Friedrich, C., Klausnitzer, A., Lasch, R.: Integrated slicing tree approach for solving the facility layout problem with input and output locations based on contour distance. Eur. J. Oper. Res. **270**(3), 837–851 (2018)
14. Liu, J., Liu, J.: Applying multi-objective ant colony optimization algorithm for solving the unequal area facility layout problems. Appl. Soft Comput. **74**, 167–189 (2019)
15. Anjos, M.F., Vieira, M.V.C.: Mathematical optimization approaches for facility layout problems: The state-of-the-art and future research directions. Eur. J. Oper. Res. **261**(1), 1–16 (2017)
16. Deming, W.E.: Out of the Crisis, 1st edn. Massachusetts Institute of Technology, Center for Advanced Educational Services, Cambridge, Massachusetts (1982)
17. Chen, Y., Li, H.: Research on engineering quality management based on PDCA Cycle. IOP Conf. Ser.: Mater. Sci. Eng. **490**(6), 062033 (2019)
18. Taylor, M.J., McNicholas, C., Nicolay, C., Darzi, A., Bell, D., Reed, J.E.: Systematic review of the application of the plan–do–study–act method to improve quality in healthcare. BMJ Qual. Saf. **23**, 290–298 (2013)

19. Reed, J.E., Davey, N., Woodcock, T.: The foundations of quality improvement science. Future Healthc. J. **3**(3), 199–202 (2016)
20. Kurowski, E.M., Schondelmeyer, A.C., Brown, C., Dandoy, C., Hanke, S.P., Cooley, H.L.T.: A practical guide to conducting quality improvement in the health care setting. Curr. Treat. Options Pediatrics **1**, 380–392 (2015)
21. Tahiduzzaman, M., Rahman, M., Dey, K., Kapuria, T.: Minimization of sewing defects of an apparel industry in Bangladesh with 5S & PDCA. Am. J. Ind. Eng. **5**, 17–24 (2018)
22. Realyvásquez, A., Arredondo-Soto, K., Carrillo-Gutiérrez, T., Ravelo, G.: Applying the Plan-Do-Check-Act (PDCA) cycle to reduce the defects in the manufacturing industry. A case study. Appl. Sci. **8**, 2181 (2018)
23. Hudson, M., Smart, A., Bourne, M.: Theory and practice in SME performance measurement systems. Int. J. Oper. Prod. Manag. **21**(8), 1096–1115 (2001)
24. Ollila, A.: Proposals for the implementation and improvement of ISO 9001. Global J. Bus. Res. **6**(2), 71–81 (2011)
25. Jain, S., Triantis, P.K., Liu, S.: Manufacturing performance measurement and target setting: a data envelopment analysis approach. Eur. J. Oper. Res. **214**(3), 616–626 (2011)
26. Nestic, S., Stefanovic, M., Aleksic, A., Djordjevic, A., Stojanovic, S.: Production process metric in medium manufacturing organization. In: Proceedings of 7th International Conference ICQME, pp. 231–237. Faculty of Mechanical Engineering Podgorica, Tivat, Montenegro (2012)
27. Wu, C.-W.: A Bayesian approach for measuring process performance with asymmetric tolerances. Eur. J. Ind. Eng. **6**(3), 347–368 (2012)

Areal Decomposition Methodology for a 5-Axis Milled Surface

Chun Chan(✉) and Karl Walton

University of Huddersfield, Huddersfield H1 3DH, West Yorkshire, UK
Chun.Chan@hud.ac.uk

Abstract. An optical 3D measurement system (Alicona Infinite Focus Measurement G4) has widely been used to scan a surface. This system can generate the topography of the scanned surface and calculate the surface parameters. The cutting process of using the 5-axis milling machine associated with the ball end cutting tool creates cusp marks on the surface. This study focused on decomposing the cusp marks to investigate the characteristics and parameters of the surface at different scales (cusp marks, feed marks and tool marks) by using the Alicona IFM G4 with other software. The two approaches used in this study were form removal and filtration. The result of using the form removal approach to decompose the cusp marks was a more suitable method for this study since it left all the features on the surface except the cusp marks. The filtration approach tended to filter some useful features alongside the cusp marks, which affected the calculated surface parameters.

Keywords: Surface decomposition · Multiscale · Cusp mark · Form removal

1 Introduction

The 5-axis milling machine has become widely used in industries such as automotive and aerospace where the generation of free form surfaces is required. Turbocharger impeller wheels are one of the automotive components that are now being machined from billet material by 5-axis milling, instead of being cast. The traditional casting process is by injecting the molten metal into the mold cavity. However, this process can create impurities within the component and increase the porosity, resulting in micro cracks, crack initiation and fatigue failure. The advantage of using the 5-axis milling process is to enhance the wheel's durability due to the purity and uniformity of the billet material. However, the disadvantage of using this machining process is that tool marks are left on the component surface. Many researchers have looked at the cusp marks created by the ball end cutter on different materials such as stainless steel, copper and aluminium at the macro scale [1–3]. However, there is a lack of research looking at the meso and micro scale of those surfaces in terms of the surface roughness due to marks within the cusps since the surface roughness is dominated by the cusp height. The meso scale is defined as the feed marks within the cusp generated by the kinematics of the cutting edge rotation and potential defects at that scale created by tool vibration and tool

S. Hloch et al. (Eds.): ICMEM 2021, LNME, pp. 212–225, 2021.
https://doi.org/10.1007/978-3-030-71956-2_17

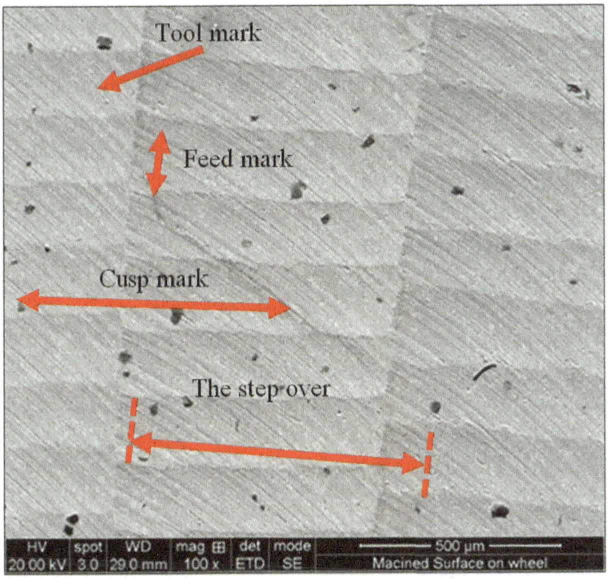

Fig. 1. Illustration of cusp, feed and tool mark.

run out. The micro scale is defined as the machining marks between each flute of the cutting tool. These marks are identified in Fig. 1.

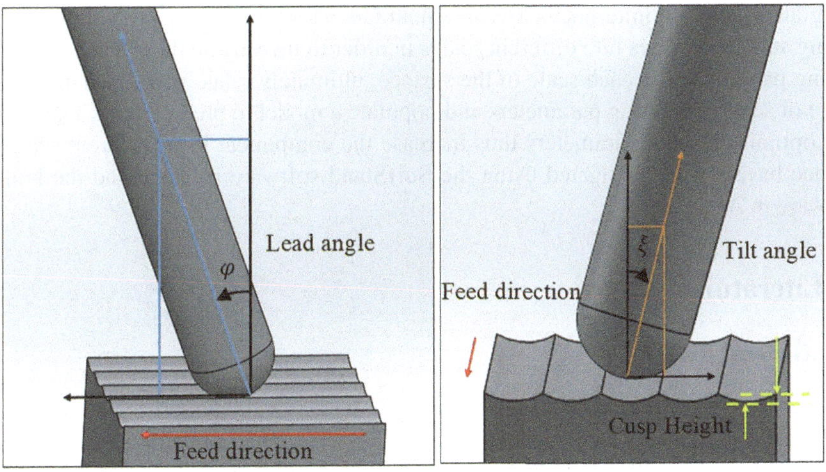

Fig. 2. Illustration of lead (φ) and tilt (ξ) angle on a work piece.

There are several parameters that can affect cutting performance. One of the commonly known parameters is the position of the cutting tool relative to the workpiece surface, defined by the lead (φ) and tilt (ξ) angles as shown in Fig. 2. As shown in Fig. 2(a), during pull cutting the cutting tool leans toward to the feed direction with a positive lead angle. In push cutting the tool leans in the opposite direction with a negative lead angle. With zero lead and tilt angle, the tip of the tool (the point on the tool axis) has no movement and no cutting can take place at that point. Figure 1 also shows the step over, which is the distance that the tool moves from one feed path to another. Cusp height is the height between the lowest and highest point of the cusp, as shown in Fig. 2.

At different scales, the surface can be represented by different wavelengths. The macro scale of the surface can be described as form, which in this study will refer to the cusp. The meso scale of the surface can be described as waviness, which can refer to the feed marks. Finally, the micro scale of the surface can be described as roughness, which can refer to the defects. These different scales are illustrated in Fig. 3.

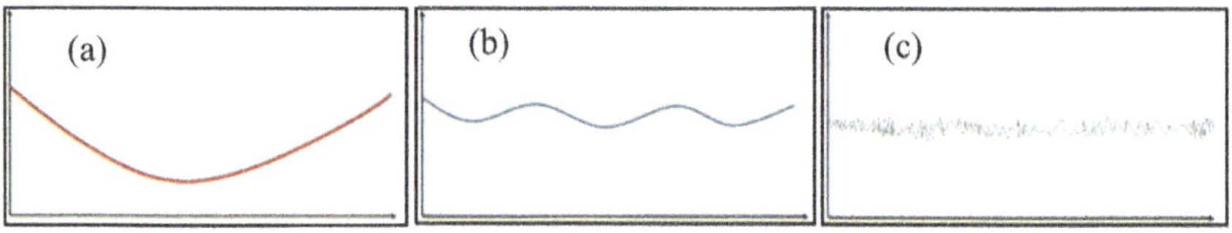

Fig. 3. The illustration of **(a)** form, **(b)** waviness and **(c)** roughness.

In this work, the meso and micro scale of a machined surfaces have been investigated using an Alicona Infinite Focus Measurement G4. This study concentrated on decomposing surface features into different scales in order to investigate the effect of different cutting parameters on each scale of the surface, ultimately relate surface character at a range of scales to cutting parameters and populate a model to predict surface character, also optimise cutting parameters thus increase the component life. Different scales of surface have been investigated using the SurfStand software package and the built-in software in Alicona.

2 Literature Review

2.1 General Review

In order to investigate turbocharger compressor wheel cycle durability, [4] repeatedly ran a wheel from 10.5% to 105% of the design speed. Investigation of the resultant fatigue failure showed cracks initiating from the machining marks. They concluded that the stress raising effect due to the machining marks was limiting wheel lifetime. [5] developed a kinematic cutting model based on the tool geometry and the lead angle, using a coordinate system with an origin at the centre of the hemispherical cutting tool tip. This assisted in defining the depth of the cutting tool's penetration into the work piece and the kinematic cutting process. They concluded that by increasing the lead angle in pull cutting style, the surface roughness in terms of the spacing and height of each feed mark can be reduced. Their study was verified by a series of experiments. [6] conducted a similar study of the lead and tilt angle effects in the 5-axis ball end milling process. As well as looking at the effect of applying a lead angle during the cutting process, this simulation also considered the marks created by the tool tip ploughing the surface and the association between cusp height and step over distance. The fundamental principle of the model was similar to the research done by [5]. However, this study also added the tilt angle into the model. The result of the simulation was validated by a series of cutting tests. This study affirmed that the tool tip marks could be avoided by changing the lead angle and cutting depth which reduced the surface roughness in terms of the height of the feed marks. They also changed the lead and tilt angle and proved that this application could achieve the same scallop height with less machining time compared to zero tilt angle.

The macro scale (cusp form) surface finish of samples machined using ball end cutters has widely been investigated. Most researchers were mainly focusing on the height and width of the cusp, and the overall surface roughness. [2] established a series of cutting experiments using a carbide ball end cutter on a Titanium alloy TC17 workpiece. The result demonstrated that the surface roughness (*Ra*) was improved when applying the lead and tilt angle during the cutting process. [7] analysed the effect of different combinations of cutting parameter, such as the feed per tooth and spindle speed and radial depth on the surface in terms of areal roughness parameters such as average roughness (*Sa*).

[8] also investigated the effect of using a ball end cutter with different combinations of spindle speed, feed rate and lead angle with cutting in the push direction. The result showed that higher spindle speeds created lower surface roughness values unless the magnitude of the cutting tool lead angle was 50° or higher, at which point tool vibration became a significant factor. [9] also carried out an experimental investigation of the influence of cutting parameters, tool run-out and deflections induced by cutting force on a 5-axis milling machine in hardened steel. Their result showed that the lead angle has a significant influence on the surface finish. The tool life can be improved and minimal cutting forces can be achieved by applying a lead angle of 15° associated with feed rate of 375 m/min. [5] found that the feed marks have a greater impact on the surface roughness than the cusp mark during the high-speed machining. The result of this study showed that the feed marks were 3–4 times higher than the cusp marks in a high feed rate cutting and the cusp height was continuously reduced while the tool lead angle was increased. A lead angle up to 10° is sufficient for most ball-end tool diameters to create lower surface roughness values compared with 0° lead angle.

The Alicona Infinite Focus Measurement G4 has widely been used for many applications for inspecting surfaces. This machine captures an image of the surface and analyses the depth of field by focus variation. Since the microscope camera of the machine can only move horizontally, the machine can only produce areal results. [10] used this machine to study the wear of coated and uncoated tools by comparing the volume of the tools before and after the cutting process. [11] used the Alicona IFM to investigate Titanium samples in order to obtain the value of the Sk (surface core height) and Svk (reduced valley depth). They used the result to characterise the finished surface on samples and provide guidance on the optimal parameters on the surface.

To achieve the desired resolution, the whole area was divided into 400 fields (a 20 × 20 grid), each scanned individually. The individual images were then overlapped until the height value of each pixel matched. This area stitching process required approximately 10% overlap.

[12] also used the Alicona IFM to scan the cutting inserts flank and rake face areas for a wear analysis. They used the scan of the new cutting insert edge as a reference to compare with the scan of the worn edge. The difference analysis function within the machine software was used to superimpose the two scans to evaluate the tool wear.

2.2 Filtration and Form Removal

Filtration is a function to separate the measured data into different scales of interest [13]. The Gaussian filter is one of the most widely used filters for surface metrology [14] and is identified within the international standard (ISO 16610-21:2012) [15]. This filter has a 50% transmission for the cut-off wavelength [16]. It has been defined as one of the standard filtering techniques for profile analysis [17]. This filter is known to determine a weighted moving average and it can be used in individual cases such as with freeform surfaces to eliminate the nominal form [13]. However, there is a drawback of this filter, since it is not robust, a large peak or valley will then be distorted and affect the final result of the surface roughness parameter [14].

Wavelet transformation is a method which avoids the distortion created by the Gaussian filter. The wavelet transform method decomposes a surface profile into a set of basic wavelets and breaks down a signal into sinusoids of different frequency [18]. This analysis technique is extensively applied in image processing, pattern recognition and some non-linear subjects, and it can be used to pick out and further evaluate the characteristics of surface structures [19]. In this study, a wavelet analysis function within the software package SurfStand has been used. The software allows the user to select from an automatically generated wavelength range in order to extract relevant data. The software calculates the longest (form) and shortest (roughness) wavelength frequency on the surface, and then divides the range into 8 levels in order the user to choose the levels to investigate [20]. However, this method can only apply to a certain range of wavelengths and it cannot accurately choose the exact maximum and minimum width of the wavelength window to investigate.

The next method, which can avoid the restriction of the wavelet transform method, is polynomial fitting. This is a smoothing method to filter the data in order to produce the necessary data. The advantages of using this method include there is no need to modify the boundary on the surface; it can be applied to various design situations [21]. The disadvantage of using this method is that it is difficult predict the outcome of the fitting procedure; many trials may be required before the system can create appropriate result [22].

Finally, the form removal function aims to remove the basic surface shape but leave other features on the surface unaffected. This will not distort the surface like the previous methods. [23] used this function to remove the cylindrical form from a measured surface, leaving a flat surface ready to be analysed for surface roughness. [24] did a study to inspect the surface on maraging steel 300 moulded by selective laser melting. They also used the form removal function to make the cylindrical surface flat. In this study, all the above methods were used to determine the most suitable for machined surface investigation.

3 Method and Material

A series of 120 mm by 25 mm by 5 mm Aluminium alloy samples were machined using a 5-axis milling machine. The cusps were machined on the middle of the sample as shown in Fig. 4.

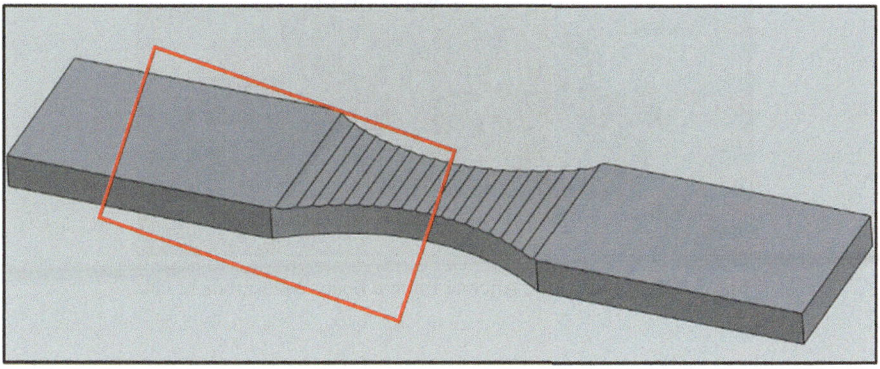

Fig. 4. CAD model of the sample (red square represents the machined area).

In this study, all samples had the same cutting lead and tilt angles of 55° and 0° respectively, and a push cutting style. The spindle speed was 12000 rpm. The feed rate and sample code numbers are shown in Table 1.

Table 1. Samples specification.

Converted feed rate (mm/min)	Quantity	Code number
972	2	BAC 1
1273	1	CAC 1
1944	2	EAC 1

The surface inspection was carried out using the optical 3D measurement system shown in Fig. 5.

This machine uses a non-contact measuring method to measure surface topography. The surface can then be reconstructed in 2.5D and the topography digitised. For this study, a magnification of 20× was used to inspect a volume of approximately 4 mm × 4 mm × 300 um. The inspected area included at least 6 cusps in each inspection. The 6 cusps were investigated in two ways. Firstly, the cusps were investigated individually and the average value of the surface roughness parameters for each cusp taken to represent

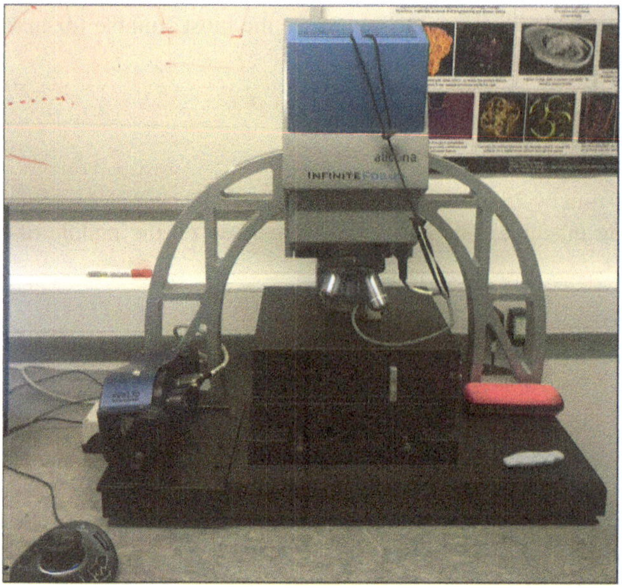

Fig. 5. The setup of the Alicona Infinite Focus Measurement G4.

the whole surface. Secondly, all cusps were investigated together to generate surface roughness parameters directly. A fixture was produced to secure the samples on the inspection bed. This step ensured that the cusps on each sample were accurately aligned to the machine x and y axis. The estimated vertical resolution was 0.21 μm and lateral resolution was 2.91 μm. The machine could not measure the full 4 mm by 4 mm area in a single field of view.

In order to decompose the surface into different scales, the post processing of the data captured by the Alicona was carried out in three ways. Firstly, there is a function proprietary in the Alicona IFM software called form removal, which allows the user to select different geometric primitives, Subsequently, the decomposed results could be used to characterise the surface at a range of scales to correlate with machining parameters. However, the cylindrical form removal function has the disadvantage of only being able to remove a single form from a data set. In this study, it was necessary to manually crop the 6 cusps on each surface into one single cusp so that the software could recognise the cylindrical shape. This step created another issue due to non-uniform crests (edges) on the cusps. These defects prevented the form removal process from working. Therefore, the edge of all the cusps had to be cropped. However, there is a chance that those features are important to this investigation. This will be investigated in future experiments. An additional function in the Alicona is the 3D-Editor. This function allows the user to select erroneous artefacts to remove and will then fill in the surface according to the surface roughness around that region. This function has been used in this project in order to lower the chance of getting errors due to artefacts that were not created during the cutting process and were not relevant to this study. However, this function relies on user judgement of how to define areas as containing defects or errors.

Secondly, the SurfStand software package was used to perform a wavelet filtration (Multi-Scale Analysis). In this case, it was assumed that the longer wavelength range sets were the cusp marks, the middle sets were the feed marks and the shorter sets were the tool marks. The middle and shorter wavelength range sets were selected to illustrate the feed and tool marks. With the wavelength ranges selected, the surface roughness parameters were calculated. However, a disadvantage of this approach was that the user did not have control of the wavelength interval in each level. Therefore, this approach might not be able to select the exact wavelength interval that the user prefers. Also, features with similar wavelet ranges (in this case the cusps and the feed marks) cannot be isolated from each other using this wavelet analysis technique.

The third approach was the other approach available in the SurfStand software, polynomial fitting. This feature can remove polynomial curves from second to twentieth order from the data. In this study, the cusp can be approximated as a second order feature. The limitation of this approach was that multiple cusps could not be defined as a polynomial of any order. Therefore, it was only possible to process a single cusp.

In this study, all the cusps in each sample were investigated using wavelet filtration, polynomial fitting and the form removal associated with the 3D-Editor described above.

4 Results and Discussion

4.1 Scales of the Sample

Figure 6 shows one of the samples containing 6 cusps with cusp marks, feed marks, tool marks and defects. At the macro scale, defects can be seen which are potentially due to material imperfection, tool run out, tool vibration, tool wear, and contamination of the surface before inspection. In order to investigate the cusp with as little error as possible, a cusp selection was performed. Figure 6(b) shows the detail of a selected area in one of the cusps in Fig. 6(a). The meso scale of this sample showed the feed marks on the surface, which are equally distributed throughout the cusp. These feed marks vary depending on feed rate. Figure 6(c) shows the detail in Fig. 6(b) illustrating the tool marks. The micro scale of this sample shows the tool marks which are different from the feed marks. The most significant difference between feed marks and tools marks is feed marks are regularly laid on the surface but tool marks appear randomly within the feed marks. Those marks were created potentially because of tool wear which leads to a defect on the tool edge and due to aluminium oxide, created during the cutting process, adhering to the tool edge.

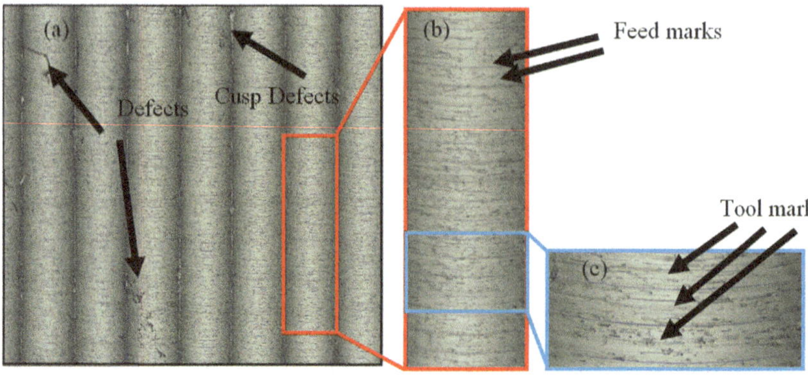

Fig. 6. (a) Alicona inspection result (Sample BAC 1-1) (b) Meso scales of the Sample (c) Micro scales of the Sample.

4.2 Result Analysis

By using the cylindrical form removal function in the Alicona to decompose each sample cusp, the areal roughness parameters in terms of *Sa* (Arithmetical mean height) and *Sz* (Maximum height), were found as shown in Fig. 7 and 8. The error bars on each sample in Fig. 7 were relatively small, as BAC 1-1 was 0.0278 μm, CAC 1-1 was 0.0459 μm and EAC 1-1 was 0.0541 μm. However, the error bar on BAC 1-1 in Fig. 8 was 27.136 μm, which is a huge number compare with 1.718μm for CAC 1-1 and 0.429 μm for EAC 1-1. Therefore, it is almost certain that an error occurred during the inspection process.

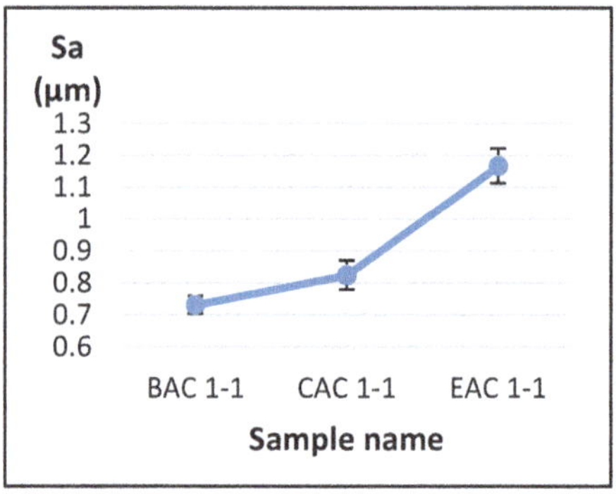

Fig. 7. The average *Sa* value with the error bar of each cusp on each sample.

These two figures represented the average value of six cusps on all three samples that were inspected. The *Sa* value gradually increased with feed rate in Fig. 7. This is potentially due to a fast feed rate creating a vibration on the tool during the cutting process. Also, the combination of a spindle speed at 12000 rpm and a feed rate of 1944 mm/min leads to insufficient cut and the effective depth of cut may have been excessive. The *Sz* value has suddenly dropped from the first sample to the second in Fig. 8. This phenomenon happened because there was a defect on the first cusp of the

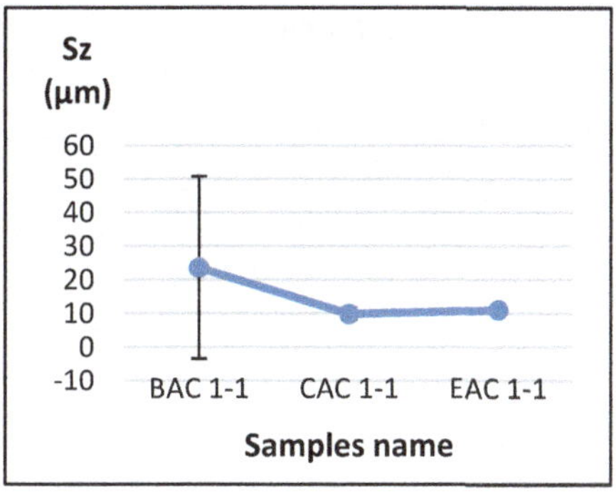

Fig. 8. The average *Sz* value with the error bar of each cusp on each sample.

sample and the *Sz* value is an extreme parameter to only pick up the maximum depth of the valley on the surface. It is not a good practice to ignore cusps that contained defects since those defects may be caused during the cutting process, rather than being post-cutting artefacts. Using the 3D-Editor function in the Alicona to eliminate the defect, the adjusted *Sz* values shown in Fig. 9 were obtained.

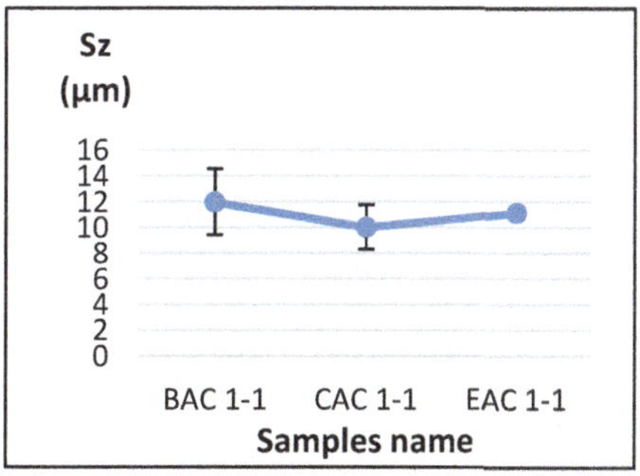

Fig. 9. The adjusted average *Sz* value.

The results show no significant changes in *Sz*. The error bar on BAC 1–1 was 2.565 µm, which is more realistic compared to the other data points. This is to be expected because the lead and tilt angle were not changed, and these are the main parameters affecting the height within the cusp.

Figure 10 shows the profile along feed direction one of the cusps in each sample (Blue – BAC 1-1, Green - CAC 1-1, Red – EAC 1-1). The sample with the slowest feed rate created a steadier line compared with others. This is potentially because the faster feed rate generates more vibration and tool run out during the cutting process. This also explains why the Sa value increased with the feed rate.

On the EAC 1-1 sample in Fig. 10, 25 peaks on the profile can be identified. This result matches the theoretical result of 25.67 marks over a 4 mm cutting path. However, the resolution of this setup could not pick up the peak marks on BAC 1-1 and CAC 1-1 since both samples had a slower feed rate.

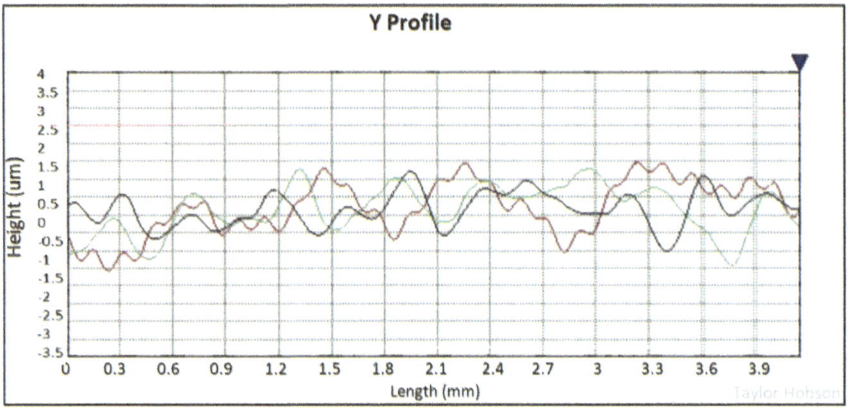

Fig. 10. 2D profile in feed direction of all samples (Blue – BAC 1-1, Green - CAC 1-1, Red – EAC 1-1).

The wavelet filtration function in SurfStand appears to effectively remove the cusp feature. In this example, it removed 4 cusps as shown in Fig. 11. However, the surface roughness parameter (*Sa*, *Sz*) based on this surface does not match the result generate in the Alicona using the cylindrical form removal. As discussed above, this is due to this process eliminating other important features from the surface.

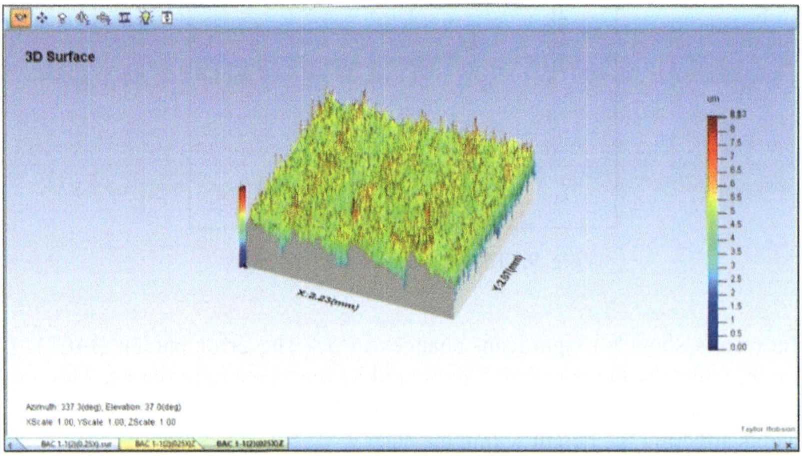

Fig. 11. The wavelet result in the SurfStand.

The result from using the polynomial fitting technique in SurfStand to remove a single cusp feature is shown in Fig. 12. This result was relatively close to the result from the Alicona in terms of the surface roughness parameters and the overall image. However, noise appeared on the surface which affected the *Sz* value.

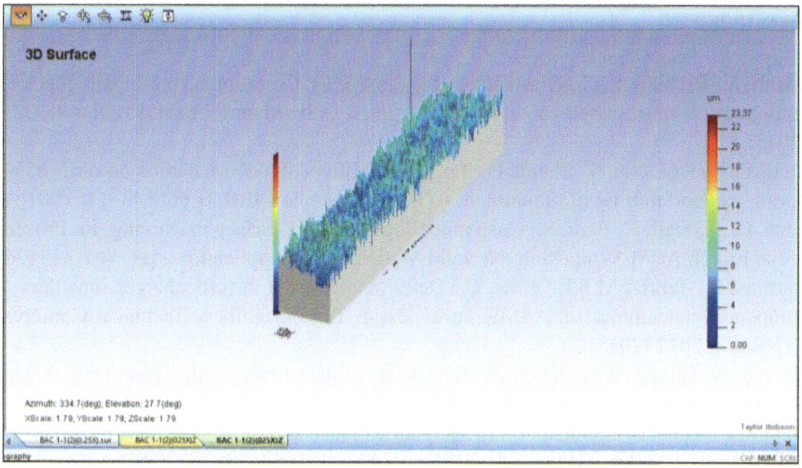

Fig. 12. Surface analysis of a single cusp using polynomial fitting in SurfStand.

5 Conclusion

- Feed rate affects the feed marks at the meso scale surface. At faster feed rates clearer feed marks appear on the surface, which also affects the surface roughness.
- Tool marks appear between the feed marks at the micro scale. These may be due to tool wear and aluminium oxide adhering to the tool edge.
- The form removal function in the Alicona worked well in this study, leaving all the features on the surface except the cusp.
- The polynomial fitting function gave a similar result to the form removal function. However, since the data was extracted from the Alicona, loss of resolution and distortion of data occurs and leads to more noise appearing in the result.
- No method is currently available to remove the cusp form from the full area scanned (i.e. all 6 cusps). Hence, manual methods must be used to combine data from multiple cusps.

6 Future Work

This paper develops the methods for decomposing the surface to isolate the various feature scales. Future work will characterise those features at the different scales to correlate them with machining parameters and ultimately with surface 'function', which in this case will be fatigue performance. The result will also be used to populate an existing surface generation model of the 5-axis milling process, in order to predict the surface roughness and optimise cutting parameters before the actual cutting process occurs.

References

1. Hauth, S., Richterich, C., Glasmacher, L., Linsen, L.: Constant cusp toolpath generation in configuration space based on offset curves. Int. J. Adv. Manuf. Technol. **53**(1–4), 325–338 (2011)
2. Yang, P., Yao, C., Xie, S., Zhang, D., Tang, D.X.: Effect of tool orientation on surface integrity during ball end milling of titanium alloy TC17. Procedia CIRP 31 **56**, 143–148 (2016)
3. Gray, P.J., Ismail, F., Bedi, S.: Cusp modeling for 5-axis surface machining. In: Proceedings of the Eighth ACM Symposium on Solid Modeling and Applications, pp. 340–345 (2003)
4. Barrans, S., Tabriz, M.S.E., Ellis, C.: Determining stress in turbocharger impellers due to component machining. In: ASME Turbo Expo: Turbomachinery Technical Conference & Exposition 2017 (2017)
5. Chen, J.-S., Huang, Y.-K., Chen, M.-S.: A study of the surface scallop generating mechanism in the ball-end milling process. Int. J. Mach. Tools Manuf. **45**(9), 1077–1084 (2005)
6. Ozturk, E., Tunc, L.T., Budak, E.: Investigation of lead and tilt angle effects in 5-axis ball-end milling processes. Int. J. Mach. Tools Manuf. **49**(14), 1053–1062 (2009)
7. Buj-Corral, I., Vivancos-Calvet, J., Domínguez-Fernández, A.: Surface topography in ball-end milling processes as a function of feed per tooth and radial depth of cut. Int. J. Mach. Tools Manuf. **53**(1), 151–159 (2012)
8. Tabriz, M.S.E., Barrans, S.M., Bills, P.J.: Effect of machining parameters on surface roughness in Al 2618 alloy subject to multi-axis machining process using ball nose cutting tools. In: 2017 8th International Conference on Mechanical and Aerospace Engineering (ICMAE), pp. 16–20. IEEE (2017)
9. Wojciechowski, S., Maruda, R., Barrans, S., Nieslony, P., Krolczyk, G.: Optimisation of machining parameters during ball end milling of hardened steel with various surface inclinations. Measurement **111**, 18–28 (2017)
10. Seid Ahmed, Y., Paiva, J.M., Covelli, D., Veldhuis, S.C.: Investigation of coated cutting tool performance during machining of super duplex stainless steels through 3D wear evaluations. Coatings **7**(8), 127 (2017)
11. Walton, K., Blunt, L., Fleming, L.: The topographic development and areal parametric characterization of a stratified surface polished by mass finishing. Surf. Topogr. Metrol. Prop. **3**, 11 (2015)
12. Polishetty, A., Alabdullah, M., Littlefair, G.: Tool wear analysis due to machining in super austenitic stainless steel. In: MATEC Web of Conferences, vol. 95: EDP Sciences, p. 05006 (2017)
13. Leach, R.: Characterisation of areal surface texture. Springer (2013)
14. Muralikrishnan, B., Raja, J.: A Brief History of Filtering. Computational Surface and Roundness Metrology, pp. 7–11 (2009)
15. ISO, ISO 16610-21, Geometrical Product Specifications (GPS)–Filtration–Part 21: Linear Profile Filters: Gaussian Filters (2011)
16. ISO16610-21, Geometrical product specifications (GPS)—filtration—Part 21: linear profile filters: Gaussian filters, ed. (2011)
17. Podulka, P.: The effect of valley location in two-process surface topography analysis. Adv. Sci. Technol. Res. J. **12**(4), 6 (2018)
18. Whitehouse, D.J.: Handbook of Surface and Nanometrology, p. 86. CRC Press, Boca Raton (2010)
19. Chen, Q., Yang, S., Li, Z.: Surface roughness evaluation by using wavelets analysis. Precis. Eng. **23**(3), 209–212 (1999)

20. Blunt, L., Jiang, X.: Advanced Techniques for Assessment Surface Topography: Development of a Basis for 3D Surface Texture Standards "Surfstand," p. 18. Elsevier, Amsterdam (2003)
21. Fan, J., Gijbels, I.: Data-driven bandwidth selection in local polynomial fitting: variable bandwidth and spatial adaptation. J. Roy. Stat. Soc. Ser. B (Methodol.) **57**(2), 371–394 (1995)

Tool Wear of Carbide Cutting Inserts Coated with TiAlN and AlTiSiN in Finish Turning of Inconel 718

Joanna Lisowicz, Witold Habrat(✉), Krzysztof Krupa, and Mateusz Fiedeń

Faculty of Mechanical Engineering, Rzeszów University of Technology,
12 Powstancow Warszawy, 35-959 Rzeszow, Poland
witekhab@prz.edu.pl

Abstract. Inconel 718 due to its good mechanical properties is widely used in industry, however the same properties make them also hard-to-cut material. The choose of proper cutting tools is really important in machining heat-resistant super-alloys. In this study tool wear and its influence on the cutting force components and on the roughness were investigated in finish turning of Inconel 718 superalloy using two coated carbide cutting inserts. The coatings selected for the tests were TiAlN and AlTiSiN. Cutting insert with AlTiSiN coating showed better machining performance in terms of tool life and cutting forces. However the surface roughness was better after machining with the cutting insert coated with TiAlN. The significant increase of passive force with the increase of the tool wear factor was noted for both cutting inserts, however the greater increase was noted for the cutting insert with TiAlN coating.

Keywords: Inconel 718 · Coated carbide inserts · Finish turning

1 Introduction

Inconel 718 is a nickel-based superalloy. Its extraordinary thermal resistance, high shear strength, and the capability to retain its mechanical properties at temperatures over 700 °C make it widely used in many industries, for example in the aircraft and nuclear industry [1]. At the same time, the good properties of nickel-based alloys, make them also hard-to-cut material. The main problems occurring during the machining of nickel-based alloys generally are large cutting forces, short tool lives, poor chip shape/swarf disposal, high power consumption, and a limited metal removal rate [2]. Moreover, the tool wear is accelerated due to the poor thermal conductivity of nickel-based alloys, which causes the rise of temperature at the tool-workpiece interface during machining [3].

In recent years, much attention has been paid to the performance of ceramic cutting tools in the machining of heat-resistant superalloys. Nevertheless, the cemented carbide cutting tools are still frequently used for machining the nickel-based superalloys, especially for the finishing of the Inconel 718. Some works studied the effect of different cutting parameters on the tool wear, cutting forces, and surface roughness in machining

S. Hloch et al. (Eds.): ICMEM 2021, LNME, pp. 226–235, 2021.
https://doi.org/10.1007/978-3-030-71956-2_18

of Inconel 718 with cemented carbide cutting tools [4, 5]. Coated carbide cutting tools have been also used in experiments comparing the performance of different cooling conditions in finish turning of Inconel 718: Ezugwu and Bonney [6] compared the tool wear, cutting forces, and surface roughness under flood and high-pressure cooling conditions, Yazid et al. [7] studied the surface integrity different cutting parameters in dry and minimum quantity lubrication (MQL) conditions.

Devillez et al. [8] analyzed the tool wear mechanism and cutting forces in the orthogonal turning process of Inconel 718 in dry machining conditions. The performance of tools with different coatings was investigated and compared to the performance of uncoated cemented carbide cutting tool. The coatings used in this study were TiAlN, AlTiN, TiAlN+WC/C, TiAlN+MoST (MoS2+Ti). Based on the experimental results, it was concluded that the AlTiN is the best coating.

Ducros et al. [9] investigated the machining performance of multilayer CrN/TiN and TiN/AlTiN PVD coated cutting tools in turning of Inconel 718 superalloy. The best results were obtained for TiAlN - based coating. TiN/AlTiN coatings presented better resistance to sticking phenomenon and better resistance to welding phenomenon therefore longer tool life was obtained than for CrN/TiN coatings.

Thakur et al. [10] conducted an experiment of the high-speed turning of Inconel 718 with multicoated carbide inserts. Tungsten carbide tools coated with multilayered TiN/TiAlN (PVD) and TiN/Al2O3/TiCN (CVD) were used for the experiment. Better results were achieved for TiN/Al2O3/TiCN (CVD) coated tool – the cutting forces were lower around 7% to 9% for the low feed rate and the cutting temperature was lower around 8% to 10% comparing with PVD coated tool. Moreover, it was observed that the tool wear of CVD coated tool was lower compared to PVD coated tool.

Li et al. [11] investigated the tool life of various coated carbides and ceramics in rapid face-turning in dry conditions. Three coated carbide inserts were selected: TiN/TiCN/TiN CVD/PVD coated (grade KC732), TiAlN PVD coated (grade 7310) and Al2O3/TiC/TiCN CVD coated (grade KC935). The three kinds of sialon ceramics tools were also chosen with different geometries and cutting angles (grades: KY2000, KY2100, KY3000). The best results in high-speed turning of Inconel 718 were achieved for KY2000 with a negative rake angle and KY2100 of round type. PVD-coated carbide inserts KC7310 were found to be a better choice for cutting Inconel 718 than CVD-coated carbides KC935 and were recommended on account of their lower price.

Arunachalam et al. [12] carried out an experiment of turning age hardened Inconel 718. The surface finish components of surface integrity and the residual stresses have been investigated. TiAlN (PVD) and TiCN/Al2O3/TiN (CVD) coated carbide cutting tools were used for the tests. At the basis of the research, the guidelines for the selection of appropriate tool (nose radius, cutting edge preparation, insert shape) and cutting conditions (coolant) for machining heat resistant super alloys with coated carbide cutting tools were prepared. It was assumed that coated carbide cutting tool inserts of round shape, small nose radius, negative type with a chamfered cutting edge in wet cutting conditions will generate primarily compressive residual stresses.

Kamata and Obikawa [13] conducted an experiment of finish turning of Inconel 718 with three different types of coated carbide tools in minimum quantity lubrication conditions. TiAlN (PVD), TiN/AlN superlattice (PVD), and TiCN/Al2O3/TiN (CVD) were the three selected coatings for the tests. The best performance in MQL cutting conditions exhibited TiCN/Al2O3/TiN coating, while the second performance was achieved by TiN/AlN superlattice coating in MQL conditions. TiCN/Al2O3/TiN coatings attained the longest tool life in wet conditions, however, the surface finish was poor.

The aim of the study is to analyze the wear of carbide cutting inserts coated with TiAlN and AlTiSiN in finish turning of Inconel 718 superalloy and the impact of this wear on selected machinability indicators. The influence of tool wear on the components of the cutting force and surface roughness parameters was determined.

2 Materials and Methods

2.1 Workpiece Material

The workpiece material used in machining tests was Inconel 718 alloy. Heat-resistant superalloy based on nickel and chromium, has a complex chemical composition (Table 1). Inconel 718 has very good weldability, high strength, corrosion resistance and excellent creep resistance at temperatures up to 700 °C. However it is considered as a hard-to-cut material.

Table 1. The chemical composition of Inconel 718.

%	Al	B	C	Cr	Co	Cu	Fe	Mn	Mo	Ni	Nb	P	Si	S	Ti
Min.	0.2	0	0	17	0	0	17	0	2.8	50	4.75	0	0	0	0.65
Max.	0.8	0.006	0.08	21	1	0.3	17	0.35	3.3	55	5.5	0.015	0.35	0.015	1.15

2.2 Cutting Tools

During machining tests VCMT 160408 cutting inserts were used with two different coatings.

- VCMT 160408 SM IC907 – is an insert made from a cemented carbide. The grade with hard submicron substrate, coated with TiAlN by PVD (Physical Vapor Deposition) method. It is characterized with high heat resistance. It is suitable for turning heat resistant alloys, austenitic stainless steels and hard steels at low to medium cutting speeds.
- VCMT 160408 NSU AC 5015S – is an insert made of a cemented carbide, produced with new technologies. The use of Absotech™, a new technology of PVD coating, allows to obtain highly heat-resistant ultra multi-layered and thin-film AlTiSiN structure. The new sintering process enables to maintain hardness while improving toughness and achieving good resistance to notch and chipping. This grade is intended for finish turning of exotic alloys at high cutting speeds.

2.3 Test Stand

The machining tests were conducted on the NEF 600 lathe. The measurements of the components of cutting forces were performed with Kistler 9257B piezoelectric three dimensional dynamometer, mounted in the turret of the lathe with a VDI holder Fig. 1.

Fig. 1. Testing stand.

2.4 Cutting Conditions

The experiment was performed using cutting parameters that ensure normal abrasive wear without the risk of chipping or cracking of the blade. These parameters were established on the basis of previous studies. The cutting parameters were within the range commonly used in the aerospace industry (Table 2).

Table 2. Cutting parameters.

Cutting speed v_c [m/min]	Depth of cut a_p [mm]	Feedrate f [mm/rev]
80	0.25	0.1

The accepted value of the tool wear index $VB_{B\,max}$ was assumed as 0.3 mm. Flood cooling with 7–8% solution of Fuchs Ecocool Global 10 cutting fluid was applied.

3 Results

3.1 Tool Wear

Table 3 shows the comparison of the effects of tool wear for cutting inserts made of grade IC 907 and AC 5015S. The cutting insert of grade IC 907 worked 9 min, during

Table 3. Effects of tool wear tests for cutting insert IC 907 and AC 5015S.

	Time of work, [min]	Volume of removed material, [cm^3]
T01	9	18.1
T02	24.68	49

which the material with a total volume of 18.1 cm^3 was removed, however the cutting insert worked 24.68 min and the total volume of removed material was 49 cm^3.

Figure 3 shows the tool wear $VB_{B\,max}$ in time function for IC907 and AC 5015S cutting inserts. It can be seen that the increase of tool wear is significantly faster for the tool with cutting insert of grade IC 907 (Fig. 2).

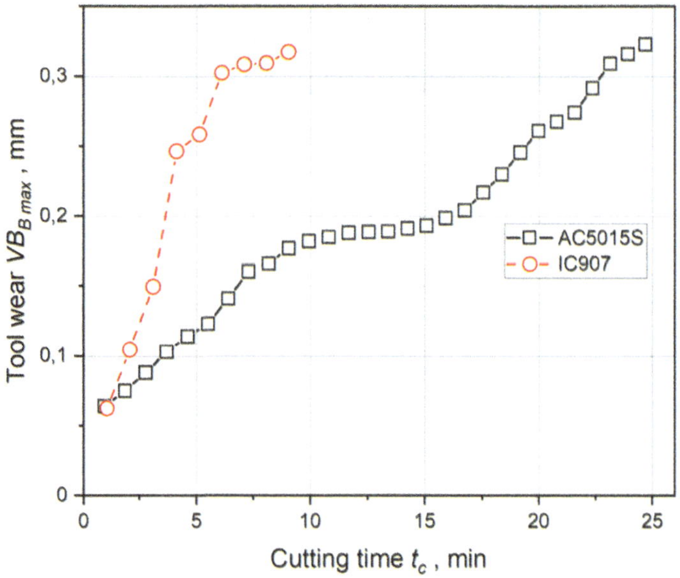

Fig. 2. Tool wear in time function for cutting inserts IC907 and AC 5015S.

3.2 The Influence of the Tool Wear on the Components of Cutting Force

Figure 3 shows the dependence of the main cutting force F_c, passive force F_p and feed force F_f of the tool wear for cutting insert IC 907. Figure 4 shows the photos of the tool at each point marked on the Fig. 3. The passive force is the dominant force, its value increased slightly with the increase of the tool wear. The rapid growth of the passive force at point 4 has been reflected in the photo by the appearance of the of clearly visible notch.

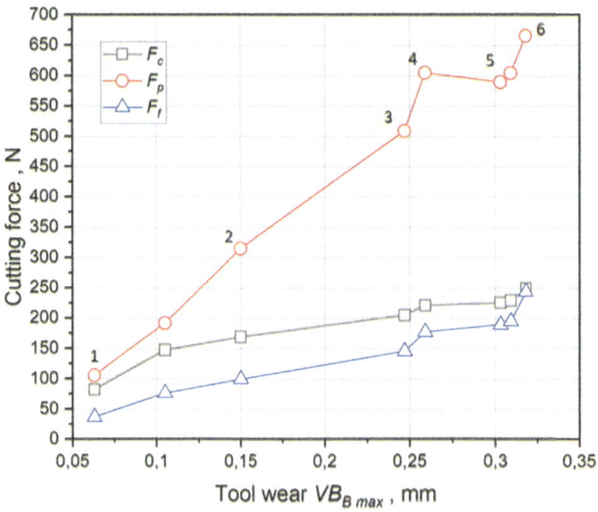

Fig. 3. Cutting force in the function of tool wear $VB_{B\,max}$ for cutting insert IC 907.

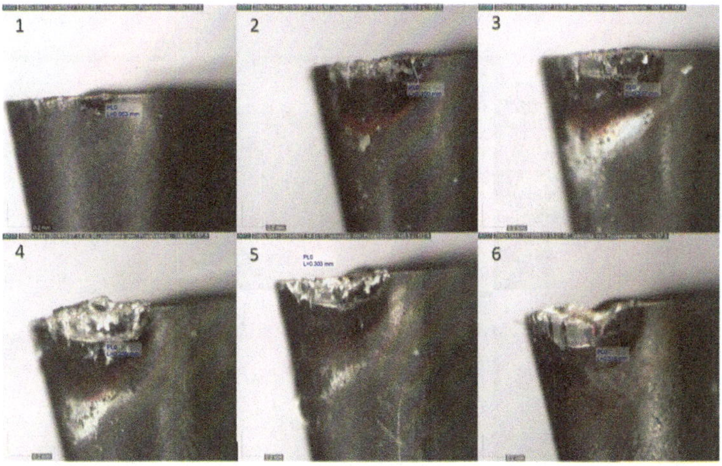

Fig. 4. Photos of the IC 907 cutting insert at different points of tool wear.

The dependence of the cutting force components of the tool wear $VB_{B\,max}$ for AC 5015S is presented in the Fig. 5. Similar to the IC 907 the passive force is the dominant force. The rapid growth of the cutting force components from point 2 to point 4 is caused by the appearance of the notch wear (Fig. 6).

It was noticed that at the beginning of the process the values of the components of cutting force were higher for the AC 5015S cutting insert. However the increase of the components of the cutting force for the IC 907 was faster and the maximum values were higher compared to the AC 5015S.

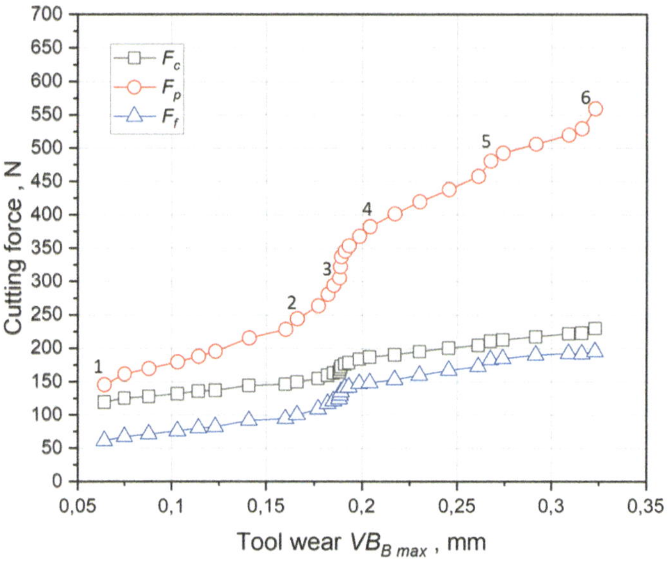

Fig. 5. Cutting force in the function of tool wear $VB_{B\,max}$ for cutting insert AC 5015S.

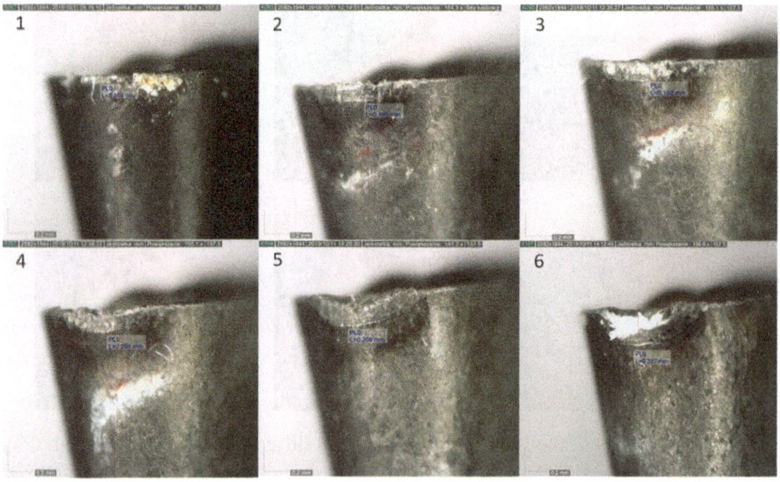

Fig. 6. Photos of the AC 5015S cutting insert at different points of tool wear.

3.3 The Influence of the Tool Wear on the Workpiece Surface Roughness

The *Ra* and *Rz* roughness parameters in the function of the tool wear $VB_{B\,max}$ are presented in the Fig. 7 and Fig. 8, respectively. For both parameters the higher values were obtained after machining with AC 5015S cutting insert. It was noticed that for IC 907 cutting insert the values of *Ra* and *Rz* remained approximately constant with the increase of tool wear till the value of 0.3 of $VB_{B\,max}$, when they started to increase. For the AC 5015S cutting insert the values of *Ra* and *Rz* parameters increased slightly at the very beginning and with the increase of the tool wear they decreased close to the values obtained after machining with IC 907 cutting insert.

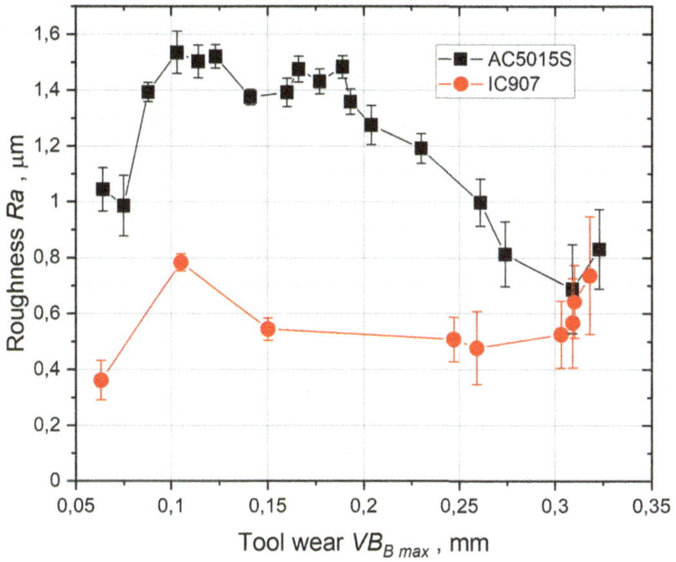

Fig. 7. *Ra* roughness parameter in the function of tool wear while machining Inconel 718 with IC 907 and AC 5015S cutting inserts.

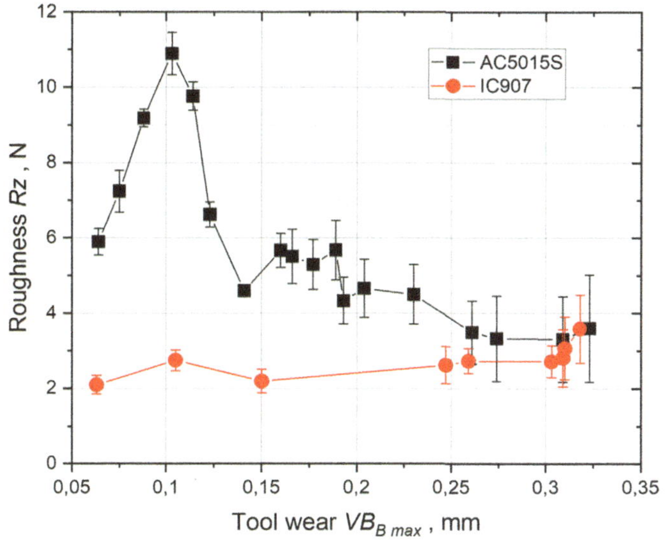

Fig. 8. *Rz* roughness parameter in the function of tool wear while machining Inconel 718 with IC 907 and AC 5015S cutting inserts.

4 Conclusions

The tool wear and its influence on the cutting force components and workpiece surface roughness has been studied in finish turning of Inconel 718 superalloy using cutting inserts with two different coatings: TiAlN and AlTiSiN. The following conclusions can be drawn:

- The longer tool life was attained for the cutting insert with AlTiSiN coating (tool T02). The total volume of removed material was over 2.7 times greater compared to the cutting insert with TiAlN coating.
- The dominant force was the passive force for both, the IC 907 and AC 5015S cutting inserts. The increase of the passive force with the increase of tool wear factor VB was faster for cutting insert with the TiAlN coating. For both cutting inserts, the rapid growth of the components pf cutting force has been noticed. This was caused by the notch wear appearance.
- The lower values of Ra and Rz roughness parameters were obtained for the IC 907 cutting insert.

In conclusion, it can be said that the AlTiSiN coating provides better wear resistance in comparison to IC907 grade in finish turning of Inconel 718, while the surface roughness is better after machining with IC 907 grade.

References

1. Altin, A., Nalbant, M., Taskesen, A.: The effects of cutting speed on tool wear and tool life when machining Inconel 718 with ceramic tools. Mater. Des. **28**, 2518–2522 (2007). https://doi.org/10.1016/j.matdes.2006.09.004
2. Ezugwu, E.O., Wang, Z.M., Okeke, C.I.: Tool life and surface integrity when machining Inconel 718 with PVD- and CVD-coated tools. Tribol. Trans. **42**, 353–360 (1999). https://doi.org/10.1080/10402009908982228
3. Ezugwu, E.O., Bonney, J.: Effect of high-pressure coolant supply when machining nickel-base, Inconel 718, alloy with coated carbide tools. J. Mater. Process. Technol. **153–154**, 1045–1050 (2004). https://doi.org/10.1016/j.jmatprotec.2004.04.329
4. Devillez, A., Schneider, F., Dominiak, S., Dudzinski, D., Larrouquere, D.: Cutting forces and wear in dry machining of Inconel 718 with coated carbide tools. Wear **262**, 931–942 (2007). https://doi.org/10.1016/j.wear.2006.10.009
5. Rahman, M., Seah, W.K.H., Teo, T.T.: The machinability of Inconel 718. J. Mater. Process. Technol. **63**, 199–204 (1997). https://doi.org/10.1016/S0924-0136(96)02624-6
6. Yazid, M.Z.A., Ibrahim, G.A., Said, A.Y.M., CheHaron, C.H., Ghani, J.A.: Surface integrity of Inconel 718 when finish turning with PVD coated carbide tool under MQL. Procedia Eng. **19**, 396–401 (2011). https://doi.org/10.1016/j.proeng.2011.11.131
7. Ezugwu, E.O., Bonney, J.: Finish machining of nickel-base Inconel 718 alloy with coated carbide tool under conventional and high-pressure coolant supplies. Tribol. Trans. **48**, 76–81 (2005). https://doi.org/10.1080/05698190590899958
8. Hao, Z.P., Fan, Y.H., Lin, J.Q., Yu, Z.X.: Wear characteristics and wear control method of PVD-coated carbide tool in turning Inconel 718. Int. J. Adv. Manuf. Technol. **78**, 1329–1336 (2015). https://doi.org/10.1007/s00170-014-6752-0

9. Ducros, C., Benevent, V., Sanchette, F.: Deposition, characterization and machining performance of multilayer PVD coatings on cemented carbide cutting tools. Surf. Coat. Technol. **163–164**, 681–688 (2003). https://doi.org/10.1016/S0257-8972(02)00656-4
10. Thakur, D.G., Ramamoorthy, B., Vijayaraghavan, L.: Some investigations on high speed dry machining of aerospace material Inconel 718 using multicoated carbide inserts. Mater. Manuf. Process. **27**, 1066–1072 (2012). https://doi.org/10.1080/10426914.2011.654158
11. Li, L., He, N., Wang, M., Wang, Z.G.: High speed cutting of Inconel 718 with coated carbide and ceramic inserts. J. Mater. Process. Technol. **129**, 127–130 (2002). https://doi.org/10.1016/S0924-0136(02)00590-3
12. Arunachalam, R.M., Mannan, M.A., Spowage, A.C.: Surface integrity when machining age hardened Inconel 718 with coated carbide cutting tools. Int. J. Mach. Tools Manuf. **44**, 1481–1491 (2004). https://doi.org/10.1016/j.ijmachtools.2004.05.005
13. Kamata, Y., Obikawa, T.: High speed MQL finish-turning of Inconel 718 with different coated tools. J. Mater. Process. Technol. **192–193**, 281–286 (2007). https://doi.org/10.1016/j.jmatprotec.2007.04.052

The Effect of Magnetic Assisted Ball Burnishing on the Corrosion Resistance of C45 Steel

Zsolt Ferenc Kovács[1,3(✉)], Zsolt János Viharos[1,2], and János Kodácsy[1]

[1] GAMF Faculty of Engineering and Computer Science, Department of Vehicle and Faculty of Economics and Business, Department of Economics and Law, John von Neumann University, Kecskemét 6000, Hungary
kovacs.zsolt@gamf.uni-neumann.hu

[2] Research Laboratory on Engineering and Management Intelligence, Intelligent Processes Research Group, Institute for Computer Science and Control, Budapest 1111, Hungary

[3] Department of Manufacturing Science and Technology, Budapest University of Technology and Economics, Budapest 1111, Hungary

Abstract. The development of modern science requires developing more economical, more environmentally friendly and more efficient procedures in the area of machine parts production as well. Corrosion, as a natural process, shortens the lifetime of our structures and machines. Although there are corrosion-resistant materials and various ways of protections against corrosion, all of them result in more expensive and more complex solutions. The increase in the corrosion resistance of metals can be achieved not only by applying surface protective coatings, but also by the usage of more economical and more efficient manufacturing processes. The developed, novel technology - Magnetic Assisted Ball Burnishing (MABB) - helps to increase the hardness of the surface and to reduce the surface roughness in the same time. This article investigates the effects of MABB machined workpieces against corrosion. The corrosion rate and the related technological parameters were optimized by the well-known Taguchi method. Corrosion tests were performed with a BioLogic SP-150 potentiostat in a conventional three-electrode corrosion cell. The recorded Tafel curves of the different specimens enable the calculation of the corrosion rate and time. The results prove that the corrosion resistance can be slightly increased by MABB machining, but it depends on the tool feed rate and the pre-machined surface roughness. The machining speed and the burnishing strategies show significant effects on the corrosion rate as well. The introduced MABB tool enables the reduction of the costs of further working-surface postprocessing, the increase in the hardness and lifetime of the treated surfaces, while surface roughness can be significantly reduced.

Keywords: Ball burnishing · Edge rounding · Magnetic assisted machining

Nomenclature

A	Corroded surface (cm^2)
CR	Corrosion rate (mm/year)
fi	Mass of the i-th element of the alloy

S. Hloch et al. (Eds.): ICMEM 2021, LNME, pp. 236–248, 2021.
https://doi.org/10.1007/978-3-030-71956-2_19

h	Gap between the tool and workpiece (mm)
pre. Ra	Pre-machined *Ra* roughness (μm)
i_{cor}	Corrosion current density (yA/cm^2)
I_{cor}	Total anodic current (yA)
n I	Number of electrons required to oxidize an atom in the corrosion process
ni	Valence of the *i*-th element of the alloy
v_b	Burnishing speed (m·min^{-1})
v_f	Burnishing feed (mm·min^{-1})
W	Atomic mass of the element
Wi	Atomic mass of the *i*-th element of the alloy
ρ	Density of materiel (g/cm^3).

1 Introduction

Numerous attempts and solutions have been made to slow down and prevent corrosion, which are able to protect the equipment exposed to corrosion. These methods are usually based on the application of coating or protective layers. However, up-to-date engineering is looking for those processes where, instead of the subsequent application of surface protection materials, the manufacturing technology itself can provide the solution for corrosion protection, such a result can be achieved by methods of causing surface plastic deformation.

In the paper, application of Magnetic Assisted Ball Burnishing (MABB) tool is examined, which machining is not yet widespread in production, and which, based on the data so far, provides an extremely promising method in corrosion protection as well beyond other positive effect on the machine workpiece.

According to the ISO 8044 standard, corrosion is a physico-chemical interaction between the metal and its environment, as a result of which the properties of the metal change and often the functional characteristics of the metal, the environment and the technical system consisting of them deteriorate [1].

According to the reaction procedure, the corrosion processes can be classified into the following two groups [2, 3]:

- chemical corrosion and
- electrochemical corrosion.

It is a generally accepted fact in the engineering sciences that surface roughness values are closely related to corrosion resistance [4–6].

With surface burnishing, during machining the tool presses down on the material surface, so that the surface layer start to flow and so the roughness peaks begin to fill-in the valleys, as shown in Fig. 1.

As the surface layer deforms plastically, it causes hardness increases, too. The principle of magnetic assisted ball burnishing is that the tool travels along the surface of the workpiece at a given speed in a straight line or in another given path. The MABB tool contains a neodymium permanent magnet, it can be clamped easily in a CNC

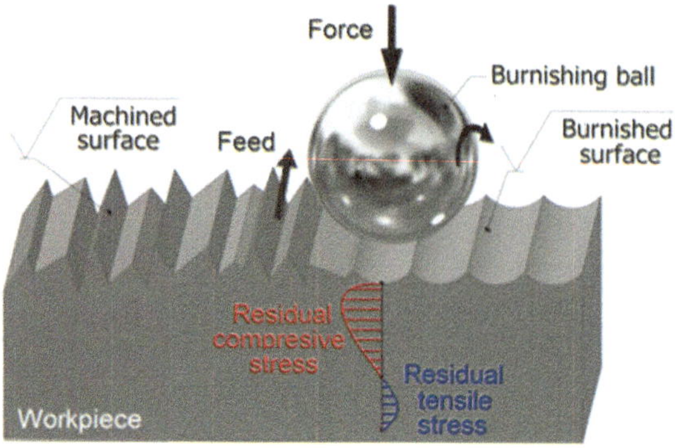

Fig. 1. Principles of ball burnishing process.

milling machine, so that the finishing operation of the workpiece can be performed in the milling machine immediately after machining, which saves significant time and money. Its construction is simple (see in Fig. 2) and it does not require any special maintenance.

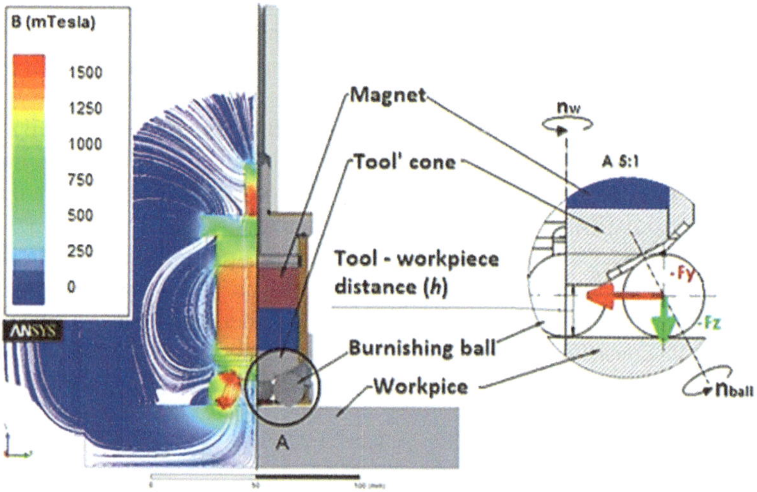

Fig. 2. Principles and parts of the MABB tool.

Honess et al. [4] published on surface quality and corrosion resistance of stainless steels. According to their results, if the average roughness is greater than 1 μm, the grooves and pits deepness of the surface can hold enough chloride ions, which leads to corrosion. If the average roughness on the surface can be reduced below 0.5 μm, the chance that the above-mentioned chloride ions are significantly reduced in the pits, so, this phenomenon is almost eliminated [4].

Asma et al. [5] investigated the relationship between surface roughness values and corrosion resistance capabilities on C22E steel in a carbon dioxide environment. During their experiment, the specimens were finished with silicon carbide abrasive papers of different particle sizes. It has been shown that specimens machined with increasingly finer abrasive papers showed improved corrosion resistance capabilities, however, the differences in corrosion are very small and are therefore not considered significant [5].

In Hryniewicz and Rokosz's [9] analyzed the relationship between surface roughness and corrosion resistance for C45 type steel. The surface of the specimens was sanded and polished with different grain sized sandpapers. The prepared samples were tested in 3% and 0.03% sodium chlorides. The difference between the coarsest 180 sandpaper, and the finer 4000 sandpaper was nearly three times greater for the corrosion resistance in favor of the finer one, after the treatment with the 3% chlorides. Also, with 0.03% chloride, a significant, 1.6-fold improvement was achieved with the finer abrasive one [6].

The previously presented studies and experiments concerned surfaces handled by abrasive machining. Burnishing process significantly differs from these. A fundamental difference is that no chips are separated during rolling, so that the grain structure of the material is not damaged as during abrasive machining.

Salahshoor [7] investigated the possibility of increasing the corrosion resistance of cylindrical MgCa0.8 alloy by ball burnishing. This study is so important because this alloy, which has very good properties among implants, corrodes very quickly in the human body, so the current usability of the material is economically unsatisfactory and also exposes the patient further surgery. In the test, the author made three flat milled and three burnished specimens with different rolling forces at different cutting speeds. The corrosion resistance of the specimens was compared by Hank's balanced salt solution test. Without wishing to be exhaustive, its results were as follows [7]:

- much less and more uniform corrosion products were produced on the rolled samples than on the machined samples,
- as the burnishing force increased, more and more corrosion occurred on the specimens.

Saldaña-Robles et al. [8] studied the effect of rolling on surface roughness, surface hardness, and corrosion resistance. The material in each case was C45 cylindrical steel, the surfaces were pre-machined by turning. Corrosion resistances were compared on a turned and two rolled specimens, one of them had the lowest average surface roughness and the other one had the highest hardness. Based on their conclusions, the corrosion resistance of the rolled specimen was higher than at the turned one [8].

Hryniewicz and Rokosz [9] conducted a similar experiment when investigated the effect of roller burnishing of C45 steel specimens on corrosion resistance. They compared the rolled pieces with the polished ones. Roller burnishing not only resulted a significantly improved surface roughness, but it significantly improved the corrosion resistance as well. The polished samples were paired with rolled samples, ensuring that the surface roughness are similar. By comparison, it can be stated that the corrosion resistance was much better on rolled surfaces than on the polished one [9].

Based on the analysis of the state-of-the-art publications, it can be stated that the conventional burnishing has significant effect on corrosion behaviour. However, there are scientific result only for cylindrical workpieces, so that, it is unknown, that in the case of flat surfaces one may had a similar positive surface improvement. In the case, if it has positive effect, another important question arises: what extent can the MABB improve it? So, the main aim of this paper is to get the exact answer for these assumptions and challenges.

2 Description of the Experimental Work

The aim is not to create or develop a corrosion-resistant material, but to test the corrosion resistance of the machined workpieces with different burnishing strategies and machining conditions. The selected material was C45, since this is a general purpose, non-alloy material.

The dimensions of the specimens were: 200 mm long, 100 mm wide and 15 mm high. The experiments were performed on the basis of an experimental design by Taguchi [11] with different machining parameters and procedures (a total of 9 rolled plus 3 pre-machined samples). Each sample was pre-machined by milling or grinding. For face milling SECO APMX160408TR-M14 MP2500 inserts were used, while the grinding was performed on a surface grinder. The goal was to produce surfaces with different roughnesses.

After the preparatory processes, for the ball burnishing phase a CNC milling machine was suitable. An experimental design method (DoE) according to Taguchi was used to define the burnishing tests. The DoE table was designed by the MiniTab18 software, which consists of 3 levels and 4 factors. Table 1 shows the change in each factor, where the variables were:

A: Pre-machined roughness, *pre. Ra* (μm): ~0.8; ~1.6; ~2.4.
B: Feed, v_f (mm/min): 10; 30; 50.
C: Burnishing speed, v_b (m/min): 20; 50; 80.
D: Rate of burnishing, (%): 100; 150; 200.

Table 1. Design of Experiments (DoE) table for MABB corrosion test.

Levels	Factors			
	A – *pre. Ra* (μm)	B - v_f (mm/min)	C - *vb* (m/min)	D (%)
1	0.8	10	20	100
2	0.8	30	50	150
3	0.8	50	80	200
4	1.6	10	50	200
5	1.6	30	80	100
6	1.6	50	20	150
7	2.4	10	80	150
8	2.4	30	20	200
9	2.4	50	50	100

The rate of rolling means the width of the tool path according to the width of the surface to be rolled by the tool (path diameter is approximately 47 mm). Thus, 100% means the straight-, 150% the small-, and 200% the large-cycloid path. The h distance between the workpiece and the tool was constants, its value was 10 mm [10].

Traces of each rolling strategy are clearly visible on the burnished specimens, as shows the Fig. 3.

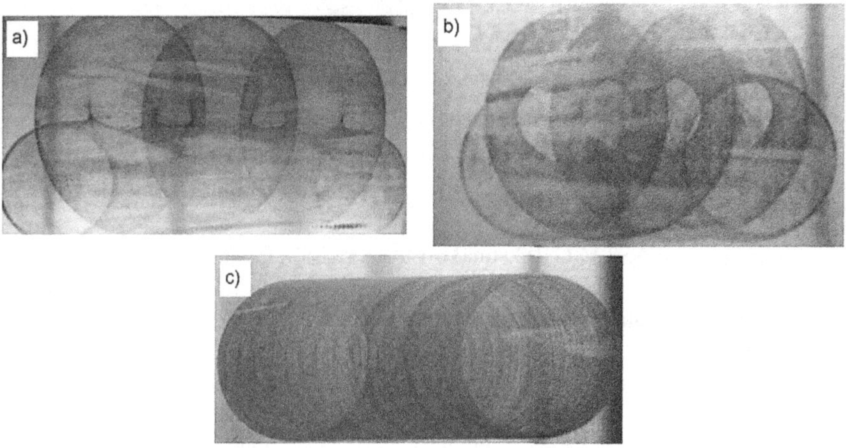

Fig. 3. Traces of burnishing strategies on the samples, a) Large-cycloid path (D-3 = 200%), b) Small-cycloid path (D-2 = 150%) and c) Straight path (D-1 = 100%).

3 Results of the Burnished Surface Analysis

The corrosion tests on the machined surfaces were performed by a BioLogic SP-150 potentiostat and ECLab software, furthermore, a Zeiss EVO MA10 scanning electron microscope was also used to evaluated the corrosion behavior on the surface; during the related scanning the voltage was 20 kV. Roughness was measured on a MITUTOYO Formtracer SV-C3000 roughness tester.

3.1 Results of Roughness Analysis

The *Ra* average surface roughness was measured before and after ball burnishing. Based on the burnished *Ra* average roughness results, the bar graph shows *Ra* values in Fig. 4, whereat blue color marks the burnished surfaces having originally $Ra = 0.8$ μm, green color marks the originally $Ra = 1.6$ μm and yellow ones the originally $Ra = 2.4$ μm surface roughness values. The Fig. 4 clearly represents that the *Ra* roughness values were significantly reduced compared to the original milled values.

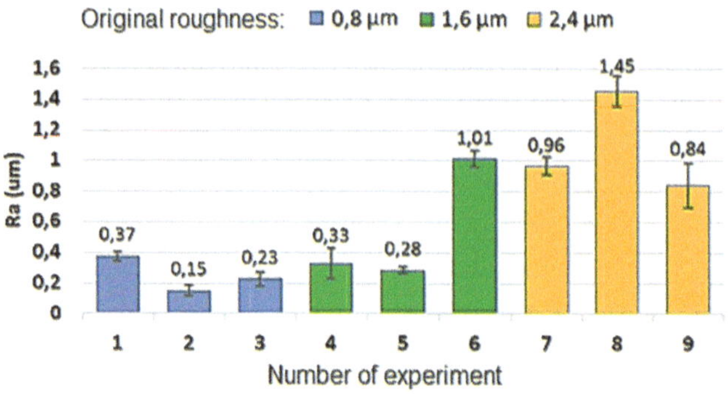

Fig. 4. Results of surface roughness measurement after ball burnishing.

3.2 Results of Corrosion Analysis

The guidelines of the ASTM G102 standard for corrosion testing were used for evaluation. Corrosion values can be measured using the galvanic cell and polarization processes, including Tafel extrapolation or polarization resistance measurements. The first step is to convert the measured or estimated current to a special current density. This can be obtained as the quotient of the total current to the geometric surface size of the electrode dipped in the solution, with the assumption that the current is evenly distributed over the aforementioned area. In the case of galvanic pairs, the immersed surface of the anode should be used. The related calculation is (1):

$$i_{cor} = \frac{I_{cor}}{A} \tag{1}$$

where: i_{cor} is the corrosion current density (yA/cm^2), I_{cor} is the total anodic current (yA) and A is the corroded surface (cm^2).

Equivalent mass has to be calculated for clean cells (2):

$$EW = \frac{W}{n} \tag{2}$$

where: W is the atomic mass of the element and n is the number of electrons required to oxidize an atom in the corrosion process, which is the valence of the element.

Relationship to equivalent mass to be used for alloys (3):

$$Q = \sum \frac{ni \cdot fi}{Wi} \tag{3}$$

where: fi is the mass of the i-th element of the alloy, Wi is the atomic mass of the i-th element of the alloy and ni = valence of the i-th element of the alloy.

Thus, the equivalent mass is a reciprocal (4):

$$EW = \frac{1}{\sum \frac{ni \cdot fi}{Wi}} \tag{4}$$

Faraday's law can be used to calculate the corrosion rate *CR* (mm/year) by (5):

$$CR = K_1 \frac{i_{cor}}{\rho} EW \tag{5}$$

where: $K_1 = 3.27 \cdot 10^{-3}$ (mm g/μA cm year) and ρ the density of materiel (g/cm^3).

Table 2 shows the calculated equilibrium mass, the calculated density, the corrosion current obtained from the evaluation, and Fig. 5 shows the measured exact surface area after corrosion from the stereomicroscopic images.

Table 2. Corrosion rate values after one hour - measurement.

Number of experiments	*Ew* (-)	ρ (g/cm^3)	I_{cor} (μA)	*A* (mm^2)	i_{cor} (μA/cm^2)	*CR* (mm/year)
1	19.293	7.462	6977.533	239.829	29.094	0.254
2			7270.618	207.698	35.006	0.267
3			7647.384	194.627	39.293	0.236
4			3914.594	194.164	20.161	0.246
5			6348.125	179.147	35.435	0.296
6			6731.190	196.744	34.213	0.332
7			4534.413	217.317	20.865	0.170
8			6986.236	235.616	29.651	0.300
9			5830.740	194.261	30.015	0.289

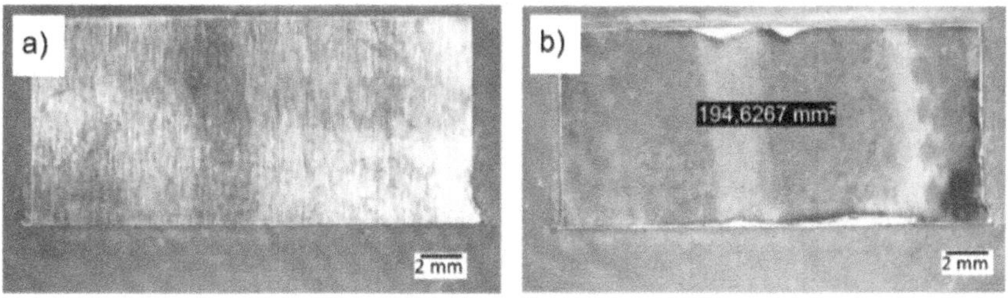

Fig. 5. Surface of sample 3 with stereomicroscope measurement, a) before corrosion and b) after corrosion.

The corrosion current can be obtained from the Tafel curves (Fig. 6 according to the ASTM G102 standard).

It was measured after a soaking time of 1 h in electrolyte. The measurement was performed at a scanning speed of 90 mV/min, so one measurement takes 10 min and 33 s. A tangent shall be fitted to the straight part of the descending and rising stems of the curve and the point intersected by them shall give the corrosion current.

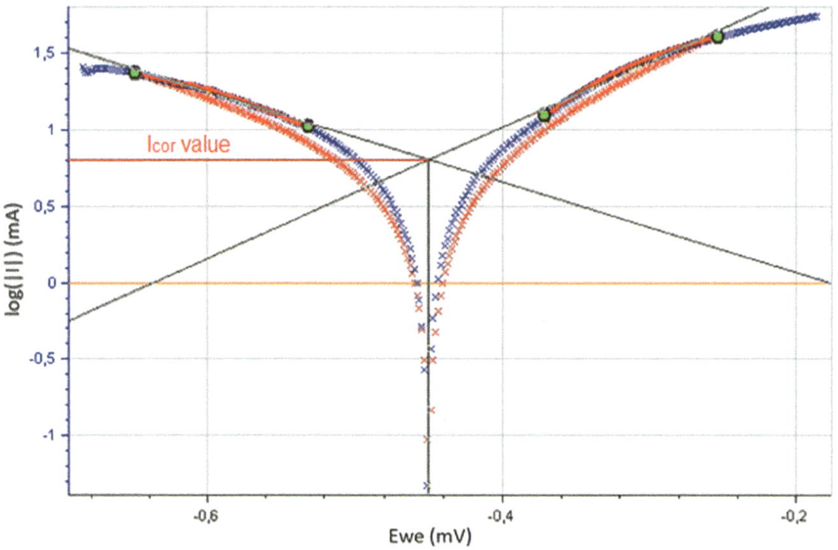

Fig. 6. Tafel curve of sample 5 after 1 h of soaking.

3.3 Roughness and Corrosion Common Effects

Based on the tests performed, results can be obtained in which the corrosion rate values improve as a result of burnishing. The corrosion rates are shown in Fig. 7. (after a hour measurement).

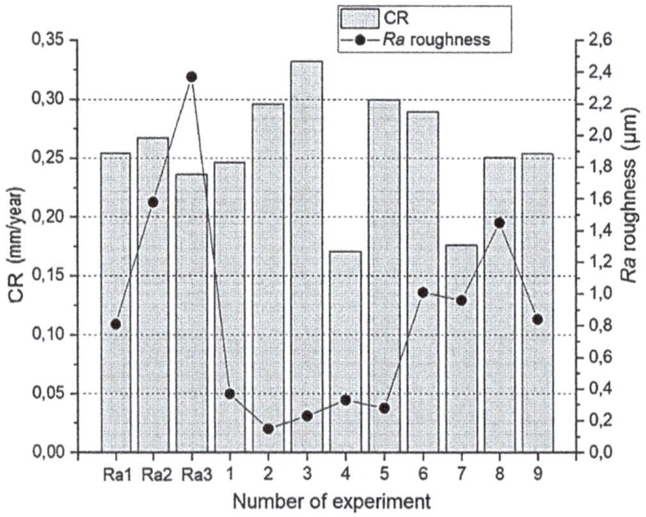

Fig. 7. Relationship between the corrosion rate and *Ra* surface roughness after MABB.

According to Fig. 7, comparing the unburnished surfaces (Ral, Ra2 and Ra3) to sample 1 (*pre. Ra* = 0.8 μm; v_f = 10 mm/min; v_b = 20 m/min) when the rate of burnishing was 100%, the MABB had lower effect, while in the cases of sample 4 (*pre. Ra* = 1.6 μm; v_f = 10 mm/min; v_b = 50 m/min; rate of burnishing 150%) and sample 7 (*pre. Ra* = 2.4 μm; v_f = 10 mm/min; v_b = 80 m/min; rate of burnishing 150%) they succeed in a higher corrosion resistance. Furthermore, it can be seen that the sample 2 (*pre. Ra* = 0.8 μm; v_f = 30 mm/min; v_b = 50 m/min; rate of burnishing 150%) and sample 3 (*pre. Ra* = 0.8 μm; v_f = 50 mm/min; v_b = 50 m/min; rate of burnishing 200%) there were a significant deterioration in the corrosion resistance. What these samples have in common is that they were made by grinding. During grinding, higher amount of heat was given to the surface, tensile stress was generated from it, and possible grinding microcracks may have contributed to the experienced deterioration.

3.4 Technology Set-Up for the Highest Corrosion Resistance

To obtain an accumulated result on the effects of each factors, the $(S/N)_S$ values have be calculated according to the Taguchi method [11] e.g. in Minitab 17 software. The "Smaller is better" analysis was chosen because the goal was to achieve the lowest possible corrosion resistance value. The effects of each factor on the corrosion rate are shown in Fig. 8.

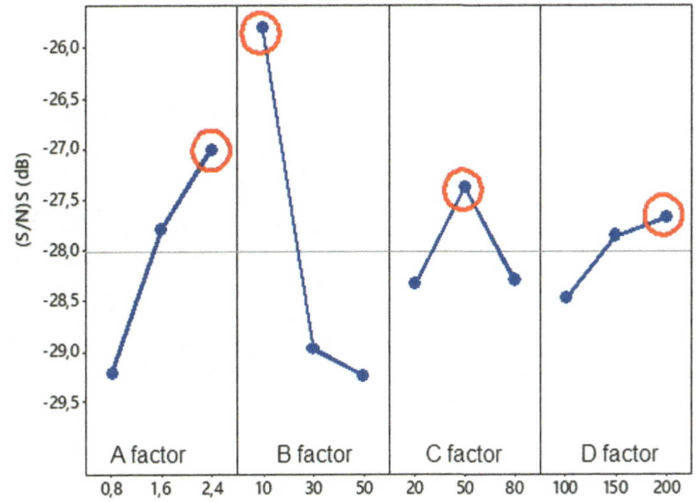

Fig. 8. Effect of burnishing factors on corrosion rate, A factor: Pre-machined roughness, B factor: Burnishing feed, C factor: Burnishing speed, D factor: Rate of burnishing: Rate of burnishing.

Based on Fig. 7 and 8, it can be identified that surfaces when higher MABB burnishing feed was applied (3; 6 and 9) show even higher corrosion rate compared to the pre-machined surfaces, while those which were machined by lower burnishing feed, resulted in lower corrosion rates. Thus, the lower the feed is, the better are the corrosion properties of the resulted workpiece. As shown in Fig. 8, the pre-machined surface roughness has the most significant effect followed by the burnishing feed on the corrosion. All this means that the corrosion resistance of surfaces depends not only on the low ball burnished surface roughness, but also on the microgeometric structure of the surface, which is affected by the low feed.

The highest level in the corrosion resistance can be achieved by (A3-B1-C2-D3), namely pre-machined roughness: 2.4 μm, feed rate of 10 mm/min, rolling speed of 50 m/min and with 200% high cyclic path.

4 Conclusions

Numerous attempts and solutions have been made to slow down and prevent corrosion, which are able to protect the equipment exposed to corrosion. In the paper, application of Magnetic Assisted Ball Burnishing (MABB) tool is examined, which machining is not yet widespread in production, and which, based on the data so far, provides an extremely promising method in corrosion protection as well beyond other positive effect on the machine workpiece.

The experimental results mirrored that there was less corrosion on the burnished surfaces and also the overall corrosion had much more even, and so, better distribution on the surface. Furthermore, looking at the less MABB burnished sample 1 and the highly MABB burnished sample 7, one might see that large parts of sample 7 stayed still resistant to corrosion, while uniform corrosion was observed over the entire surface of sample 1 (Fig. 9 and Fig. 10).

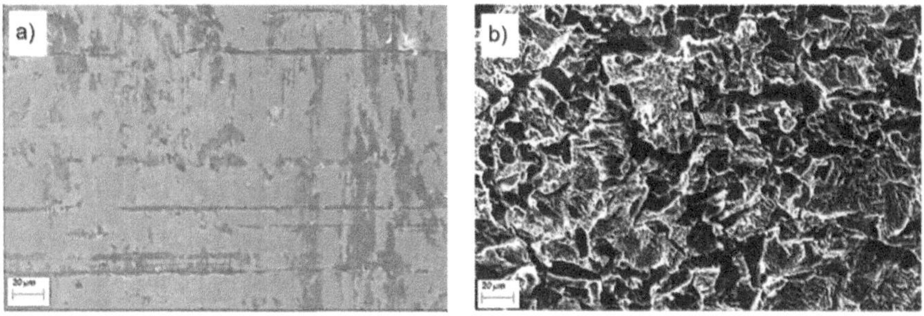

Fig. 9. SEM images of samples 1 (less MABB burnished) a) before and b) after corrosion test (N = 1000×).

The surface corrosion resistance of non-alloy structural steels (e.g. C45) can be improved by the introduced magnetic assisted ball burnishing:

- In terms of increasing the corrosion resistance, the burnishing feed has the greatest effect, so the lower burnishing feed can produce higher corrosion resistance surface and
- the studied cycloid burnishing strategy has also positive effect on corrosion resistance.

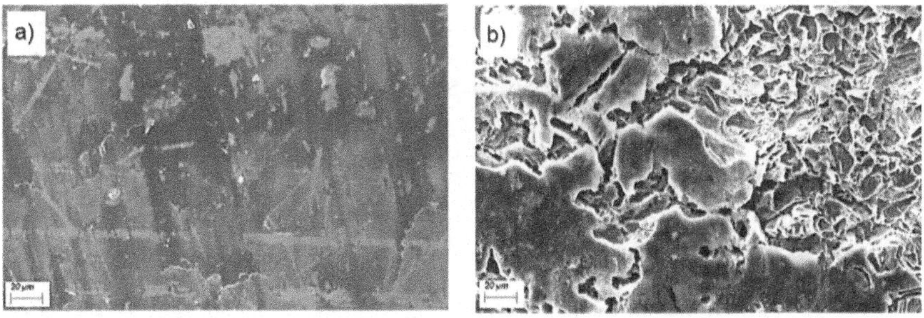

Fig. 10. SEM images of samples 7 (highly MABB burnished) a) before and b) after corrosion test (N = 1000×).

From these statements it can be concluded that the degree of burnishing (number of rolls) has an important effect on the corrosion resistance of the surface. Furthermore, as a result, thanks to the MABB burnishing, for example surfaces of workpiece parts can be exposed to moisture or to a corrosive medium for a short time because MABB prevents the starting of the corrosion.

Overall, MABB technology can, under certain conditions, be an effective tool in short term for corrosion protection of surfaces.

Acknowledgement. This research is party supported by EFOP-3.6.1-16-2016-00006 "The development and enhancement of the research potential at John von Neumann University" project. The Project is supported by the Hungarian Government and co-financed by the European Social Fund.

The research in this paper was partly supported by the Hungarian ED_18-2-2018-0006 grant on a "Research on prime exploitation of the potential provided by the industrial digitalization".

References

1. ISO 8044:2015: Corrosion of metals and alloys—Basic terms and definitions
2. Alcantara, A.S., Fábián, E.R., Furkó, M., Fazakas, É., Dobránszky, J., Berecz, T.: Corrosion resistance of TIG welded joints of stainless steels. In: Materials Science Forum, vol. 885, pp. 190–195 (2017)
3. Abbas, A.S., Fazakas, É., Török, T.I.: Corrosion studies of steel rebar samples in neutral sodium chloride solution also in the presence of a bio-based (green) inhibitor. Int. J. Corros. Scale Inhib. **7**(1), 38–47 (2018)
4. Hones, C., Harison, A.: Importance of surface finish in the design of stainless steel. Engineering Utilities, pp. 14–15 (2006)
5. Nor Asma, R.B.A., Yuli, P.A., Mokhtar, C.I.: Study on the effect of surface finish on corrosion of carbon steel in CO2 environment. J. Appl. Sci. **11**, 2053–2057 (2011)
6. Hryniewicz, T., Rokosz, K.: Corrosion behaviour of C45 carbon steel after mechanical surface finishing. In: Proceedings of 13th International Metallurgical & Materials Conference, Metal, pp. 18–20 (2004)
7. Salahshoor, M., Guo, Y.B., Li, C.: Surface integrity and corrosion performance of biomedical magnesium-calcium alloy processed by hybrid dry cutting-finish burnishing. Procedia Manuf. **10**, 467–477 (2017)

8. Saldaña-Robles, A., Plascencia-Mora, H., Aguilera-Gómez, E., Saldaña-Robles, A., Marquez-Herrera, A., Diosdado-De La Peña, J.A.: Influence of ball-burnishing on roughness, hardness and corrosion resistance of AISI 1045 Steel. Surf. Coat. Technol. **339**, 191–198 (2018)
9. Hryniewicz, T., Rokosz, K.: Corrosion behaviour of C45 carbon steel after roller burnishing. J. Hradec nad Moravici **5**, 24–26 (2005)
10. Kovács, Zs.F., Viharos, Zs.J., Kodácsy, J.: Determination of the working gap and optimal machining parameters for magnetic assisted ball burnishing. Measurement **118**, 172–180 (2018)
11. Roy, R.K.: A Primer on the Taguchi Method. Van Nostrand Reinhold, New York (1993)

Characterization of Spheroidized Alumina Powders for Ceramic 3D Printing Application

Ram Krishna[1(✉)], Mohan Sreenivasan[2], Mykola Lyutyk[3], and Mykola Barosh[3]

[1] Madanapalle Institute of Technology and Science, Madanapalle 517325, AP, India
krishnamme@gmail.com
[2] Pace Institute of Technology and Sciences, Ongole 523272, AP, India
[3] Klakona System LLC, Kyiv, Ukraine

Abstract. Alumina powders with requisite morphology and properties are required for the application in ceramic 3D printing. In this research, the Radio Frequency Plasma Spheroidization technique has been used for obtaining the requisite alumina powders. In this article, the alumina powders have been investigated for morphology, crystallinity, chemical analysis using different techniques such as scanning electron microscopy, energy dispersive X-ray and fluorescence spectroscopy, and X-ray diffraction. The experimental results showed that the obtained alumina powder particles have been almost globular ranging from 2 μm to 100 μm in size and observed as 100% sphericity. The obtained alumina powders are 98.9% ultra-pure. The morphology of the obtained powder particles' surfaces shows a rugged surface and cellular structure. The results have shown the presence of only well-crystallised α-Al_2O_3 phase in the powders. Due to excellent physical and mechanical properties, ultra-pure powders are widely preferred for ceramic 3D printing and related promising applications.

Keywords: Radio frequency plasma spheroidization · 3D Printing Spheroidized Al_2O_3 powders · Surface morphology

1 Introduction

Ceramics three-dimensional (3D) printing is an Industry 4.0 revolution in advanced manufacturing technology. It is generally known for quick improvement in the manufacturing of ceramic 3D products within a short period of time and with better precision. The components that cannot be manufactured utilizing conventional material removal techniques can be manufactured using this technique [1]. This endorsed technique uses powder materials of necessary morphology and properties as feedstock in layer-by-layer building up approach to manufacture 3D objects from a computer-assisted design model [2]. The powder-based 3D printing process mainly uses powder as a layered bed that usually contains dispersed ceramic powder particles. The powder-based routes are better and suitable to manufacture complex 3D forms and shapes with a lesser number of manufacturing steps compared to other traditional methods. Basically, a layer at a time generates directly from a computer-aided model for producing a high-value component

S. Hloch et al. (Eds.): ICMEM 2021, LNME, pp. 249–257, 2021.
https://doi.org/10.1007/978-3-030-71956-2_20

with complicated internal features [3]. The fabrication process involves several tool-less manufacturing steps which produces dense components with better precision. The final components properties resulting from the endorsed technique depend on composition, microstructure, morphology and particle distribution in the feedstocks and used process parameters [4].

Three-dimensional printing (3DP)/indirect ceramic inkjet 3D printing, selective laser sintering (SLS) and selective laser melting (SLM) are a range of important commercially viable powder-based manufacturing techniques [2, 4–6]. Among these approaches, the process operation is associated with solid-state sintering which involves melting of surface particles followed by a controlled grain growth depending upon the process parameters. The lasers, plasma torch and electron beams are the heat sources widely used in such techniques for shaping fusible bed layers [7]. The SLS/SLM techniques have shown the ability to generate temperatures in the melting range of materials to initiate solid-state sintering processes. Here, to note that sintering is an extremely sensitive step, because it determines the final microstructure, mechanical and metallurgical properties, and often influences the geometry of the components [8]. Reportedly, bonding of the particles in powder bed occurs either by particle cohesion due to available laser energy or binder phase mixing, depending up on the technologies used [9]. Other binding mechanisms, such as chemical bonding, partial or absolute melting, are also been reported. Due to higher melting point range (>2000 °C) of ceramics, bonding of binders/fluidizers using thermal energy sources is difficult to achieve. Ceramic components of simple geometries may be generated through conventional powder compaction, but SLS/SLM technique are more promising to fabricate complicated geometries [6, 8].

The success of these techniques depends on the nature of the powders consisting of the necessary morphology and properties of the particles [10]. Therefore, feedstocks with requisite characteristics need to be used in the ceramic 3D printing to obtain robust products with less porosity and better properties. However, the variability of powder materials used as a feedstock and absence of standardize method for testing and characterization poses challenges in the success of the technique. Uniformity in the composition and morphology of the particles minimizes the distribution of fine to coarse particles, and it allows the creation of a proper surface finish and a complex internal geometry of 3D printed structures. Despite the potential advantages, ceramic 3D printing has not been widely utilized in industrial production, which uses powder as a feedstock material. Therefore, powder with distinctive features and quality are necessary for the development of a successful powder-based ceramic 3D printing technology. Many researchers have reported challenging issues related to powder's characteristics such as properties, microstructure and morphology, which can be overcome with the utilization of proper selection of powders in the ceramic 3D printing. Travitzky et al. [4] has observed that fine distribution of particles reduce the thickness of the powder bed layers during additive manufacturing process, which results in difficulty during uniformly spreading of particles across the workspace. Zhang et al. [8] has found that reducing the powder bed thickness effectively decreases the porosity in the product and thus, improves the properties of the final 3D components.

The details on the consistency of the composition, quality, variability, versatility and properties of ceramic powders should be accessible so that powder can be easily selected with confidence. In fact, to maximize the final 3D products density and property, the powder bed density has to be increased. It can be achieved through using the powders with particles having spherical morphology and narrow size range distribution. Interestingly, ceramic powder-based manufacturing has not been well investigated mostly due to lack of sophisticated utilities for melting and sintering compared to the other advanced materials. The ceramic 3D products can be used as critical components for sophisticated applications where the requirements are less, but internal geometries are complex, such as those can found in aerospace, automobile, tooling, electrical, electronic, medical and design industries [7].

Alumina is produced in a variety of purities and is characterised by lesser density compared to metallic materials [5]. It has several polymorphic metastable phases such as α, γ, κ, ρ, η, θ and χ. Among these phases, α-Al2O3 is a pure and the most thermodynamically stable phase. It is one of the hardest and densest phases that is produced by reheating other polymorphic variety of alumina to a temperature over 1200 °C as a result of thermal transformation [11–13]. The α-phase has hexagonal structure and lattice parameters of $a = 4.7602$ Å, $c = 12.993$ Å, and $\gamma = 120°$[14]. The first order transformation of alumina phase is represented in Eq. (1).

$$\gamma - Al_2O_3 \xrightarrow{750\,^\circ C} \delta \xrightarrow{900\,^\circ C} \theta \xrightarrow{1200\,^\circ C} \alpha \quad (1)$$

Alumina powders are synthesized by a number of methods like sol-gel method, laser ablation, mechanical attrition, microwave combustion, high-energy ball milling, chemical and physical vapor deposition, aluminium isopropoxide hydrolysis, liquid aerosols and flame pyrolysis [13]. Obtaining an absolute structural crystallinity is one of the major drawbacks of these methods. Further, processing criteria in these approaches are very complex, and a number of complicated procedures have to be followed for a limited production. Due to their complexity, these methods have made no contribution to the development of a commercial scale.

Ceramic powders with better structural crystallinity have better properties and can be used in a broad spectrum of applications based on material properties such as rheological, mechanical, chemical and physical properties [2, 15]. The required properties and industrial application of microcrystalline alumina have been discussed by Nadolny [16]. The excellent chemical and physical behaviour of microcrystalline alumina and their characteristics can be used for numerous interesting applications. In addition, for innovative technical application submicron- to nanocrystalline alumina are expected to expand rapidly. In fact, crystalline structure determines the properties of alumina powder and helps to understand the evolution of 3D components' characteristics to obtain optimum performance during the service condition. Only a few studies are available on microstructural design of crystalline ceramics and there is a need to develop and design future generation ceramics to compete with other materials [8, 17].

The ceramic powders are known for their diverse as well as variety of other potential uses such as catalysis, gas separation, abrasive polishing, optical application, abrasives, cutting and stamping devices, biomaterial, thermal insulation in fission and fusion nuclear reactors, filters at elevated temperature, fuel cells, sensors and refractories [6]. As an excellent chemical corrosion resistant and wear resistant behaviour, it is widely used for protection coating and inner liner material for extreme condition applications.

Recent emerging trends show a better efficiency of additive manufacturing with the advancement of computer programming and software technologies aimed at sustainable manufacturing industry. The relevant machine and materials, fabrication process, optimization of powder production and origins of material lumps to be processed into fines are the squeezing parameters for the growth of powder-based ceramics manufacturing. Therefore, the quality design model and the efficient creation of 3D products with superior strength and better surface finish need to be mapped and understand on the basis of the microstructural relationship and properties of the powder materials and their effect on the final product performance. In this correspondence, alumina powders have been obtained from radio frequency plasma spheroidization technique and the properties of alumina powders have been studied in order to address the reliability problems in their 3D printing application.

2 Materials and Experimental Methods

Radio Frequency Plasma Spheroidization (RFPS) technique is used to manufacture alumina microbeads powder from crushed alumina ingots of size ranging from 3 to 20 mm. The process consists of in-flight heating and melting of lump materials accompanied by their subsequent cooling and crystallization under regulated conditions [18]. The next section discusses the RFPS technique.

2.1 Radio Frequency Plasma Spheroidization

RFPS is one of the novel techniques uses plasma technology to produce requisite quality material powder. This technique is more effective than other methods such as vacuum gas atomization [19], plasma atomization [20] and plasma rotating electrode [21].

Figure 1 illustrates schematic of RFPS process for obtaining ultra-pure spheriodized powders from material lump of irregular shapes. The technique has been successfully used in producing material powders of uniform morphology, structure, geometry, composition, purity for usage in tool-less additive manufacturing [14]. The technique utilizes the presence of a elevated temperature zone in high-frequency plasmatron and a rapid cooling that allows uniform nucleation resulting in spherical particles ranging from nanometres to micrometres sizes [22].

The details of high-frequency plasmatron have been discussed in Ref. [23]. RFPS has unique advantages such as temperature up to 10,000 °C, fast quenching, high-power density for an effective nucleation, condensation and tailored chemistry of particles in nanometric range, and can easily melt various types of materials of extreme melting point range. The technique is effective for commercial production of ultra-pure fine powders in a size range up to 100 μm or more, which is difficult or impossible to obtain by other available techniques [24].

The Based on the size and density of the crushed alumina ingots, the time of downward movement is regulated in such a way that melted material would have adequate time for complete crystallization before reaching the bottom of the reaction chamber [23]. The purity of alumina powder was measured using the analytical energy dispersive X-ray fluorescence spectrography (Make: Rigaku) technique. X-ray diffractometer

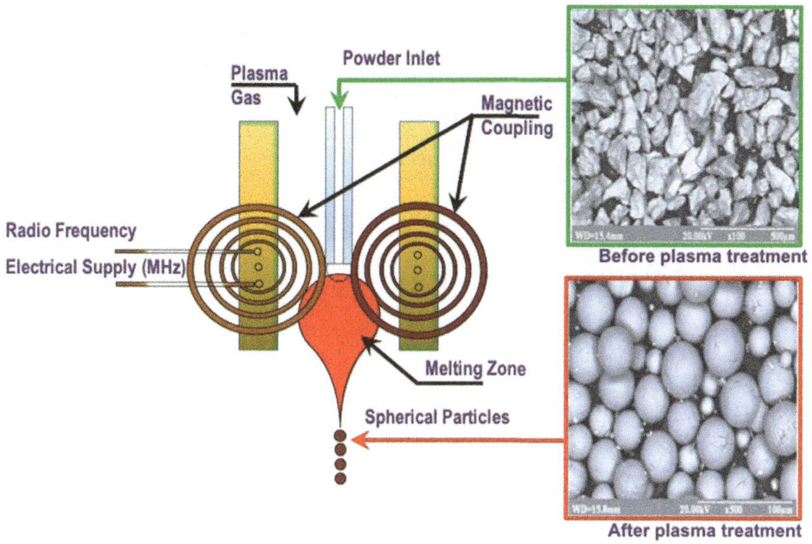

Fig. 1. Schematic of Radio Frequency Plasma Spheroidization technique for obtaining ultra-pure powder from the material lumps of irregular shapes.

(PANalytical Empryean) was used to determine the structural crystallinity and lattice strain of the powders. The experimental parameters for XRD was Cu Kα (λ = 1.54056 Å) radiation at a voltage of 45 kV and a current of 40 mA in the 2θ range varying from 10° to 90°. The crystallite average size (Å) was estimated from the experimental X-ray line broadening using the Debye-Scherrer's equation. The detail procedure was discussed in Ref. [25]. The interplanar spacing $<d_{hkl}>$ for the prominent (012) and (110) peak positions were used to calculate the lattice parameters 'a' and 'c' of alumina phase. Scanning Electron Microscopy (SEM) TESCAN VEGA 3 was used to inspect the morphology and elemental analysis of alumina powders at an accelerating voltage of 15 kV.

3 Results and Discussion

The purity and compositional analysis of alumina powders are shown in Table 1. The compositional analysis confirms 99.98% pure alumina powders obtained from the RFPS technique.

Experimental XRD pattern of alumina microbeads powder after plasma spheroidization is shown in Fig. 2.

The XRD results show the formation of the crystallized phase of well-defined diffraction peaks. In the XRD spectrum, the indexed peaks confirmed the formation of a single-phase of pure α-Al2O3. The identified phase in feedstock alumina ingots was γ-Al2O3. The result of XRD from initial material is not shown here.

The XRD pattern did not show any additional peaks to confirm other alumina phases except the α-phase, indicating the formation of only α-Al_2O_3 and other important alumina phases are faded away as a result of thermal transformation. Most of the alumina

Table 1. Compositional analysis of Al_2O_3 powders within the detection limit of the EDXRF.

Phases	$Al_2 O_3$ Powders (Obtained from RFPS technique)
	Average Size (μm): 1–100
	Concentration (percent)
$Al_2 O_3$	98.98
Na_2O	0.266
CaO	0.007
MgO	0.017
Fe_2O_3	0.021
SiO_2	0.14

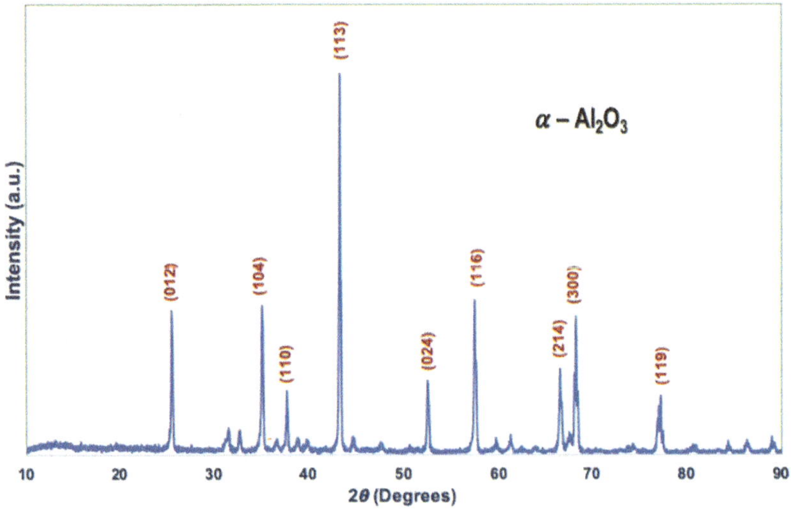

Fig. 2. Experimental XRD spectrum of alumina microbeads powder obtained from RFPS.

phases disappeared at a temperature above 1200°C and only thermodynamically stable α-Al_2O_3 phase is observed.

The analysed result is consistent with the previous literature where the thermal transition of phases have been reported from γ-phase to α-phase [12]. The crystallite average size values of α-Al_2O_3 obtained from Debye-Scherrer relation are shown in Table 2. A particle size refinement can also be noted from the observed diffraction data. This is also consistent with the results obtained from SEM micrographs.

Presence of the residual impurity or uncrystallized phase was well below the limit of XRD detection. The powder shows a better degree of crystallinity at different planes of (113), (104), (012), (300) and (116). The average crystallite size and lattice strain

Table 2. Structural crystalline parameters of alumina microbeads powders obtained from the experimental XRD pattern.

α-Al_2O_3	Scherrer crystallite size (nm), corresponding to interplanar distances (d_{hkl})			
	$d_{(012)}$ = 3.479 Å	$d_{(104)}$ = 2.550 Å	$d_{(113)}$ = 2.081 Å	Average crystallite size (nm)
	42.56	36.28	29.78	36.21 ± 6.4
	Lattice Strain (percent)			
	0.38	0.33	0.32	0.35
	Lattice Parameters (Å)			
	c = 13.037		a = 4.751	$\frac{c}{a} = 2.74$

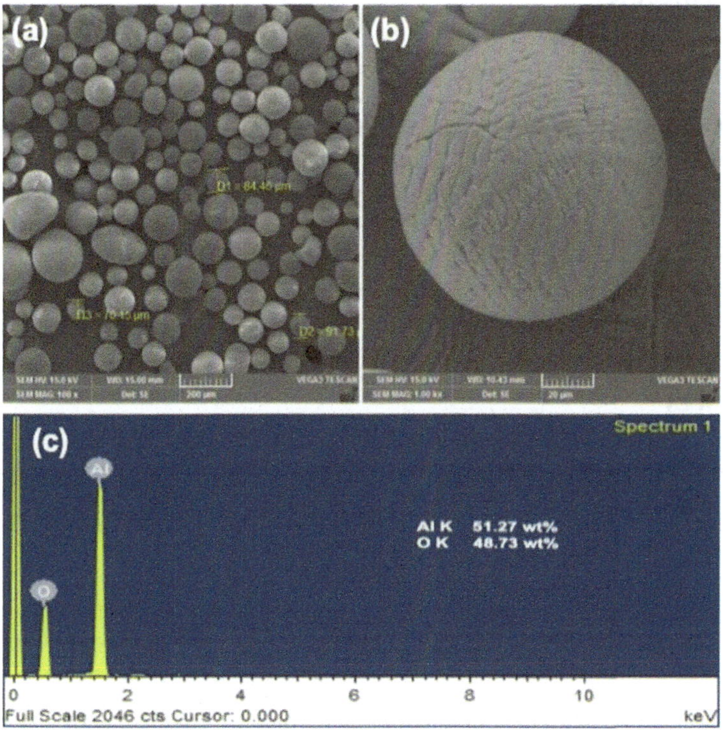

Fig. 3. SEM micrographs show (a) alumina powder particles obtained from RFPS technique, (b)globular alumina particle with 100% sphericity and surface roughness, and (c)EDX analysis of alumina globular particle.

of alumina powder detected as 36.21 ± 6.4 nm and 0.35%, respectively. The experimental average lattice parameters (c = 13.037 Å and a = 4.751 Å) were comparable to the available literature data [14]. Powder particles' morphology was evaluated from the micrographs obtained from scanning electron microscope. Figure 3 shows the representative SEM micrographs of alumina powders. All the powder particles were observed to be spherical with unique surface roughness. The particle surface was rugged with a cellular structure. The diameter of alumina particles was varying from 2 μm to 100 μm and an average size of 8.7 μm. Such particles are potentially a perfect starting feedstock material for ceramic additive manufacturing. The globular shape of particles in a specific range potentially packs tightly the powder bed before sintering. This helps to attain a homogeneous distribution of particles to obtain uniform dispersion within the layered bed in order to achieve a high-performance 3D final product. Quality powder particles help to develop a compact and smooth 3D products and have better properties [23, 26]. According to the elemental SEM-EDX analysis, the powder consists of Al and O in a respective weight ratio, confirming the presence of Al_2O_3, with no turbidity.

4 Conclusions

Radio frequency plasma spheroidization technique was used for the production of alumina powder from the lumps. The process was completely controlled to obtain an essential morphology of the particles for use in a powder-based ceramic 3D printing. The obtained alumina powder was used for evaluating chemical composition, particles morphology and structural crystallinity. High-temperature zone in plasmatron converts alumina lumps into spheroidized and well-crystallised powder. The alumina particles comprised of only a single and stable α-Al_2O_3 phase of uniform chemistry. The powders were ultra-pure and of finer crystallites. The morphology of alumina particles was almost perfect globular shape.

References

1. Deckers, J., Vleugels, J., Kruth, J.P.: Additive manufacturing of ceramics: a review. J. Ceram. Sci. Technol. **5**, 245–260 (2014)
2. Chen, Z., Li, Z., Li, J., Liu, C., Lao, C., Fu, Y., Liu, C., Li, Y., Wang, P., He, Y.: 3D printing of ceramics: a review. J. Eur. Ceram. Soc. **39**, 661–687 (2019)
3. Bourell, D.L.: Perspectives on additive manufacturing. Annu. Rev. Mater. Res. **46**, 1–8 (2016)
4. Ngo, T.D., Kashani, A., Imbalzano, G., Nguyen, K.T.Q., Hui, D.: Additive manufacturing (3D printing): a review of materials, methods, applications and challenges. Compos. Part B Eng. **143**, 172–196 (2018)
5. Moritz, T., Maleksaeedi, S.: 4 - Additive manufacturing of ceramic components. In: Zhang, J., Jung, Y.-G. (eds.) Additive Manufacturing, pp. 105–161. Butterworth-Heinemann (2018) (2018)
6. Zocca, A., Colombo, P., Gomes, C.M., Günster, J.: Additive manufacturing of ceramics: issues, potentialities, and opportunities. J. Am. Ceram. Soc. **98**, 1983–2001 (2015)
7. Tofail, S.A.M., Koumoulos, E.P., Bandyopadhyay, A., Bose, S., O'Donoghue, L., Charitidis, C.: Additive manufacturing: scientific and technological challenges, market uptake and opportunities. Mater. Today **21**, 22–37 (2018)
8. Bertrand, P., Bayle, F., Combe, C., Goeuriot, P., Smurov, I.: Ceramic components manufacturing by selective laser sintering. Appl. Surf. Sci. **254**, 989–992 (2007)

9. Grau, J., Moon, J., Uhland, S., Cima, M., Sachs, E.: High green density ceramic components fabricated by the slurry-based 3DP process. In: Solid Freeform Fabrication Symposium, pp. 371–378 (1997)
10. Yeong, W., Yap, C., Mapar, M., Chua, C.: State-of-the-art review on selective laser melting of ceramics. In: High Value Manufacturing: Advanced Research in Virtual and Rapid Prototyping: Proceedings of the 6th International Conference on Advanced Research in Virtual and Rapid Prototyping. CRC Press, Leiria (2013)
11. Zielinski, P.A.: Structural transformations of alumina by high energy ball milling. J. Mater. Res. **8**, 2985–2992 (1993)
12. Wang, Y., Suryanarayana, C., An, L.: Phase transformation in nanometer-sized γ-alumina by mechanical milling. J. Am. Ceram. Soc. **88**, 780–783 (2005)
13. Liu, H., Ning, G., Gan, Z., Lin, Y.: A simple procedure to prepare spherical α-alumina powders. Mater. Res. Bull. **44**, 785–788 (2009)
14. Ruys, A.: 4 - Processing, structure, and properties of alumina ceramics. In: Ruys, A. (ed.) Alumina Ceramics, pp. 71–121. Woodhead Publishing (2019)
15. Singh, B.P., Menchavez, R., Takai, C., Fuji, M., Takahashi, M.: Stability of dispersions of colloidal alumina particles in aqueous suspensions. J. Colloid Interface Sci. **291**, 181–186 (2005)
16. Nadolny, K.: State of the art in production, properties and applications of the microcrystalline sintered corundum abrasive grains. Int. J. Adv. Manuf. Technol. **74**, 1445–1457 (2014)
17. Zhang, W., Melcher, R., Travitzky, N., Bordia, R.K., Greil, P.: Three-dimensional printing of complex-shaped alumina/glass composites. Adv. Eng. Mater. **11**, 1039–1043 (2009)
18. Sreenivasan, M., Krishna, R., Skulskyi, M., Lyutyk, M.: Opportunities and challenges for radio frequency plasma spheroidized nanocrystalline powders in additive manufacturing. Mater. Today Proc. **33**, 1021–1025 (2020)
19. Flinn, J.E., Smolik, G.R.: Intrinsic features of Fe-40wt.%Ni powders prepared by centrifugal or vacuum gas atomization. Mater. Sci. Eng. A **124**, 39–48 (1990)
20. Entezarian, M., Allaire, F., Tsantrizos, P., Drew, R.A.L.: Plasma atomization: a new process for the production of fine, spherical powders. JOM **48**, 53–55 (1996)
21. Ozols, A., Sirkin, H.R., Vicente, E.E.: Segregation in Stellite powders produced by the plasma rotating electrode process. Mater. Sci. Eng. A **262**, 64–69 (1999)
22. Ananthapadmanabhan, P.V., Thiyagarajan, T.K., Sreekumar, K.P., Venkatramani, N.: Formation of nano-sized alumina by in-flight oxidation of aluminium powder in a thermal plasma reactor. Scripta Mater. **50**, 143–147 (2004)
23. Skulskyi, M., Lyutyk, M., Sreenivasan, M., Krishna, R.: High Frequency Plasmatron, in: Indian Patent, 202041004694, Mumbai (Eds.) (2020). https://www.ipindia.nic.in/India
24. Skulskyi, M., Lyutyk, M., Sreenivasan, M., Krishna, R.: Synthesis of Nano-Powders Using High-Frequency Plasma Technology for Use in Next Generation Additive Manufacturing, in: Indian Patent, 201941053287, Mumbai (Eds.) (2019). https://www.ipindia.nic.in/India
25. Krishna, R., Wade, J., Jones, A.N., Lasithiotakis, M., Mummery, P.M., Marsden, B.J.: An understanding of lattice strain, defects and disorder in nuclear graphite. Carbon **124**, 314–333 (2017)
26. Yves-Christian, H., Jan, W., Wilhelm, M., Konrad, W., Reinhart, P.: Net shaped high performance oxide ceramic parts by selective laser melting. Phys. Procedia **5**, 587–594 (2010)

Layered Composite Materials and the Use of Dynamic Model Analysis

Gabriela Kuchtová(✉)

Faculty of Manufacturing Technologies with a Seat in Prešov, Technical University of Košice, 080 01 Prešov, Slovakia
gabriela.kuchtova@tuke.sk

Abstract. Each product consists of a certain material. In order to determine the specific material of a component, it is first necessary to know what load will be exerted on the component and what effect the environment in which the component will be located will have (temperature, humidity, chemical environment, etc.). At the same time, it is necessary to have an overview of the available materials, their properties, price and possibilities of their application in practice. With the ever-increasing volume of technical production, the consumption and quality requirements of technical materials also increase. Natural material resources are depletable and non-renewable. For this reason, new and affordable technologies for the production and processing of technical materials are currently being formed. Composites are currently widely used in various fields, especially for their excellent mechanical properties and ability to fully replace the "basic" resp. Original material in order to optimize the production process, reduce costs and modernize production and the products themselves. Experimental measurement of mechanical and viscoelastic properties of composites under stress (tension, compression, bending, compression) is performed by various methods. One of them is DMA analysis (Dynamic Model Analysis), which is part of the methods of thermal analysis. The purpose of the analysis is to quantify the relationships between deformations and deformation forces (as a function of temperature) in situations where these quantities change over time, which allows the measurement of highly elastic materials such as rubber and plastics. The article offers a brief overview of the basic types of reinforcement and core materials used for layered composite structures, examples of use in practice and experimental measurements using DMA analysis.

Keywords: Composite material · Layered structure of composite materials · Laminates and sandwich structure · Fibrous composite materials · DMA analysis

1 Introduction

The development of the first composite materials dates in the distant past, although at that time they were not yet known under this name as we know them today.

A composite is defined as an artificially created material which consists of two or more components of different chemical composition and at the same time different

S. Hloch et al. (Eds.): ICMEM 2021, LNME, pp. 258–269, 2021.
https://doi.org/10.1007/978-3-030-71956-2_21

physical and mechanical properties. The individual components must be macroscopically evenly distributed throughout the volume of the composite material and the resulting properties must be different from the properties of the components of which it is composed and are synergistic.

The primary impulse for the creation of new materials is the fact that natural resources are depletable and non-renewable.

The production of composites can be defined as a purposeful and at the same time experimental activity based on the intention to create a different material from natural materials with even better properties. During the formation, the matrix material (primary phase), the material and the geometry of the reinforcement (secondary phase) are combined and other factors are taken into account in order to achieve an ideal composite that overcomes the shortcomings of the original materials.

Countless types of composite materials are already known today, such as metallic, non-metallic (ceramic, inorganic, organic, polymeric), and structures (fibrous - single-layer and layered, particulate).

The aim of layered composite structures (laminates and sandwich structures) is to create a material understood as a whole and thus to achieve infinitely thin, shear-non-deformable and perfect joints between the individual layers.

2 Layered Structure of Fibrous Composite Materials

To increase the strength and stiffness, the base materials are reinforced with fibres placed in the matrix - fibre composite materials.

The following Fig. 1 shows examples of fibre arrangements in composite materials.

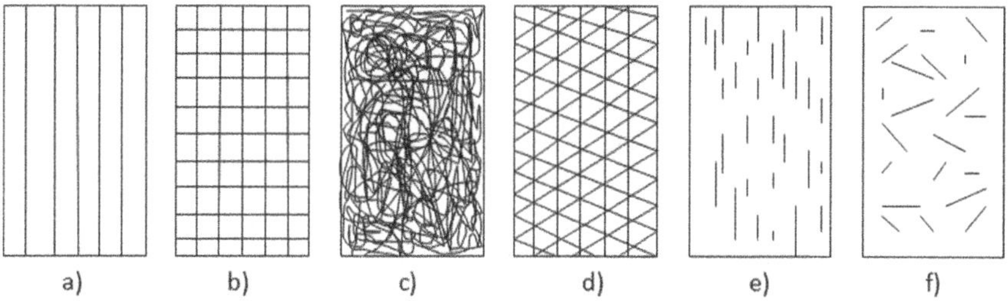

Fig. 1. Structure of the arrangement of fibre reinforcements in composite materials: a) unidirectional arrangement, b) fabric, c) mat, d) multiple reinforcements of continuous fibres, e) short fibres - unidirectionally oriented, f) short fibres - with random orientation [1].

Types of Fibre Reinforcement Materials and their Application

Table 1 shows an overview of the mechanical properties of selected types of fibres reinforcing the material in comparison with steel.

Carbon fibres are characterized by a good ratio of tensile/weight resistance, thermal and chemical resistance. The disadvantage is the relatively expensive production. Fibres

Table 1. Table of mechanical properties of selected fibres.

Material	Modulus of elasticity in tension [GPa]	Tensile strength [MPa]	Density [kg.m^{-3}]
Carbon	300–800	2 700	1 860
Boron	420	3 500	2 650
Aramid Kevlar 29	60	2 700	1 450
Aramid Kevlar 49	130	2 900	1 450
Aramid Kevlar 149	190	2 600	1 450
Glass	70–85	2 200–3 500	2 540
Polyester (PC)	1.3–4.5	45–85	1 100–1 500
Polyamide (PA)	3.0–3.1	80–190	1 200–1 900
Polypropylene (PP)	1.1–1.5	28–41	900
Steel (for comparison)	210	340–2 500	7 800

form a reinforcing component of polymers; in non-polymer applications it is the use of carbon-graphite composites. They are applied mainly in the missile, military, transport and sports industries, e.g. production of individual parts of aircraft, racing vehicles, bicycles, wind generators, etc. [2].

The advantage of boron fibres is good adhesion to the matrix and the disadvantage is high density and low specific strength, the fibres cannot be bent and therefore not woven. They are used in the production of planar structures [3].

Aramid fibres, are divided into meta-aramids (MPIA), which excel in thermal resistance and electro-insulating properties (e.g. Nomex) and para-aramids (PPTA) with higher tensile strength and higher modulus of elasticity (e.g. Kevlar). In practice, these fibres are used in the aerospace industry for aircraft components, for modern balloons in combination with aluminium foil and a protective layer (e.g. silicone or polyurethane coating), in the textile industry for the production of protective fireproof clothing, in the ballistic industry for the production of bulletproof vests., etc. [4].

Glass fibres are characterized by very good tensile strength, which is 2 to 3 times greater than for steel. They are divided into optical fibres used for data transmission and glass fibres intended for textile products made of glass, for insulation and for reinforcing other materials (e.g. fiberglass) [5].

Synthetic fibres are various types of fibres (e.g., polyester, polyamide, polypropylene, etc.). Positive characteristics include good flexibility, strength, resistance to mould and moths, non-shrinkage. On the contrary, their negative properties are flammability, ability to charge with static electricity, low ability to absorb sweat and are non-degradable, which burdens the environment. They are mainly used in the textile industry for the production of clothing, home and industrial textiles [6].

And other types of fibres (ceramic, natural, etc.). Physical and mechanical properties of different types of composites using natural fibres are described in [7].

Layered Structure. The simple sandwich construction of composites consists of three layers of different materials, namely a core which is closed by two outer layers. The core is characterized by low weight. A resin or a resin is applied to the outer layers. an adhesive, a binder whose job is to bond the core to the outer layers [8].

The multilayer sandwich construction of composites is thus also formed by three layers - the outer layers between which there is a core. However, in a multilayer structure, the core of the composite consists of several layers with further intermediate layers.

The layered structure of composites is also considered to be the layering of several types of materials or one type of material with different orientation of the fibre reinforcement - laminates to be a layered structure of composites.

2.1 Laminates

Laminates consist of several unidirectional layers stacked together in different orientations, cf. Fig. 2. The individual layers of the laminate are called laminates. The laminates can have different thicknesses and can consist of different materials [9].

The laminate can also be formed from layers of different materials. Such a laminate is referred to as an interply hybrid composite. An inner layer hybrid composite is formed when different types of fibres are used within one layer. (intraply hybrid composite.) [9].

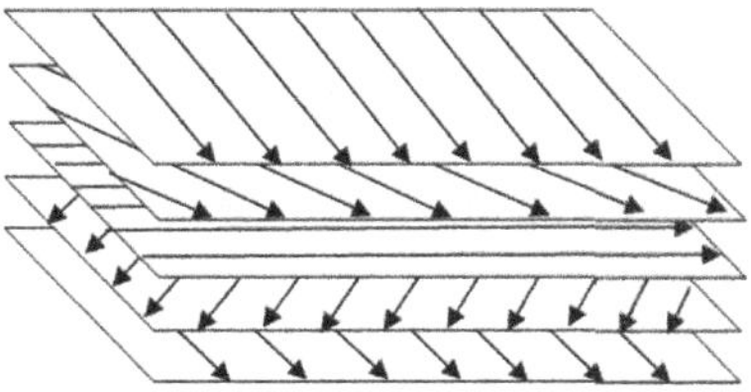

Fig. 2. Laminate layers.

2.2 Sandwich Structures of Composite Materials

Sandwich structures of composite materials are often used in practice due to their high flexural strength and stiffness at low weight.

The following Fig. 3 shows the individual layers of a single and multilayer sandwich structure of a composite material.

In a multilayer sandwich structure, the core is made of two or more kinds of materials. An intermediate layer is applied between each layer of material, the task of which is the perfect connection of the individual layers and thus create a solid material understood as a whole.

The choice of a suitable material is adapted to the properties that the resulting sandwich structure should have.

Examples of materials used for the outer layer of the sandwich structure are mainly thin and solid materials - e.g. aluminium, steel, laminate, wood.

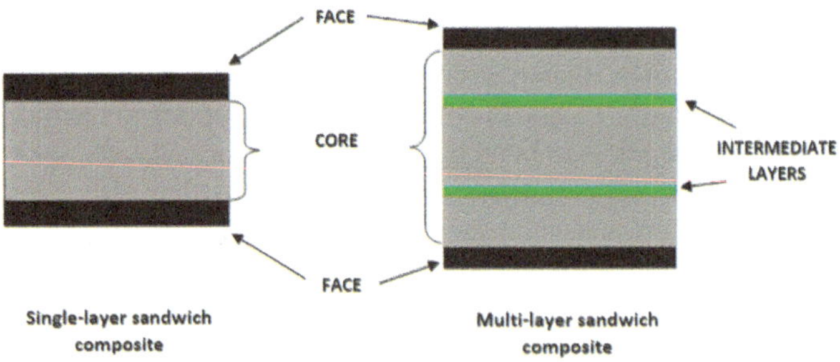

Fig. 3. Scheme of simple and multilayer sandwich structure of composite materials.

Different materials with different thicknesses and shapes are used for the core. By increasing the thickness of the core, we can achieve higher stiffness of the composite without a substantial increase in weight and cost.

Types of Materials for Cores of Sandwich Structures of Composite Materials and their Application in Practice. The honeycomb consists of a series of cells inserted into each other to form panels similar in appearance to the cross-section of a beehive plate. Depending on the material used, different types of honeycombs are on the market, e.g. aramid, aluminium, kevlar honeycombs. The advantage of this type of core is relatively good non-flammability, flexibility, lightness and impact resistance. It offers the best ratio of strength to weight of core materials, which can be influence by adjusting the wall thickness and changing the volume of the honeycomb cells. It is used mainly in the transport industry.

Aramid honeycomb, see Fig. 4, is used wherever reduced flammability is required, especially in train or aircraft interiors. They are also used for special types supplemented with phenolic foam, which reduces and dampens vibrations - an insulating function [8].

Fig. 4. Aramid honeycomb [11].

The problem with aluminium honeycombs is the formation of corrosion, which can be treat with a suitable coating. It is also not advisable to deposit aluminium fibres directly on the aluminium structure due to the risk of galvanic corrosion [8].

Kevlar honeycombs are non-metallic honeycombs and represent a combination of the properties of aramid and aluminium honeycombs. Compared to aramid honeycomb, this type is lighter by 40% and up to 4 times stronger and is resistant to corrosion compared to aluminium honeycomb [8].

One of the frequently used materials for the core of the sandwich structure is also universal polyvinyl chloride (PVC) foams, cf. Fig. 5. A characteristic feature is closed pores, they have balanced mechanical properties and resistance to water absorption and aggressive chemicals. Examples of applications are in particular ship constructions such as ship hulls or the production of interior parts in the transport industry [8].

Fig. 5. PVC foams of different thicknesses [12].

Balza, Fig. 6, is a type of wood with relatively high compressive strength (depending on the size of the stressed material), high elasticity, insulating properties. At the microscopic level, it has the structure of bee plastic. It is used mainly in model making, in the automotive and construction industries and in sports, e.g. for the production of models, furniture, skis, ping-pong rackets, surfboards, etc. [13].

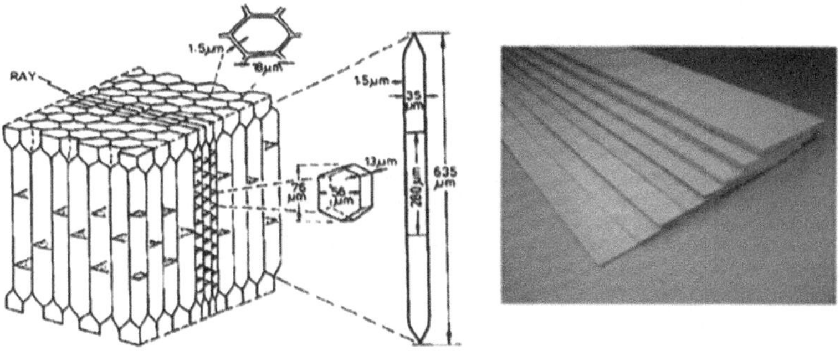

a) b)

Fig. 6. a) layered arrangement of balsa cells with fibrous structure, b) balsa - boards [14].

Non-woven polyesters, see. Fig. 7 represent another possibility for forming an efficient sandwich core, which is deposited by hand during the production of the composite. The advantage is its affordability compared to PVC foams and the possibility of using all types of resin. Applicability is found in various areas, e.g. in healthcare for the production of surgical clothing, wipes, cosmetic products, etc., in agriculture as a fabric providing crop protection, industrial nonwovens e.g. filter materials, insulating materials, etc. [8].

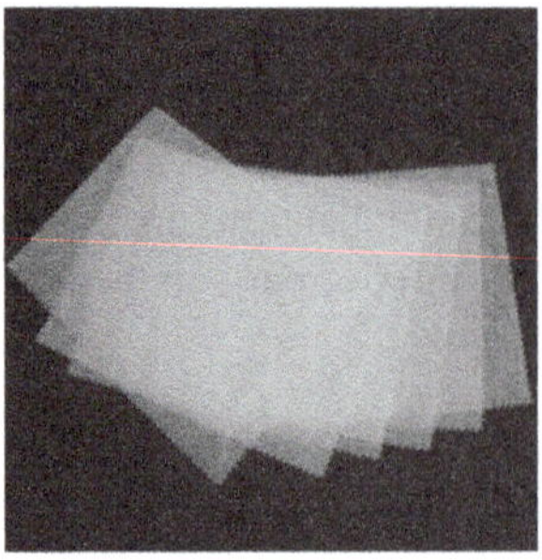

Fig. 7. Non-woven polyesters [14].

3 DMA (Dynamic Model Analysis)

For a suitable choice of composite, the materials are subjected to various tests for introduction and subsequent comparison of their mechanical and physical properties. One of the methods for determining the mechanical properties of a material is DMA (Dynamic Model Analysis). The principle of DMA analysis consists in monitoring the viscoelastic response of a material subjected to a small oscillating stress and includes the elastic modulus E' [MPa] (real part), the loss modulus E'' [MPa] (imaginary part) and the loss factor $tan\ \delta = (E'')/(E')$ characterizing the damping capacity of the material.

Sample Preparation. As a practical example, a sample of a carbon fibre reinforced composite was tested. A research sample of 40 × 10 × 1 mm for DMA analysis was cut from the original sample, Fig. 8. It is a 3-layer composite material - a twill-woven carbon fabric (200 g/m^2) with a polymer matrix. The warp angle of the fabric in each layer is 0°.

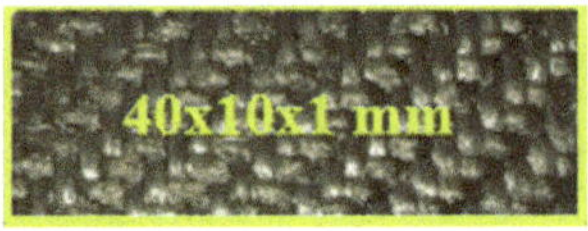

Fig. 8. Sample of carbon twill weave fabric with polymer matrix (40 × 10 × 1 mm) for DMA analysis.

Equipment and Measuring Conditions. The experiment was performed using a DMA 242 device from NETZSCH with a holder for three-point bending, see. Fig. 9 a). Figure 9 b) represents a measuring device with an inserted sample. The vibration frequency was $f = 1$ Hz and $f = 10$ Hz. The obtained results are shown in the graph, Fig. 9.

Results of DMA Analysis of the Test Sample. Using DMA analysis, the dependence of the mechanical strength of the sample E' and the value of the loss factor $tan\ \delta$ depending on the temperature changes was investigated, at the oscillation frequency $f = 1$ Hz, $f =$ 10 Hz, Fig. 10.

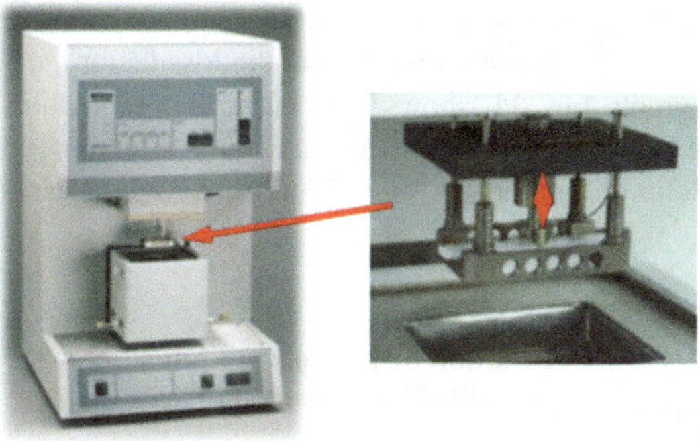

a) b)

Fig. 9. DMA 242 from NETZSCH, b) sample anchored in the device holder [16].

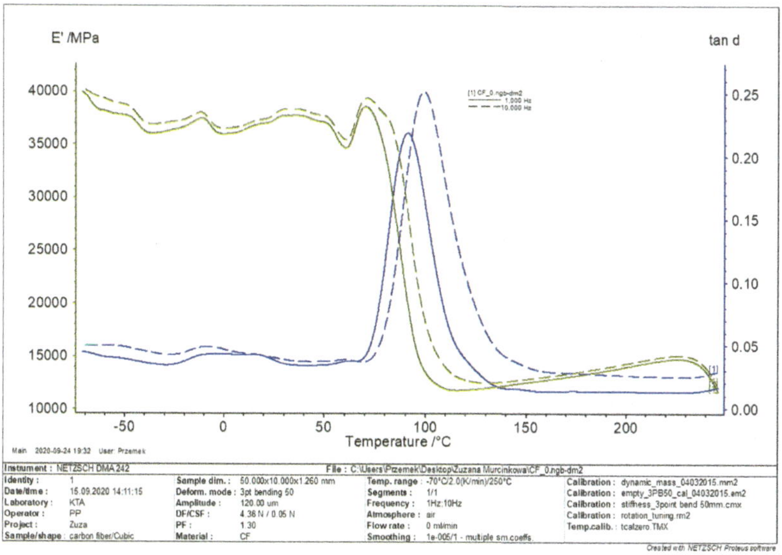

Fig. 10. DMA analysis for carbon fabric in twill weave with 0° fibre orientation.

In the temperature range −70 °C to 70 °C, $tan\ \delta = 0{,}03$ to 0,05. With increasing temperature, 70 °C to 100 °C, the loss factor $tan\ \delta$ increases. At 100 °C, frequency $f = 10$ Hz and loss factor $tan\ \delta = 0{,}25$, the glass transition temperature T_g occurs, at which mechanical and physical changes in the properties of the sample occur. At higher temperatures, from 100 °C, the damping capacity of the composite decreases. At a frequency of 1 Hz, the temperature T_g already occurs at 90 °C and $tan\ \delta = 0{,}22$, which is a decrease of 30%. Sufase modulus E' (elastic modulus of elasticity) at temperatures from −70 °C to 70 °C reaches the range of 35 000 MPa–40 000 MPa. As the temperature increases, the strength of the composite sample decreases. The temperature range 70 °C–100 °C can be considered the maximum for a given sample in order to maintain a relatively good strength and damping ability of the composite. At higher temperatures there is a decrease in the Sufase modulus E' and the loss factor *tan* δ.

The following graph, Fig. 11, shows how the Young's modulus changes with the change of the angle between the applied tension and the direction of the fibre ($E_f = 210$ GPa (Young's modulus of the fibre) at $V_f = 50\%$ (fibre volume fraction), $E_f = 3$ GPa at $V_f = 70\%$) [17].

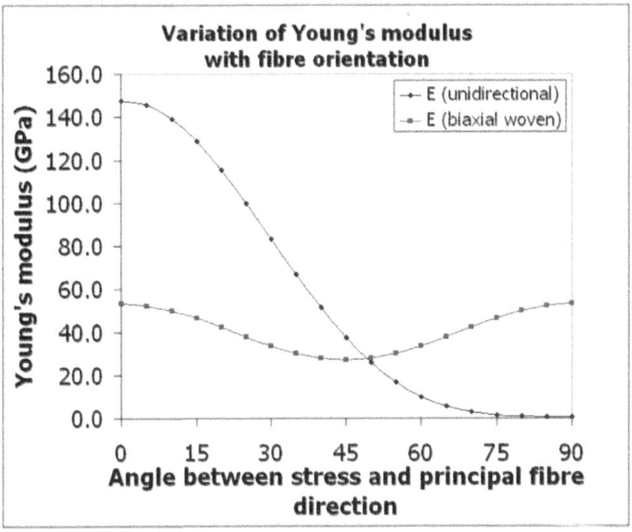

Fig. 11. Changes of Young's modulus with fibre orientation [17].

It can be seen that in the horizontal orientation of the fibres the modulus of elasticity is higher than in the vertical orientation of the fibres, resp. in other directions of the fibres. It is true that the higher the E', the greater the stress (with the same cross-section the greater the force) to achieve the same deformation (elongation) [17]. It is also true that the larger the volume fraction of fibre, the stronger the material. For the highest fibre content, the rate of change is always highest, regardless of the angle and the rate of deformation [17]. The expected course of elasticity for a sample of carbon fabric in a twill weave with a fibre orientation of 0° can also be seen in Fig. 11.

4 Applications of Layered Composite Structures

Composite materials offer more durable, flexible and resilient material options and thus form an integral part of use in various industrial areas and industries. Here are a few examples of the use of composite materials in practice.

Defence and Army Industry. The representation of composites in the defence and military industries is generally extremely important due to their durability and lightness. Uses are found e.g. in the production of ballistic vests, helmets, ballistic goggles, use for combat equipment, weapons, etc.

Transport Industry. The transportation industry represents a wide application of composites in all areas, such as automotive, aviation, shipping and rail transport. In the automotive industry, it is an application for dashboards, door insulation inserts, brake pads, brake discs (discs), shafts, glazing, etc. In addition to the already mentioned application possibilities, the aviation industry uses composites for the production of aircraft components, for compressor blades, turbine blades, and the like. These are, in particular, ship constructions for ship transport and, for railway transport, use for train constructions and the like. The reason for choosing composites is to lighten the carcass of the vehicle and increase the impact resistance.

Building Industry. The service life of the material is a very important property, especially for the construction industry. Examples of use are e.g. for building facades, kiosks, cold stores, building melds, etc. An example of a laminated composite material in construction is hardened paper, so-called umakart. It is a combination of cellulose paper and modified epoxy resin and is used as a covering or insulating substance. One of the most used layered composite materials is also glass laminate. It is a composite that consists of glass fibres and a resin, most often a polyester resin, that holds them together.

Engineering Industry. The principle of layering is also used in the mechanical engineering.

In recent years there is a significant physical development of new technologies allowing to create a thin film on the surface and thus significantly improve the useful properties of the instrument. A special group of cutting materials represent coated cutting materials. In addition to the monolayer coatings it is nowadays also applied the multilayer coating for to better quality of the surface modification of cutting tools. Especially in case of multilayers, the layered composition of deposited materials, thickness of individual layers is key to durability. By means of the experiments and numerical analyses the problem of under surface stress and stress jumps is researched to find the optimal composition for specific conditions. Numerical analysis in [18] showed that the lower stress was observed in the case of multilayers comparing to monolayers. However, in multilayer is on the interface of layers which causes the formation of micro-cracks between the layers of the coating. According to [18] the most preferred multi-layered coating composition type appears accumulating the last (outside) layer. In that case, the maximum stress is the smallest and maximum is not on the boundary substrate – coating. Moreover, the structure of individual coating layers as fibre-like structure in one direction (along with surface or perpendicular to surface) is important as microstructure is able to directs heat flow to wanted direction [19].

Conventional use of laminates, layered sandwich structures made of aramid honeycomb, PVC foam etc. are in airplanes to make lightweight structure. That principle is applied for machinery, especially for movable parts of machine tools as lowering the weight is significant up to 45% [20]. Thus, the velocity of movement can be 150% higher, even the acceleration (up to 300%) [20] as the moment of inertia is significantly lower.

Very successful are applications of hybrid and layered structures for lowering the vibrations and noise. Combination of weld structure and mineral casting significantly improve damping of machine tools [23]. Damping properties of composite materials are very promising analysing results of experimental measurements (more in [21, 22]).

5 Conclusion

Modern composite materials push original materials out of the market, especially for their better mechanical and physical properties. They significantly affect the environment due to their lower weight. They have excellent thermal insulation and acoustic properties, better resistance in aggressive environments, high modulus of elasticity and strength, etc.

Like any material, composite ones have several disadvantages, e.g. higher price, service life of the material, for some types it is necessary to use protective coatings, worse workability due to high density, etc.

The technology of composite materials makes it possible to create a structure which corresponds to deformation (stress) and high strength at the point of stress. A special group of composite materials are sandwich composite materials, the great advantage of which is their high rigidity and at the same time low weight.

When developing composites, emphasis is also placed on the structure of the material itself. One of the thermal and at the same time structural analyses is DMA analysis. It is one of the most sensitive techniques by which we can characterize the mechanical behaviour of a wide range of materials under thermal and mechanical loading. Thanks to this method we can find out e.g. glass transition temperature T_g, softening and pour point, mechanical losses in materials, crystallization, crosslinking, presence of fillers and additives in polymeric materials, etc.

Based on an experimental measurement, a carbon sample in twill weave with a fibre orientation of 0° in the temperature range −70 °C to 250 °C at $f = 1$ Hz and $f = 10$ Hz, on a DMA instrument, we can say that the maximum temperature of the sample load is in the range from 70 °C to 100 °C, to maintain the good strength and damping ability of the composite. At higher temperatures, both the sulfase modulus and the loss factor decrease. In the region of the glass transition temperature, the material reaches the maximum value of the loss factor and the relatively rapidly decreasing value of the modulus of elasticity as a function of increasing temperature. There is a transition to a rubbery state and thus changes in the mechanical and physical properties of the material.

Acknowledgement. The author would like to thank the Slovak Research and Development Agency for supporting this research through the grant under Contract No. APVV-18–0316.

References

1. Rendek, M.: Kompozitní materiály používané pri výrobe automobilů. Bakalárska práca. Vysoké učení technické v Brne, Fakulta strojního inžinírství, Brno (2016)
2. https://sk.wikipedia.org/wiki/Uhl%C3%Akov%C3%A9_vl%C3%A1kno. Accessed 28 Apr 2020
3. https://www.talentica.sk/vlaknite-kompozity-vyroba-a-vyuzitie/. Accessed 30 Apr 2020
4. https://www.veba.cz/cs/materialy. Accessed 01 May 2020
5. https://sk.wikipedia.org/wiki/Sklen%C3%A9_vl%C3%A1kno. Accessed 22 Apr 2020
6. https://cs.wikipedia.org/wiki/Syntetick%C3%A1_vl%C3%A1kna. Accessed 21 Mar 2020

7. Lalta, P., Kumain, A., Patel, R.V., Yadav, A., Winczek. J.: Physical and mechanical behavior of hemp and nettle fiber-reinforced polyester resin-based hybrid composites. J. Nat. Fibers 1–16 (2020)
8. Kender, Š.: Sendvičové kompozity. In: Transfer inovácií (36/2017), pp. 52–55. Technická univerzita v Košiciach, Košice (2017)
9. Jones, R.: Mechanics of Composite Materials. Taylor & Francis, Milton Park (1999)
10. Allen, H.: Analysis and Design of Structural Sandwich Panels. Pergamon Press, New York (1969)
11. https://m.sk.haxcorematerial.com/honeycomb-core/nomex-honeycomb/aramid-honeycomb-core-sheet.html. Accessed 19 Mar 2020
12. https://www.chinaexpress.cz/cs/item/pvc-foam-board-plastic-model-pvc-foam-sheet-board-white-color-foamboadrd-model-plate-300x200mm/32807272921.html. Accessed 19 Mar 2020
13. https://www.fibreglast.com/product/end-grain-balsa/Vacuum_Bagging_Sandwich_Core. Accessed 19 Mar 2020
14. Grossová, O.: Analýza mechanických vlastností balzového dřeva na mesoúrovni pomocí zkušebního stroje iErudio. Diplomová práce. Mendelova universita v Brne, Lesnická a dřevařska fakulta, Brno (2015)
15. https://sk.nonwovenfabricchina.com/nylon-non-woven-fabric/insulation-materials/thermal-cable-insulation-materials-non-woven.html. Accessed 19 Mar 2020
16. Postawa, P., Stachowiak, T., Jaruga, T.: Influence of the processing conditions on the dynamic mechanical properties of gas assisted injection moulded parts. Arch. Mater. Sci. Eng. **44**(2), 104–111 (2010)
17. Summerscales, J.: Composites Design and Manufacture (Plymouth University teaching support materials). Young´s moduli, Poisson´s ration, shear and bulk moduli. https://www.fose1.plymouth.ac.uk/sme/MATS347/MATS347A2%20E-G-nu.htm. Accessed 11 Sep 2019
18. Murčinková, Z., Baron, P., Tiňo, L., Pollák, M., Murčinko, J.: Research and analysis of stress distribution in multilayers of coated tools. Int. J. Mater. Res. **108**(6), 495–506 (2017)
19. Kompiš, V., Murčinková, Z., Očkay, M.: Temperature fields in short fibre composites. In: Computational Modelling and Advanced Simulations, pp. 99–116. Springer, Dordrecht (2011)
20. Mráz, P., Talácko, J.: Konstrukce strojů s kompozitními materiály. ČVUT v Praze, Praha (2006)
21. Murčinková, Z., Vojtko, I., Halapi, M., Šebestová, M.: Damping properties of fibre composite and conventional materials measured by free damped vibration response. Adv. Mech. Eng. **11**(5) (2019)
22. Murčinková, Z., Zajac, J.: Damping ratio of the fibre and particle composites and the time decaying and resonance curves in measurement of damping. Modelling of mechanical and mechatronic systems. Technická univerzita v Košiciach. s. 164–174 (2019)
23. Murčinková, Z.: Composite materials: Micro and Macro Behavior, Numerical Simulation and Experiment, Applications in Mechanical Engineering, 1st edn. RAM-Verlag, Lüdenscheid (2018)

A Comparative Simulation Analysis of Conventional and Composite Leaf Spring

Dinesh Kumar Madheswaran[1,2(✉)], Mohanraj Thangamuthu[3], Suresh Gopi[4], Ram Krishna[2,5], Sreenivasan Mohan[5], Arunkumar Jayakumar[1], Yogesh Chinnampalli[5], and Shaik Tappa Nayab Hussain[5]

[1] Department of Automobile Engineering, SRM Institute of Science and Technology, Kattankulathur, Chennai 603203, India
mdineshautomobile@gmail.com

[2] Department of Mechanical Engineering, PACE Institute of Technology and Sciences, Ongole 523272, India

[3] Department of Mechanical Engineering, Amrita School of Engineering, Amrita Vishwa Vidyapeetham, Coimbatore, India

[4] Department of Mechanical Engineering, Rajalakshmi Institute of Technology, Kuthambakkam, Chennai 600124, India

[5] Department of Mechanical Engineering, Madanapalle Institute of Technology and Science, Angallu, Chitoor 517325, India

Abstract. Conventional materials used for an automotive suspension system are one of the key parameters which influence the vehicle weight. Hence the use of composite material for suspension system would offer a significant benefit of being light in weight also have a high life span compared to existing materials, without compromising the load-carrying capacity and stiffness. The primary emphasis of this paper is on gaining a finite element-based performance analysis of advanced composite materials as a substitute for conventional materials used for the application of leaf spring suspension of an automobile. The leaf springs were modeled using CATIA, to replace the existing material of steel alloy and carbon alloy leaf springs with a composite material (Carbon Fiber-reinforced polymer) for a pickup truck and Finite Element Analysis (FEA) was performed using ANSYS for static load condition and the results were compared with conventional leaf spring materials to determine the structural characteristics. Furthermore, vibrational characteristics were also presented by modal analysis for the designed leaf springs. It was inferred from the results that advanced composite materials are light in weight, optimal, and ideal material for the leaf spring offering better structural & mechanical behavior.

Keywords: ANSYS · CATIA · Composite materials · Finite Element Analysis · Leaf springs · Natural frequency · Static analysis

1 Introduction

In modern times, automobile manufacturers focus extensively on vehicle weight reduction, as it is the potential talk on the ground to enhance vehicle performance on a weight

S. Hloch et al. (Eds.): ICMEM 2021, LNME, pp. 270–281, 2021.
https://doi.org/10.1007/978-3-030-71956-2_22

basis [1]. The modern automobile has progressed from being a metal heap in the 1950s to the vast use of substitute lighter materials [2]. This drastic transition over through the ages is often due to rising mandates to reduce vehicle weight to provide high functionality, fuel savings, and comply with legislation and regulations [3]. Design optimization techniques, component consolidation & material substitution will facilitate light-weighting. Progressively, thermoplastics composite materials such as carbon-fiber-reinforced plastic (CFRP) & glass-fiber-reinforced plastic (GFRP) have been seen in automobile structural components, which including body-in-white (BiW), leaf springs, crash frames, etc. [4]. This research work intent on substitution of conventional materials for leaf spring composites used on a commercial pick-up truck.

The leaf spring suspension is a basic and simple form of suspension system widely used in most goods vehicles all around. Typically, a leaf spring comprises multi-layers or a single layer of metals called leaves that are straight or twisted into an arc. The number of leaves is based on the requirements. Leaves are simply thin and long, slightly narrow or curved plates, clamped together as a single system, rigidly mounted directly or mounted between a moving shackle to the chassis of the vehicle, which the system of leaves are rested over or under the axle [5].

When the vehicle is said to travel in different road conditions, the road shocks are transferred on the axle through road wheels via the leaves, which the leaves compress and retain the accelerated force instead of transferring all the energy straight to the vehicle chassis or body. This results in minimizing the vibrations and direct impacts on vehicle body and chassis, which offers a comfortable drive and improved handling of the vehicle [6]. However, the advent in the automobile industry was providing advanced suspension systems, leaf spring suspensions still have a potential market value due to their advanced load-bearing characteristics [7].

In the majority of the cases, steel alloys, and carbon alloys are conventionally used materials for leaf springs which provide better performance over a long decade [8]. Even though conventional materials are providing good performance, in recent days, several research works are concentrating on the transition of conventional leaf spring materials into modern composites due to the insufficient performance of conventional materials and improving the efficiency of the vehicle [9]. As leaf springs are subjected to a countless number of varying stress cycles, brake torque, which leads to fatigue failure of the leaves, results in a crack, fracture, or breakage of a leaf. Also, conventional materials are not much durable, which the material may easily corrode, leads to decreased life span or breakage when they are not maintained frequently. Functionally graded materials can be used for enhanced performances [10, 11]. The dimensions of the leaf spring can be optimized with any optimization techniques like Taguchi's design of experiments [12], response surface methodology [13], and bio-inspired algorithms [9].

Despite the structural parameters of leaf springs, automotive industries are concentrated highly on replacing the existing components with lightweight components without compromising the performance and safety aspects. Composite fibers are a desirable material option over commonly used materials for leaf springs, as they have enhanced material properties [14]. The primary emphasis of this work is to introduce advanced materials over the conventional existing materials, which provides enhanced performance compared to the conventional material associated with significant weight reduction, less maintenance, and increased life span. Among various materials chosen, comparative results of Finite Element Analysis were obtained using ANSYS for the static and dynamic conditions to determine the optimal material for automotive leaf spring.

2 Methodology

2.1 Design of Leaf Springs

The present research work designed the leaf spring for a cargo truck [15]. The design model is a semi-elliptical type multi-leaf spring comprised of seven leaves in number. This type is the oldest one and has largely been used in automotive, due to their increased spring action range and high life span compared with other types. Specifications of the designed leaf spring are given in Table 1.

Table 1. Specifications of leaf spring.

Description	Data
Number of leaves	7
Diameter of the spring eyes	50 mm
Length of the master leaf (eye to eye)	1175 mm
Length of the bottom leaf	525 mm
Thickness of leaves	10 mm
Width of leaves	68 mm
Graduation of leaves	15°

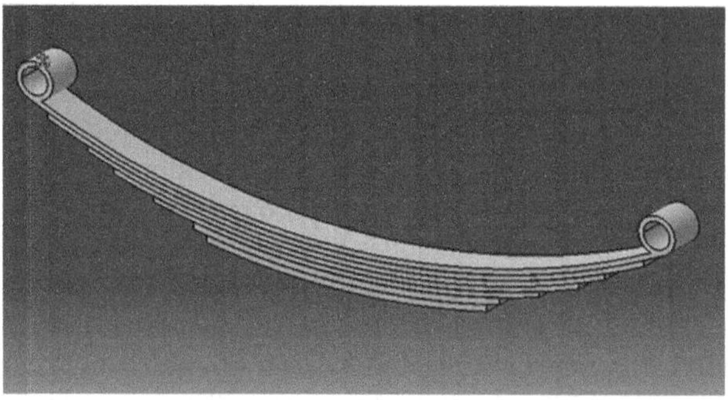

Fig. 1. An isometric view of the designed leaf spring obtained from the CAD model using CATIA V5 software.

Figure 1 shows the isometric view of the CAD model of the designed leaf spring. Gradually decreasing length from master leaf to bottom leaves of constant thickness and constant width [16] of leaves are chosen. The master leaf extends to form the eye of the spring on both sides for mounting. The designed leaf spring was said to be fixed to the shackle at the front end for action and the rear end is rigidly mounted to the chassis of the vehicle. Semi-elliptical multi-leaf springs are modeled using CATIA V5 software because of its easy and compact usage. The theoretical calculation provided by the manufacturer has been selected for providing the dimensions for modeling the springs.

2.2 Material Selection

Material selection for leaf springs must be highly concentrated towards their strength characteristics and better load tolerable character [17]. Steel alloys and Carbon-based steel are commonly used for automotive leaf springs. The conventional leaf spring materials are failed easily and cause improper directional stability on rough and hard terrain conditions [18]. Silicon Manganese Steel and Chrome Vanadium Steels are advanced leaf springs materials [19, 20], nevertheless, they do give only a very small variation in performance over the traditional materials [21].

Composite springs offer an extremely good strength-to-weight ratio, fatigue resistance, and natural frequency, with durability about five times than that of steel springs [22]. In this work, on account of its extremely good properties, Tairyfil Carbon (TC 33 6K) fiber reinforced polymer is selected in contrast to the conventional material. Reinforced carbon fibers when properly engineered in their appropriate metal matrix, significant weight reduction, and extremely rigid composite structure would be achieved [23, 24]. Carbon fiber reinforcements offer an excellent fatigue tolerance in addition to their high strength to weight ratio, they are rich in strength, having lower weight, offering excellent corrosion and wear resistance, high load tolerance is the other key characteristics to consider for automotive leaf spring fabrication [25].

Therefore, in the application of leaf springs, reinforced carbon fiber polymer is selected as ideal and superior material [26]. The material properties were provided in Table 2. The weight of a leaf spring modeled was calculated using the volume and density of the material. Table 3 provides the weight of the leaf spring for different materials. The density of the carbon fiber is very low, but it has high strength over the other materials, would be able to tolerate high loads and impact forces, which makes it highlighted among the others by satisfying all the needs that necessary for the leaf spring applications.

Table 2. Properties of materials.

Properties	Material			
	AISI 1090 carbon steel	6150 chrome vanadium steel	ASTM A299-B silicon manganese steel	TC 33 6K carbon fibre reinforced polymer
Density (g cm^{-3})	7.85	7.85	7.82	1.8
Poison's ratio	0.29	0.29	0.29	0.48
Brinell's hardness	250	270	190	1015
Tensile strength (MPa)	860	940	630	3450
Yield Strength (MPa)	490	620	360	3110
Elastic modulus (GPa)	202	210	190	230
Elongation at break (%)	11	22	21	1.5

Table 3. Weight of the leaf spring.

Material	Weight (Kg)
Carbon steel	35.91
Chrome vanadium steel	35.84
Silicon manganese steel	35.61
Carbon fiber reinforced polymer	8.27

Table 3 shows that Carbon fiber is significantly lightweight for the modeled leaf spring, which is 75% to 76% less compared to other materials chosen.

2.3 Analysis

The finite element analysis tool, ANSYS 16.0 was used to study the structural nature of the materials in the application of leaf springs. A fine mesh geometry that comprises a finite number of nodes and elements was generated to achieve precise results. The mesh model of the designed leaf spring is given in Fig. 2.

To determine the structural nature of the leaves, material properties are initially described. Boundary conditions are applied by assuming the leaves are said to be fixed at the front end and the rear end is allowed to move along X- coordinate and rotation along with Y- coordinate. Based on the overall payload capacity of the vehicle, a factor

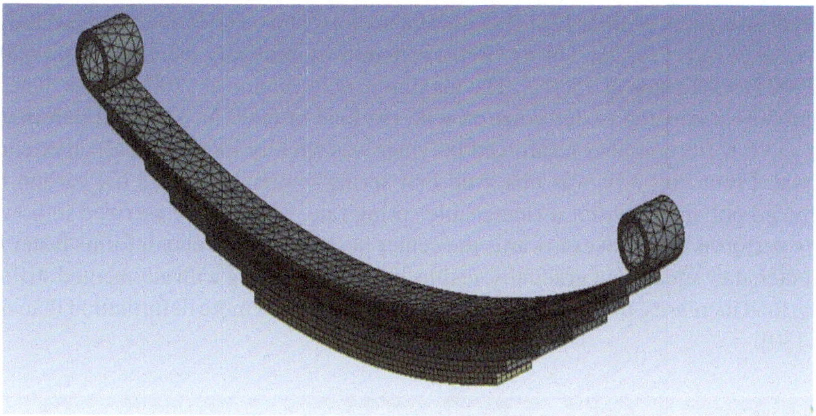

Fig. 2. A fine mesh geometry with a finite number of nodes and elements for the designed leaf spring from a Mesh model.

of safety (FOS), and using the quarter-car model calculation, loads varying from 6000 N to 12000 N were applied on the center of the bottom spring to determine the total deformation, stress, and strain of the springs [27]. Modal analysis was also performed to determine the natural frequency for different modes varying from 1 to 6.

3 Results and Discussions

3.1 Structural Analysis

Static condition results for total deformation, von-Mises stresses, and strains for varying loads are given in Tables 4, 5, and 6 respectively.

Table 4. Total deformation.

Load (N)	6150 Chrome vanadium steel		ASTM A299B silicon manganese		AISI 1090 carbon steel		TC 33 6K carbon fibre reinforced polymer	
	Min (mm)	Max (mm)	Min (mm)	Max (mm)	Min (mm)	Max (mm)	Min (mm)	Max (mm)
6000	0.0434	0.0438	0.0481	0.0484	0.0451	0.0455	0.0392	0.0396
7500	0.0543	0.0548	0.0605	0.0605	0.0564	0.0569	0.0491	0.0495
9000	0.0652	0.0657	0.0721	0.0726	0.0677	0.0683	0.0589	0.0590
12000	0.069	0.0876	0.0961	0.0969	0.0903	0.0911	0.0785	0.0792

From Table 4, it was observed that total deformation was gradually increased with an increase in load. A similar trend was found in composite leaf spring reinforced with different structures of fiber [28]. Load-deformation characteristics were measured as a performance index for leaf spring [29]. The load was gradually increased from 6000 N to 12000 N with a step of 1500 N. The maximum deformation at 12000 N was increased around 50% compared to deformation with the load of 6000 N. The total deformation of TC 33 6 K Carbon fiber reinforced polymer was slightly lesser than all other chosen material. From Fig. 3, it was observed that spring materials (TC 33 6K carbon fiber reinforced polymer) exhibit a changeable spring rate. This can be ascribed to how the spring performs as it moves towards the center position. As the end deforms faster than the center, they squash out gradually, results in a shorter spring with augmented stiffness. As the load increases, the center portion eventually exhibits more deformation than outer ends [30].

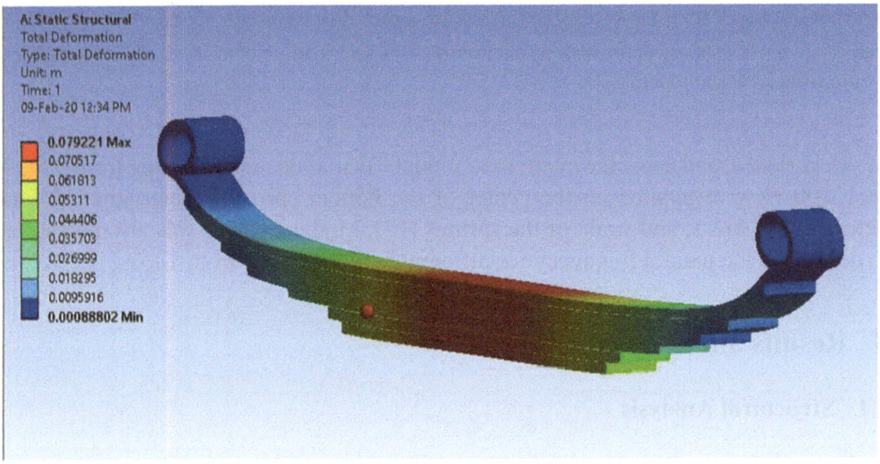

Fig. 3. Load deformation characteristics of carbon fiber reinforced polymer, showing maximum deformation at the center position under the load of 12000 N.

Table 5. Von-Mises stress.

Load (N)	6150 chrome vanadium steel		ASTM A299B silicon manganese		AISI 1090 carbon steel		TC 33 6K carbon fibre reinforced polymer	
	Min (MPa)	Max (MPa)	Min (MPa)	Max (MPa)	Min (MPa)	Max (MPa)	Min (MPa)	Max (MPa)
6000	18.7	21.1	18.7	24.8	18.7	21.1	18.8	20.8
7500	23.4	26.4	23.4	26.4	23.4	26.4	23.5	26.0
9000	28.1	31.7	28.1	31.7	28.1	31.7	28.2	31.2
12000	37.5	42.3	37.5	42.3	37.5	42.3	37.6	41.6

From Table 5, it was found that von-Mises stress values were increased with an increased load. It was evident that the linear relationship between load and stress. In general, applied load induces stress in the material. A similar trend was found in [31]. The stress values for all the selected materials except TC 33 6K carbon fiber reinforced polymer was in the range of 42.3 MPa. The stress value for TC 33 6K was slightly lesser than the remaining materials. The yield strength of the TC 33 6K (3110 MPa) was very high and it exhibited the stress value of 41.6 MPa. The factor of safety is around 74. The material can withstand up to a maximum load of 888 KN. On the other hand, the factor of safety for 6150 Chrome Vanadium steel, ASTM A299B Silicon Manganese, and AISI 1090 Carbon steel was 11.5, 14.6, and 8.5 respectively. Compared to all the selected materials, the factor of safety is very high for TC 33 6K carbon fiber reinforced polymer due to its higher yield strength. Figure 4 shows the von-Mises stress for TC 33 6K carbon fiber reinforced polymer at 12000 N load conditions. The produced stress value was very low compared to its yield strength. This revealed that the design for the leaf spring application is safe [32].

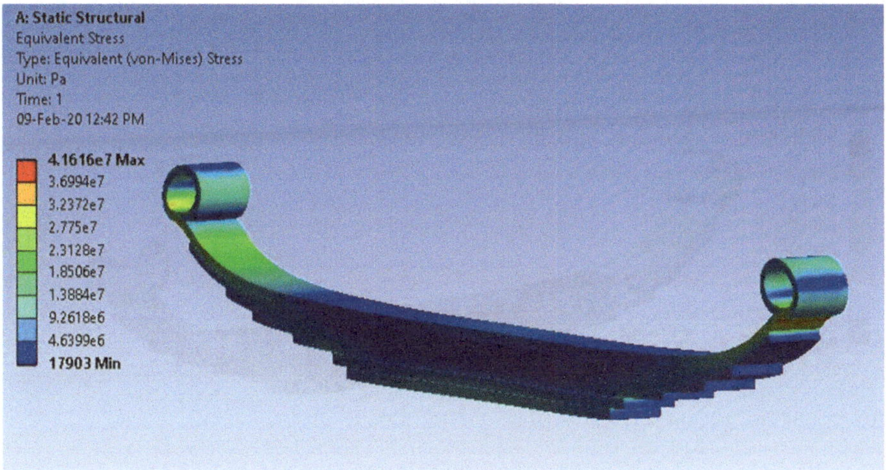

Fig. 4. Von-Mises stresses for the designed leaf spring made from carbon fiber reinforced polymer under the load of 12000 N, illustrating a safer design criterion.

The strain value for different load conditions was presented in Table 6. The strain values were gradually increased with the applied load. When the load increases, the material tends to yield and exhibits a strain. From Table 6, the strain value for TC 33 6K carbon fiber reinforced polymer was very low compared to other selected materials. Figure 5 shows the strain values for TC 33 6K carbon fiber reinforced polymer 0.18166 which is equivalent to a minimum strain of 6150 Chrome Vanadium Steel. The maximum strain produced at 12000 N was almost 1.13 to 1.23 times lesser than that of selected materials. From the results achieved, for all loading conditions, carbon fiber reinforced polymer is less deformed over other materials. Also, stress and strain were found minimum for carbon fiber reinforced polymer. The deformation, stress, and strain were found maximum at the center of the leaves, which shows that it gradually decreasing the vibration absorbed to the end of springs mounted to the vehicle.

Table 6. Von-Mises strain.

Load (N)	6150 chrome vanadium steel		ASTM A299B silicon manganese		AISI 1090 carbon steel		TC 33 6K carbon fibre reinforced polymer	
	Min (m/m)	Max (m/m)	Min (m/m)	Max (m/m)	Min (m/m)	Max (m/m)	Min (m/m)	Max (m/m)
6000	0.91e−4	1.01e−4	0.99e−4	1.11e−4	0.93e−4	1.05e−4	0.82e−4	9.08e−4
7500	1.12e−4	1.26e−4	1.24e−4	1.39e−4	1.17e−4	1.31e−4	1.03e−4	1.13e−4
9000	1.35e−4	1.51e−4	1.49e−4	1.67e−4	1.41e−4	1.57e−4	1.24e−4	1.36e−4
12000	1.81e−4	2.05e−4	1.99e−4	2.23e−4	1.87e−4	2.11e−4	1.65e−4	1.81e−4

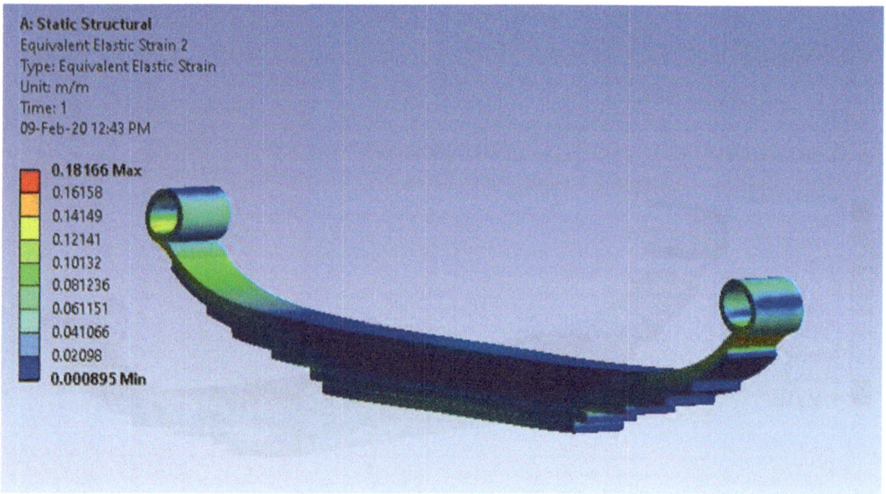

Fig. 5. Von-Mises strain of the designed leaf spring made of carbon fiber reinforced polymer at 12000 N loading condition, illustrating the deformation is less compared to other analyzed materials.

4 Modal Analysis

The road irregularities typically have a maximum frequency of 12 Hz [33], so the designed leaf spring should own a natural frequency higher than 12 Hz to circumvent the resonance. The stiffness of TC 33 6K polymer is higher than all the materials. Yet, the weight of the TC 33 6K polymer is lesser than the other selected material. Hence, the first natural frequency of TC 33 6K polymer is 31 Hz which is higher than all materials considered in this study. Using ANSYS, the first six natural frequencies of selected leaf springs are listed in Table 7.

From the results, carbon fiber reinforced polymer has a low natural frequency, comparatively. The first natural frequency of TC 33 6K leaf spring is nearly three times the road frequency and resonance will not occur [34]. The results obtained show that TC 33

Table 7. Natural frequency on model analysis.

Frequency (Hz)				
Modes	6150 Chrome Vanadium Steel	ASTM A299B Silicon Manganese	AISI 1090 Carbon Steel	Tairyfil Carbon Fibre TC 33 6K
1.	21.863	20.835	21.442	35.3659
2.	1148.5	1094.5	1126.4	1173.725
3.	3352.9	3195.4	3288.4	3528.95
4.	4842.1	4614.5	4748.9	4930.24
5.	6044.9	5760.8	5928.6	6441.76
6.	7085.1	6752.1	6948.7	7466.25

6K Carbon fiber reinforced polymer is an optimal material for leaf springs having better structural characteristics and less weight compared to the conventional materials.

5 Conclusion

The leaf springs were designed using CATIA V5. Static structural and modal analysis was carried out using ANSYS 16.0. The materials chosen for analysis were Silicone manganese steel, chrome vanadium steel, carbon steel, and carbon fiber-reinforced polymer. To find the most appropriate material for leaf springs, a comparative analysis was carried out. Based on the analytical results, TC 33 6K carbon fiber reinforced polymer was found to be relatively efficient, which has low deformation, less stress and strain values in contrast to other selected materials used for leaf springs. Carbon fiber leaf springs have high strength and weight is four times lesser than other selected materials. This infers that ideal and superior materials for leaf springs are Carbon fiber reinforced polymer.

6 Future Scope

In nearly all light and heavy commercial vehicles, leaf springs are used as primary suspension. The reduction of weight of these leaf springs by substituting traditional materials for composite material could well lead to a near reduction of 10 to 15% of the total vehicle weight, which in turn leads to optimized performance, improved mileage, and enhanced comfort of the vehicle as well. In addition to that, these composites materials having alluring aspects such as relatively low-density and lower weight, high strength of compression, viable and flexible manufacturing nature, and corrosion resistance would undeniably lead to a wide range in the future automobile world eventually.

References

1. Khatkar, V., Behera, B.K.: Experimental investigation of composite leaf spring reinforced with various fiber architecture. Adv. Compos. Mater. **29**(2), 129–145 (2020)
2. Mann, G.S., Singh, L.P., Kumar, P., Singh, S.: Green composites: a review of processing technologies and recent applications. J. Thermoplast. Compos. Mater. **33**(8), 1145–1171 (2020)
3. Karali, N., Abhyankar, N., Sharpe, B., Bandivadekar, A.: Improving Fuel Efficiency for Heavy-Duty Vehicles of 3.5–12 Tonnes in India: Benefits, Costs, and Environmental Impacts. Lawrence Berkeley National Lab (LBNL), Berkeley (2019)
4. Pradeep, S.A., Iyer, R.K., Kazan, H., Pilla, S.: Automotive applications of plastics: past, present, and future. In: Applied Plastics Engineering Handbook, pp. 651–673. William Andrew Publishing (2017)
5. Rai, V., Saxena, G.: Development of a composite leaf spring for a light commercial vehicle (Tata Magic). Int. J. Eng. Res. Appl. **1**, 110–114 (2013)
6. Tanabe, K., Seino, T., Kajio, Y.: Characteristics of carbon/glass fibre reinforced plastic leaf spring. SAE Trans. **91**, 1628–1636 (1982)
7. Dishant, E., Sharma, M.: Suspension systems: a review. Int. Res. J. Eng. Technol. **4**, 148–160 (2017)
8. Jun, K., Zhenyu, W., Zhiping, C.X.Y.: A review on material selection, design method and performance investigation of composite leaf springs. Compos. Struct. **226**, 111277 (2019)
9. Rajendran, I., Vijayarangan, S.: Optimal design of a composite leaf spring using genetic algorithms. Comput. Struct. **79**, 1121–1129 (2001)
10. Radhika, N., Sam, M.: Tribological and wear performance of centrifuge cast functional graded copper based composite at dry sliding conditions. J. Central S. Univ. **26**, 2961–2973 (2019)
11. Radhika, N., Sam, M., Thirumalini, S.: Experimental studies on mechanical and wear behaviour of TiC reinforced Cu-Sn-Ni functionally graded composite. Tribol. Ind. **41**, 537-547 (2019)
12. Mohanraj, T., Shankar, S., Rajasekar, R., Deivasigamani, R., Arunkumar, P.M.: Tool condition monitoring in the milling process with vegetable based cutting fluids using vibration signatures. Mater. Test. **61**, 282–288 (2019)
13. Shankar, S., Mohanraj, T., Pramanik, A.: Tool condition monitoring while using vegetable based cutting fluids during milling of inconel 625. J. Adv. Manuf. Syst. **18**, 563–581 (2019)
14. Tripathi, G.: Application and future of composite materials: a review. Int. J. Compos. Constituent Mater. **3**, 1–4 (2017)
15. Moon, I.D., Oh, S.H., Oh, C.Y.: Modeling of a multi-leaf spring for dynamic characteristics analysis of a large truck. J. Korean Soc. Precis. Eng. **21**(10), 147–153 (2004)
16. Kueh, J.-T.J., Faris, T.: Finite element analysis on the static and fatigue characteristics of composite multi-leaf spring. J. Zhejiang Univ. Sci. A **13**, 159–164 (2012)
17. Bhandarkar, D.K., Kushwaha, A: Weight and material optimization of mono leaf spring for light weight vehicle. Am. J. Art Des. **2**, 16–20 (2017)
18. Nataraj, M., Thillikkani, S.: Failure analysis of leaf spring suspension system for heavy load truck vehicle. Int. J. Heavy Veh. Syst. **27**, 1–7 (2020)
19. Suresh, J., Muneendra, K., Mohammed, R.: Design and analysis of a steel and composite material leaf spring. Int. J. Mod. Trends Sci. Technol. **5**, 12–20 (2019)
20. Thangapazham, P., Kumaraswamidhas, L., Muruganandam, D.: Evolution of 52CrMoV4 from 51CrV4 material to withstand field severity of parabolic leaf spring suspension in heavy-duty commercial vehicles. Trans. Can. Soc. Mech. Eng. **43**, 387–397 (2019)

21. Dinesh, M., Rubanrajasekar, B., Asokan, R., Vignesh, S., Rajesh, S.: Study on carbon, glass, and flax hybrid composites using experimental and computational techniques. In: Innovative Design, Analysis and Development Practices in Aerospace and Automotive Engineering (I-DAD 2018), pp. 131–143. Springer, Singapore 2019.
22. Yang, S., Shi, W., Chen, Z., Qian, C., Yang, C., Hang, L.: Composite mechanics and energy method based stiffness prediction model for composite leaf springs. Mech. Des. Struct. Mach. **47**, 375–386 (2019)
23. Suresh, G., Srinivasan, T., Chidambaram, C.T., Santhoshpriya, K., Vijayakumar, K.R., Kumar, M.D.: A study on sliding wear behaviour of carbon fiber reinforced IPN composites. Mater. Today: Proc. (2020)
24. Sreenivasan, M., Kumar, M.D., Krishna, R., Mohanraj, T., Suresh, G., Kumar, D.H., Charan, A.S.: Finite element analysis of coil spring of a motorcycle suspension system using different fibre materials. Mater. Today: Proc. **33**, 275-279 (2020)
25. Karataş, M.A., Gökkaya, H.: A review on machinability of carbon fibre reinforced polymer (CFRP) and glass fibre reinforced polymer (GFRP) composite materials. Defence Technol. **14**, 318–326 (2018)
26. Shokrieh, M.M., Rezaei, D.: Analysis and optimization of a composite leaf spring. Compos. Struct. **60**, 317–325 (2003)
27. Wang, S., Hua, L., Yang, C., Tan, X.: Nonlinear vibrations of a piecewise-linear quarter-car truck model by incremental harmonic balance method. Nonlinear Dyn. **92**(4), 1719–1732 (2018)
28. Khatkar, V., Behera, B.: Experimental investigation of composite leaf spring reinforced with various fibre architecture. Adv. Compos. Mater. 29(2), 1–17 (2019)
29. Subramanian, C., Senthilvelan, S.: Development and preliminary performance evaluation of discontinuous fibre reinforced thermoplastic leaf spring. Proc. Inst. Mech. Eng. Part L: J. Mater.: Des. Appl. **223**, 131–142 (2009)
30. Sancaktar, E., Gratton, M.: Design, analysis, and optimization of composite leaf springs for light vehicle applications. Compos. Struct. **44**, 195–204 (1999)
31. Shamim, S., Anwer, J.: Design and optimization of automotive multi-leaf spring by finite element method. Int. J. Res. Aeronaut. Mech. Eng. **2**, 46–54 (2014)
32. Zheng, S., Cheng, K., Wang, J., Liao, Q., Liu, X., Liu, W.: Failure analysis of frame crack on a wide-body mining dump truck. Eng. Fail. Anal. **48**, 153–165 (2015)
33. Yu, W.J., Kim, H.C.: Double tapered FRP beam for automotive suspension leaf spring. Compos. Struct. **9**, 279–300 (1988)
34. Kumar, M.D., Teja, P.S., Krishna, R., Sreenivasan, M. : Design optimization and simulation analysis of formula SAE frame using chromoly steel [Текст] // Журнал інженерних наук. - 2019. - Т. 6, № 2. - С. D9-D13 (2019). https://doi.org/10.21272/jes.2019.6(2).d2

Research of the Effect of Macrogeometric Structures on the Melt Front Using Simulation

Krisztian Kun(✉) and Zoltan Weltsch

GAMF Faculty of Engineering and Computer Science, Department of Vehicle Technology, John Von Neumann University, Kecskemét 6000, Hungary
kun.krisztian@gamf.uni-neumann.hu

Abstract. In plastics processing, sizing procedures in many cases form a compound system, which cannot described in analytical form. Nowadays, interacting flow behaviours of complex geometries are almost solely solved in simulation software. These are capable of mapping phase transitions in injection moulding. During the process, the tool cavity is filled with an expansion flow. As a result, a stretching flow develops on the surface of the melt front, resulting a high degree of orientation. In addition, the shear stress in the layers next to the wall also changes the confrontation of the molecules. Inside the melt front, on the other hand, there is a much smaller degree of orientation. In addition to the velocity gradient, there is also a temperature gradient. However, it should not be forgotten that the flow of the melt front is also greatly affected by the quality of the tool surface. Based on all this, a complex shell-core structure is formed in injection moulded products. The aim of this research is to investigate the effect of the macro geometric structure formed on the mould insert on the progress of the melt front. The simulation results were tested in practice with the same parameters.

Keywords: Injection moulding simulation · Surface structure · Flow front

1 Introduction

The development of plastic parts and their tools is one of the most complex engineering tasks today. In plastics processing, sizing procedures in many cases form a complex system that cannot be described in analytical form. Typically, interacting flow problems are solved using some kind of simulation software. The main objective of the research is to investigate the effect of different surface structures on the progress of the melt front, using simulations. The simulation results were tested in practice. Since the quality of the tool surface changes the filling process and the cooling efficiency, it is assumed that the ratio of the layers with different molecular orientations changes, so that the mechanical properties of the injection moulded products also changes. The present research examines two types of surface structures and aims to investigate the changes in mechanical properties that can be related to the results of the research.

S. Hloch et al. (Eds.): ICMEM 2021, LNME, pp. 282–289, 2021.
https://doi.org/10.1007/978-3-030-71956-2_23

2 Optimization of Injection Moulding Process Through Simulation

The central unit of the simulation programs is an algorithm that mathematically describes and solves physical phenomena. The analytical solution is based on continuity-, momentum-, energy- and rheological equations. The geometric CAD models, the material properties, the injection-moulding machine and the associated technological parameters provide the required input. Autodesk Moldflow Adviser 2019 software was used to simulate the tested processes [1].

2.1 Geometric Boundary Conditions

For the research, an experimental injection moulding tool were used, containing two symmetrically placed mould cavities. The product manufactured by the tool is two bending specimens. The main objective was to create two surface structures on the A plate's cavity that would affect the melt flow differently during injection moulding. Perpendicular (vertical) and parallel (horizontal) groove geometries with a depth of 100 μm were formed (Fig. 1). During the simulation, 3D mesh was used. This type is based on a volumetric mesh design [4, 5].

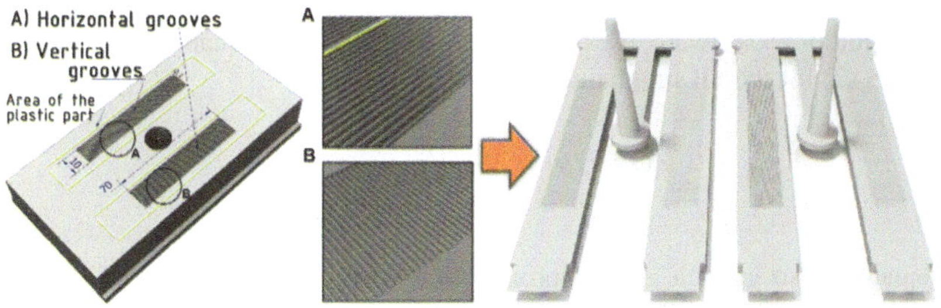

Fig. 1. CAD model of the cavity with the created surface structure (left) and two injection moulded products (right).

2.2 Material Properties, Technological Parameters and Solving Algorithms

After recorded the geometric boundary conditions, the raw materials were selected (Table 1). Due to the different behaviour, the use of an amorphous and a semi-crystalline material was obvious. During the phase transition, the density changes: the crystalline phase has a higher density than the amorphous phase or the melt, so the shrinkage of the semi-crystalline polymers is higher [2]. The choice fell on two materials used by the university's laboratory.

The viscosity of polymers is a time-, pressure- and shear rate-dependent behaviour. The applied simulation software use the Cross Eq. (1) because it calculates viscosity even at high shear rates [3, 4]

$$\frac{\eta_0 - \eta_\infty}{\eta - \eta_\infty} = 1 + k \cdot \dot{\gamma}^{\frac{2}{3}} \tag{1}$$

Table 1. The examined polymers and their properties.

Properties	Lexan 141R	Dowlex 2027G	Unit
Material type and degree of crystallinity			
	PC (Amophous)	LLDPE (Crystalline)	
Recommended processing temperatures			
Tool surface temperature	100	16	°C
Melt temperature	320	230	°C
pvT properties			
Density (melt)	1,0534	0,7288	g/cm^3
Density (solid)	1,1983	0,9282	g/cm^3
MFI (Melt flow index)	11	4	g/10 min

where:
η_∞ and η_0 – viscosity values for high and low shear rates.
k – time constant.
γ ˙ – shear rate or velocity gradient.

The Moldflow software uses the shape of the Cross equation with additional parameters (2) for the fill simulations:

$$\eta(T,\dot{\gamma},p) = \frac{\eta_0(T,p)}{1+\left[\frac{\eta_0(T)\cdot\dot{\gamma}}{\tau^*}\right]^{(1-n)}} \tag{2}$$

The equation also gives a good approximation when applied over a wider range of shear rates, where the transition between Newtonian and the intermediate section is τ^*. The slope of the equation is described by the value of (1-n). In this model, the value of the zero viscosity, as well as its pressure and temperature sensitivity is specified by separate factors. Thanks to these properties, it makes the model suitable for filling process simulations [3, 4].

The technological values of the simulation were set to the recommended processing parameters given in the raw material datasheet. The parameters of the applicable injection moulding machine were also taken into consideration. Following the simulation processes, the practical manufacturing was performed at the same settings. The structures were 100 μm deep, 60 μm wide, and 120 μm apart from each other, thus, a stepped shape pattern was formed. They were machined by using an industrial femtosecond laser. The pulse frequency was 188 kHz, the pulse width was 277 fs. The maximum pulse energy is 80 μJ, which induces a peak power of 280 MW [7, 8].

2.3 Simulation Results in Moldflow

It can be seen in Fig. 2 that the effect of the two surface structures on flow was significant: the grooves, which were parallel to the flow direction (A - horizontal) do not significantly slow down the flow velocity. The surface elements perpendicular to the flow (B - vertical) provide more resistance. The deceleration occurs during the filling of the structures. The colours indicated in the figures show the cavity filling time and the pressure drop in the cavity, which indicates how many MPa the injection pressure dropped after filling each range. It can be clearly seen that filling the vertical orbit was a more difficult task for the melt compared to the horizontal one, where a lower pressure drop can be expected [5].

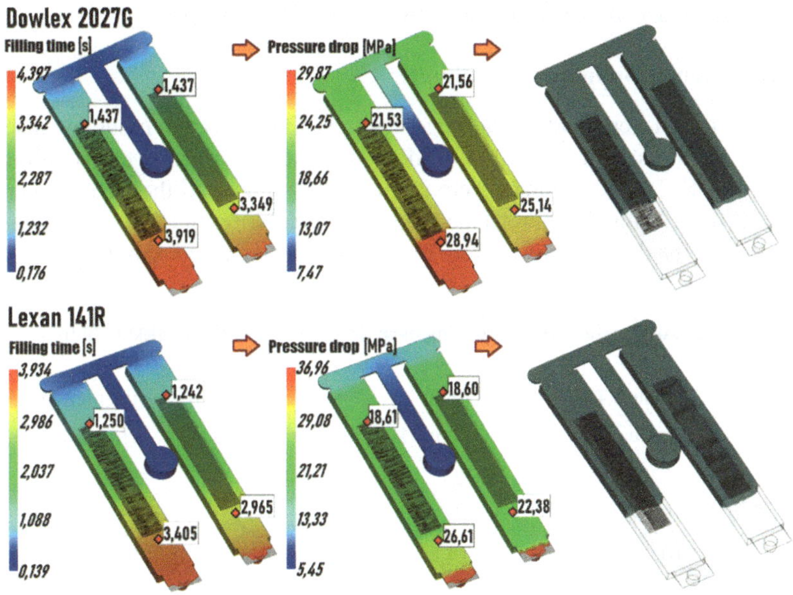

Fig. 2. Cavity filling time (left), pressure drop (middle) and filling process (right).

As shown in Fig. 3, during injection moulding, there is always some degree of molecule orientation in the product, which differs in cross sections. If a Newtonian fluid flows in a channel under isothermal conditions, the velocity of the layers next to the wall due to friction, will be lower than that of those flowing in the middle of the channel. It can be clearly seen that the formed surfaces also have an effect on the molecular orientation.

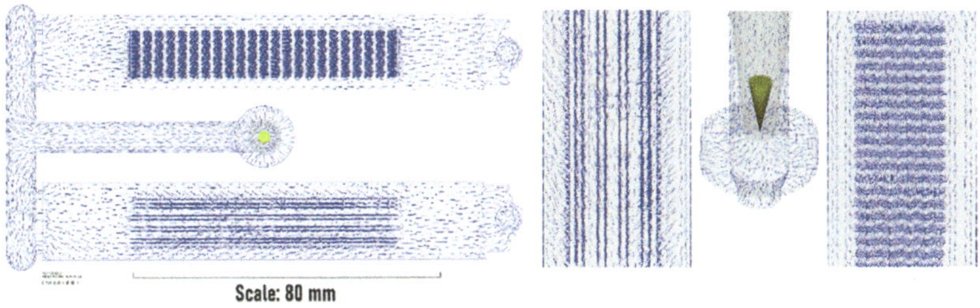

Fig. 3. The molecular orientation at the surface boundary layer.

As a result of the flow velocity profile, the orientation at the wall is greater than at the center of the channel. The layer covering the flow front expands on the tool wall and cools further. The speed of tool filling influences the formation of orientation. Since the quality of the tool surface changes the filling process and the cooling efficiency, it is assumed that the ratio of the layers in different orientations changes, so that the mechanical properties of the injection product also change [6].

3 Investigation of the Progress of the Melting Front

3.1 Experimental Boundary Conditions

The test specimens were made on an Engel emac 310/100 injection moulding machine. By varying the amount of injection moulded dose, the difference in cavity filling was examined (Table 2). The products were cut from the runner and then using a certified Mitutoyo 570–227 digital height gauge (Fig. 4) for recording data. The deviations after the injection of the different doses were measured through several samples.

Table 2. Average lengths of specimens at different injection moulded volume.

Injected dose [cm^3]	Lexan 141R product length		Dowlex 2027G product length	
	Vertical [mm]	Horizontal [mm]	Vertical [mm]	Horizontal [mm]
20	62.78	66.29	60.09	63.07
25	88.63	92.17	78.54	81.60
28	102.56	106.19	88.49	91.63
30	111.09	114.95	94.57	97.90

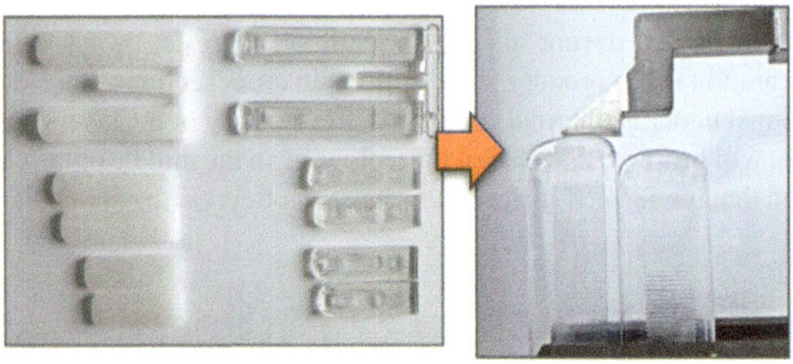

Fig. 4. Test specimens from the two materials (left) and measurement process (right).

3.2 The Effects of the Surface Structures on the Melt Front

After an injection of 20 cm^3. The melt passed head-to-head in the two cavities and then the filling of the structured surface begins. The changes in length between each products form three distinct measuring sections: 20 to 25. 25 to 28 and 28 to 30 cm^3. The diagram on Fig. 5 shows the path taken by 1 cm^3 of melt in the two mould cavities. The results clearly show that the melt front slows down continuously in the formed structure. Furthermore. it can be stated that the filling of the perpendicular (vertically) structured cavity slows the flow melt more than the parallel (horizontal). This can be observed in both PC and LLDPE materials.

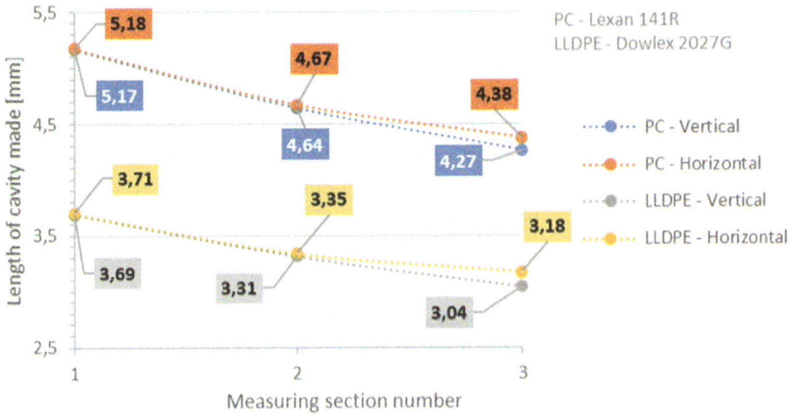

Fig. 5. Distances made by 1 cm^3 melt in each measuring sections.

A decrease in the filling speed has an effect on the pressure distribution. The temperature distribution and through them the deformation (shrinkage and warping). It is important to note that the values shown are calculated results. Which is why a dotted line represents the function. The difference between the two materials (Fig. 6) is also due to their flow ability. The results depicted were also supported by simulation.

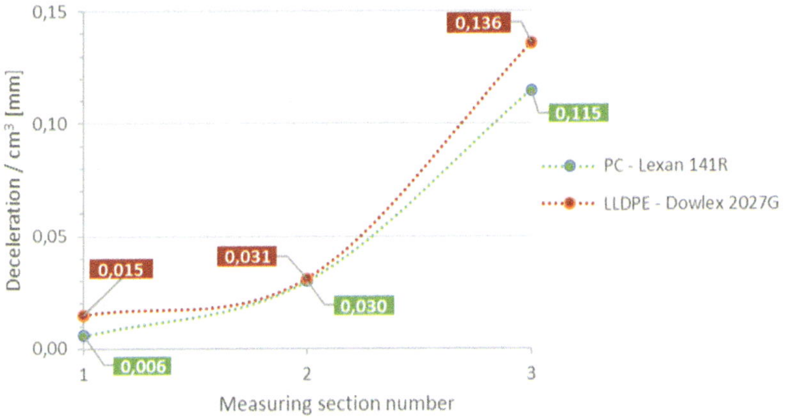

Fig. 6. Vertical path deceleration for the two material types.

The shell-core structure formed in polymer products can be influenced not only by the processing parameters but also by the quality of the forming surface. The effect is very complex because the surface quality also determines the adhesion of the melt as well as the heat transfer conditions. In addition. of course. it also has an aesthetic effect on the product.

4 Summary

In this research. The effect of surface structures on polymer melt flow were investigated by simulation and practical measurements. In the two cavity. The structures were formed parallel and perpendicular to the flow direction of the melt. Which affected the filling differently. The filling time. The pressure drop. And the melt front deceleration were examined. The research also investigated the molecular orientation formed in the product. As for quantifiable results. The effects of the designed surface structures on the flow properties of the applied materials were proved by simulations and verified practical measurements.

These phenomena may affect the mechanical properties of polymer products. Examining interactions and searching for regularities requires further research. The next step in the research could be to investigate the effects of surfaces on heat transfer that are directly related to molecular orientation. The long-term goal is to form surface structures that. Through the flow of the melt front. With a guided molecular orientation. Could shape the emerging mechanical properties according to different needs.

Acknowledgement. The project has been supported by the European Union. co-financed by the European Social Fund. EFOP-3.6.1–16-2016–00014.

References

1. Eladl, A.E., et al.: Effect of process parameters on flow length and flash formation in injection moulding of high aspect ratio polymeric micro features. Micromachines **9**(2), 58 (2018). https://doi.org/10.3390/mi9020058
2. Bánhegyi, G.: Kristályos és amorf polimerek termikus vizsgálata differenciális pásztázó kalorimetriával (DSC). In: Műanyagipari szemle (2014)
3. Kovács, J.G.: Fröccsöntési szimulációs programok elemzése. Műanyag és Gumi. **38**(9), 350–358 (2001)
4. Autodesk Knowledge Network – Moldflow Adviser. https://help.autodesk.com/view/MFAA/2019/ENU/?guid=GUID-7BC3A8F0-8B41-4FCB-BDF1-F1159E4DD175. Accessed 05 Jan 2020
5. Kun, K., Boza, P.: Investigation of inserts surface structures on injection moulded parts. In: IOP Conference Series: Materials Science and Engineering, vol. 448, p. 012052 (2018). https://doi.org/10.1088/1757-899X/448/1/012052
6. Sienz, J.: Applied mathematical modelling. In: Huang-Ya, L., Wen-Bin, Y. (eds.) Analysis of the Filling Capability to the Microstructuresin Micro-Injection Molding. Elsevier (2008). https://doi.org/10.1016/j.apm.2008.12.012

7. Temmlera, A., Willenborg, E., Wissenbach, K.: Design surfaces by laser remelting physics procedia lasers in manufacturing 2011. In: Proceedings of the 6th International WLT Conference on Lasers in Manufacturing. 2011, Part A, vol. 12, pp. 419–430. https://doi.org/10.1016/j.phpro.2011.03.053
8. Preußner, J., Oeser, S., Pfeiffer, W., Temmler, A., Willenborg, E.: Microstructure and residual stresses of laser structured surfaces. Trans Tech Publications Ltd. Switzerland (2014). https://doi.org/10.4028/www.scientific.net/AMR.996.568

Replication of Microstructures Formed by Femtosecond Laser During Injection Moulding

Krisztian Kun(✉), Janos Líska, and Zoltan Weltsch

GAMF Faculty of Engineering and Computer Science, Department of Vehicle Technology, John Von Neumann University, Kecskemét 6000, Hungary
kun.krisztian@gamf.uni-neumann.hu

Abstract. Due to increased customer needs in automotive industry, the quality of plastic parts is an increasingly relevant issue. Besides processing parameters, it is also crucial how to design the mould. However, it is important to know how the processing parameters affect the mould surface mapping. During injection moulding in the tool cavity, the velocity of the layers next to the wall will be lower than that of those flowing in the middle of the channel, due to friction. In the case of plastics, due to the pseudoplastic behaviour, the velocity profile between the layers flowing parallel in the middle of the channel flattens and the macromolecules are oriented parallel to the direction of flow. The shear stress in the layers next to the wall also changes the clashes of the molecules and forming a complex shell-core structure. Therefore the shell-core structure and the flow of the melt front is influenced by the processing parameters and also the quality of the tool surface. Based on these, the cavity surface mapping is also depends not only the geometric conditions, but also on the technological parameters. During the research, two microstructured surface created, with different effects on melt flow formed by femtosecond laser. For examining the effect of technological variables on surface mapping an experimental design were used.

Keywords: Surface replication · Femtosecond laser · Microstructures injection moulding

1 Introduction

During injection moulding, there is always some degree of molecular orientation in the product, which differs in the cross sections. The surface structure of the tool changes the filling process and the cooling efficiency. For this reason, it can be observed that the ratio of the layers with different molecular orientations changes. A related phenomenon of polymers can be that mechanical properties can also change as a result of molecular changes. However, to understand the effect of surface structures on orientation, we must first understand the filling process. This research provides a basis for further investigations by give a technology recommendations for filling.

S. Hloch et al. (Eds.): ICMEM 2021, LNME, pp. 290–300, 2021.
https://doi.org/10.1007/978-3-030-71956-2_24

2 Geometric Design of the Experimental Mould

For the research, a special injection mould was designed and manufactured, which include two symmetrical cavities. The products manufactured by the tool were two bending specimens. The aim was to create two surface structures on the A plate's cavity that would affect the melt flow differently during injection moulding. Perpendicular (vertical) and parallel (horizontal) groove-geometries with a depth of 100 μm were formed (Fig. 1). These were machined by an industrial femtosecond laser.

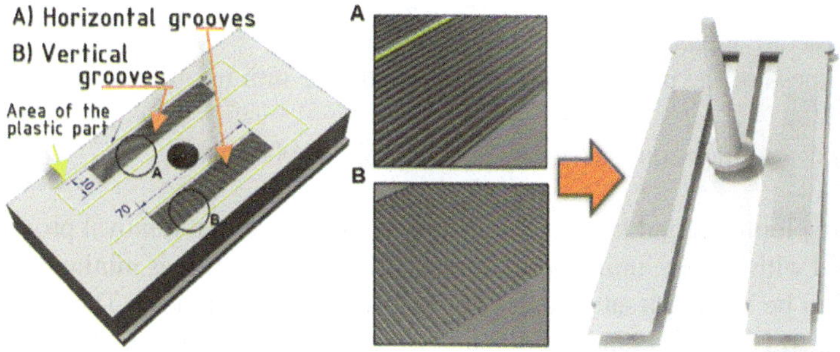

Fig. 1. CAD model of the cavity with the created surface structure (left) and two injection moulded products (right).

The goal was to apply a manufacturing technology that can achieve a uniform microstructure (Fig. 2) on the injection mould insert. The process parameters set to make the geometries as regular as possible and the heat zone as small as possible. Particular attention was paid to avoiding technology-specific errors such as plasma formation, possible melt residues and oxidation [1].

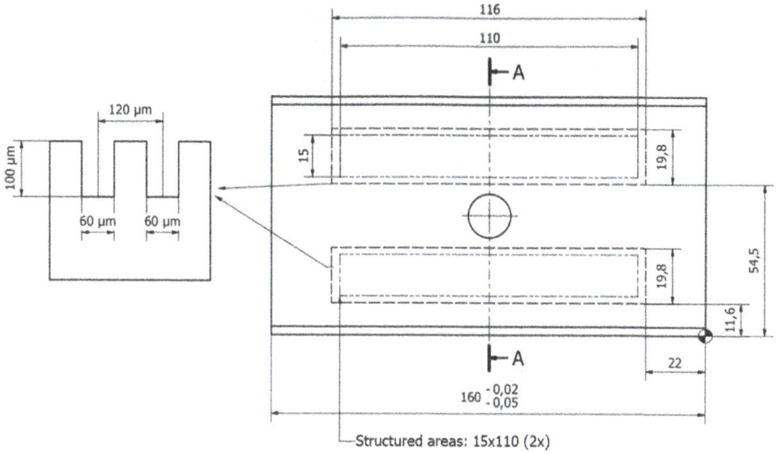

Fig. 2. The drawings of the A plate's cavity with the structured areas.

2.1 Structuring of the Cavity Using a Femtosecond Laser

According to several literature, the result of femtosecond laser machining is influenced by several factors. Researchers have been able to achieve significant results using the technology. Prominent surface quality can be achieved on thin (>0.5 mm) materials. However, for thicker materials it is much more difficult to achieve a similar result. As far as the parameters are concerned, there is no consensus among researchers: they have studied different pulse widths, frequencies and power, with various results [5–7].

As in previous research [2, 3]:

- microstructure means surfaces below 200 μm,
- the thickness of the product must be greater than 2 mm in order to obtain a suitable filling,
- the expected dimensional accuracy is 0.127 mm (0.005 inch)

By using femtosecond laser, a controlled, accurate material removal processing can be expected with quality, uniform surface, where burr formation is minimal. The cavity insert had to be surface treated under argon gas during machining. The used machine was a Monaco 1035-80-40 industrial femtosecond laser. The laser beam using a LINOS F-Theta-Ronar type scanner optics with $f = 254$ mm focal length lens (Fig. 3). The used type of material was C45 structural steel.

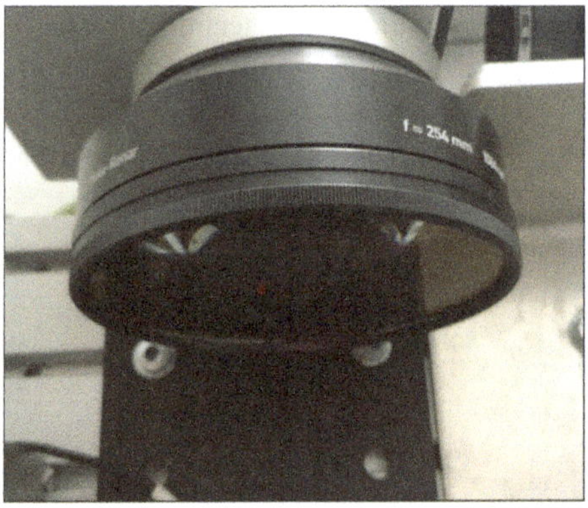

Fig. 3. LINOS F-Theta-Ronar type scanner optics with $f = 254$ mm focal length lens.

The technological parameters used were as follows: The pulse frequency was 188 kHz, the pulse width was 277 fs, and the laser operated at 25% power compared to the average power of 60W. The interesting thing about the technology is that the laser is able to generate very short pulses of a few hundred femtoseconds in length, so even though the average power remains low, the peak power can be up to 280 megawatts. In our case, the maximum pulse energy was 80 μJ. The geometry thus formed did not prove to be deep enough compared to the design, so had to scan 30 times in a row for more accurate depth (Table 1).

Table 1. The technical parameters used of the Monaco 1035-80-40 type femtosecond laser.

Wavelength (nm)	1035 ± 5
Average power (W)	60
Energy (μJ)	80 (at 500 kHz)
Frequency range	188 kHz
Pulse width (fs)	270 fs
Modus	TEM00, M2 < 1.2

2.2 Analysis of the Micro Structure Formed on the Cavity's Surface

After forming micro-grooves parallel and perpendicular to the melt flow on the surface of the mould cavity, the surface was examined using a Keyence VHX-2000 digital microscope. The main question was whether the planned depth of 100 μm was reached. It can be clearly seen in the figure that the average structure depth was 83.1 μm, which approximated the planned depth of 100 μm [1] (Fig. 4).

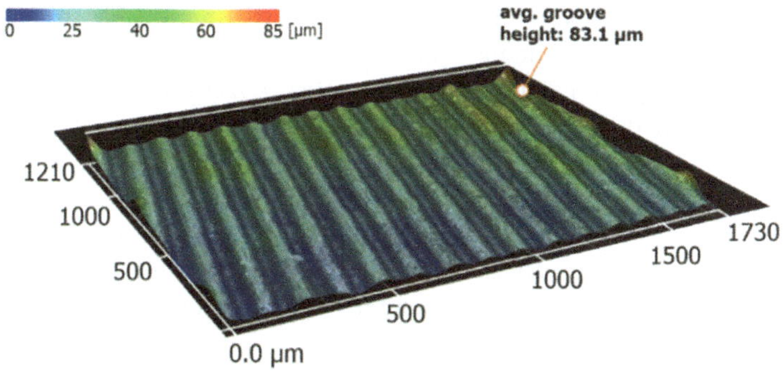

Fig. 4. Microscopic image of the cavity's surface with the designed microstructure dimensions.

2.3 The Polymer Material Used for Injection Moulding

The specification of the geometric boundary conditions was followed by the choice of polymer material. It was obvious to use a material that typically occurs in an area of industry where microstructures can be encounter. The choice of raw material was made of a polypropylene (PP), which is a thermoplastic crystalline polymer. Manufacturer name: TATREN RM 85 82 Clear. This material is suitable for research because of its high flow ability and excellent processing stability. Recommended for the production of thin-walled injection moulded products where clear transparency is required. Typical applications are transparent cases, caps, ribbed lids, etc. Its physical property is in the Table 2 [8].

Table 2. The properties of the examined polymer [8].

Properties	TATREN RM 85 82 Clear	Unit
Material type and Degree of crystallinity		
PP (Crystalline) Random copolymer		
Recommended processing temperatures		
Melt temperature	190–230	°C
pvT properties		
MFI (Melt flow index)	85	g/10 min
Mechanical properties		
Tensile stress at yield	30	MPa
Tensile strain at yield	12	%
Flexural modulus	1250	MPa

3 The Experimental Method

The Taguchi experimental design is a design method with a traceable structure and an efficient one that allows us to find the optimal combination of process parameters. By using it, the right technology can be found for quality products which includes the most fulfilled depth of microstructures.

3.1 Experimental Boundary Conditions

Mean values were determined based on the performance of the injection moulding machine and the properties of our raw material. The machine settings are summarized in Table 3.

Table 3. Machine settings for injection moulding. The three variables used in the experimental design (A, B, C) are highlighted.

Dose		**Filling parameters**	
Peripheral speed	200 mm/s	Switching pressure	350 bar
Ram pressure	50 bar	Injection speed	**(A)** 80 cm^3/s
Injected dose	40 cm^3	Pressure	2000 bar
Decompression		**Compression**	
Speed	3 cm^3/s	Switching point	14 cm^3
Injected dose	3 cm^3	Back pressure (pressure)	**(B)** 350 bar
Injection unit		Back pressure (time)	30 s
Nozzle	200 °C	Cooling time	**(C)** 30 s

Following Taguchi's philosophy, a strategy developed at which the variables do not interact with each other. The experimental design was created in Minitab 19 software. Based on experience, three factors had to considered, namely injection speed (A), back-pressure (B), and cooling time (C). In order to create three variables, a deviation of – 30% and + 30% from the mean value was set. The experimental design is shown in Table 4.

Table 4. Taguchi experimental design, generated in Minitab 19 [7].

#	Injection speed [cm^3/s]	Backpressure [bar]	Cooling time [s]
1	56	245	20
2	56	350	30
3	56	455	40
4	80	245	30
5	80	350	40
6	80	455	20
7	98	245	40
8	98	350	20
9	98	455	30

During injection moulding, a robot was used to remove the part after ejection. A total of nine different experiments were performed and six specimens were injection moulded for each experiment to ensure that the product-specific setting was correct (Fig. 5).

Fig. 5. Injection moulded products are sorted by experimental number (left) and the microstructured tool for injection moulding in the machine workspace (right side).

3.2 Microscopic Analysis and Evaluation of the Mapped Structures

A ZEISS Axio Imager.M2m microscope was used to evaluate the filling and to examine the microstructures (Fig. 6). The depths were able to measure using autofocus: Focused first on the deeper surface and then on the upper surface. The difference between the two gave the exact groove depth. The parallel and perpendicular microstructures were separated, taking 3-3 measurement points at the beginning, middle and end of the track on each surface.

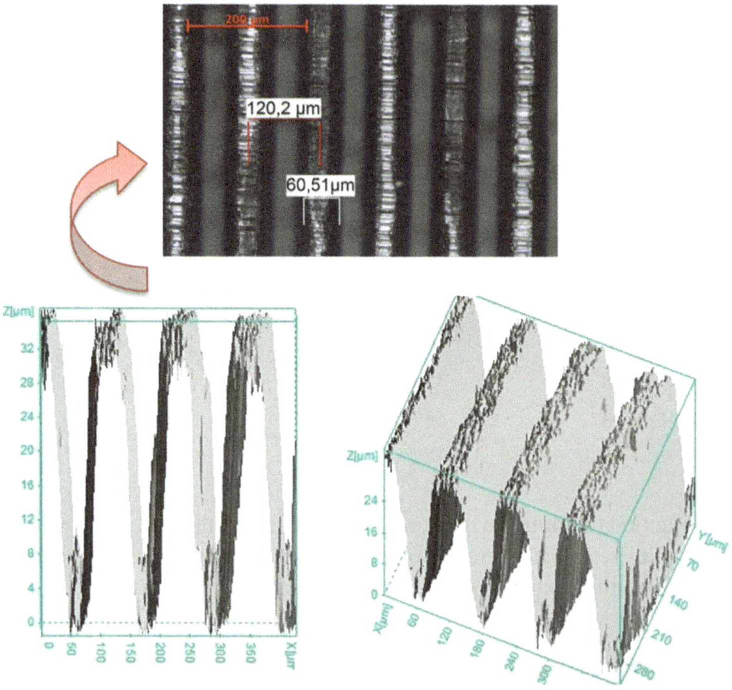

Fig. 6. Microscopic view: Test specimen with No.3 experiment parameters, parallel grooves.

4 Evaluation of Measurement Results

The results showed that, under some technological parameters, the parallel grooves guided the plastic melt as opposed to the perpendicular. The phenomenon can be explained by the fact that the melt is more difficult to replicate perpendicular microstructures during moulding due to the orientation of the giant molecules. The movement of the melt in parallel grooves was in the same direction as the microstructures orientation, which made the replication much more uniform. At the perpendicular surfaces the furthest part of the grooves from the inlet showed incomplete filling at certain settings. On the parallel side, this phenomenon was not observed. The experimental design generated in the Minitab software could be evaluated after recording the measurement results (depths). The values of the depths measured on the specimens containing perpendicular and parallel microstructures were recorded in an Excel table (Table 5). Of the nine studies, six products were measured per experiment from parallel and six from perpendicular pieces. Three measurement points per product were examined, at the beginning, middle, and at the end of the structured surface. This resulted in a total of 324 measurements. The means shown the average numbers of the three measurement points per products belonging to each experiment [4].

Table 5. Mean values of microscopic measurements on the two different surface structures.

Number of experiments	Perpendicular structures (vertical) [mm]	Parallel structures (horizontal) [mm]
1	0.0141	0.0238
2	0.0151	0.0376
3	0.0223	0.0362
4	0.0279	0.0397
5	0.0298	0.0482
6	0.0347	0.0493
7	0.0412	0.0558
8	0.0403	0.0510
9	0.0457	0.0580

The purpose of the evaluation is to determine which factors affect the expected depth values and which affect their variance. The Taguchi experimental design calculates signal-to-noise ratios (SN) closely related to robustness. The signal-to-noise ratio filters out the increase or decrease from the standard deviation. Caused by the mean. Having the average values summarized in Table 5 the Taguchi experimental design can be evaluated using Minitab. The method used for evaluation was the Larger is better (1) [7].

$$SN_L = -10\log\left(\Sigma_i\left(\frac{1}{y_i^2}\right)/n\right) \tag{1}$$

After ranking the factors. The greatest influence was the injection speed. 30% decrease from the mean value led to a drastic decrease in the filling. While a 30% increase resulted good surface mapping and replication. This phenomenon can be observed both in parallel and in perpendicularly structured cavities. In the perpendicularly structured cavity. it can be observed that. There was no significant change by reducing the backpressure by 30%. But by increasing the value. The results showed significant improvements. However. in the parallel structured cavity. Reducing the backpressure by 30% significantly reduces the filling. While increasing the value does not cause a significant improvement. The cooling time had the least effect in both cases. But for parallel structured surfaces it is not negligible. The effect diagrams generated by the Minitab software is shown in Fig. 7.

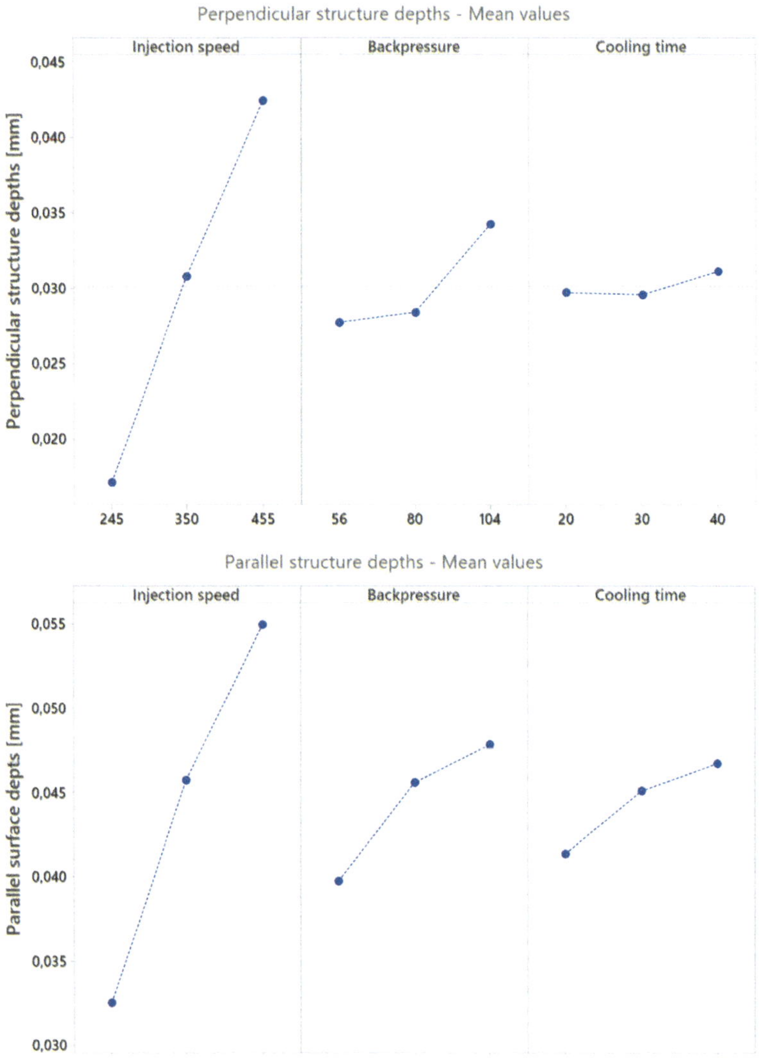

Fig. 7. The effect diagrams generated by the Minitab software.

Based on the results. The recommended technology for injection moulding for the best replicated microstructures:

- Injection speed: 104 cm^3/s
- Backpressure: 455 bar.
- Cooling time: 30 s.

The regression equation was used to describe the relationship between the response and the terms in the model. The regression equation is an algebraic representation of the response contours (Fig. 8) [3, 7]. The regression equation was used with coded units.

The low level of a factor is −1 and the high level of a factor is + 1. If the response surface contains curvature. Then a polynomial model of higher degree is used.

The second-order model is (2):

$$Y = \beta_0 + \sum_{i=1}^{k} \beta_i x_i + \sum_{i=1}^{k} \beta_{ii} x_i^2 + \sum\sum_{i<j} \beta_{ij} x_i x_j + \varepsilon \tag{2}$$

- Y is the response variable.
- β_0 is the constant and $\beta_1. \beta_2. \ldots \beta_k$ are the coefficients.
- $x_1. x_2. \ldots x_k$ are the values of the terms.
- ε is the error term [7].

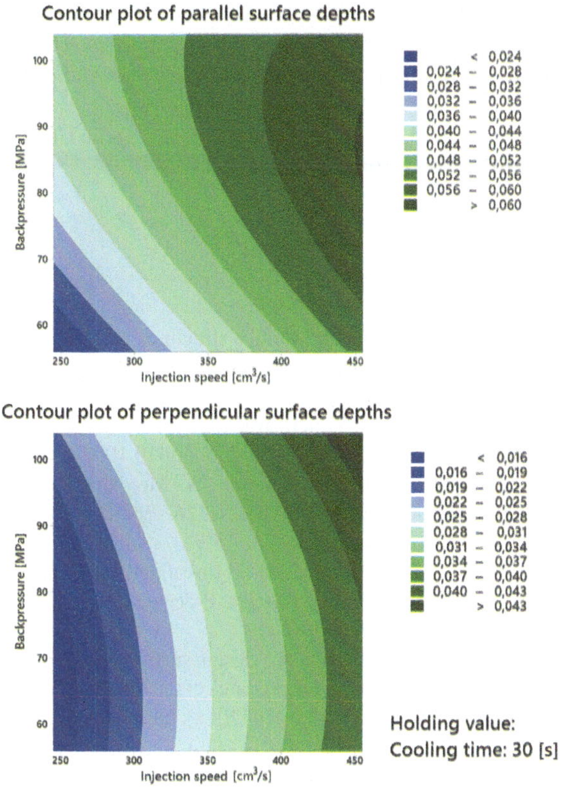

Fig. 8. Contour plot diagrams generated by the Minitab software.

Using the diagrams in Fig. 8. The quality of the fill can be predicted from the combination of the two most influential factors. Taking into account the interactions.

5 Summary

In this research polymer test specimens with femtosecond laser formed microstructures were examined during injection moulding. In the two cavity. The structures were formed parallel and perpendicular to the flow direction of the melt. Which affected the filling differently. Laser-formed structures were examined with laboratory microscopes. The designed surface structures were completed with a good approximation due to the applied process parameters. Thereafter the filling of the cavity's structures and the surface replication were examined. To achieve this. A Taguchi experimental design was compiled. Following the completion of the experimental design. The evaluation resulted an injection moulding process technology recommendation for a detailed surface replication. The result of the research are two effect diagrams (contour plot) that can be used to predict the filling.

The future plans are to build on the foundations of these results to investigate the effects of different surface structures on mechanical properties. The right technological parameters defined in this research will be essential to perform the future experiments.

Acknowledgement. The project has been supported by the European Union. co-financed by the European Social Fund. EFOP-3.6.1-16-2016-00014.

References

1. Kun, K., Boza, P.: Investigation of inserts surface structures on injection moulded parts IOP Conf. Ser.: Mater. Sci. Eng. **448** (2018). https://doi.org/10.1088/1757-899X/448/1/012052. Paper: 012052
2. Theilade, U.A., Hansen, H.N.: Surface microstructure replication in injection molding. Int. J. Adv. Manuf. Technol. **33**, 157–166 (2007). https://doi.org/10.1007/s00170-006-0732-y
3. Holthusena, A.K., Riemera, O., Schmützb, J., Meiera, A.: Mold machining and injection molding of diffractive microstructures. J. Manuf. Process. **26**, 290–294 (2017). https://doi.org/10.1016/j.jmapro.2017.02.014
4. Sha, B., Dimov, S., Griffiths, C., Packianather, M.S.: Investigation of micro-injection moulding: factors affecting the replication quality. J. Mater. Process. Technol. **183**, 284–296 (2007). https://doi.org/10.1016/j.jmatprotec.2006.10.019
5. Temmlera, A., Willenborg, E., Wissenbach, K.: Design surfaces by laser remelting. In: Physics Procedia Lasers in Manufacturing 2011 - Proceedings of the 6th International WLT Conference on Lasers in Manufacturing, Part A, pp. 419–430 (2011). https://doi.org/10.1016/j.phpro.2011.03.053
6. Preußner, J., Oeser, S., Pfeiffer, W., Temmler, A, Willenborg, E.: Microstructure and residual stresses of laser structured surfaces, pp. 568–573. Trans Tech Publications Ltd., Switzerland (2014). https://doi.org/10.4028/www.scientific.net/AMR.996.568
7. Minitab Software Support – Regression Equation. https://support.minitab.com/en-us/minitab/. Accessed 10 May 2020
8. Technical Datasheet of Polypropylene TATREN RM 85 82 C. https://www.molgroupchemicals.com/userfiles/products/80/80_tds_en.pdf. Accessed 05 May 2020

Design of a Rig to Assess the Structural Performance of Rotors for High Speed Electric Machines

Levi Mallin(✉)

University of Huddersfield, Huddersfield HD1 3DH, UK
Levi.Mallin@hud.ac.uk

Abstract. The usage of high-speed electric machines (HSEM) has increased due to the performance gains over mechanical transmission. However, the permanent magnets (PM) used within the electromagnetic rotor have low tensile strength and are susceptible to failure. For high-speed applications, surface-mounted PM rotors are most suitable, consisting of three interfering components. Accurate mechanical analysis of the rotor is required to predict a successful rotor design. Generalised plane strain (GPS) theory has been developed for three-cylinder rotors and requires experimental validation. In this paper, a test rig and suitable rotor are successfully designed to experimentally verify the GPS theory, with reduced complexity when compared to those in existing literature. The test rig enables the use of digital image correlation (DIC) to analyse the rotor surface strain, requiring a rotor design that produced enough outer surface strain to be detected. The result can then be compared to the GPS theory prediction for validation.

Keywords: Test rig · Digital image correlation · High-speed electric machines · Permanent magnet rotor

1 Introduction

1.1 Motivation

When exploring rotors in HSEMs, the existing literature rarely uses testing to validate the theoretical or FEA stress results. Using a pass or fail criteria on rotor test success, appears to be the extent of validating stress predictions, as shown in [14]. This paper will investigate existing test rig designs and determine a suitable rig for validating various stress analyses.

1.2 Background

HSEM usage has increased over the last decade due to the performance benefits over mechanical transmission, as discussed by [1]. Benefits include efficiency [1–3], reliability and power density [4, 5] due to the electromagnetic forces inducing rotation [6, 7]. Electromagnetic forces result in less parts in contact within the machine, improving

S. Hloch et al. (Eds.): ICMEM 2021, LNME, pp. 301–323, 2021.
https://doi.org/10.1007/978-3-030-71956-2_25

efficiency and reliability while the PMs increase the power density. An increase in power density enables a reduction in rotor size and weight, thus increasing the operating speed [8]. Samarium Cobalt or Neodymium are the high-energy sintered PMs that are typically used in HSEMs according to [9]. Applications of HSEMs include compressors, turbine generators and medical drills [8], while [7] adds turbochargers and fuel pumps. All operate over 10,000 rpm.

'Interior' (IPM) and 'surface-mounted' (SPM) PM rotors are the two configurations used within HSEMs. IPM rotors have the PMs buried within the rotor core, whereas SPM rotors have the PMs bonded to the surface of the rotor core. [10, 11] recommend SPM rotors as most suitable for high-speed applications, due to the stress concentrations generated by the magnet slots in IPM rotor designs, which limit operating speed. A comprehensive review conducted by [12] examined the stress analysis methods for SPM rotors undertaken across the field, including hand calculations and finite element analysis (FEA). [12] concluded that the common theoretical approaches of plane stress and plane strain were unsuitable and recommended the development of generalised plane strain (GPS) theory for three-cylinder SPM rotors. The GPS theory was developed in [13] and was shown to have a much stronger alignment with FEA simulations, specifically for long, thin rotors. However, testing was not discussed in [12] or [13], which is required to validate stress prediction models. A suitable test rig must be designed to enable the validation of the GPS theory and other stress analyses.

2 Literature Review

2.1 Power Supply

Of the reviewed literature, over 75% of conducted tests were completed with HSEMs used as the power supply. [14–37] all use electromagnetic forces to rotate their test rotors. Designing a HSEM to test a rotor provides very realistic operating conditions and performance while it also utilises the power density gains to produce a smaller test rig with a higher speed range. HSEM rigs are beneficial when the electromagnetic performance is also of importance to the author. Aspects such as windage losses [29], back EMF [17] and magnet losses [26] can be explored when testing HSEMs. However, when the focus is the mechanical performance of the rotor, HSEMs may be unnecessarily complex. To analyse the accuracy of stresses produced in the rotor, the rotor will need to be accessible to measuring devices. HSEM based test beds used in the literature typically have the rotor surrounded by the stator, obstructing the rotor.

The second most common approach was an air turbine, as shown in [7, 29, 36, 38]. Air turbines can be utilised to rotate rotors by attaching the rotor to the air turbine shaft. [7, 38] tested an electrically assisted turbocharger which included the complexities discussed above for HSEMs. A cold gas stand was used to rotate the turbine in [38], while [7] showed engine exhaust gases being used from a Cummins engine test bed. The turbine was only used by [36] to identify air friction losses, while the rotation was created via a HSEM setup. Only [29] focussed on the mechanical aspects of the rotor design, using the turbine to rotate the rotor and reducing the complexity of the rig. The expansion of the rotor was measured so a stator was not required, and the magnets could be swapped for stainless steel. Air turbines reduce the complexity requirement of a test rig when compared to HSEMs and open the possibility of accessing the rotor to obtain mechanical analysis measurements.

Another method of powering a rotor is to use a standard air motor. [39] shows an air motor shaft coupled to the rotor. However, to overcome the inertia of the rotor, an electric motor with a pulley system across the rotor was employed. This method also reduces complexity against HSEMs and enables the rotor to be seen while testing.

2.2 Stress Measurement Methods

Measuring rotor stresses during operation is required to validate the GPS theory. It is important to predict accurately due to the magnets used in HSEMs being a limiting factor in the design. As stated by [9], the sintered magnets are weak in tension and become the critical factor in a mechanical design, as the operating speed increases. Magnet failure is a high risk when assembling and testing electromagnetic rotors. The current practice used to validate HSEM rotor stresses appears to be a pass or fail criteria based on if the rotor fails during operation, as shown by [14, 29, 37, 40]. This does not confirm whether the theoretical or FEA analyses were accurate. Most studies into HSEMs do not tend to focus on the mechanical aspect of the rotor.

Methods of analysing stresses from rotating components are available despite the lack of exploration in the HSEM literature. Strain gauges coupled with a slip ring can extract the values produced by the gauges, as used in [41]. Strain gauges require attaching to the rotor surface and only measure strain at their placed location, while [42] states that slip rings are problematic and suffer from measurement noise.

Non-intrusive methods of analysing stresses are available, as described in [43]. Using dynamic photogrammetry, optical target points on components can be tracked via a software. This enables the displacement of those targets to be analysed when compared to a reference stage. However, this method is typically used for a small number of targets on a component, which can be seen in use by [44] on a rotating helicopter blade.

Digital image correlation is a similar technique, but provides analysis across whole surfaces to show continuous strain patterns. [43] states that the component surface is prepared with a high contract random speckle pattern which is then tracked by the software. Each random speckle can be identified by the software to provide continu-ous results patterns across the component surface. [45] uses DIC to analyse a rotor blade by incorporating a strobe light to ensure the same blade is captured with every image. [44] states that due to the surface preparation requirements and the need for high resolution cameras, DIC is not commonly used for rotating components.

[43] mentions the availability of photogrammetry options that do not use optical targets like DIC and dynamic photogrammetry. Instead component defects or features are tracked, removing the need to prepare difficult surfaces. However, they are not as accurate as DIC and dynamic photogrammetry.

2.3 Bearings

Only a relatively small percentage of the reviewed literature discussed the bearings used in the test rig. [14, 17, 23, 29, 31, 32, 35, 36, 38, 39] use a wide range of bearings including: ball bearings; magnetic bearings, angular contact bearings; foil air bearings; and static air bearings.

As discussed in [17, 46], foil air bearings levitate using air pressure which considerably reduces friction on the rotor and requires no lubrication. Static air bearings used by [32, 35] require an external air supply integrated into the test rig design [17]. [17] adds that air bearings for this application are also not commercially available.

Magnetic bearings used by [14, 23, 31, 36] utilise a magnetic force to stabilise the rotor and give similar advantages to air bearings. However, active magnetic bearings were most commonly used amongst [14, 23, 31, 36] and they require a complex de-sign procedure as they use sensors and actuators [17].

Angular contact bearings are ball bearings designed to withstand high axial loads, used by [38]. According to [17], ball bearings provide simplicity and a small size, but they are limited by temperature, lubrication and speed.

3 Instrumentation/Components

3.1 Air Turbine

As discussed in Sect. 2.1, an air turbine greatly reduces the complexity of a test rig design over a HSEM, due to a reduction in components, and allows the mechanical behaviour to be measured. An air turbine is appropriate for the test rig design in this paper as the mechanical aspect of the rotor is the focus, therefore only rotor rotation is required. The performance of a HSEM power supply is not relevant to this test rig.

To ensure expenditure remained low, an air turbine was sourced within the university. Spare turbochargers were available so the choice of utilising the turbine stage was made. A suitable turbocharger was sourced that was designed to operate at over 100 krpm so a large range of results could be tested. A redesign of the bearing housing was required to allow the rotor to be observed however, the turbine housing and turbine shaft would be utilised.

3.2 Air Compressor

To power the test rig, inspiration was taken from [38], where a cold gas stand was used within the design. This was deemed a simpler than sourcing and utilising an internal combustion engine, as shown in [7]. This would also enable the control of thermal testing rather than being constrained by the exhaust gas temperature.

The university had access to two electric air compressors and a mounting stand was already designed for the operation of the compressor. The utilisation of both the compressor and stand reduced the design complexity greatly and only required the mounting of the test rig to the compressor stand. However, as discussed in Sect. 5, issues with the manufacturing process required a change in air turbine that provided more commercial access to individual components. The primary gas turbine was only available as a full turbocharger unit at an excessive cost. Therefore, a lower cost alternative was required.

A similar size gas turbine was sourced with access to individual components. To ensure this replacement air turbine matched the compressor, turbine and compressor flow charts were compared. It was important to match the components to extract maximum operating performance. Fig. 1 and 2 show the compressor map and turbine flow chart, respectively.

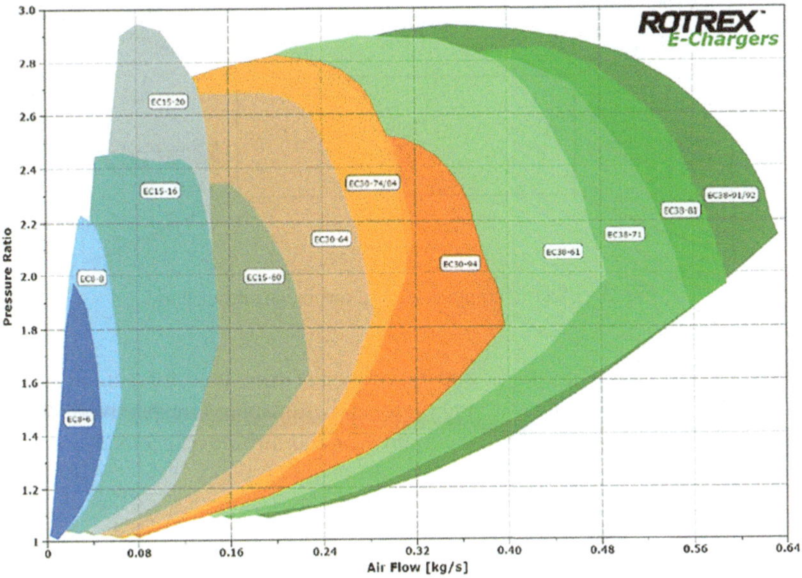

Fig. 1. Compressor map for the university owned electric air compressor [47].

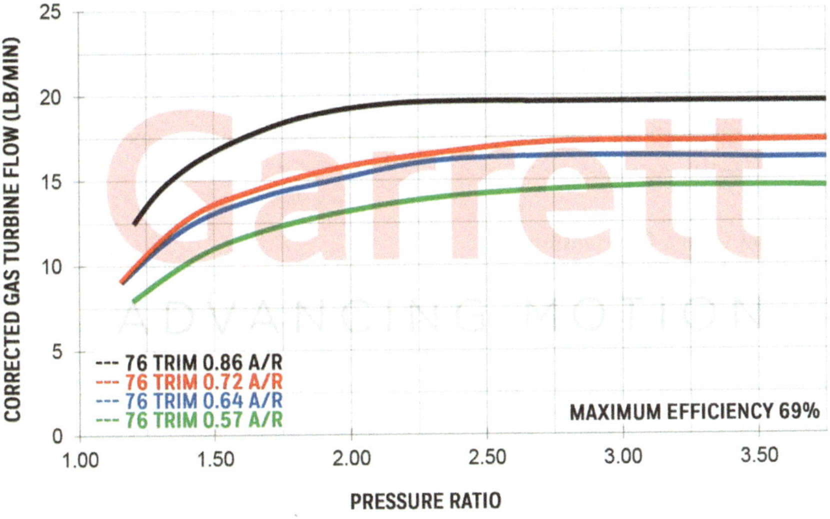

Fig. 2. Air turbine flow chart [48].

From Fig. 1, the EC15 compressor wheel was used, while the 0.57 A/R line was the air turbine flow from Fig. 2. A schematic of the airflow between the compressor and air turbine is shown in Fig. 3.

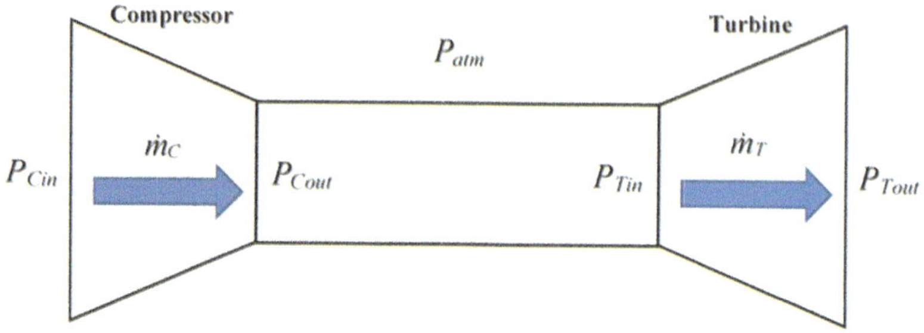

Fig. 3. Electric compressor and turbine air flow chart.

Where P_{Cin}, P_{Cout}, P_{Tin}, P_{Tout} and P_{atm} are the pressures for compressor inlet, compressor outlet, turbine inlet, turbine outlet and atmospheric respectively, while $\dot{m}_C$ and $\dot{m}_T$ are the mass flow rates for the compressor and turbine respectively.

$$P_{Cin} = P_{Tout} = P_{atm}$$

$$P_{Cout} = P_{Tin}$$

$$\dot{m}_c = \dot{m}_T$$

Since the compressor inlet and turbine outlet pressures are the same, while the compressor outlet and turbine inlet are the same, both the compressor and turbine have the same pressure ratio. Mass flow rate through the compressor must match the flow rate through the turbine. Operating at a pressure ratio of 2.25 will be within both the compressor and turbines capabilities, producing high performance and enabling high rotational speeds.

3.3 Digital Image Correlation

As discussed in Sect. 2.2, DIC enables the extraction of strain results without any equipment required to be attached to the rotor or test rig. It provides a full surface pattern of the strain and can be used for rotating components with the introduction of a strobe light. Fig. 4 shows a typical DIC set up.

The DIC system to be used on this test rig will be set up as shown in Fig. 4. The use of two cameras enables a 3D analysis by providing a depth of view. Before setting the cameras, the test specimen will require a speckle pattern. A base later is used to coat the surface, followed by a contrasting spray to create the speckles. This must be a random pattern so the cameras can identify unique speckles to reference.

When the test specimen and cameras are in place, a reference image is taken. As the test progresses, surface strain will affect the size and position of the speckles. Images

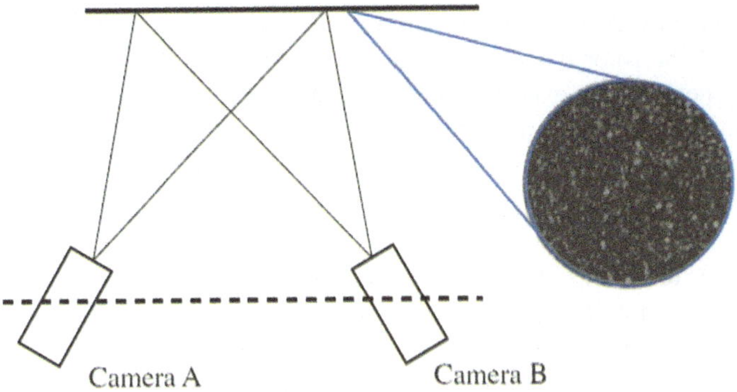

Fig. 4. DIC setup with speckle pattern on test specimen [45].

taken throughout the test are references against the reference image to determine the level of strain taking place. The software then converts this into a visual pattern of surface strain.

With such high rotor speeds, a light strobe will be required to capture the same section of the rotor in each image. Each strobe will illuminate the specimen to be visible to the cameras when the image is recorded. To ensure the timing of the strobe is correct, a rotor speed sensor output can be input into the system during testing. The collection of images will then undergo post processing to determine surface strain value which can then be compared against the GPS predictions, to validate the theory.

4 Three Cylinder Rotor Design

4.1 Rotor Concept

Figure 5 shows a three-cylinder concept rotor to be tested, as discussed earlier. The carrier is a sliding fit onto the shaft. The interference locations between the carrier and

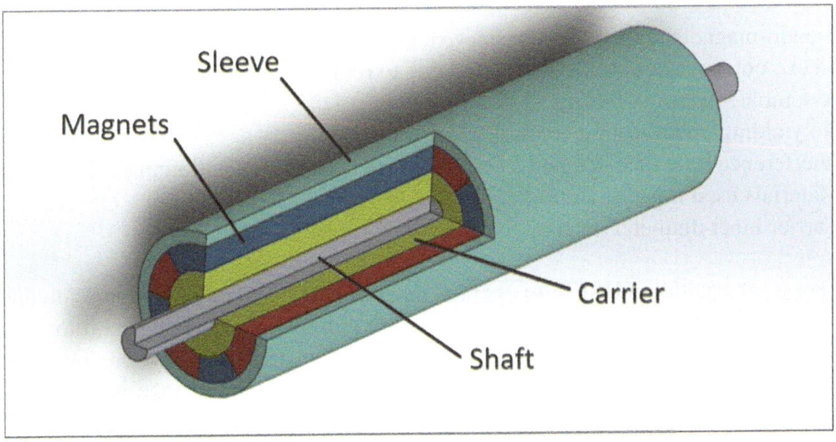

Fig. 5. Annotated model of a three cylinder rotor [13].

magnets; and the magnets and sleeve, are known as radii c and d respectively [13]. As the purpose of the rotor being designed was to allow experimental determination of the strains generated, the expensive rare earth magnets were replaced with much less expensive, metal 'pseudo- magnets'.

4.2 Rotor Manufacturing and Assembly

To manufacture the rotor within university capabilities, the diameter, length, and tolerance of the rotor is important. The machinists do not have the capability to measure finer than a hundredth of a millimetre. Therefore, any tolerance must be specified to at least 10 microns. This could particularly affect the interference levels, which are the smallest measurements of a rotor design.

The diameter and length are related when discussion machining limitations. The machining would be limited by the aspect ratio of the rotor. If the rotor is long and thin, the machines used to manufacture the component are likely to chatter as the machining moves further from the rotor ends. This is due to the component flexing under the pressure applied during cutting. Excessive flexing would destroy the parallelism and tolerances of a rotor component.

When assembling a rotor, [10] explores axial pressing and shrink fitting, both capable of being conducted at the university. [10] observes a rotor failure and lists the axial pressing assembly process as harmful to the rotor sleeve strength, although they admit it is difficult to quantify. [49] also states that in HSEMs, the PMs have low tensile strength and the tensile load during assembly can be reduced via shrink fitting. The rotor for the design in this paper will enable the assembly method of shrink fitting to avoid the negative effects of axial pressing.

4.3 Design Constraints

There are many difficulties in designing a successful rotor for this rig that requires delicate balancing. The rotor design constraints were:

- Pseudo-magnets retained by the sleeve under testing conditions.
- Sleeve outer surface strain differential due to rotation greater than 300 $\mu\varepsilon$.
- Assemble via shrink fitting.
- No yielding under testing condition.
- Interference tolerance at c and d must not have a lower resolution than 10 microns.
- Materials used must be isotropic.
- Carrier inner diameter must be 6 mm.

For a rotor within a HSEM to operate, magnets must remain in compression due to the low tensile strength. However, this condition did not constrain the test rotor design as the pseudo-magnet material could withstand tensile stresses. The focus of the rotor design was to achieve a measurable outer surface strain. The change in sleeve outer surface strain from stationary to operating conditions had to be large enough for the DIC system to identify.

To increase the sleeve outer surface strain, the overall rotor diameter can be increased, thus increasing the centrifugal force. However, an increase in centrifugal force may reduce the interference between the cylinders and risk rotor separation. To maintain retention, the interference levels between the cylinders can be increased. However, if the interference is increased excessively the rotor may be impossible to assemble by shrink fitting as the temperature differential required to generate the required thermal expansion differential, may be larger than the materials can withstand. Another risk of excessive interference is the rotor collapsing under its own compression when stationary.

As discussed in Sect. 4.2, the level of interference between each component must have tolerances limits of at least 10 microns. Based on the turbocharger discussed in Sect. 3.1, the maximum rotor length was 60 mm due to the fixed turbine shaft length. The turbine shaft also fixed the inner diameter of the inner cylinder at 6mm. All materials had to be isotropic to correlate with the assumption made in the GPS theory in [13]. Specifying the use of thin, long rotors for testing was contemplated as that application is where GPS theory produces the most accurate results. However, if the testing results correlate with FEA for thick rotors, then the FEA set up must be correct, thus providing confidence in the theoretical analysis as this agreed with the FEA.

4.4 Detailed Design

The developed GPS theory in [13] was utilised to predict whether a rotor design would be successful during testing. Material specifications, dimensions and operating conditions were the inputs while the key outputs were component stresses, interference pressures S_1 and S_2 at radii c and d respectively, and the sleeve outer surface strain.

The first decision was to select materials in order of stiffness, with the sleeve the stiffest and the carrier the least stiff. It was predicted that as the load increased on the rotor, the outer material would deform less than the interfering inner material. Therefore, both c and d should remain in compression. Despite the risk of separation, high density materials were selected to attempt to strain the outer sleeve as the high stiffness sleeve material would resist deformation. The chosen materials and the corresponding specifications are shown in Table 1.

Table 1. Material selection for first rotor design iteration.

Cylinder component	Material	Poisson's ratio	Density (kgm^{-3})	Young's modulus (GPa)	Coefficient of thermal expansion	Yield strength (MPa)	Ultimate strength (MPa)
Inner	Aluminium 6061-T6	0.33	2700	68.9	2.36×10^{-5}	276	310
Middle	C464 Brass	0.28	8410	100	2.12×10^{-5}	317	476
Outer	304L stainless steel	0.29	8000	193	1.73×10^{-5}	210	505

Table 2. Dimension selection for first rotor design iteration.

Cylinder component	Inner surface diameter (mm)	Outer surface diameter (mm)	Interference (mm)
Inner	6	12.02	0.01
Middle	12	22.02	0.01
Outer	22	26	

Table 3. First design iteration initial tests.

Test Speed (krpm)	Rotor separation?	Sleeve outer surface strain (με)	Maximum inner cylinder Von Mises stress (MPa)	Maximum middle cylinder Von Mises stress (MPa)	Maximum outer cylinder Von Mises stress (MPa)
0	No	792	167	93	205
50	No	869	145	102	220
80	No	989	111	117	245

Another input, the dimensions, must then be determined. Assuming a maximum diameter value of 50 mm, based on the size of the turbine, a medium thickness rotor was chosen. This provided a sensible starting point and ensured the machining was within the university's capacity. The dimensions are specified for the first design iteration in Table 2.

As shown by the initial calculations shown in Table 3, the rotor remained in compression from stationary to 80 krpm, well above the 50 krpm requirement. Most of the rotor stresses remained within the yield limits of the material, but the outer sleeve was predicted to plastically deform at both 50 krpm and 80 krpm. With the rotor stationary the sleeve would already be close to yielding. However, the most significant problem was the lack of range for the sleeve outer surface strain. The 77 με change, from stationary to 50 krpm would not be reliably detectable by the DIC system.

For the second design iteration, the rotor diameter was substantially increased to maximise the range of sleeve outer surface strain being produced. To combat the extra stresses this would put on the sleeve, the brass and aluminium materials were switched to the inner and middle cylinders, respectively. The minimum interference remained as a baseline test. Tables 4, 5 and 6 show the design and test details of the second design iteration.

For this second iteration, the sleeve outer surface strain increased from 77 to 165 με. Whilst this did not meet the design criteria, the intention of the second iteration was successful. There was also a substantial reduction in the stresses produced in each component, eliminating any yielding at stationary or at an operating speed of 50 krpm.

Table 4. Material selection for second rotor design iteration.

Cylinder component	Material	Poisson's ratio	Density (kgm^{-3})	Young's modulus (GPa)	Coefficient of thermal expansion	Yield strength (MPa)	Ultimate strength (MPa)
Inner	C464 Brass	0.28	8410	100	2.12×10^{-5}	317	476
Middle	Aluminium 6061-T6	0.33	2700	68.9	2.36×10^{-5}	276	310
Outer	304L Stainless steel	0.29	8000	193	1.73×10^{-5}	210	505

Table 5. Dimension selection for second rotor design iteration.

Cylinder component	Inner surface diameter (mm)	Outer surface diameter (mm)	Interference (mm)
Inner	6	20.02	0.01
Middle	20	40.02	0.01
Outer	40	48	

Table 6. Second design iteration initial tests.

Test speed (krpm)	Rotor separation?	Sleeve outer surface strain (με)	Maximum inner cylinder stress (MPa)	Maximum middle cylinder stress (MPa)	Maximum outer cylinder stress (MPa)
0	No	457	96	64	122
50	No	622	33	69	153
80	Yes	-	-	-	-

However, the rotor would separate before reaching 80 krpm as the centrifugal forces of the sleeve pulled it from the middle cylinder. The speed limit for the second design iteration was found to be 60 krpm. To combat this in the third design iteration, different material grades were selected. A reduction in sleeve density and an increase in stiffness would reduce centrifugal forces to delay rotor separation, but also may hinder sleeve surface strain. An increase in density and a reduction in stiffness for the inner cylinder would also reduce rotor separation and increase sleeve outer surface strain. Due to supply issues the aluminium was swapped for a similar grade. Tables 7, 8 and 9 show the design and test details of the third design iteration.

Table 7. Material selection for third rotor design iteration.

Cylinder component	Material	Poisson's ratio	Density (kgm^{-3})	Young's modulus (GPa)	Coefficient of thermal expansion	Yield strength (MPa)	Ultimate strength (MPa)
Inner	CZ121 Brass	0.31	8470	97	2.09×10^{-5}	369	495
Middle	Aluminium 6082-T6	0.33	2700	70	2.40×10^{-5}	260	310
Outer	EN16T Steel	0.30	7800	210	1.20×10^{-5}	680	850

Table 8. Dimension selection for third rotor design iteration.

Cylinder component	Inner surface diameter (mm)	Outer surface diameter (mm)	Interference (mm)
Inner	6	20.02	0.01
Middle	20	40.02	0.01
Outer	40	48	

Table 9. Third design iteration initial tests.

Test speed (krpm)	Rotor separation?	Sleeve outer surface strain ($\mu\varepsilon$)	Maximum inner cylinder stress (MPa)	Maximum middle cylinder stress (MPa)	Maximum outer cylinder stress (MPa)
0	No	451	101	66	132
50	No	608	38	71	164
80	Yes	-	-	-	-

The results for the third iteration in Table 9 appear very similar to the second iteration results. There are slight increases in the stresses throughout all three-cylinder components and the sleeve outer surface strain range has slightly decreased. The rotor also separates before reaching 80 krpm. However, the third design reached 64 krpm instead of the 60 krpm achieved by the second iteration. As the sleeve outer surface strain range has not met the design criteria, the operating range must be increased. However, the third design iteration showed that these criteria offset each other. To identify further potential rotor designs, the third design iteration was further explored using different interference levels for δ_c and δ_d at radii c and d respectively. The results are shown in Fig. 6, 7, 8, 9, 10, 11, 12 and 13.

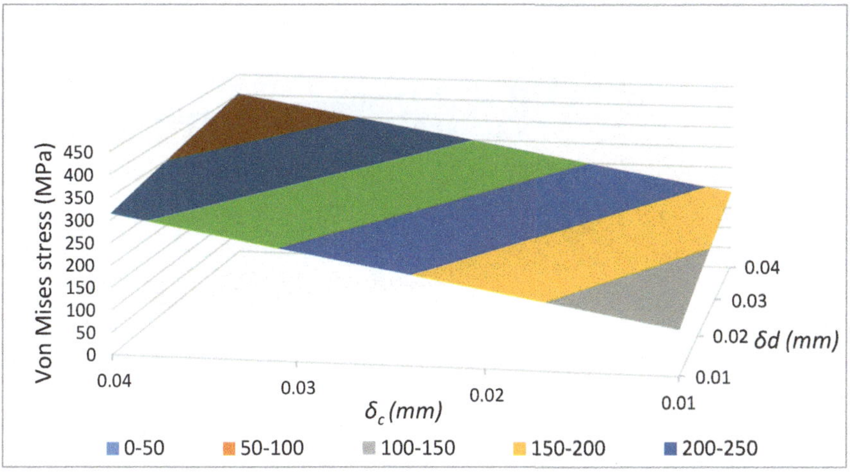

Fig. 6. Inner cylinder peak Von Mises stress across various interferences when stationary.

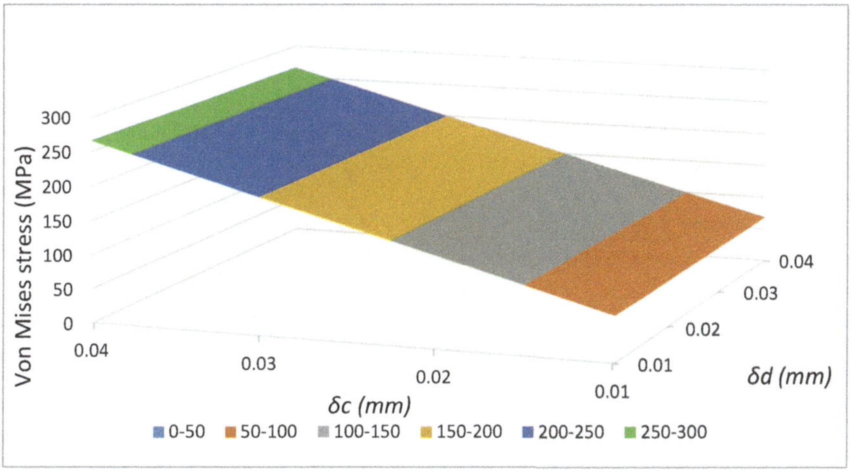

Fig. 7. Middle cylinder peak Von Mises stress across various interferences when stationary.

From Fig. 6 and 7, to avoid yielding of the inner and middle cylinders, the following design constraints are required:

$$\delta_d \leq 0.04 - 2(\delta_c - 0.03)$$

$$\delta_c \leq 0.04$$

From Fig. 8, the outer cylinder remains elastic for all levels of interference.

As shown in Table 9, a 50 krpm operating speed did not produce enough change in surface strain to be measured by the DIC system. Therefore, an operating speed of 80 krpm was examined to determine a suitable rotor design. As shown in Fig. 9, yielding

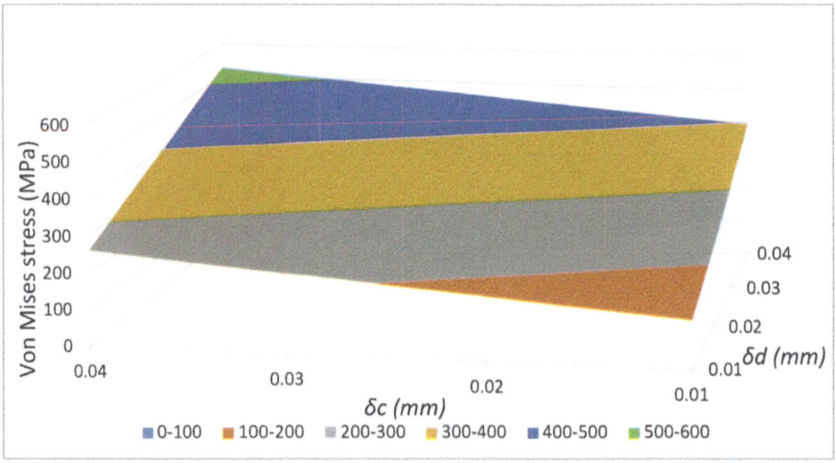

Fig. 8. Outer cylinder peak Von Mises stress across various interferences when stationary.

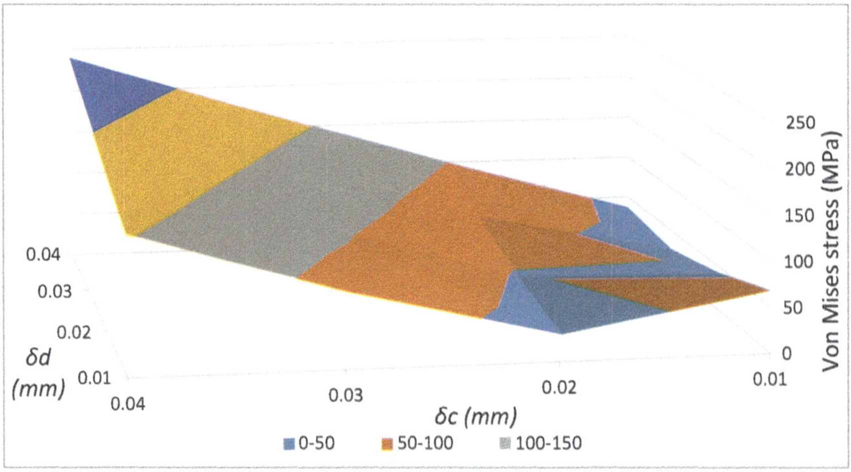

Fig. 9. Inner cylinder peak Von Mises stress across various interferences at 80 krpm.

was not predicted in the inner cylinder for any combination of interferences. For the middle cylinder in Fig. 10, yielding was predicted when the interference, δ_c was 0.04 mm with any δ_d value. The outer cylinder in Fig. 11, again shows that all levels of interference remained within the cylinder stress limit. However, rotor separation was predicted by the theory when interference, δ_d was 0.01 mm with δ_c range of 0.01–0.02 mm.

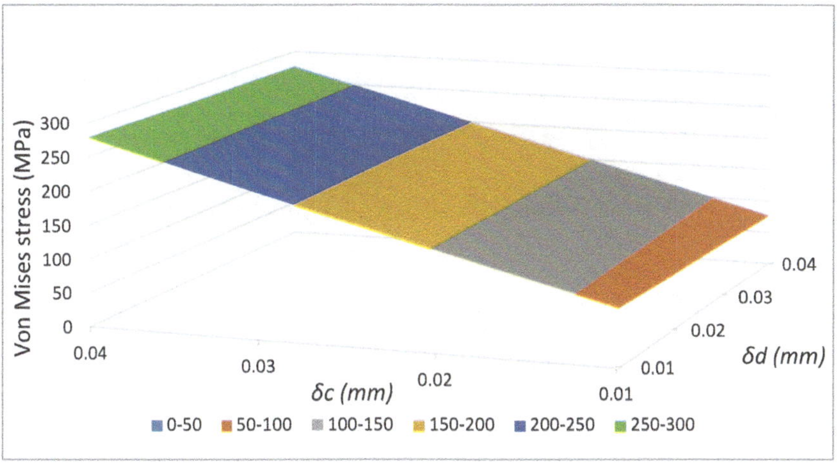

Fig. 10. Middle cylinder peak Von Mises stress across various interferences at 80 krpm.

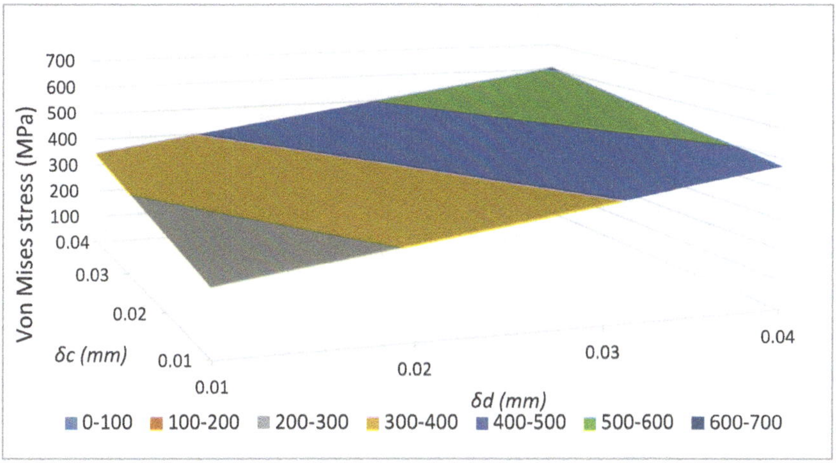

Fig. 11. Outer cylinder peak Von Mises stress across various interferences at 80 krpm.

Figure 13 shows the strain values produced during rotor assembly without an impact from rotation. All values are over the 300 $\mu\varepsilon$ limit required to be detected by the DIC system. This allows strain testing to be achieved at stationary and operating conditions and could be utilised to conduct stationary thermal tests too. Figure 12 shows that the change in sleeve outer surface strain is constant for all interference topologies. Circa 400 $\mu\varepsilon$ is consistently produced, enabling the DIC system to detect these results. The added strain due to rotation is the same for all interference levels because the materials, rotor size and rotational speed are the same. Based on the presented results, the final rotor design dimensions are shown in Table 10. The materials from Table 7 will be used.

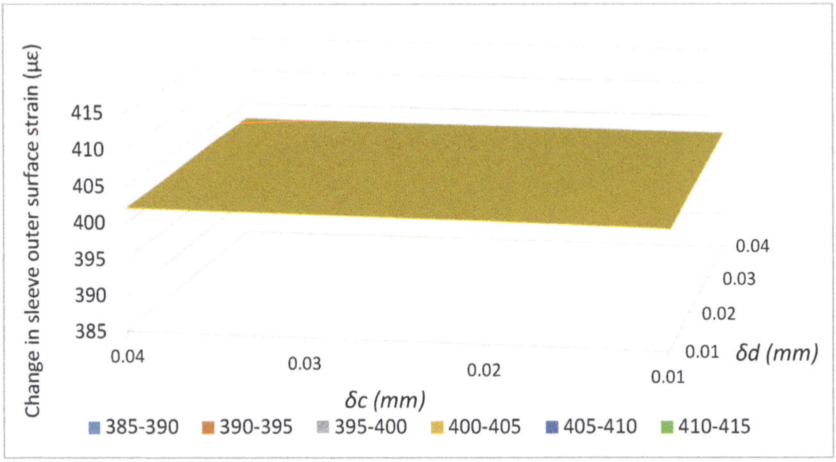

Fig. 12. Change is sleeve outer surface strain from stationary to 80 krpm for various interferences.

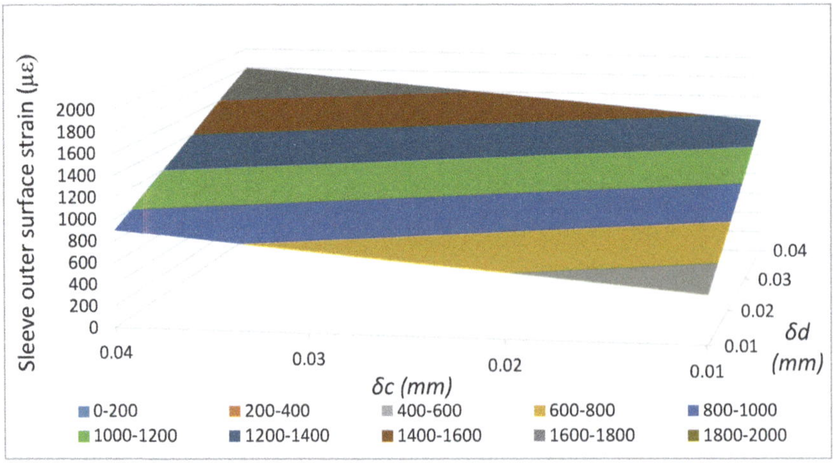

Fig. 13. Sleeve outer surface strain for various interferences when stationary.

The final rotor design uses the minimum tolerance resolution at δ_d and achieves the required sleeve strain range suitable for the DIC system.

For the shrink fitting process, it is assumed the materials are at 20 °C room temperature and are cooled with liquid nitrogen to −196 °C. Only the inner and middle cylinders will require shrinking to slide into their larger outer cylinder. Table 11 shows that the rotor can be assembled via shrink fitting and therefore meets all criteria set in Sect. 4.1.

Table 10. Dimension selection for final rotor design iteration.

Cylinder component	Inner surface diameter (mm)	Outer surface diameter (mm)	Interference (mm)
Inner	6	20.02–20.04	0.01–0.03
Middle	20	40.04–40.06	0.02–0.03
Outer	40	48	

Table 11. Shrink fit calculations for final rotor design iteration.

Cylinder Component	Material	Coefficient of Thermal Expansion	Temperature change (°C)	Starting diameter (mm)	Final diameter (mm)	Change in diameter (mm)	Required change in diameter (mm)
Inner	CZ121 Brass	2.09×10^{-5}	-216	20.02-20.06	19.93–19.97	0.09	0.03-0.07
Middle	Aluminium 6082-T6	2.40×10^{-5}	-216	40.04-40.06	39.83–39.85	0.21	0.05-0.07

5 Test Rig Design

5.1 Design Requirements

After reviewing the existing literature, the following constraints were applied to the design of the testing rig.

- Rotor must be visible for the DIC system to measure surface strain.
- Test rig must operate over 50,000 rpm.
- Any machining and testing should be completed in-house at the University.
- Existing parts available at the University should be utilised.

5.2 Detailed Design

As discussed in Sect. 3, the test rig design must incorporate the use of the turbocharger, which is to be powered by an electric compressor fixed to an existing rig. The rotor designed in Sect. 4 must be visible to the DIC system so strain analysis can be captured to validate the GPS theory. The constraints in Sect. 5.1 must also be met, however the rig must now operate at 80 krpm due to the design of the rotor from Sect. 4.

The gas turbine in Sect. 3.1 operates at over 100 krpm which enables the rotor to be used up to 80 krpm as intended. To simplify the test rig, the compressor end of the turbocharger was removed as it was unnecessary to the testing, leaving an air turbine. However, this removed the compressor load acting against the turbine making it more difficult to slow the rotor down. The operating speed would have to be tightly controlled to avoid over-speed which could cause rotor failure.

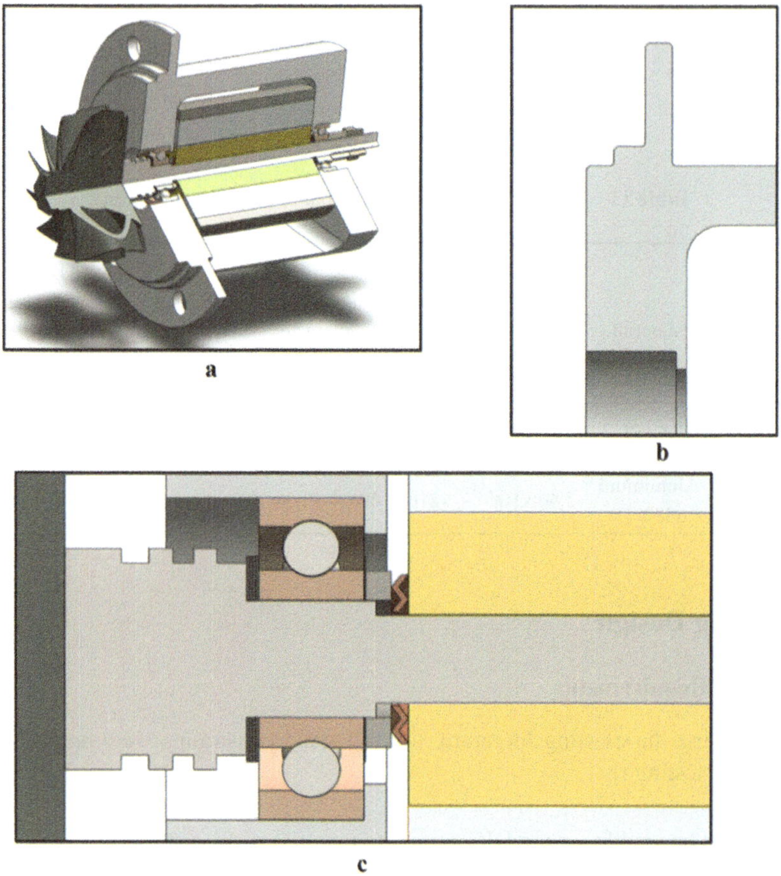

Fig. 14. Final test rig design a. Assembly excluding turbine housing. b. Bearing housing detail. c. Spring washer integration detail.

To visualise the rotor, the bearing housing had to be redesigned. The internal turbine housing detail was measured to accurately create a close-fitting flange on the bearing housing, shown in Fig. 14b, locating the component and retaining concentricity. A pocket was then designed to be milled, where the rotor could be visualised, shown in Fig. 14a.

A central bore diameter was used throughout the component to increase manufacturing ease. Two shallow diameters would then be bored at either end to create the press fit bearing locations. A lip remained before the milled pocket to press the bearings against and easily locate them.

To reduce tight tolerances that would be difficult to manufacture with the university's capabilities, shims were input into the design as shown in Fig. 14c. A series of shims enabled the positional control of the turbine wheel to the turbine housing and the rotor within the milled pocket. This would improve turbine performance as air leakage could be minimised. A major challenge was reducing turbine shaft machining tolerances. To reduce the tolerances along the middle section of the turbine shaft, a step was created at the turbine end, as shown in Fig. 14a. The bearing at the turbine end was therefore larger and cleared the shaft during assembly. Only tight tolerances were required at the bearing locations on the turbine shaft.

Based on Sect. 2.3, ball bearings were chosen due to their simplicity and low cost. Bearings were identified that could withstand speeds of 80 krpm when appropriately cooled. This test rig would not run long enough for the temperature within the components to be an issue. These bearings were chosen over angular contact bearings once the axial load capabilities were identified.

To enable rotation, the rotor was designed to be a sliding fit onto the turbine shaft which was then gripped between the bearings via the compressor nut. However, to ensure the ball bearings operate smoothly, preloading must be applied. To ensure the ball bearings always had an axial load present on them, a spring washer was introduced into the assembly. Having shims enabled the spring washer to be easily placed without affecting much of the design. As shown in Fig. 14c, this ensured an axial load was present across the bearings while the components along the turbine shaft were still compressed together.

As discussed, these decisions produced a test rig that met all requirements listed in Sect. 5.1.

6 Conclusion

A lack of exploration into the experiment testing of high-speed rotors, focussing on the mechanical analysis, was identified after examining the existing literature. Most of the experiments in the literature were focussed on the electromagnetic performance of the motor, leading to the use of complex HSEM testing rigs. However, papers that explored the mechanical aspect of the rotor showed that test rig complexity could be greatly reduced using air turbines or compressors as the power source, whilst also making it more accessible to extract test results. Few papers attempted to extract stress results from a high-speed rotor and alternatively opted for post-test inspections. However, non-intrusive strain testing methods have been successfully used in other rotor applications, such as DIC.

Findings from the literature, combined with constraints on the designs, led to a rotor design and an inexpensive testing rig that can be used to obtain accurate strain data. Materials for the rotor design were identified to provide enough outer surface strain to be detectable by the non-intrusive DIC system whilst also removing the use of fragile permanent magnets, simplifying the design for testing purposes. The rotor was also suitable for shrink fitting, eliminating any residual stress build up that could affect testing results. The test rig utilised spring washers and shims to enable tolerances to be increased and manufacturing costs reduced. Further costs were reduced by identifying suitable ball bearings and enabling the rig to be powered by an external compressor, also reducing the complexity of the rig compared to the HSEMs used in existing literature. The rotor is exposed when assembled in the rig to enable results to be extracted via DIC. The results that can be extracted from the test rig, can be used to verify theoretical or FEA analysis. This greatly improves the current situation of examining a rotor, post-test, and determining success based purely on if the rotor failed or not.

The next steps are to manufacture the rotor and test rig to conduct the mechanical analysis. The results will then be compared against GPS theory predictions to verify the accuracy of the GPS analysis.

Appendix A: List of Symbols and Abbreviations

See Table 12.

Table 12. List of Symbols and Abbreviations

Symbol or abbreviation	Definition
HSEM	High-speed electric machine
PM	Permanent magnet
GPS	Generalised plane strain
DIC	Digital image correlation
IPM	Interior permanent magnet
SPM	Surface-mounted permanent magnet
FEA	Finite element analysis
P_{Cin}	Compressor inlet pressure
P_{Cout}	Compressor outlet pressure
P_{Tin}	Turbine inlet pressure
P_{Tout}	Turbine outlet pressure
P_{atm}	Atmospheric pressure
$\dot{m}_C$	Compressor mass flow
$\dot{m}_T$	Turbine mass flow
Radii c	Interference location between carrier and magnets
Radii d	Interference location between magnets and sleeve
S_1	Interference location pressure carrier and magnets
S_2	Interference location pressure magnets and sleeve
δ_c	Interference level between carrier and magnets
δ_d	Interference level pressure magnets and sleeve

References

1. Tenconi, A., Vaschetto, S., Vigliani, A.: Electrical machines for high-speed applications: design considerations and tradeoffs. IEEE Trans. Industr. Electron. **61**(6), 3022–3029 (2014)
2. Rahman, M.A., Chiba, A., Fukao, T.: Super high speed electrical machines – summary. In: IEEE Power Engineering Society General Meeting, Denver, CO, USA 2004, vol. 2, pp. 1272–1275. IEEE (2004)
3. Fang, H., Qu, R., Li, J., Zhang, P., Fan, X.: Rotor design for high-speed high-power permanent-magnet synchronous machines. IEEE Trans. Ind. Appl. **53**(4), 3411–3419 (2017)
4. Li, S., Sarlioglu, B.: Assessment of high-speed multi-megawatt electric machines. In: 2015 Electric Machines and Drives Conference, Coeur d'Alene, ID, USA. IEEE (2015)
5. Kolondzovski, Z., Arkkio, A., Larjola, J., Sallinen, P.: Power limits of high-speed permanent-magnet electrical machines for compressor applications. IEEE Trans. Energy Convers. **26**(1), 73–82 (2011)
6. Gerada, C., Boglietti, A., Cavagnino, A.: High-speed electrical machines and drives. IEEE Trans. Ind. Electron. **61**(6), 2943–2945 (2014)
7. Gerada, D., Mebarki, A., Brown, N., Gerada, C., Cavagnino, A., Boglietti, A.: High-speed electrical machines: technologies, trends and developments. IEEE Trans. Ind. Electron. **61**(6), 2946–2959 (2014)
8. Burnand, G., Araujo, D.M., Perriard, Y.: Very-high-speed permanent magnet motors: mechanical rotor stresses analytical model. In: Electric Machines and Drives Conference, Miami, FL, USA. IEEE (2017)
9. Barrans, S., Al-Ani, M., Carter, J.: Mechanical design of rotors for permanent magnet high-speed electric motors for turbocharger applications. IET Electr. Syst. Transp. **7**, 278–286 (2017)
10. Binder, A., Schneider, T., Klohr, M.: Fixation of buried and surface-mounted magnets in high-speed permanent-magnet synchronous machines. IEEE Trans. Ind. Appl. **42**(4), 1031–1037 (2006)
11. Smith, D.J.B., Mecrow, B.C., Atkinson, G.J., Jack, A.G., Mehna, A.A.A.: Shear stress concentrations in permanent magnet rotor sleeves. In: International Conference on Electrical Machines, Rome, Italy. IEEE (2010)
12. Mallin, L., Barrans, S.: A review of the high-speed permanent magnet rotor stress analysis used for automotive air-handling machines. Eur. J. Eng. Technol. Res. **5**(4), 448–456 (2020). https://doi.org/10.24018/ejers.2020.5.4.1814
13. Mallin, L., Barrans, S.: Comparison of theoretical approaches to determine the stresses in surface mounted permanent magnet rotors for high speed electric machines. J. Strain Anal. Eng. Des. (2021)
14. Chen, L.L., Zhu, C.S., Zhong, Z., Liu, B., Wan, A.: Rotor strength analysis for high-speed segmented surface-mounted permanent magnet synchronous machines. IET Electr. Power Appl. **12**(7), 979–990 (2018)
15. Tsao, P., Senesky, M., Sanders, S.R.: An integrated flywheel energy storage system with homopolar inductor motor/generator and high-frequency drive. IEEE Trans. Ind. Appl. **39**(6), 1710–1725 (2003)
16. Bianchi, N., Bolognani, S., Luise, F.: Analysis and design of a pm brushless motor for high-speed operations. IEEE Trans. Energy Convers. **20**(3), 629–637 (2005)
17. Zwyssig, C., Kolar, J.W.: Design of a 100 W, 500000 rpm permanent-magnet generator for mesoscale gas turbines. In: 2005 Industry Applications Conference, Kowloon, Hong Kong, China, vol. 1, pp. 253–260. IEEE (2005)
18. Atkinson, G.J., Mecrow, B.C., Jack, A.G., Atkinson, D.J., Sangha, P., Benarous, M.: The analysis of losses in high-power fault-tolerant machines for aerospace applications. IEEE Trans. Ind. Appl. **42**(5), 1162–1170 (2006)

19. Pfister, P., Perriard, Y.: A 200 000 rpm, 2 kW slotless permanent magnet motor. In: 2008 International Conference on Electrical Machines and Systems, Wuhan, China, pp. 3054–3059. IEEE (2008)
20. Laskaris, K.I., Kladas, A.G.: High torque internal permanent magnet wheel motor for electric traction applications. In: 2008 18th International Conference on Electrical Machines, Vilamoura, pp. 1–4. IEEE (2008)
21. Zwyssig, C., Kolar, J.W., Round, S.D.: Megaspeed drive systems: pushing beyond 1 million r/min. IEEE/ASME Trans. Mechatron. **14**(5), 564–574 (2009)
22. Pfister, P., Perriard, Y.: very-high-speed slotless permanent-magnet motors: analytical modeling, optimization, design, and torque measurement methods. IEEE Trans. Industr. Electron. **57**(1), 296–303 (2010)
23. Mitterhofer, H., Amrhein, W.: Design aspects and test results of a high speed bearingless drive. In: 2011 Power Electronics and Drive Systems (PEDS), Singapore, pp. 705–710. IEEE (2011)
24. Thomas, A.S., Zhu, Z.Q., Jewll, G.W.: Comparison of flux switching and surface mounted permanent magnet generators for high-speed applications. In: Power Electronics, Machines and Drives (PEMD), IET, pp. 111–116 (2011)
25. Wang, T., Du, G., Yu, Z., Zhang, F., Bai, Z.: Design and DEVELOP of a MW direct drive highspeed permanent-magnet machine for compression. In: 2013 International Conference on Electrical Machines and Systems (ICEMS), Busan, South Korea, pp. 892–895. IEEE (2013)
26. Uzhegov, N., Kurvinen, E., Nerg, J., Pyrhönen, J., Sopanen, J.T., Shirinskii, S.: multidisciplinary design process of a 6-slot 2-pole high-speed permanent-magnet synchronous machine. IEEE Trans. Industr. Electron. **63**(2), 784–795 (2016)
27. Ahn, J., Choi, J., Park, C.H., Han, C., Kim, C., Yoon, T.: Correlation between rotor vibration and mechanical stress in ultra-high-speed permanent magnet synchronous motors. IEEE Trans. Magn. **53**(11), 1–6 (2017)
28. Hong, D.K., Lee, T.W., Jeong., Y.H.: Design and experimental validation of a high-speed electric turbocharger motor considering variation of the **L D** ratio. IEEE Trans. Magn. **54**(11), 1–4 (2018)
29. Sanchez, R., et al.: Mechanical validation of a high power density external cantilevered rotor. IEEE Trans. Ind. Appl. **54**(4), 3208–3216 (2018)
30. Wang, Y., Feng, Q., Li, X., Ma, W.: Design, analysis, and experimental test of a segmented-rotor high-temperature superconducting flux-switching generator with stationary seal. IEEE Trans. Industr. Electron. **65**(11), 9047–9055 (2018)
31. Ismagilov, F., Vavilov, V., Miniyarov, A., Urazbakhtin, R.: Super high-speed electric motor with amorphous magnetic circuit for the hydrogen fuel cell air supply system. Int. J. Hydrogen Energy **43**(24), 11180–11189 (2018)
32. Gong, C., Li, S., Habetler, T.G.: Rotor dynamic analysis of ultra-high speed switched reluctance machines over 1 million rpm. In: 2018 IEEE Energy Conversion Congress and Exposition (ECCE), pp. 1704–1709. IEEE (2018)
33. Du, G., Xu, W., Zhu, J., Huang, N.: Rotor Stress analysis for high-speed permanent magnet machines considering assembly gap and temperature gradient. IEEE Trans. Energy Convers. **34**(4), 2276–2285 (2019)
34. Burnand, G., Perriard, Y.: Very-high-speed miniaturized permanent magnet motors: modeling and experimental validation. In: 2019 IEEE Energy Conversion Congress and Exposition (ECCE), pp. 5251–5257. IEEE (2019)
35. Gong, C., Li, S., Habetler, T.: High-strength rotor design for ultra-high speed switched reluctance machines. IEEE Trans. Ind. Appl. **56**(2) 1432–1442 (2020). https://doi.org/10.1109/TIA.2020.2966184

36. Huang, Z., Fang, J.: Multiphysics design and optimization of high-speed permanent-magnet electrical machines for air blower applications. IEEE Trans. Industr. Electron. **63**(5), 2766–2774 (2016)
37. Zhang, F., Du, G., Wang, T., Liu, G., Cao, W.: Rotor retaining sleeve design for a 1.12-MW high-speed pm machine. IEEE Trans. Ind. Appl. **51**(5), 3675–3685 (2015)
38. Terdich, N., Martinez-Botas, R.: Experimental efficiency characterization of an electrically assisted turbocharger. SAE Technical Paper Series (2013)
39. Bonello, P., Hassan, M.F.B.: An experimental and theoretical analysis of a foil-air bearing rotor system. J. Sound Vib. **413**, 395–420 (2018)
40. Xu, D., Wang, X., Li, G.: Optimization design of the sleeve for high speed permanent magnet machine. In: Industrial Electronics and Applications (ICIEA), Hefei, China, pp. 2531–2535. IEEE (2016)
41. Ha, S.K., Kim, M.H., Han, S.C., Sung, T.-H.: Design and spin test of a hybrid composite flywheel rotor with a split type hub. J. Compos. Mater. **40**(23), 2113–2130 (2006)
42. Warren, C., Niezrecki, C., Avitabile, P.: Applications of digital image correlation and dynamic photogrammetry for rotating and non-rotating structures. In: Proceedings of the 7th International Workshop on Structural Health Monitoring, Stanford, CA (2009)
43. Baqersad, J., Poozesh, P., Niezrecki, C., Avitabile, P.: photogrammetry and optical methods in structural dynamics–a review . Mech. Syst. Signal Process. **86**, 17–34 (2017)
44. Baqersad, J., Niezrecki, C., Avitabile, P.: Extracting full-field dynamic strain on a wind turbine rotor subjected to arbitrary excitations using 3D point tracking and a modal expansion technique. J. Sound Vib. **352**, 16–29 (2015)
45. Sirohi, J., Lawson, M.S.: Measurement of helicopter rotor blade deformation using digital image correlation. Opt. Eng. **51**(4), 043603 (2012)
46. Agrawal, G.L.: Foil air/gas bearing technology—an overview. In: ASME 1997 International Gas Turbine and Aeroengine Congress and Exhibition. American Society of Mechanical Engineers Digital Collection (1997)
47. Rotrex E-charger, Rotrex, pp. 1–7 (2015)
48. Garrett. (2020, 2020). GT2871R Turbocharger
49. Wang, J.-Q., Wang, F.-X., Bao, W.-B., Guan, E.-L.: Rotor design and strength analysis of high speed permanent magnet machine. Proc. Chin. Soc. Electr. Eng. **25**(15), 140 (2005)

A Commentary on the Role of Noise Measurement Uncertainty in the Industrial Context

Consolatina Liguori, Alessandro Ruggiero(✉), Domenico Russo, and Paolo Sommella

Department of Industrial Engineering (DIIn), University of Salerno, Fisciano, SA, Italy
ruggiero@unisa.it

Abstract. This commentary highlights the importance of uncertainty and estimation associated with the measurement of noise levels in the industrial field.

As example evaluating noise workers' exposure in a workplace has to take into account levels of uncertainty. The authors investigated the effect of the variability of the measurand with respect to the uncertainty of the measurement and proposed a novel data-driven sampling approach which consists of three phases.

Firstly, because the reliability of the noise indicators assessment depends significantly on the temporal variability of the noise, a statistic procedure is proposed to locate a corresponding minimum number of sound pressure levels necessary for having statistical significance of the initial data set.

Next, a numerical algorithm finds and deletes the anomalous values (outliers) from a population of real data.

Finally, the inherent variability of uncertainty of acoustic noise is calculated using the normal bootstrap method.

Keywords: Measurement uncertainty · Noise control · Bootstrap method

1 Introduction

Nowadays there is a growing attention in acoustic noise in every area of human activity. In the industrial field, for example, phonometric measurements are widely used for various purposes, in order to verify the conformity of a product with reference to the specifications on noise emission or to evaluate the exposure of workers to noise in a workplace [1]. In the first case, maximum suitable limits are fixed for the equivalent sound pressure level [2], that is the time average-sound level with the same energy of the original time-varying signal

$$L_{eq} = 10Log\frac{1}{T}\int_{T}\left[\frac{p(t)}{p_{rif}}\right]^2 dt \qquad (1)$$

where $p_{rif} = 20$ mPa is the reference sound pressure, corresponding to the hearing threshold at a frequency of 1 kHz. In the other case, the reference is daily noise exposure

S. Hloch et al. (Eds.): ICMEM 2021, LNME, pp. 324–331, 2021.
https://doi.org/10.1007/978-3-030-71956-2_26

level, that is normalized to an 8 h working day and takes into account both the A-weighted noise level and the duration of exposure [3]:

$$L_{EX,8h} = L_{p,A,eqTe} + 10Log\left(\frac{T_e}{T_0}\right) \tag{2}$$

where:

$L_{p,A,eqTe}$ is the A-weighted equivalent sound pressure level occurred during T_e;
T_e is the actual duration, in hours, of the working day;
T_0 is the working day duration (8 h).

Unfortunately, several factors can influence a measurement result, so the associated risks of incorrect decisions have to be considered before establishing a measurement decision rule.

2 Decision Rules

In industrial context, many are the applications which contemplate the comparison of a general dataset to specified requirements [4] from technical standards and/or from laws. This activity is in general not always straightforward, because it consists of a comparison between a measurement result (a mere approximation of measurand true value) and a threshold: thus, it is not a comparison between two numbers. Many decision-making processes depend on the results of these comparisons and they may also have economic consequences. So, it is necessary not to ignore the measurement uncertainty, as highlighted in [5–7], because it plays a significant role in quantifying the incompleteness of the knowledge related to the measurement finding [8]. The higher the measurement uncertainty the smaller the rate of compliance, but current technical international standards don't exactly demonstrate how to calculate the confidence level in this area, increasing the risks of a wrong decision.

The lower values of the established upper limit belong to a specification zone, that is the interval of tolerated levels. The measured value uncertainty is called the confidence zone [9].

The evaluation of compliance with near limit values is a complex matter, so it is mandatory to make a choice among the available decision-rules, which determine a range of risks of incorrect decisions of compliance (guardband). The fundamental rules are known as simple acceptance and simple rejection [10].

The first states that for values within the specification zone the product is declared compliant, while the second states that for values outside the specification zone it is declared non-compliant.

Generally, in professional acoustics environments, the decision rules are selected on the basis of indications derived from economic, health or other sectors. In more detail, decision rules based on stringent acceptance and relaxed rejection is in favour of the receiver (Fig. 1), decision rules based on relaxed acceptance and stringent rejection is in favour of the source (Fig. 2).

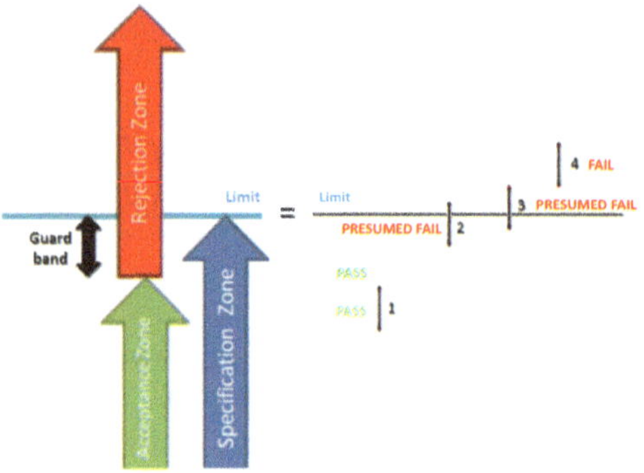

Fig. 1. Conformity assessment with an upper limit using the decision rule based on stringent acceptance and relaxed rejection.

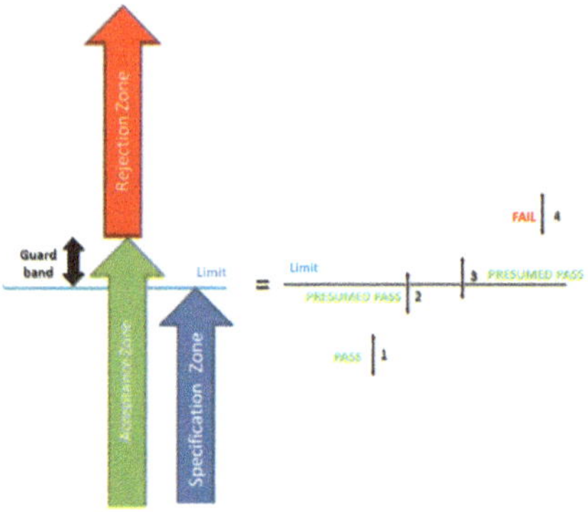

Fig. 2. Conformity assessment with an upper limit using the decision rule based on relaxed acceptance and stringent rejection.

3 Sources of Uncertainties in Noise Measurement

Nowadays, an adequate technical and procedural reference standard about the estimation of noise measurement uncertainties has not been made available yet, therefore, lately the scientific community has focalized the attention on this theme. In particular, the possible sources of uncertainty have been identified:

- instrumentation of measurement ($u_{instrum}$)
- measurement position (u_{pos})
- other sources, for example wind, alarm signals (u_{other})

where $u_{instrum}$ is the uncertainty introduced by the measurement instrumentation, which is evaluated for class 1 noise measuring equipment equal to 0,49 dB (on the basis of empirical data), u_{pos} is the standard uncertainty due to positioning of the measuring instruments and the last contribution u_{other} is commonly neglected in the professional field.

Thus, assuming that the contributions of inherent uncertainty are uncorrelated, the global uncertainty can be expressed as:

$$u_{Leq}=\sqrt{u_{instrum}^2 + u_{pos}^2 + u_{other}^2} \tag{3}$$

In order to meet industrial and regulatory requirements, in any measurement scenario the expanded uncertainty $U = ku$ is calculated, in which the parameter k (coverage factor) is the function of the confidence interval.

Nevertheless, to adequately estimate the uncertainty of noise measurement (u_{LAeq}), the intrinsic variability of the measurand (u_{var}) has to be considered.

4 The Proposal

The authors, as part of a research activity carried out over the last few years at the Department of Industrial Engineering (DIIn) of the University of Salerno -Italy-, have proposed an integration of three different steps and the creation of a new procedure (Fig. 3) for the evaluation of the uncertainty of measurement of noise, with reference to measurand variability.

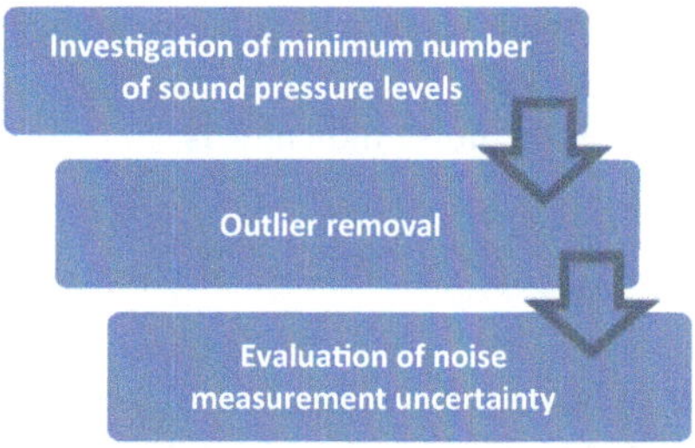

Fig. 3. Block diagram of the proposal.

The algorithm is based on the bootstrap method which does not require knowledge of the probability density function of considered statistics, and is, therefore, more correct for the analysis of noise, whose population is not strictly Gaussian [11].

From the acquired dataset this approach generates new datasets by sampling with replacement (bootstrap samples). Starting from a random variable X and a sample (x_1, x_2, ..., x_n), which is a realization of $X = \{X_i, i = 1, 2, \ldots, n\}$, a certain statistics $R(x, F)$ *is* determined on the sample space. The standard steps of the basic bootstrap procedure are the following:

1. scheming the distribution of probability using the function P $(X_B = x_K) = \frac{1}{n}$ for $k = 1,\ldots,n$ named bootstrap distribution from sample and indicated by $\hat{F}$, where n is the sample size;
2. sampling on the basis of the distribution of values $(x_1^*, x_2^*, \ldots, x_n^*)$, representing the realisation of variable $x^* = (X_1^*, X_2^*, \ldots, X_n^*)$ and it is named the bootstrap sample;

3. distribution of statistics R is estimated with the bootstrap distribution: $\vartheta^* = R\left(x^*; \hat{F}\right)$.

Referring to the scheme in Fig. 4, in the first phase, since the reliability of the noise indicators depends on the temporal variability of the noise, a data-driven sampling strategy is employed, in order to detect a corresponding minimum number of sound pressure levels necessary for statistical significance of the acquired dataset [12–15]. In the proposed approach the bias corrected percentile bootstrap method (CPER) is used for determining the Confidence Interval of the short time statistics Leq, with a fixed Confidence Level [16]. The interval endpoints are function of the bias correction, determined by the bootstrap sampling distribution.

This model is also effective for predicting the noise indicator by analysing a measurement window shorter than the observation period.

The focus of the next stage is the study of the influence of the spot events occurrence on the measurement uncertainty: for this purpose, a numerical algorithm based on K-neighbors' distance is applied [17] and experimentally verified using real data to ascertain so as to remove these anomalous outliers ("Outlier detection") from the acquired population of data [18]. In the last step of the study, the uncertainty associated with the measured noise is calculated. In particular, the authors, for determining the contribution of the variability of measurand to the measurement uncertainty, have selected the normal bootstrap method, in which bootstrapped bias and standard error approximate the confidence interval.

The bootstrap estimation of the noise indicator is:

$$\overline{L_{eq}^*} = \frac{\sum_{b=1}^{n} L_{eqb}^*}{n} \tag{4}$$

where n is the number of bootstrap replicas and L_{eqb}^* is the level of the bth bootstrap estimate of index L_{eq}, and bootstrap estimate of standard uncertainty is:

$$u\left(\overline{L_{eq}^*}\right) = \left(\overline{L_{eq}^*}\right)\sqrt{\frac{\sum_{b=1}^{n}\left(L_{eqb}^* - \overline{L_{eq}^*}\right)^2}{n-1}} \tag{5}$$

Experimental results, concerning the application of the proposal, have been matched with the evaluation of the expected value of the noise indicator and the relative uncertainty by using the classical approach [5].

Table 1 shows, as practical example, the contribution of the variability of workers' exposure to noise in a workplace on measurement uncertainty, taking into account the residual value of the standard deviation of outlier detection algorithm too (s). In particular, the global uncertainty is evaluated considering both the estimated standard deviation and the residual uncertainty of the outlier detection algorithm.

This reference is evaluated by means of the classical method to the whole population consisting of 14.400 sound pressure levels because the time history logging step was fixed to 0.125 s and observation period was equal to 30 min.

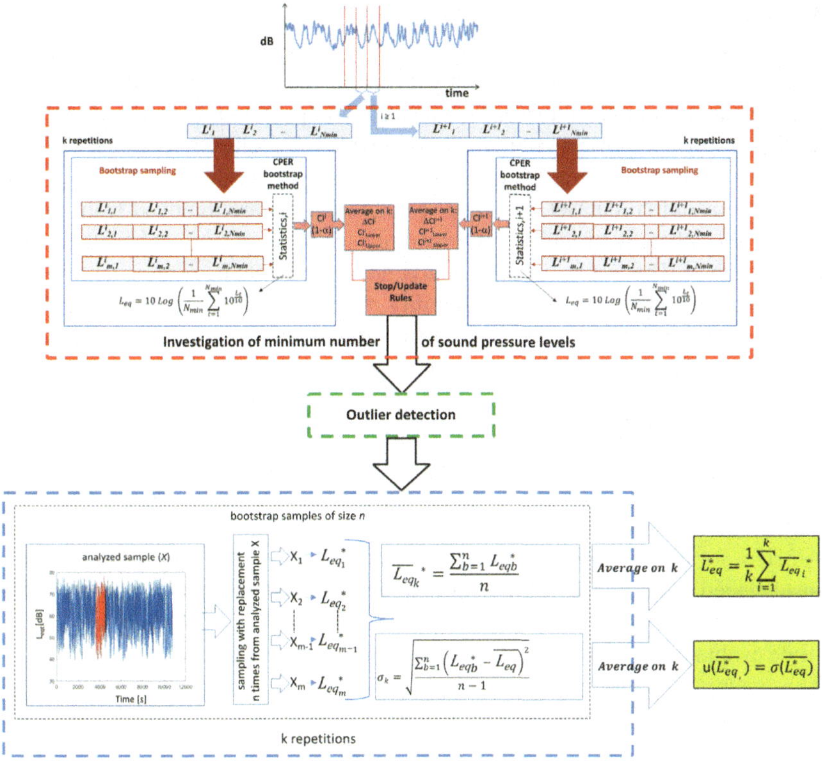

Fig. 4. Scheme of the procedure (first stage = dash red, second stage = dash green, third stage = dash blue).

As may be noted, the classical method and the proposal approximate the expected value for the short-term noise indicator very well when few samples are available. Moreover, the reduced number of samples leads to an overestimation of the true standard uncertainty, especially by applying the classical method.

Table 1. Estimation of the equivalent sound pressure level and uncertainty in measurement time and observation period.

Bootstrap method					Classical method				
		Measurement time interval			Measurement time interval			Observation period	
Noise source	Sample size (n)	Expected value [dB]	δ [dB]	Global uncertainty [dB]	Expected value [dB]	δ [dB]	Global uncertainty [dB]	Expected value [dB]	Global uncertainty [dB]
Workers' exposureina workplace	1200	61.040	0.016	0.101	61.520	0.027	0.115	61.100	0.005
	2400	61.160	0.020	0.074	61.160	0.022	0.085		
	3600	61.130	0.013	0.056	61.130	0.016	0.067		

5 Conclusions

The paper has covered a procedure for estimating the uncertainty of noise measurement as practical example of a variable dataset to compare with such imposed threshold value. The proposal represents a summary of a wide research activity of the authors and it essentially consists of three phases: first an algorithm assures the statistical significance of the acquired dataset, then spot events (outlier) are removed from the signal and finally the inherent variability of uncertainty measurement is calculated using the bootstrap method. The results are particularly interesting both from the scientific point of view, because they demonstrate the reliability of the procedure for predicting the noise indicator when considering a reduced dataset, and from a professional point of view, because they allow the experts working in this field to estimate the contribution of the measurement uncertainty of the variability of the measurand.

The novel procedure has been validated on real data acquired during several measurement campaigns and in this paper the tangible values related to two noise measurement have been mentioned: the results have confirmed the reliability, the stability and the robustness of the proposal.

A patent for an industrial invention has been granted to the authors, in more detail, for an advanced electronic measurement system of equivalent sound pressure level, with automatic real-time determination of the minimum measurement time and automatic real-time evaluation of the inherent contribution to measurement uncertainty [19].

This integrates some additional blocks in the phonometric chain, enhancing the traditional functions of the sound level meter.

References

1. Liguori, C., Ruggiero, A., Russo, D., Sommella, P.: Task-based measurements for estimating the workers' exposure to acoustic noise. In: IEEE International Instrumentation and Measurement Technology Conference (I2MTC), pp. 1062–1067 (2018)
2. Ruggiero, A., Russo, D., Sommella, P.: Determining environmental noise measurement uncertainty in the context of the Italian legislative framework. Measurement **93**, 74–79 (2016)
3. ISO I: 9612: Acoustics-Determination of Occupational Noise Exposure-Engineering Method, Geneva, Switzerland: International Organization for Standardization (2009)
4. Liguori, C., et al.: Proposal for the automatic evaluation of workers' exposure to acoustic noise following task-based approach. Measurements. https://doi.org/10.1016/j.measurement.2020.108534
5. BIPM, IEC, IFCC, ILAC, ISO, IUPAC, IUPAP and OIML: Evaluation of measurement data—the role of measurement uncertainty in conformity assessment. JCGM, **106**, 1–106 (2012)
6. ISO 14253-1:2013: Geometrical product specifications (GPS) -- Inspection by measurement of workpieces and measuring equipment -- Part 1: Decision rules for proving conformity or nonconformity with specifications
7. Ferrero, A., Salicone, S.: Measurement uncertainty. IEEE Instrum. Meas. Mag. **9**(3), 44–51 (2006)
8. ASME B89.7.3.1–2001: Guidelines to decision rules in determining conformance to specifications
9. ISO 10576–1:2003: Statistical methods – guidelines for the evaluation of conformity with specified requirements, Part 1: General principles. ISO, Geneva

10. Liguori, C., Ruggiero, A., Sommella, P., Russo, D.: Choosing bootstrap method for the estimation of the uncertainty of traffic noise measurements. IEEE Trans. Instrum. Meas. **66**(5), 869–878 (2017)
11. Liguori, C., Ruggiero, A., Russo, D., Sommella, P.: Innovative bootstrap approach for the estimation of minimum measurement time interval in road traffic noise evaluation. Measurement **98**, 237–242 (2017)
12. Liguori, C., Ruggiero, A., Russo, D., Sommella, P.: Influence of measurement time on acoustic noise uncertainty. In: IEEE International Instrumentation and Measurement Technology Conference (I2MTC), pp. 1–6 (2017)
13. Liguori, C., Ruggiero, A., Russo, D., Sommella, P.: Estimation of the minimum measurement time interval in acoustic noise. Appl. Acoust. **127**, 126–132 (2017)
14. Liguori, C., Ruggiero, A., Russo, D., Sommella, P.: Accurate estimation of the environmental noise through sampling approach: selection of the measurement time. IEEE Trans. Instrum. Meas. **67**(5), 1006–1013 (2018)
15. Stępień, B.: Bootstrap confidence intervals for noise indicators. Acta Acustica united with Acustica **102**(2), 389–397 (2016). https://doi.org/10.3813/AAA.918955
16. Sheng, B., Li, Q., et al.: Outlier detection in sensor networks. In: Proceedings of MobiHoc (2007)
17. Liguori, C., Paolillo, A., Ruggiero, A., Russo, D.: Outlier detection for the evaluation of the measurement uncertainty of environmental acoustic noise. IEEE Trans. Instrum. Meas. **65**(2), 234–242 (2016)
18. Pietrosanto, A., Ruggiero, A., Russo, D., Sommella, P.: Electronic measurement system of equivalent sound pressure level, with automatic real-time determination of the minimum measurement time and automatic real-time evaluation of inherent measurement uncertainty contribution, Italian patent for an industrial invention n. 102016000124340 granted on 17th June 2019

Attendants at MMC Composite Milling

János Líska(✉), Krisztián Kun, and Zsolt Kovács

Department of Innovatice Vehicles and Materials, John von Neumann University, Izsáki út 10, Kecskemét, Hungary
liska.janos@gamf.uni-neumann.hu

Abstract. The usage of composite materials is showing a growing tendency in the industry, especially in the field of automotive engineering. The goal to improve emissions and safety requires eco-friendly solutions, reducing the weight of vehicles. The efficient and economical use of these materials requires many cutting investigations. To achieve the best results and tool life well-chosen technology and technological parameter are needed.

Keywords: Milling · Composite material · Tool wear

1 Introduction

During the research of goodness by cutting, it is found worthy to compare the two examined materials. The used aluminium alloys are M6061 with modified silicon and magnesium content and C355, which contains small portions of ceramic particles. These are known as raw material for aluminium foam. A number of technological parameters were examined and evaluated. Two types of cutting tools and the two types of composite aluminum alloy mentioned earlier were used. A milling machine was used for the machining process and chip removal was free cutting. After machining the surface roughness, the chips and tool wear were evaluated, and the forces were also measured during the milling process.

The results of the experiments may help in further research as a cutting guide for the industry. For example, in the first part of the experiment, it was proved that there are also 4 input parts. Of course, the force generated during complex machining is much higher than the reference material, but it does not show a noticeably large difference when using a tool specifically designed for steel or cast iron.

The machining of aluminum composite materials is a complex process, and the quality of the tool used greatly influences the output values and results of the experiment (Fig. 1).

S. Hloch et al. (Eds.): ICMEM 2021, LNME, pp. 332–342, 2021.
https://doi.org/10.1007/978-3-030-71956-2_27

Fig. 1. C355 aluminium composite with ceramic particle.

As we know, these examined materials are not ordinary aluminum alloys, so I would first describe the base material of it, which is the aluminum. In addition, it is important to know that both are casting plates.

The following Table 1, summarizes the mechanical and chemical properties of C355 and M6061 alloys, both in T6 heat treatment condition [1]:

Table 1. Mechanical and chemical properties of the tested aluminum alloys.

	C355	*M6061*
Tensile strength [MPa]	290	310
Yield strength [MPa]	200	270
Tensile shear strength [MPa]	190	210
Elastic modulus [GPa]	70	69
Allay content [%]		
Al	91.7–94.1	95.9–98.6
Si	4.5–5.5	0.5–1
Mg	0.4–0.6	1–1.5
Fe	0–0.2	0–0.7
Ti	0–0.2	0–0.15
Mn	0–0.1	0–0.15
Zn	0–0.1	0–0.25
Cr	-	0.04–0.35
Cu	-	0.15–0.4
Contamination	0–0.15	0–0.15

It is important to know the mechanical properties and alloy content of the materials. Because both data are important in machining aspect. We know that C355 contains ceramic granules size of from 80 nm to 20 μm. Maximum up to a six percent. Which are added for the molten aluminium by a high temperature unification technology. This means, we are talking about composite materials. In this case, the ceramic granules are aluminum oxide (Al_2O_3). Thereby increasing the strength of the finished product. i.e. metallic foam. Which appears to have lower characteristics than M6061.

2 Methods

We can evaluate the experiment so long, so I tried to determine those parameters that are relevant to the future project.

I chose two different workpieces and two types of tools from the input parameters. During the manufacturing which was a free cutting. The tangential force (*Ft*) was measured by a Kistler 9257B dynamometer. While the input parameters were changed. I measured the surface roughness among the output parameters and collected separately the chips.

The workpieces were C355 and M6061 alloys. I chose two different qualities of tools. Because to the differences are clearly visible. In addition, it helps to determine which direction is worth to choose the right tool. I didn't find any similar topic in my literature research, so I chose the most optimal option for me. In the technological process, the cutting speed was set to constant. The value of it was $v_c = 700$ m/min. The feed rate per tooth was $f_{z1} = 0.1$ and $f_{z2} = 0.15$ mm/tooth. For composite materials the applied value is usually $f_z = 0.84$ or 0.17 tooth [6]. The cutting depth was 1 and 3 mm. The above mentioned parameters are the variables. The applied machine was a NCT Toomill 250 CNC machining center with NCT 204 controller (Fig. 2).

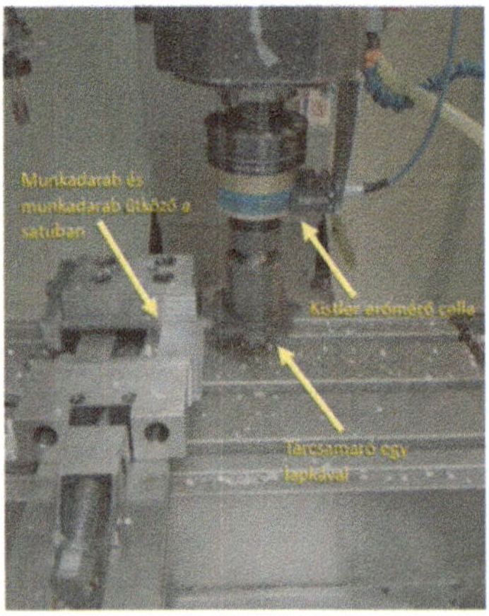

Fig. 2. Cutting force measuring layout.

The surface roughness was measured by Mitutoyo C-3000 roughness and contour measuring machine, it helps to determine several roughness profiles. The separated chips were examined by a Dino-Lite Digital microscope. The abrasion during cutting was evaluated using a Mitutoyo Quick Vision Elf Pro microscope (Fig. 3). The tool wear was evaluated by Mitutoyo Quick Vision Elf Pro microscope. This equipment can perform high-precision measurements, even with nanometer precision.

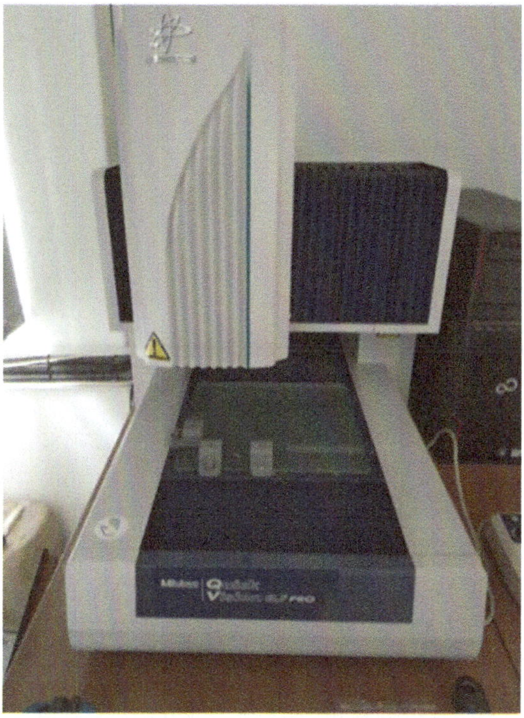

Fig. 3. Mitutoyo Quick Vision Elf Pro microscope.

During the experiment the type of robust design experimental methodology was used. This experimental design was created in the Minitab software. The generated experimental design is visible in Table 2.

Table 2. Technological parameters and factors.

	Cutting speed. V_c (m/min)	Feed per tooth. f_z (mm/teeth)	Cutting depth. a_e (mm)	Tool	Material
1	700	0.1	1	1	0
2	700	0.1	1	1	1
3	700	0.1	1	2	0
4	700	0.1	1	2	1
5	700	0.1	3	1	0
6	700	0.1	3	1	1
7	700	0.1	3	2	0
8	700	0.1	3	2	1
9	700	0.15	1	1	0
10	700	0.15	1	1	1
11	700	0.15	1	2	0
12	700	0.15	1	2	1
13	700	0.15	3	1	0
14	700	0.15	3	1	1
15	700	0.15	3	2	0
16	700	0.15	3	2	1

The first column of the table clearly shows that the main parameter is the cutting speed. The recommended cutting speed for the carbide insert was $v_{c1} = 315$ m/min for the insert which deigned for steel and $v_{c2} = 1100$ m/min for the insert which deigned for aluminum. Therefore was set up an intermediate value of $v_c = 700$ m/min. The recommended value for feed per tooth according to the first insert $f_{z1} = 0.15$ mm/tooth and for the second insert $f_{z2} = 0.19$ mm/tooth. Before the start of each experiment. The insert was changed and also the work-piece. The tools are marked with 1 and 2. While the work-piece with 0 and 1. The dimension of applied work-piece is presented in Fig. 4.

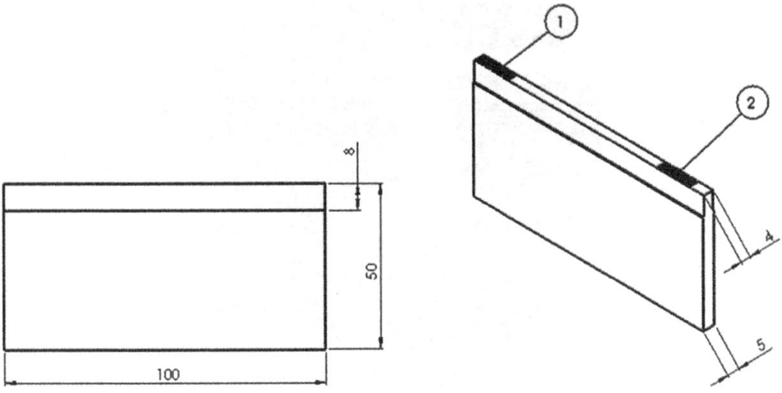

Fig. 4. The work-piece and its dimensions.

The surface roughness was evaluated at two points and program of the measurement instrument examined a total length of 24 mm. First, the measurement was completed at 12 mm on stage 1, and then about 5 mm before the end of the piece, was again measured 12 mm on stage 2. This long measuring distance so important because to be filtered the measurement inaccuracy and the standard deviation.

After preparation of the work-pieces for the test. I chose the right tool. Since there is not many literature on the subject, so I tried to choose a monolithic carbide tool which can machined most materials. The selected tool was a Sandvik 1P222–1200-XA-1630, it has 4-edge, PVD AlCrN coating and diameter of 12 mm.

The spindle of the selected machine can be rotated up to 12000 rpm. so the maximum available cutting speed is slightly more than 450 m/min. This is a sufficient enough cutting speed range to perform wear tests.

The tool wear was measured the above mentioned Dyno-lite microscope. In order to repeat the measurement several times. The carbide tool was fixed in an own designed magazine. The measurement is shown in Fig. 5.

The applied edge geometry of tool is positive so it suitable aluminum cutting. Table 3 shows the technological parameters.

3 Results

In this capture we can see the results of the measured datas. We was measured the tangential cutting forces (*Ft*) and surface roughness also. In the Table 4. we can see the analytical evaluation of the values. The table clearly shows the factor. Which we was changed (f_z. a_e. Cutting tool, material). For the better understanding at the two last factor we was used numerical marking.

We was made a regression for the data in Table 4. The measured data mean different points in a plane. If a curve attached to these points, we can get the goodness of the research by the Excel software. The equation for the curve determines each component how affects the test factor. In the present case, the following Eq. (1) can be described by

Fig. 5. Tool wear measurement set-up.

Table 3. Technological parameters for tool life measurement.

	Revolution (RPM)	v_f (mm/min)	Cutting speed. v_c (m/min)	a_e (mm)	a_p (mm)	Feed per tooth. f_z (mm/tooth)
1	7958	1592.4	300	6	1	0.05
2	6632	1326	250	6	1	0.05
3	4775	995	180	6	1	0.05

examining the tangential force:

$$F_t = 114.625 + 390 \cdot f_z + 25.5 \cdot a_e - 55 \cdot tool + 9.75 \cdot material \quad (1)$$

It can be seen that the greatest effect for the cutting force is the change of the feed per tooth. Then come the chosen tool number 2. The depth of the cut changes the power output by 25.5 times. Finally, the material has the least impact on the force. I made a diagram (Fig. 6) to shows the results by Minitab.

In this case. The goodness was 95.3%, which represents the r-square. The F-significant value is 3.2×10^{-7}, and the p-value is less than 0.05% in most cases.

From this equation, we can determine the tangential force. From the tangential force can be determined the specific cutting force and from cutting force we can conclude the machine performance.

Figure 7 shows how the coefficients influence the change in surface roughness based on the Rz profile. The following equation can be written:

$$R_{zst} = 1.45 \cdot f_z + 0.2125 \cdot a_e - 0.475 \cdot tool + 1.05 \cdot material \quad (2)$$

Table 4. Results of the cutting force and surface roughness.

	fz (mm/tooth)	a_e (mm)	Cutting Tool	Material	F_t (n)	*Ra* Start (μm)	*Ra* Exit (μm)	*Rz* Start (μm)	*Rz* Exit (μm)
1	0.1	1	1	0	132	0.235	0.28	1.25	1.5
2	0.1	1	1	1	128	0.325	0.35	2.45	1.8
3	0.1	1	2	0	73	0.2	0.23	1.86	1.9
4	0.1	1	2	1	90	0.25	0.22	2.03	1.8
5	0.1	3	1	0	175	0.34	0.355	1.99	1.8
6	0.1	3	1	1	178	0.41	0.27	2.92	1.6
7	0.1	3	2	0	115	0.21	0.17	1.41	1.2
8	0.1	3	2	1	125	1	0.6	5.44	4.2
9	0.15	1	1	0	151	0.2	0.18	1.17	1
10	0.15	1	1	1	150	0.3	0.3	1.61	2.2
11	0.15	1	2	0	64	0.18	0.2	1.89	1.1
12	0.15	1	2	1	102	0.3	0.28	3.04	2.3
13	0.15	3	1	0	200	0.21	0.17	1.25	0.9
14	0.15	3	1	1	200	0.35	0.315	1.79	2.2
15	0.15	3	2	0	145	1.5	1.75	1.10	0.9
16	0.15	3	2	1	160	0.225	0.37	1.88	2.7

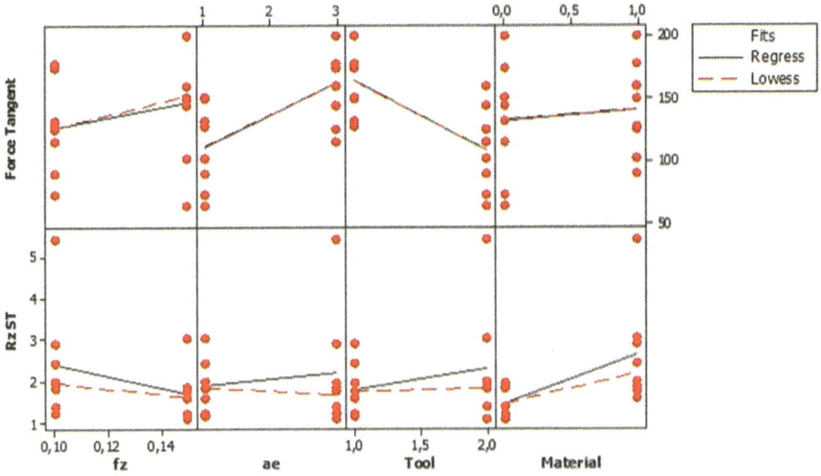

Fig. 6. Main effect plot of the cutting force and surface roughness.

The blue curve represents the first measurement. It can be seen that at $v_c = 300$ m/min there is no abrasion phase. The tool is worn out linearly. Despite the small amount of

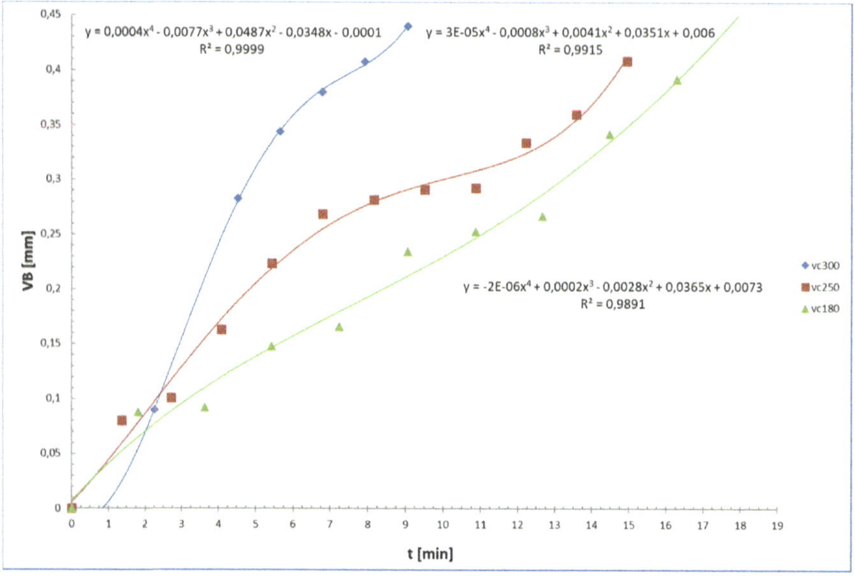

Fig. 7. Tool life plot.

data. The curve can be fitted. The manufacturer's recommended cutting speed of v_c = 700 m/min for aluminum, in this case it is very high because of the containing ceramic particles. The microscopic images clearly show that the coating has worn off the surface of the tool despite the short section.

The orange curve shows edge wear measured at a cutting speed of $v_c = 250$ m/min. Here. The wear Sect. (1) can be clearly distinguished. Followed by a shorter. Even wear Sect. (2). And soon enough the over-wear (3) arrives. This wear is called progressive wear.

The third green curve. The last measured tool wear curve. This is the closest to the ideal. Here the above mentioned three stages can be well distinguished. The wear Sect. (1) is rather short. The uniform wear Sect. (2) is flatter and difficult to distinguish from the over-wear Sect. (3). This curve is degressive.

The black line indicates the allowable wear value.

4 Conclusion

Based on the series of experiments. it can be concluded that composite cutting is a complex process. Tests on carbide tools provide guidance for aluminum composite cutting. Which helps you to choose the right tool and parameters.

In the first part of the experiment we explored the evolution of the force and it was proved here that the 4-phase force run is also present. We gave a clearer picture of what's happening when you start cutting by quick camera.

The generated force in case of composite materials was much higher than that during of the M6061 reference material. However, it does not show a marked difference when using machining tools for steel or cast iron.

Feed rate has the most effect for forces. Then comes the tool and the depth of cut. From the derivative equation. The tangential force factors were determined. Which provides guidance for cutting materials. After running the regression. The goodness of the experiment became 95.3%. Surface quality is influenced by the tool and feed rate. Drastic changes can be achieved by varying the cutting speed. So it is recommended to reduce this and use a tool with sharper edge geometry to increase tool life. Typically. The aluminum-specific chip produced better results. With the increase in feed rate the difference was no longer significant. The difference between the average surface roughness and the roughness height is not four times higher than in the literature.

The composite is capable of abrasion of inserts and milling tools in a short time due to their abrasive properties. Sharp geometry and lower cutting speeds are recommended. If possible. Use PCD or CBN tools. Recommended coating types are TiN. TiCN and Al_2O_3.

Edge burr can be eliminated with sharper tool geometry and less material removal. Coolant and climb machining reduce the likelihood of leading edge burr.

Increasing the cutting depth did not significantly affect the force development or surface roughness. So deeper depth of cut can be used for productivity.

From a coating point of view. a higher hardness. a lower coefficient of friction. And a higher heat resistance are recommended.

The chip formed was fractured with a matte surface in the case of composite material. Ideal for transport and storage. it can be washed off by coolant lubricant.

The Taylor curve and its calculated coefficients help to select the machining parameters of aluminum composite.

Further experiments:

- Test other types of inserts. Milling tools and coatings.
- Take high-speed shots of the shear plane.
- Taking thermal recordings of the tool.

In the future we will plan the new experiments. Which orientated to the chip formation and cutting temperature during this kind of materials.

Acknowledgements. The project has been supported by the European Union. co-financed by the European Social Fund. EFOP-3.6.1–16-2016–00014.

References

1. Unknow – mailto. https://www.szfki.hu/~konczos/tanfolyam/9.pdf
2. Ismeretlen. https://www.cnc.hu/g-kod-kereso/
3. Sandvik Coromant, A.B.: Training Handbook. Metal Cutting Technology (2017)
4. Békés, J.: A fémforgácsolás tervezése (1984)
5. Paulo Davim, J.: Machining of Metal Matrix Composites. Springer (2012)
6. Líska, J.: Chip breaking phenomenon during cutting steel and polymer materials (2014)
7. Ligeti, G., Miskolczi, I.: Formaadó szerszámok bevonatolása - Oerlikon Balzers Magyarország (2017)

8. Babcsán, N., Beke, S., Makk, P., Szamel, G., Kadar, C.: Pilot production and properties of ALUHAB aluminium foams. Procedia. Mater. Sci. **4**, 127–132 (2014)
9. Li, Y.: Comparison of aluminium foams produced by melt foaming and gas injection processes. Spec. Cast. Nonferrous Alloy. **31**, 1097–1099 (2011)
10. Zhang, H., Chen, X., Fan, X., Li, Y.: Compressive properties of aluminum foams by gas injection method. China Foundry **9**, 215–220 (2012)

Design and Verification of an Acoustic Mitigation Intervention for an Air-Cooled Inverter Heat Pump: A Case Study

Alessandro Ruggiero[1](✉), Domenico Russo[1], Roberto D'Amato[2], and Nicolae Ungureanu[3]

[1] Department of Industrial Engineering, University of Salerno, Via Giovanni Paolo II 132, 84084 Fisciano, SA, Italy
ruggiero@unisa.it
[2] Departamento de Ingeniería Mecánica, Química y Diseño Industrial, Technical University of Madrid (UPM), Madrid, Spain
[3] Faculty of Engineering, Department of Engineering and Technology Management, Technical University of Cluj-Napoca, Baia Mare, Romania

Abstract. An acoustic mitigation intervention with reference to an air conditioner outdoor fan motor (heat pump and liquid chiller) CLIVET WSAN-XIN 141 is explored. On the basis of a measurement session conducted ante operam, the paper focuses on the phases of design activity of an innovative structure aimed at reducing the equivalent sound pressure level on the façade to the most sensitive receiver, according to the current Italian legislation. In these cases resistive or dissipative silencers with a circular or rectangular duct, internally lined with sound absorption material, are widely used, but the axial fan under examination has a residual prevalence very low to allow the installation of suitable silencers. So in order to guarantee the adequate air flow to the radiator heat exchanger along the lateral surface of the heat pump, avoiding to reduce the efficiency of the system, an original structure of appropriate geometry, able to achieve high-performance sound mitigation, is presented. The phonometric campaign post operam shows the results achieved at the end of the realization of the proposed acoustic mitigation intervention and this approach may be adopted as guideline by the practitioner.

Keywords: Acoustical design · Experimental verification · Reverberation time integrated impulse response

1 Introduction

Nowadays noise reduction is one of the key issues for improving the quality of life. High level noise for a long time can cause to people exposed several health issues, for example stress and even mental health problems. In literature, there is an important scientific knowledge to take up the challenge of monitoring acoustic waves to face all technological requirements. In acoustical engineering noise levels are caused by a mixture of air-borne and structure-borne sound transmission [1, 2]. Generally material insulators are used to

S. Hloch et al. (Eds.): ICMEM 2021, LNME, pp. 343–354, 2021.
https://doi.org/10.1007/978-3-030-71956-2_28

reduce mechanical vibrations to avoid factor weakening [3]. Moreover, in literature there are several designs of sound proofs in order to reduce the sound insulation and vibration, for example the socalled sandwich panels [4, 5], consisting of two thin plates [6–8]. This work presents an acoustic design approach and the experimental verification of a mitigation intervention for an air-cooled inverter heat pump. In particular, the authors illustrate the result of an acoustic reclamation project for an air conditioner outdoor fan motor (heat pump and liquid chiller) CLIVET WSAN-XIN 141 (Fig. 1), located in Salerno (Italy), whose technical characteristics are reported in Table 1.

Table 1. Main parameters of the air conditioner outdoor fan motor.

WSAN-XIN 141	Heating Capacity [kW]	Cooling Capacity [kW]	Sound Power Level (dB)								
			Octave bands (Hz)								
	32.36	33.19	63	125	250	500	1000	2000	4000	8000	dB(A)
			85.0	83.0	80.0	79.0	83.0	75.0	72.0	61.0	85.0

Phonometric measurements are carried out *ex ante* and *ex post* with respect to the implementation of the intervention, allowing the authors to compare the data obtained with the legal limits [9] and to evaluate the obtained acoustic results through project realization.

Fig. 1. Position of air-cooled inverter heat pump.

Following an analysis of the acoustic current regulation of the zone of interest and the consequent determination of the maximum acceptable limits of environmental noise established by the law, on the basis of the system operating period, the attention has been focused on the diurnal reference time (6.00–22.00).

A phonometric campaign has been carried out (Fig. 2) and the measured equivalent sound pressure level at the facade of the nearest sensitive receiver was greater than corresponding legal threshold; so an acoustic mitigation intervention has been necessary.

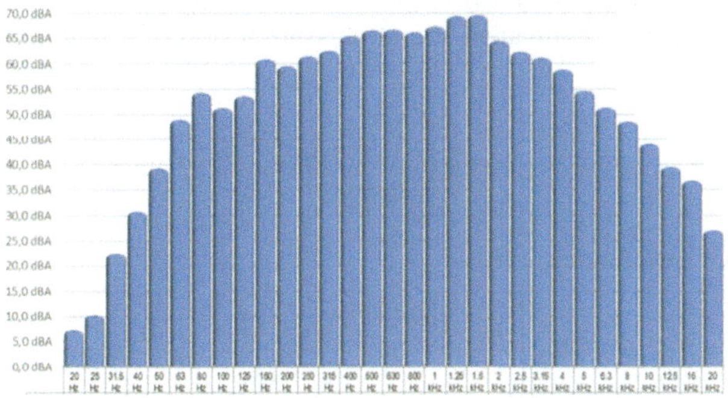

Fig. 2. Spectral representation in octave thirds of the A weighted equivalent sound level measured at 1 m distance (pump side) during full operating phase of the system.

It has been revealed particularly complex due to the need to allow the system to operate properly and, in particular, the adequate air flow to the radiator heat exchanger along the lateral surface of the heat pump, avoiding to reduce the efficiency of the system.

Generally, noise reduction occurs through intervention on the sound source [10], but, in this case, a different approach has been proposed, by combining passive noise control techniques [11].

Regarding to the problem of air conditioning systems noise control, the resistive or dissipative silencers with duct are widely used [12], because these devices have a good level of sound attenuation [13].

After the choosing of the silencer compatible with the system characteristics, the authors have discarded this solution because the axial fan has a residual prevalence very low to allow the installation of suitable silencers.

So, an acoustic innovative structure located around the system, based on modular panel system comprising a layer of sound-insulation material with externally facing spherical caps on top of another layer of sound-absorbant material, has been designed. As last step, another phonometric campaign has been conducted for analysing the activity findings, after the realization of the proposal.

2 Acoustic Mitigation Intervention

The acoustic emission of the external fan motor of the air conditioner (heat pump and liquid chiller) CLIVET WSAN-XIN 141 was evaluated with a class 1 sound level meter (Larson Davis 831) by taking a measurement at a distance of 1 m from the nearest building facade and a height of 4 m [14]. Measurement campaign was performed in one day (wind speed less than 5 m/s and no rain) with acquisition taking place over 50 min accounting for two different working conditions (namely startup and modulation state) of the air conditioner fan motor. Regarding to the instrument setting, the time history logging step was fixed at 0.1 s, (dataset of 30000 consecutive sound pressure levels). The measured equivalent sound pressure level (70 dBA) was greater than corresponding threshold established by the law (60 dBA): during diurnal reference time ($T_R = 16$ h) the heat pump is operative every day for about ten hours ($T_O = 10$ h), working in startup and modulation state, so equivalent sound pressure level corresponding to the nearest building facade was [15]:

$$L_{Aeq,T_R} = 10\log\left[\frac{1}{T_R}\sum_{i=1}^{n}(T_O)_i 10^{0.1L_{Aeq,(T_O)_i}}\right] \tag{1}$$

in the case under examination [16]:

$$L_{Aeq,T_R} = 10\log\left[\frac{T_O \cdot 10^{0.1 \cdot L_{Aeq,T_M}} + (T_R - T_O) \cdot 10^{0.1 \cdot L_R}}{T_R}\right] \tag{2}$$

where $L_{Aeq,TM}$ is equivalent sound pressure level referring to measurement time T_M and L_R is residual noise level.

In order to outline the operating principle of the specific class of heat pump in question, the schemes both of the type of component and of the position on the system are shown in Fig. 3 and Fig. 4.

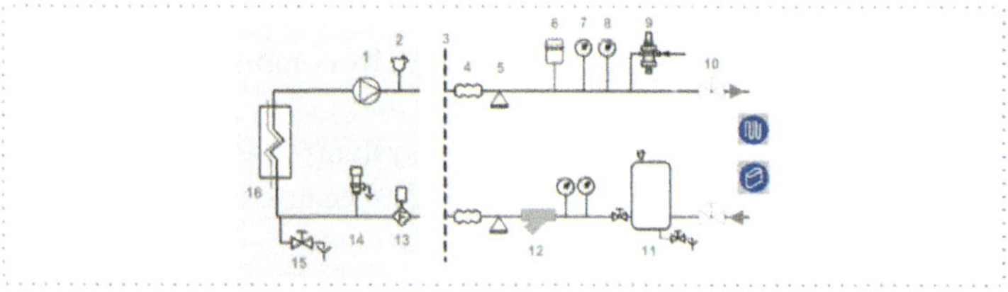

1. Pump / circulating pump
2. Vent
3. Limit unit
4. Antivibration joints
5. Piping support
6. Expansion vessel
7. Pressure gauge
8. Thermometer
9. Filling valve
10. Shut-off valve
11. Internal storage tank
12. Filter
13. Flow Switch
14. Safety valve
15. Drain valve
16. Utility side exchanger

Fig. 3. Scheme of the type of component of heat pump CLIVET WSAN-XIN 141.

In a first design phase, considering heat pump under examination, with particular attention to the axial fan currently mounted on board (Fig. 5), whose characteristic curves are shown in Fig. 6 and corresponding values are reported in Table 2, a potential installation on the top of machine of an appropriate silencer was analysed, according to the following procedure:

1. identification of the suitable type of silencer, on the basis of the system characteristics,
2. estimation of the heat pump equivalent sound pressure level at the facade of the nearest receiver following the sound attenuation due to the installation of the silencer.

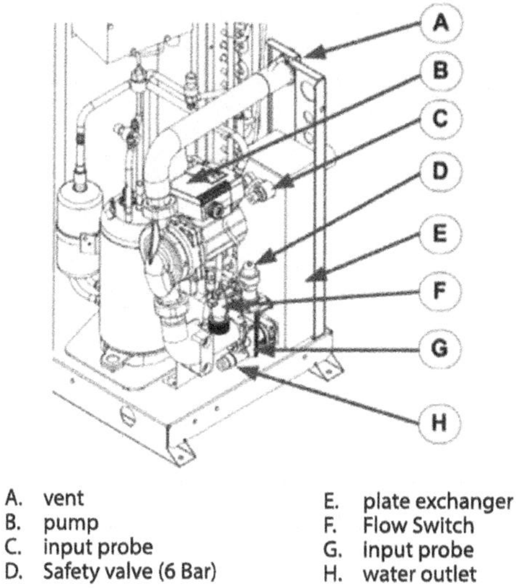

Fig. 4. Scheme of the position on the system CLIVET WSAN-XIN 141.

Generally, the rectangular absorption silencers are particularly indicated for noise abatement in air conditioning and ventilation systems, both in the civil and industrial field, because they control and reduce the noise transmitted through the system ducts. Moreover, rectangular channels are preferred to circular ones because they are more rigid and they have a lower capacity to absorb energy.

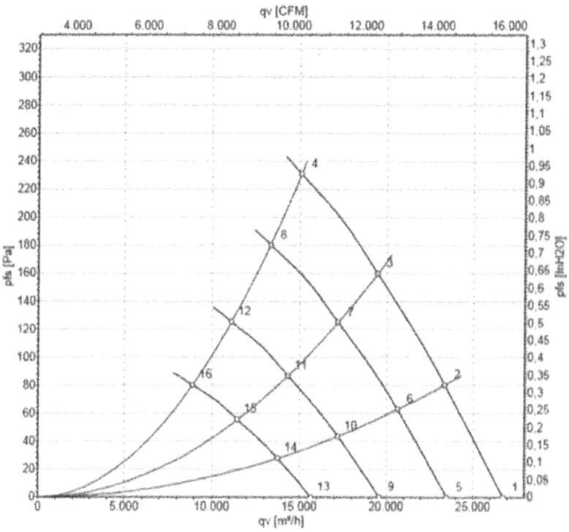

Fig. 5. Axial fan under examination drawing.

The systems with axial fans have a solid and resistant construction: they are made up of a 1 mm thick galvanized sheet steel casing with quick flanges with holes at the four corners. The sound-absorbing partitions consist of a galvanized steel frame with

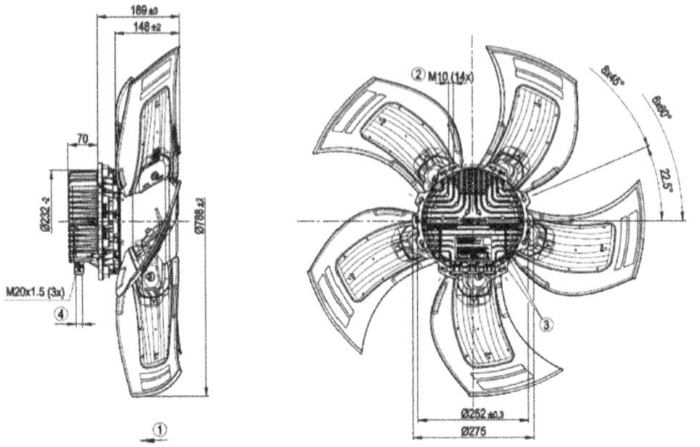

1	Direction of air flow "V"
2	Depth of screw max 25 mm
3	Tightening torque 3.5 ± 0.5 Nm
4	Cable diameter: min 4 mm, max 10 mm, tightening torque: 4 ± 0.6 Nm

Fig. 6. Characteristic curves of CLIVET axial fan heat pump WSAN-XIN 141.

variable thickness depending on the size of the partitions. The internal coating is made of high quality, non-hygroscopic and non-combustible material with a density of 55 kg/m^3, internally protected by a black glass tube. The working temperature limit for standard models is about 200 °C. The attenuation depends of the width of the air passages, the thickness of the partitions and the length of the silencer. The pressure losses depend substantially on the useful section of air passage (base and height). With the same thickness and the number of sound-absorbing partitions, in order to increase the attenuation, it is necessary to reduce the section of the air passages and increase the length of the silencer; to reduce pressure losses, simply increase the height and width of the silencer.

In order to choose a silencer it is important to consider these data: air flow, admissible loss of load, attenuation referred to a specific frequency or the value of the acceptable noise level after the installation of the silencer, installation dimensions (base, height and depth).

Another significant quantity is the free surface, which is a fictitious area that allows, on the basis of the speed of the air, to calculate the flow through the silencer.

$$Q = v_k \cdot S \cdot 3600 \tag{3}$$

Where

Q = air flow rate [m^3/h]
v_k = average measured speed [m/s]
S = free front surface [m^2]

For the dimensioning of the silencer, according to the desired acoustic attenuation, the minimum depth of the component has been identified; then, with reference to the air flow through the silencer, obtained from the axial fan datasheet, and the admissible load

Table 2. Operating conditions of CLIVET axial fan heat pump WSAN-XIN 141.

	U	f	n	P_{ed}	I	LpA_{in}	LwA_{in}	LwA_{out}	qv	p_{fs}
	V	Hz	min^{-1}	W	A	dB(A)	dB(A)	dB(A)	m^3/h	Pa
1	400	50	1020	1654	2.62	68	75	76	26760	0
2	400	50	1020	1938	3.04	68	75	74	23365	80
3	400	50	1020	2195	3.42	69	76	75	19480	160
4	400	50	1020	2560	3.90	74	82	80	15105	230
5	400	50	900	1120	1.78	65	72	73	23500	0
6	400	50	900	1340	2.10	65	71	71	20660	63
7	400	50	900	1517	2.36	66	73	72	17220	125
8	400	50	900	1702	2.64	71	78	77	13350	180
9	400	50	750	648	1.03	60	68	68	19580	0
10	400	50	750	776	1.22	60	67	67	17215	44
11	400	50	750	878	1.37	61	68	67	14350	87
12	400	50	750	985	1.53	66	74	73	11125	125
13	400	50	750	332	0.53	55	62	63	15665	0
14	400	50	600	397	0.62	54	61	61	13775	28
15	400	50	600	449	0.70	56	63	62	11480	55
16	400	50	600	504	0.78	61	68	67	8900	80

U = Supply voltage

f = Frequency

n= Speed

P_{ad} = Power input

I = Current draw

LpA_{in} = Sound pressure level inlet side

LwA_{in} = Sound power level inlet side

LwA_{out} = Sound power level outlet side

qv = Air flow

P_{fs} = Pressure increase

losses, the minimum free passage area has been calculated. Finally, the most suitable silencer model has been selected, considering the speed of the air passing through the silencer's passages (Eq. 4):

$$v = \frac{Q\left(\frac{m^3}{h}\right)/3600}{A\left(m^2\right)} \tag{4}$$

where Q represents the air flow rate and A represents the product of the number of passages for the air passage section.

Unfortunately, the best solution to the problem, in terms of rectangular silencer (base = 2450 mm, height = 900 mm, length = 2000 mm, with seven septa of thickness of 200 mm in sound-absorbing mineral wool), is not compatible with the heat pump under examination, due to a total loss of load equal to about 18 Pa: in this way the axial fan has a residual prevalence too much low.

So, in this case it is not possible to realize an acoustic mitigation intervention for the considered air-cooled inverter heat pump using a standard solution (silencer). Thus, for reducing the sound pressure level and to ensure the correct operating phase of the system, in particular the adequate air flow to the radiator heat exchangers located along the lateral surface of the heat pump, the authors have designed an innovative structure of appropriate geometry, with a diffractive element at varying inclination on the top (Fig. 7), made of sandwich panels (Fig. 8) consisting of:

- low-density expanded polyurethane foam panels with a calibrated closed cell structure, resistant to water and humidity, self-supporting, with excellent fire behaviour;
- sound insulation material comprising a polymeric base and a layer of polyethylene caps with a sound insulation index Rw about 40 dB;
- anti-vibration support sheet made of aluminium alloy panels made of Peraluman (AlMg -EN AW 3003-), resistant to weather and mechanical characteristics such as to guarantee the absence of resonant phenomena with resistance to mechanical vibrations induced by the system under examination.

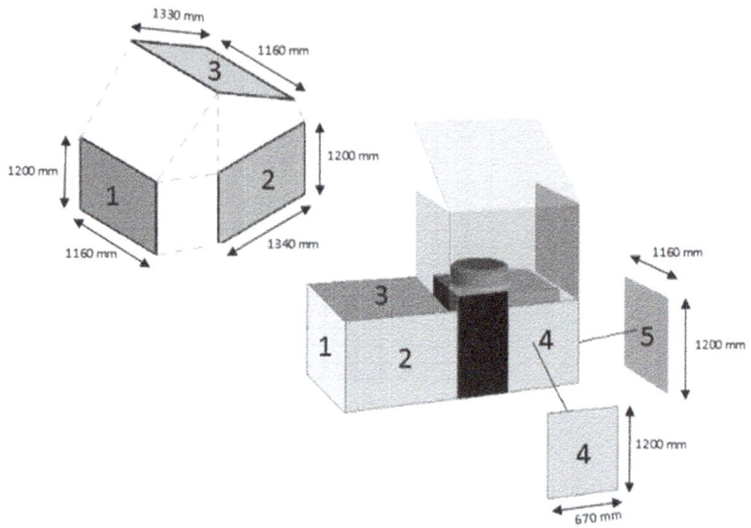

Fig. 7. View of designed structure for acoustic mitigation.

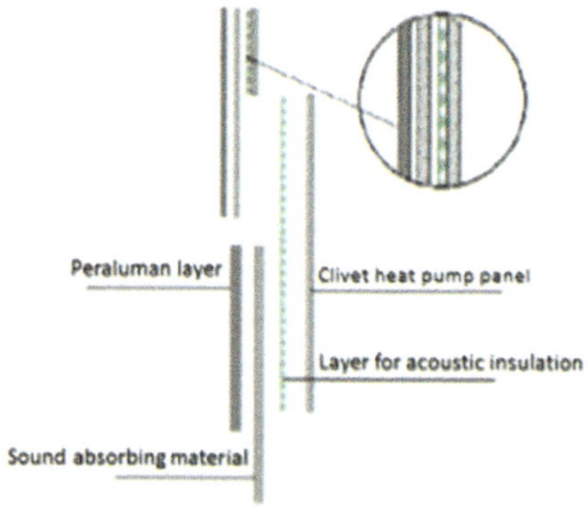

Fig. 8. Section scheme of sandwich panels.

The proposed structure, achieving high-performance sound mitigation, ensures the correct ventilation of the heat pump: in particular, regarding to axial fan, the right air buffer has been determined (Fig. 9), and, with reference to the heat exchanger (Fig. 10), a panel at distance of 100 mm has been installed on the front side (Fig. 11) and there is no closing panel on the rear side (Fig. 12).

3 Measurement Session *Post-operam* and Results

After making this acoustic correction, a new measurement campaign was conducted, on the basis ofthe standard ISO 1996–2 [14]. Also in this case the Sound Level Meter Larson Davis 831, Class 1 Environmental Noise & Building Acoustics Analyzer, was used at a distance of 1 m from the nearest building facade and a height of 4 m. Data was acquired in

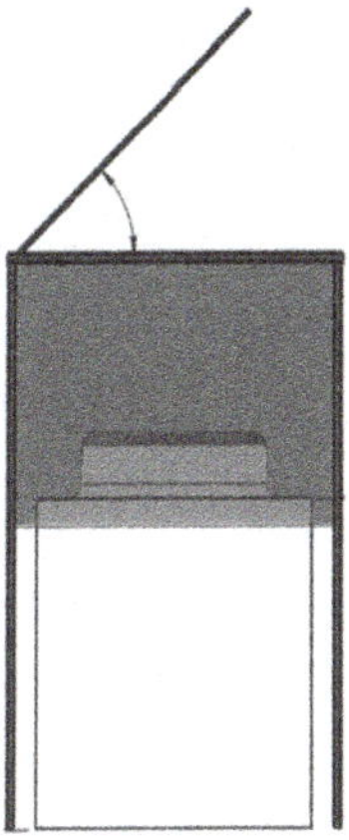

Fig. 9. View of the hat.

Fig. 10. Heat exchanger.

Fig. 11. Structure front side.

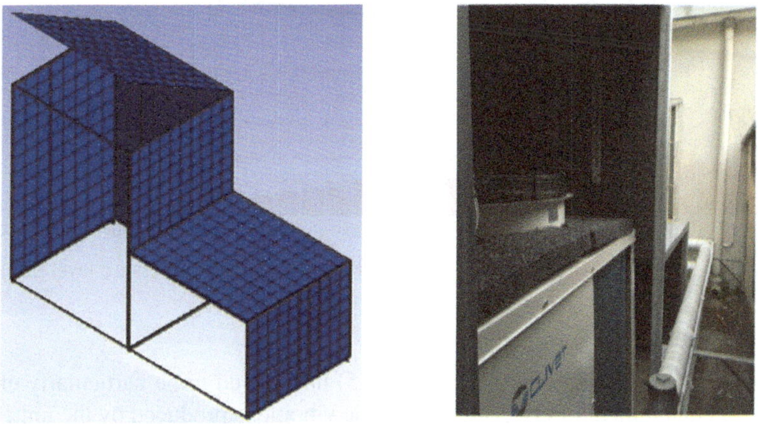

Fig. 12. Rear side of the structure.

one day (wind speed less than 5 m/s and no rain) accounting for startup and modulation state of the air conditioner fan motor. The measured equivalent sound pressure level (59 dBA) [14, 15] was smaller than corresponding threshold established by the law (60 dBA). Fig. 13 shows the sound pressure level measured at the nearest building facade during measurement time [17] after the acoustic treatment and the equivalent sound pressure level measured at the nearest building facade *post operam* at frequency varying is reported in Fig. 14.

Therefore, the mitigation intervention, whose complexity derives mainly from the need to guarantee the correct activity of the system from a thermal point of view, has allowed to limit the equivalent sound pressure level on the façade to the most sensitive receiver in the limits imposed by the current legislation.

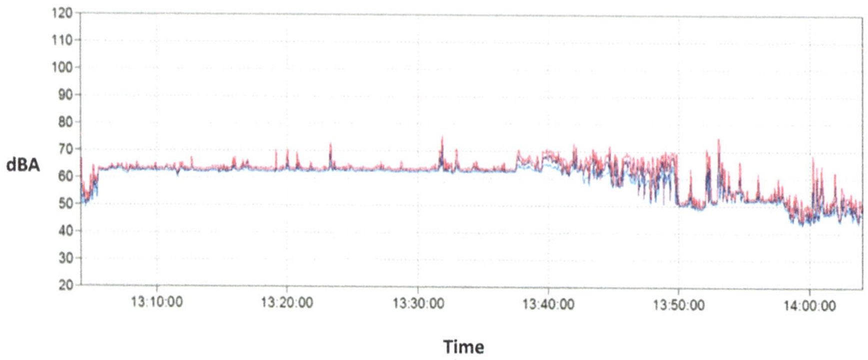

Fig. 13. Sound pressure level measured at the nearest building facade during measurement time.

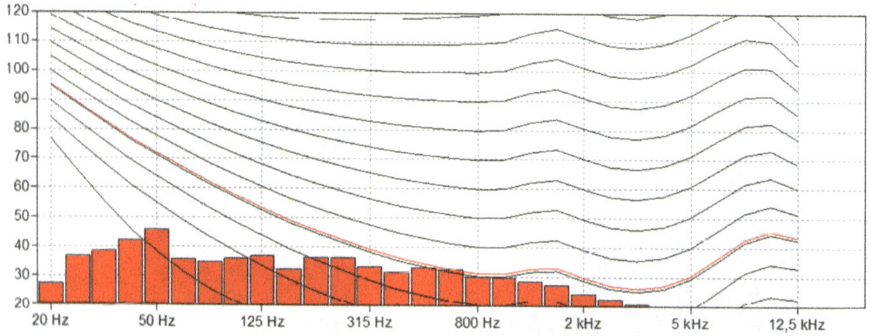

Fig. 14. Spectral representation of the A weighted equivalent sound pressure level measured at the nearest building facade *post operam.*

Moreover, the particular structure (Fig. 15) has proved to be particularly effective also in the containment of effects related to the vibrations produced by the unit.

Fig. 15. Picture of the acoustic barriers system.

The proposed innovative acoustic mitigation system reveals to be performing in the abatement of the equivalent sound pressure level (about 10 dBA), ensuring also the correct ventilation of the heat pump, an aesthetic pleasant result and a high resistance to atmospheric agents.

4 Conclusions

An original contribution has been made towards the design of acoustic mitigation intervention regarding to an air-cooled inverter heat pump. Focus has been placed on a passive noise control technique by proposing an innovative structure of appropriate geometry, based on modular panel system, with a diffractive element at varying inclination on the top for ensuring the efficiency of the system and an anti-vibration support sheet made of aluminium alloy panels made of Peraluman. It has been particularly difficult to reduce the equivalent sound pressure level on the façade to the most sensitive receiver, according to the current Italian legislation, and at the same time to guarantee the correct ventilation of the heat pump. A measurement session *ante-operam* has been conducted, in order to analyse the state of the art of the system, from an acoustic point of view, and another phonometric campaign has been carried out *post operam*, which showed an abatement of about 10 dBA relative to the equivalent sound pressure level. The tangible values based on the experiments conducted after the completion of noise reduction suggested ensure compliance with the legal limits and show great similarity with the simulated design phase, so the authors are convinced that the proposed structure is performing.

References

1. Mohanty, A.R., Pierre, B.D.S., Suruli-Narayanasami, P.: Structure-borne noise reduction in a truck cab interior using numerical techniques. Appl. Acoust. **59**(1), 1–17 (2000)
2. Ruggiero, A., Domenico, R.: Acoustical design and experimental verification of school music rooms: a case study. Appl. Acoust. **107**, 1–9 (2016)
3. Yairi, M., Sakagami, K., Morimoto, M., Minemura, A., Andow, K.: Effect of acoustical damping with a porous absorptive layer in the cavity to reduce the structure-borne sound radiation from a double-leaf structure. Appl. Acoust. **64**(4), 365–384 (2003)
4. Russo, D., Ruggiero, A.: Choice of the optimal acoustic design of a school classroom and experimental verification. Appl. Acoust. **146**, 280–287 (2019)
5. Craik, R.J.: Non-resonant sound transmission through double walls using statistical energy analysis. Appl. Acoust. **64**(3), 325–341 (2003)
6. Assouar, B., Oudich, M., Zhou, X.: Acoustic metamaterials for sound mitigation. C R Phys. **17**(5), 524–532 (2016)
7. Qian, Z., Chang, D., Liu, B., Liu, K.: Prediction of sound transmission loss for finite sandwich panels based on a test procedure on beam elements. J. Vib. Acoust. **135**(6), 061005 (2013)
8. Larbi, W., Deü, J.F., Ohayon, R.: Vibroacoustic analysis of double-wall sandwich panels with viscoelastic core. Comput. Struct. **174**, 92–103 (2016)
9. Ruggiero, A., Russo, D., Sommella, P.: Determining environmental noise measurement uncertainty in the context of the Italian legislative framework. Measurement **93**, 74–79 (2016). https://doi.org/10.1016/j.measurement.2016.07.007
10. Papini, G.S., Pinto, R.L., Medeiros, E.B., Coelho, F.B.: Hybrid approach to noise control of industrial exhaust systems. Appl. Acoust. **125**, 102–112 (2017)

11. Liguori, C., et al.: Proposal for the automatic evaluation of workers' exposure to acoustic noise following task-based approach. Measurement 108534 (2020). https://doi.org/10.1016/j.measurement.2020.108534
12. Beranek, L.L., Ver, I.L.: Noise and Vibration Control Engineering-Principles and Applications, 814 p. Wiley, Hoboken (1992)
13. Snyder, S.D.: Active noise control primer – modern acoustics and signal processing. Springer, New York (2000)
14. ISO 1996–2:2007 Acoustics -- Description, measurement and assessment of environmental noise -- Part 2: Determination of environmental noise levels
15. Ruggiero, A., Domenico, R., Sommella, P.: Determining environmental noise measurement uncertainty in the context of the Italian legislative framework. Measurement **93**, 74–79 (2016)
16. Farina, A.: Misurazioni e rilievi fonometrici. In: Proceedings of "L'acustica negli edifici e nella città", Firenze, 26 marzo 1999
17. Liguori, C., et al.: Innovative bootstrap approach for the estimation of minimum measurement time interval in road traffic noise evaluation. Measurement **98**, 237–242 (2017)

Effects of Tool Rotation Speed on Friction Stirred Material Flow Behavior of AA6061-T6 Aluminum Alloys

Anil Kumar Maurya, Anuj Jain, and Akshoy Ranjan Paul(✉)

Department of Applied Mechanics, Motilal Nehru National Institute of Technology Allahabad, Prayagraj 211004, India
arpaul2k@gmail.com

Abstract. Material flow behavior of AA6061-T6 aluminum alloy during friction stir welding (FSW) is computationally investigated in the present study. Eulerian approach is considered in the computational simulation assuming that the tool rotates at a fixed location and material flows at a constant speed. A flat cylindrical tool profile is used to investigate the material flow at different tool rotational speed during friction stir welding (FSW) of the alloy. Temperature contours for different rotational speed reveal that temperature distribution near the tool improved as tool rotational speed increases and temperature decreases as distance from center of tool increases. Velocity contours depict that the plasticized material moves around the tool and maximum velocity is observed when the tool rotates at 1200 rpm. Mixing of plastically deformed material around the tool governs the welding quality. This study suggest that tool rotation influence the material flow during FSW. Higher tool rotation gives improved temperature distribution and material mixing which leads to improved weld quality.

Keywords: Friction stir welding (FSW) · Numerical simulation · Material flow

1 Introduction

Friction stir welding (FSW) is a type of joining process having a rotating tool in which no filler material is used [1]. After the evolution of process at The Welding Institute (TWI) in 1991, FSW mostly studied on aluminum alloys because aluminum is very difficult to weld using conventional fusion welding techniques owing to formation of defects such as aluminum oxide layer, blow holes, porosity etc. Since its discovery, FSW has been used extensively in automobile, aircraft, marine and aerospace industries [2]. Material welded by this technique possess strength approximately equal to the strength of original material and great dimensional stability [3].

It is very difficult to comment on phenomenon occurring during FSW by experimental methods. Computational tool can be utilized to understand the flow of material during FSW process [4]. In this regard, some attempts have been made so far to predict the material flow behavior through numerical simulation. J. Zhang et al. [5] investigated 3D temperature and metal flow during FSW of AA6061-T6 alloy using numerical

S. Hloch et al. (Eds.): ICMEM 2021, LNME, pp. 355–360, 2021.
https://doi.org/10.1007/978-3-030-71956-2_29

simulation. G. Chen et al. [6] have performed a comparative study between boundary velocity model and boundary shear stress model in order to predict temperature and material deformation. G. Chen et al. [7] extended 3D analysis based on computational model to investigate the retractable pin tool to eliminate exit hole in FSW process. A. F. Hasan [8] studied the CFD modeling of FSW and anticipated pressure distribution at various locations of tool surface. Kadian et al. [9] examined the effect of tool design on the metal flow with the help of CFD code. R. Padmanaban et al. [10] discussed metal flow and temperature circulation during FSW of Al alloy using CFD based model.

FSW has a complex phenomenon which depends upon various parameters. Analysis of process parameters by experimental methods is very difficult due to involvement of material deformation at high temperature, heat generation and metal flow. So, it is required to develop CFD models to examine the outcome of process parameters on FSW process. In the current study an attempt has been made to study the material flow behavior at different tool rotational speed with the help of Ansys-Fluent package using the Eulerian approach.

2 Numerical Methodology

2.1 Geometrical Model and Mesh Generation

The geometry of tool comprises of a flat cylindrical tool shoulder and a cylindrical tool pin. The diameter of the shoulder is 25 mm and that of pin is 6 mm with a depth of 2.5 mm. The dimensions of the work piece used in the study are 100 mm length, 50 mm width, and 3 mm thickness. A cavity is made in the work piece of the size and shape of the pin as shown in the Fig. 1. Since material moves freely throughout the domain without any mesh distortion in the Eulerian approach, the work piece is meshed with static mesh for numerical analysis based on finite volume method. Sphere of influence is used to generate finer mesh around the cavity up to the periphery of the tool.

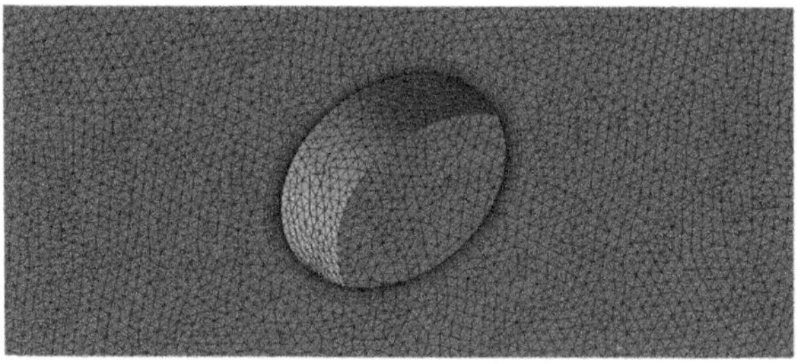

Fig. 1. Enlarged view showing cavity in the plate.

2.2 Material Parameters

The parameters like specific heat (C_p), coefficient of viscosity (μ), thermal conductivity (K) and density are the main influential material parameters. The density of AA6061 is

2700 kg/m^3, which is considered to be constant. Thermal conductivity and specific heat is temperature dependent whose values has been taken from literature [9].

As during FSW, material is exposed to plastic deformation and the temperature near the tool region is high while it decreases as increase in distance away from the tool. Therefore, viscosity must be different in different region of weldment. In this study viscosity of the plasticized material is assumed to be followed by the expression given in following equation.

$$\mu = \frac{\sigma_R}{3\dot{\varepsilon}} \tag{1}$$

where, μ is the material viscosity, σ_R is effective stress and $\dot{\varepsilon}$ is strain rate σ_R can be expressed by Zener-Hollomon equation [10] as given in Eq. 2.

$$\sigma_R = \frac{1}{\alpha}\ln\left[(Z/A)^{\frac{1}{n}} + \left(1 + (Z/A)^{\frac{2}{n}}\right)^{\frac{1}{2}}\right] \tag{2}$$

$$Z = \dot{\varepsilon}\exp(Q/RT) \tag{3}$$

where, Q = activation energy, R = gas constant, T = temperature, Z = Zener – Hollomon parameter, A, n and α are material constants.

Kadian et al. [9] expressed the values of parameters for AA6061 as: $Q = 191$ kJ/mol, $n = 5.33$, $A = 1.63 \times 10^{13}$ s^{-1}, $\alpha = 0.016$ and $R = 8.314$ J/mol-K.

2.3 Governing Equations

The process follows mass conservation, momentum conservation and energy conservation equations.

The mass conservation equation is given by:

$$\frac{\partial\rho}{\partial t} + \nabla.(\rho\boldsymbol{u}) = 0 \Rightarrow \nabla.\boldsymbol{u} = 0 \tag{4}$$

The momentum conservation equation is given by considering Newtonian fluid:

$$\frac{\partial(\rho\boldsymbol{u})}{\partial t} + \nabla.(\rho\boldsymbol{u}\otimes\boldsymbol{u}) = -\nabla p + \nabla.\left(\mu\left(\nabla\boldsymbol{u} + (\nabla\boldsymbol{u})^T\right) - \frac{2}{3}\mu\boldsymbol{I}(\nabla.\boldsymbol{u})\right) + \rho f \tag{5}$$

Considering incompressible behaviour of weldmet

$$\Rightarrow \rho\frac{\partial\boldsymbol{u}}{\partial t} + \rho(\boldsymbol{u}.\nabla)\boldsymbol{u} = -\nabla p + \nabla.\left(\mu\left(\nabla\boldsymbol{u} + (\nabla\boldsymbol{u})^T\right)\right) + \rho f$$

The energy conservation equation by neglecting viscous dissipation Φ (Conversion of mechanical energy to heat):

$$\frac{\partial(\rho c_p T)}{\partial t} + \nabla.(\rho c_p T\boldsymbol{u}) = -\nabla.(p\boldsymbol{u}) + \nabla.(K\nabla T) \tag{6}$$

Here, ρ = density, c_p = specific heat, K = thermal conductivity, μ = coefficient of viscosity, T = temperature of material.

2.4 Boundary Conditions and Solver Setup

The Eulerian approach is used in the CFD simulation. The material is supposed to be highly viscous non-Newtonian fluid and it is being stirred by the rotating tool located at a fixed position. Material is allowed to enter at inlet with velocity at which tool is supposed to move. A user defined function (UDF) is incorporated for the temperature dependent viscosity. In FSW, the value of temperature near the tool is considered to be approximately 80% of the melting point of material to be welded. 800 K is used as welding temperature and given to the pin wall. Different tool rotational speed (800 rpm, 1000 rpm and 1200 rpm) is used to investigate the material flow in FSW. A translational velocity of 2.2 mm/s is given to the material at the inlet. Double precision, pressure based solver, pressure velocity coupling as SIMPLE, and second order upwind scheme for discretization of governing equations is used. For energy, convergence criterion is selected as 10^{-6}, and for others it is selected as 10^{-5}.

3 Results and Discussion

3.1 Validation

For the validation purpose, the model is verified with the experimental result published by V. Soundararajan et al. [11]. The model taken for the validation has the same dimension as the experimental work carried out by them. In their experiment the temperature was measured by a thermocouple placed at 6 mm distance from the joint line thus in this study temperature is measured along a line at a distance of 6 mm from center of the tool. Tool rotation was 344 rpm and travel speed was 133 mm/min (Fig. 2).

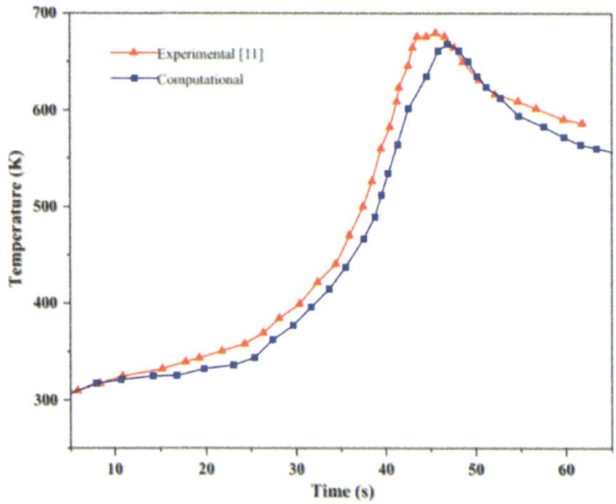

Fig. 2. Model validation with the experimental results.

The computational results have been compared and plotted against the experimentally obtained results and shows good agreement with experimental result.

3.2 Effect of Rotational Speed

Tool rotation speed is an important parameter in FSW. In this study constant work piece travel speed, 2.2 mm/s is taken and rpm of the tool is varied as 800 rpm, 1000 rpm and 1200 rpm. Temperature variation is observed at the upper surface of the work piece in each cases as shown in Fig. 3. It is depicted from the temperature contour that increasing rotational speed of the tool gives better heat generation and increase material flow area, thus improves the weld quality. The maximum temperature is observed in the highest deformation area (near the tool) and the temperature decreases as distance increases from weld line.

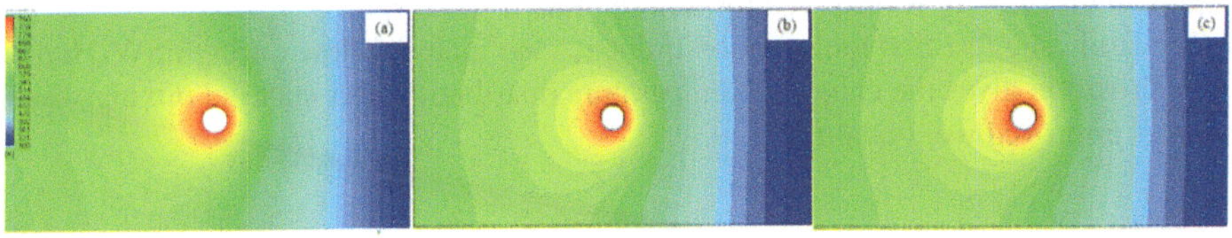

Fig. 3. Temperature contours showing variation at (a) 800 rpm, (b) 1000 rpm and (c) 1200 rpm.

In order to analyze the material flow at given tool rotation speed in the welding process, velocity contours for each case is observed as shown in Fig. 4. The observation reveals that the high tool rotation speed gives the greater deformation.

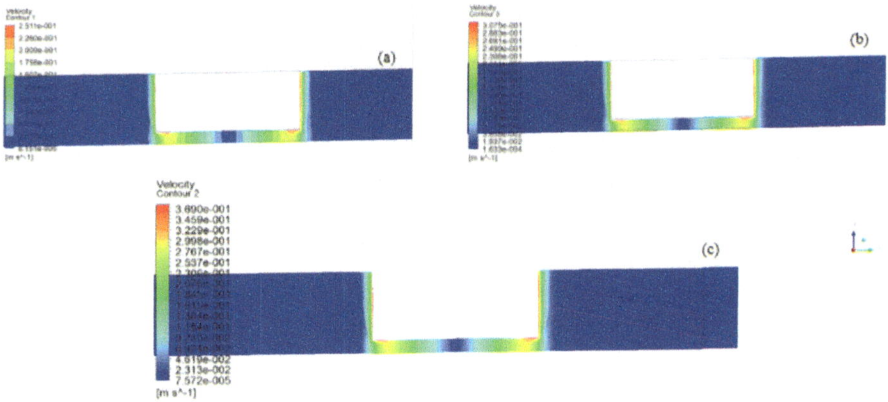

Fig. 4. Velocity contours showing velocity variation near the tool at (a) 800 rpm, (b) 1000 rpm and (c) 1200 rpm.

4 Conclusion

The major outcome of the current study is concluded as follows:

- A CFD based computational model is developed to investigate FSW of AA6061-T6 aluminum alloy for various tool rotation speed.
- The simulated temperature versus time plot shows decent settlement with the published experimental result for the validation of computational model developed.
- The temperature contours reveals that temperature distribution near the tool is improved with the increase in rotational speed.
- The velocity contours depict that the plasticized material moves around the tool and maximum velocity is observed for tool rotating at 1200 rpm.
- The study suggest that tool rotation influence the material flow during FSW. Higher tool rotation gives the enhanced temperature distribution and material mixing which gives improved weld quality.

References

1. Shankar, S., Vilaca, P., Dash, P., Chattopadhyaya, S., Hloch, S.: Joint strength evaluation of friction stir welded Al-Cu dissimilar alloys. Measurment **146**, 892–902 (2019)
2. Shariq, M., Srivastava, M., Tripathi, R., Chattopadhyaya, S., Vilaca, P., Gubeljak, N., Krolczyk, G.: Optimization and characterization of friction surfaced coatings of ferrous alloys. Mater. Test. **60**(7–8), 707–718 (2018)
3. Shankar, S., Saw, K., Chattopadhyaya, S., Hloch, S.: Investigation on different type of defects, temperature variation and mechanical properties of friction stir welded lap joint of aluminum alloy 6101-T6. Mater. Today Proc. **5**(11), 24378–24386 (2018)
4. Neto, D.M., Neto, P.: Numerical modeling of friction stir welding process: a literature review. Int. J. Adv. Manuf. Technol. **65**(1–4), 115–126 (2013)
5. Zhang, J., Shen, Y., Li, B., Xu, H., Yao, X., Kuang, B., Gao, J.: Numerical simulation and experimental investigation on friction stir welding of 6061-T6 aluminum alloy. Mater. Des. **60**, 94–101 (2014)
6. Chen, G., Ma, Q., Zhang, S., Wu, J., Zhang, G., Shi, Q.: Computational fluid dynamics simulation of friction stir welding: a comparative study on different frictional boundary conditions. J. Mater. Sci. Technol. **34**(1), 128–134 (2018)
7. Chen, G., Wang, G., Shi, Q., Zhao, Y., Hao, Y., Zhang, S.: Three-dimensional thermal-mechanical analysis of retractable pin tool friction stir welding process. J. Manuf. Process. **41**, 1–9 (2019)
8. Hasan, A.F.: CFD modelling of friction stir welding (FSW) process of AZ31 magnesium alloy using volume of fluid method. J. Mater. Res. Technol. **8**(2), 1819–1827 (2019)
9. Kadian, A.K., Puri, G., Das, S., Biswas, P.: Effect of tool geometry and process parameters on the material flow of friction stir welding. In: 5th International and 26th All India Manufacturing Technology, Design and Research Conference (AIMTDR 2014), pp. 12–14 (2014)
10. Padmanaban, R., Kishore, V.R., Balusamy, V.: Numerical simulation of temperature distribution and material flow during friction stir welding of dissimilar aluminum alloys. Procedia Eng. **97**, 854–863 (2014)
11. Soundararajan, V., Zekovic, S., Kovacevic, R.: Thermo-mechanical model with adaptive boundary conditions for friction stir welding of Al 6061. Int. J. Mach. Tools Manuf. **45**(14), 1577–1587 (2005)

Enhancement of Surface Properties of High Speed Steel Using Powder Mixed Micro-electrical Discharge Process

Shalini Mohanty(✉), Partha Bhattacharyya, Alok Kumar Das, and Amit Rai Dixit

Department of Mechanical Engineering, Indian Institute of Technology (IIT-ISM), Dhanbad 826004, Jharkhand, India
shalinimohanty5@yahoo.com

Abstract. In the present study, an approach is made for surface modification of high speed steel (HSS) using tungsten disulphide powder mixed in dielectric and tungsten tool electrode through micro-electrical discharge machining (μ-EDM). A hard and self-lubricating surface layer is formed on the work piece by considering various parameters such as duty factor, voltage, and coating time (t_m). The XRD and EDS plots confirms the presence of carbides, oxides and sulphides of iron and tungsten that owes to the improvement in surface properties. The micro-hardness of the modified samples varied in the range 613.93 HV to 1408.52 HV due to the presence of tungsten carbide and iron (III) oxide in the coated surface. The elemental diffusion of tool and powders mixed in dielectric in to the work piece surface is confirmed from EDS reports.

Keywords: Micro-EDM · Micro-hardness · XRD · Diffusion

1 Introduction

Micro-electrical discharge machining is gaining popularity among researchers as non-conventional machining process prior to its use in manufacturing of micro-components for tool, mould, and die industries. Miniaturized and complex shapes can be generated on difficult-to-cut materials by use of μ-EDM thus making it popular for roughing and finishing operations [1].

Electrical discharge machining process has not only been used as material removal technique but also as a surface modification by alloying/coating of engineering materials [2–5]. Surface alloying/coating of engineering materials has gained attention of researchers for modification of surfaces by adding up certain properties to the recast layer. Conventional methods such as vapour deposition (PVD and CVD), laser cladding, plasma deposition, ion implantation, etc. are used for increasing the service life and reducing the manufacturing costs in tool and die making industries, yet the use of high end equipment and the surface pre-treatment makes these process complicated [6]. On the other hand, electrical discharge alloying/coating is carried out by EDM wherein the

S. Hloch et al. (Eds.): ICMEM 2021, LNME, pp. 361–370, 2021.
https://doi.org/10.1007/978-3-030-71956-2_30

surface is modified in the interim of material removal process [3]. This technique follows similar spark erosion process by melting of both electrodes and the flow of molten material from the tool electrode takes place. This molten material carried away from the tool electrode either reacts with the work material to form certain inter-metallic compounds or gets deposited directly filling the vacancies on the work material that is accountable for the modification of the work piece surface properties. Hence, the recast layer developed on the work piece show enhanced properties. To this aspect, many methods have been adopted by researchers for surface modification or surface alloying/coating including use of conventional electrode [8], powder metallurgy electrode [1, 5], semi-sintered/sintered powder metallurgy electrode [3], powder mixed dielectric [4, 6], or by altering the dielectric fluid [7]. Some relevant work in context to the present study has been discussed.

According to [3, 9–11], the use of suspended particles lowers the breakdown strength of the dielectric liquid thus creating a high chance of discharge within the fluid medium. The material removal rate (MRR) is increased, tool wear rate (TWR) is decreased whereas the sparking efficiency is enhanced. In addition to that, conductive powders play an essential role in decreasing the surface roughness by enlarging the spark gap [9]. A wider spark gap lowers the capacitive effect of the electrodes, hence favoring frequent sparks [12]. In a study conducted by Mohanty et al. [2], tungsten disulphide powders suspended in deionized water was used to generate a hard and lubricious layer on a super alloy (Ti6Al4V). They found Viker's hardness ranging between 421.4 HV to 880.3 HV which was twice that of the substrate (417.6 HV). In a comparative study, Kumar et al. [13] examined thc effect of using different electrodes (Cu, Cu-W, Cu-Cr) and tungsten powders mixed in dielectric liquid on die steel work surface. Results indicated a substantial improvement in micro-hardness by around 76% whilst using Cu-W electrode and 111% while using powders in dielectric fluid. SiC powder suspension in dielectric during micro-EDM of Ti6Al4V by Kuriachen et al. [10] showed higher MRR and lower TWR is obtained at 5g/L powder concentration, gap voltage of 115 V and 0.1 μF capacitance. Vedprakash et al. [14] in their work proved that the recast layer thickness is more with kerosene as dielectric (16.2 μm to 31.6 μm) as compared to EDM oil (16.42 μm to 27.47 μm) while micro-EDM alloying of Ni plate with Ti-alloy tool electrode. Batish et al. [15] conducted experimental studies on various type of die steels (H 11, EN 31, HCHCr) using graphite and aluminium powder suspended in dielectric fluid. Two electrodes were used namely copper and tungsten-copper and results indicated the transfer of more material from tool with graphite powder suspension as compared to those with aluminium in order to increase the micro-hardness of the work specimen. In a similar work, Bhattacharya et al. [16] used Si, graphite and W powder in dielectric fluid and investigated the micro-hardness and surface roughness. Results showed that W-Cu electrode along with tungsten powder suspension gave lower surface roughness values and higher micro-hardness values.

Much of research has been accomplished using suspended powders in the dielectric medium for surface modification of engineering materials through EDM. However, less has been explored for surface modification through micro-EDM process for smaller parts. To be specific, limited study has been reported so far using solid lubricating powders in the dielectric fluid. Thus in the present study an effort is made for surface modification of HSS work material by generating a hard layer over it with tungsten disulphide powder suspended by using a tungsten micro-tool electrode.

2 Experimental Methodology

An in-house built micro-EDM set-up (Fig. 1) was used to conduct the entire set of experiments using HSS work piece, tungsten micro-tool electrode (diameter 1mm) and WS_2 powders mixed in DI water as the dielectric liquid. Initial set of pilot experiments defined the range of parameters for the actual set of experiments for the coating. Twenty seven experiments were carried out with varying voltage (V), duty cycles (DF) and coating time (t_m) as per settings given in Table 1. Reverse polarity (tool +ve and work piece −ve) was set out for conducting all experiments. The micro-EDM set-up is a CNC equipped system attached with pulse power supply (solid model shown in Fig. 1). Among three axes, longitudinal and transverse movement are given to work piece i.e., in x and y direction and last linear motion is imparted to motor-equipped rotating tool in z-direction with the help of lab-view software installed in the computer. The work piece is totally submerged in dielectric fluid and reverse polarity is preferred to enhance surface properties in PMEDM [14]. Micro-EDM works on the spark generation phenomenon amidst electrodes submerged in the dielectric fluid where material melting, vaporization, and solidification yields the formation of recast layer. Accretion of powders in dielectric fluid enhances the recast layer surface by adding up some properties to the base material [17].

Table 1. Parameters for the coating process.

Parameter	Units	Range
Voltage	V	30, 45, 60
Duty factor	%	45, 55, 65
Coating duration	minutes	5, 10, 15

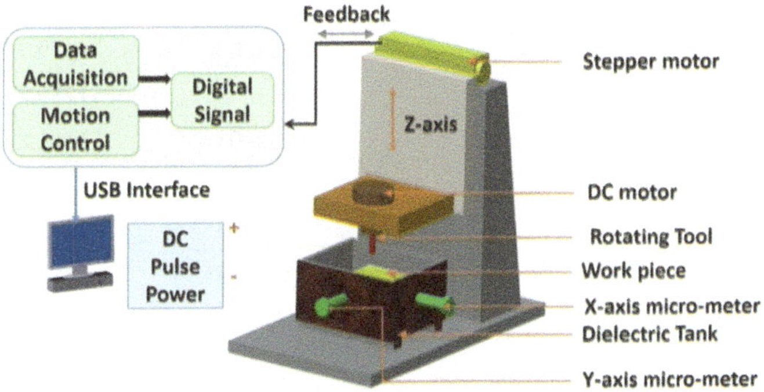

Fig. 1. Solid model of in-house developed micro-EDM set-up used for coating.

3 Results and Discussion

The modified surface were cleaned properly by dipping them in acetone filled ultrasonic bath. The micro-hardness (HV) and average surface roughness parameter (SR) are calculated and are recorded as shown in Table 2.

Table 2. Experimental observations for surface modification of HSS.

Voltage	Duty factor	Coating duration	Surface roughness	Micro-hardness
30	45	5	1.06 ± 0.02	613.93 ± 18.42
		10	1.08 ± 0.02	690.76 ± 20.72
		15	1.19 ± 0.02	770.79 ± 23.13
	55	5	1.12 ± 0.02	691.25 ± 20.74
		10	1.15 ± 0.02	730.47 ± 21.91
		15	1.29 ± 0.03	810.13 ± 24.30
	65	5	1.22 ± 0.02	742.27 ± 22.27
		10	1.29 ± 0.03	760.35 ± 22.81
		15	1.34 ± 0.03	860.41 ± 25.81
45	45	5	1.11 ± 0.02	809.26 ± 24.28
		10	1.17 ± 0.02	900.32 ± 27.01
		15	1.24 ± 0.02	951.51 ± 28.54
	55	5	1.18 ± 0.02	830.17 ± 24.91
		10	1.28 ± 0.03	910.33 ± 27.31
		15	1.42 ± 0.03	980.81 ± 29.42
	65	5	1.30 ± 0.03	900.22 ± 27.01
		10	1.37 ± 0.03	1003.83 ± 30.11
		15	1.51 ± 0.03	1157.74 ± 34.73
60	45	5	1.21 ± 0.02	990.15 ± 29.71
		10	1.26 ± 0.03	1179.73 ± 35.39
		15	1.41 ± 0.03	1259.80 ± 37.79
	55	5	1.28 ± 0.03	1102.84 ± 33.08
		10	1.41 ± 0.03	1274.85 ± 38.24
		15	1.51 ± 0.03	1321.44 ± 39.64
	65	5	1.41 ± 0.03	1187.37 ± 35.62
		10	1.57 ± 0.03	1354.37 ± 40.63
		15	1.79 ± 0.04	1408.52 ± 42.26

3.1 Micro-hardness

The micro-hardness parameter was measured using Economet VH-1 MD Viker's hardness tester using a pyramidal shaped diamond tip indenter with included angle of 136 degrees. The measuring range of the equipment is between 8 HV to 2900 HV and the readings were taken at 100 gf load with a dwell time of ten seconds. The observations presented in Table 2 show that the average of micro-hardness value is ranging between 613.93 HV to 1408.52 HV which is about thrice the base material (610.2 HV). The improvement in hardness could have been due to the accumulation of carbides and oxides of tungsten and iron. This has been confirmed by X-ray diffraction (XRD) results in later part of the study. Figure 2 indicate the transverse section of recast layer (RL) developed on the base material as per different parameter settings. The FESEM images indicate the molten material deposition in form of coated layer distinguished from the base material.

Figure 3 represent the influence of both voltage and coating time on hardness of the coated RL at various duty factors. It is indicative that the micro-hardness increases with the rise in voltage and duty factor. It is because of the high amount of spark energy developed in the discharge gap that attains to liberation of heat energy therein allowing excessive material melting from the tool, work piece, and some from the powders added in the dielectric. This forms a molten pool in the dielectric fluid which upon quenching action gets accumulated on the HSS surface to form the coated recast layer. The recast layer constitute different hard phase of intermetallic compounds that aids to the improvement in the hardness. Moreover, the continuous exposure of the surface of the work piece to the molten pool results in progressive deposition of phases of carbide on the recast surface. Therefore, as length of coating duration increases, it results in enlargement of recast layer width and hence improvement in micro-hardness. It may also be noted that the use of tungsten tool significantly affects the hardness of the recast layer. The range of hardness obtained by [2] is comparatively less than the present study owing to the use of brass tool electrode that is responsible for decreasing the hardness.

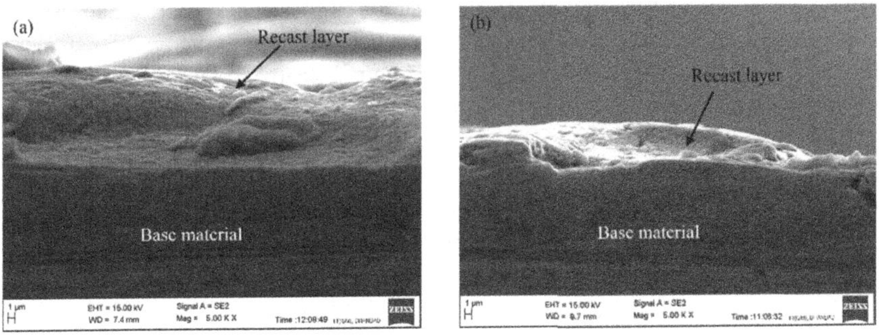

Fig. 2. Transverse sectioned FESEM images showing recast layer at (a) 30 V, 55% DF, 10 min, and (b) 45V, 55% DF, 15 min coating time.

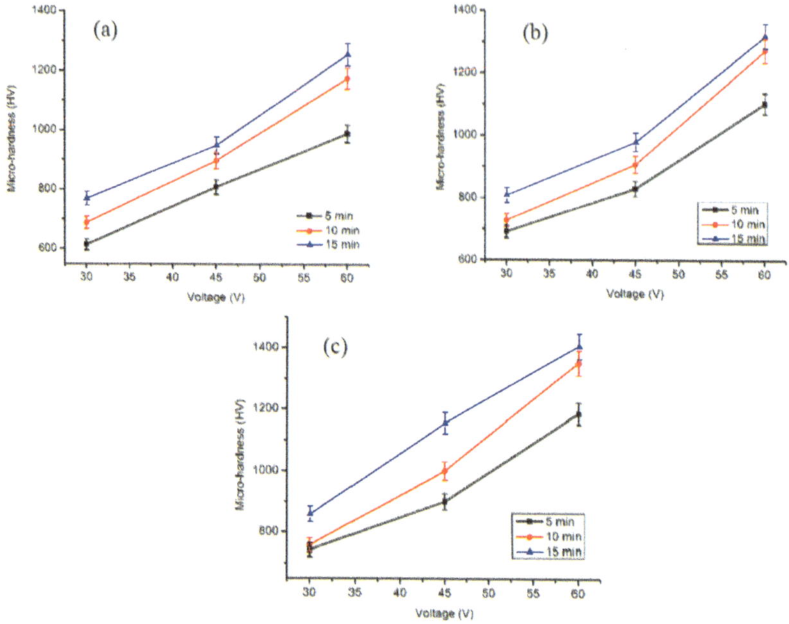

Fig. 3. Variation of micro-hardness at different voltage and coating time at duty factor (a) 45%, (b) 55%, (c) 65% respectively.

3.2 Surface Roughness

MITUTOYO surface roughness tester with 2 μm tip radius, cut-off length 0.25 mm with a measuring force of 4mN was used to evaluate the average surface roughness parameter with three readings, the mean value of which is presented in Table 2. Figure 4 illustrates the variation of average surface roughness with voltage and coating time at different duty factors such as 45% (Fig. 4 a), 55% (Fig. 4 b), and 65% (Fig. 4 c). It is seen that the average SR increases with voltage since a higher spark energy causes higher amount of material melting from the tool, work piece along with the powder suspended in the dielectric [2]. This contributes to the creation of deep and large sized craters thus giving higher average surface roughness values. Similarly, the increase in duty factor from 45% to 65% results in the increase in discharge duration, thus formation of deep craters takes place and SR is higher. Again with the increase in coating time, there is more transmission of discharge energy to the work surface thereby increasing the crater depth and hence the average surface roughness.

3.3 Diffusion by EDS Analysis

The diffusion of tool and powder particles from the dielectric liquid into the recast layer was investigated by Energy dispersive spectroscopy (EDS) reports as shown in Fig. 5. Small area (15 μm × 15 μm) portion is taken up for EDS study initiating from the HSS base surface, and then moving towards RL, covering the intermittent region. Each of the plot (Fig. 5 b, c, d, e) indicate the elemental composition for each region at parametric

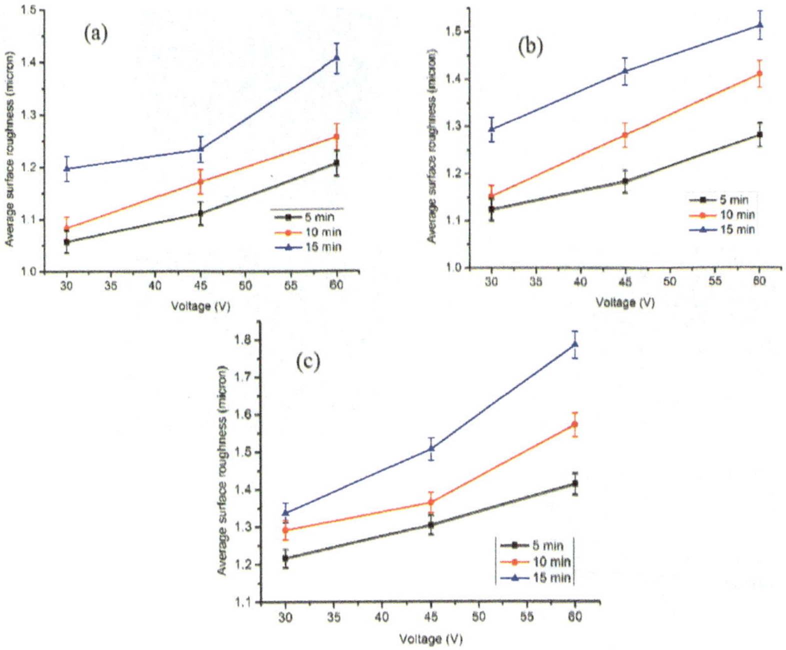

Fig. 4. Influence of voltage and coating time on surface roughness at various duty factor (a) 45%, (b) 55%, and (c) 65% respectively.

settings of 60V, 0.55 duty factor and 10 min machining duration. Figure 5(a) shows the FESEM image distinguishing the recast layer from the work piece indicating different zone of spectra. Figure 5(f) indicate the variation of elemental weight percentage of dominating elements from region 1 (HSS material) to region 4 (recast layer) where it is confirmed that the carbon and tungsten content increases from starting region towards the transformation region due to the accumulation of WC and cementite on the coated surface. Iron and oxygen content decrease from the transition region as they dissociate to form iron oxide that helps in improvement of micro-hardness of the coated specimen. Though there is transfer of material from both tool and dielectric fluid (suspended with powders) to the recast layer surface, yet the elemental weight percentage variation is insignificant. This denotes there is coating taking place in the recast region using the micro-electrical discharge process [6].

3.4 XRD Analysis

XRD analysis was done on the prepared coated surface with settings 60V, 45% duty factor and 10 min of coating time (Fig. 6). It indicate the existence of cementite (Fe_3C), tungsten carbide (WC), and iron oxide (FeO_3) that is accountable for improving the micro-hardness of the coated surface. The presence of other compounds suggest the hauling of tool and the dielectric material (powders mixed) on to the work piece surface. The traces of WS_2 seen on the plot indicate that the coated surface of HSS may inhibit some lubricity properties.

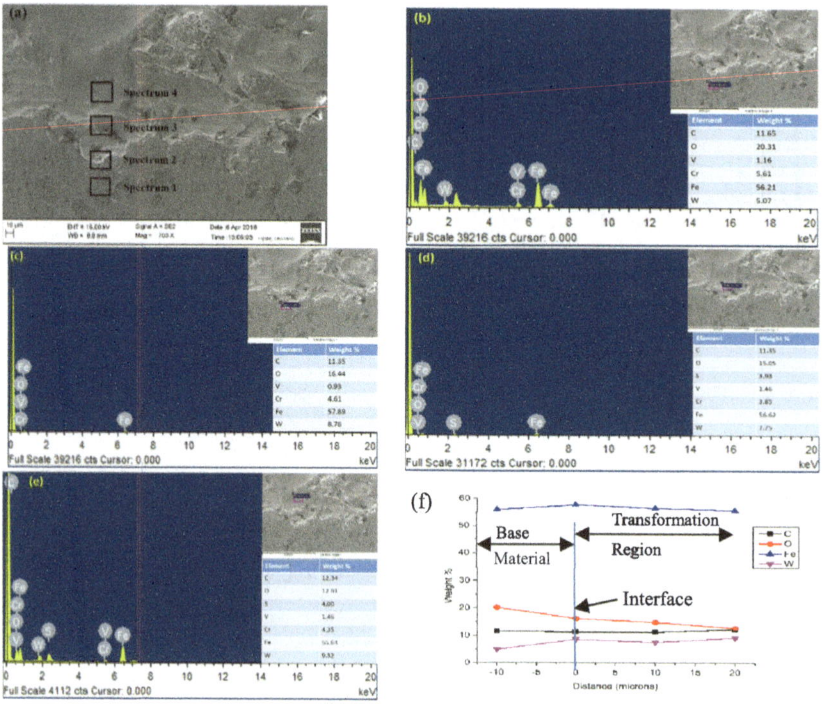

Fig. 5. Region wise elemental diffusion to reacst layer through EDS at (a) different region, (b) spectrum 1, (c) spectrum 2, (d) spectrum 3, (e) spectrum 4, and (f) variation of weight % along distance from transformation region.

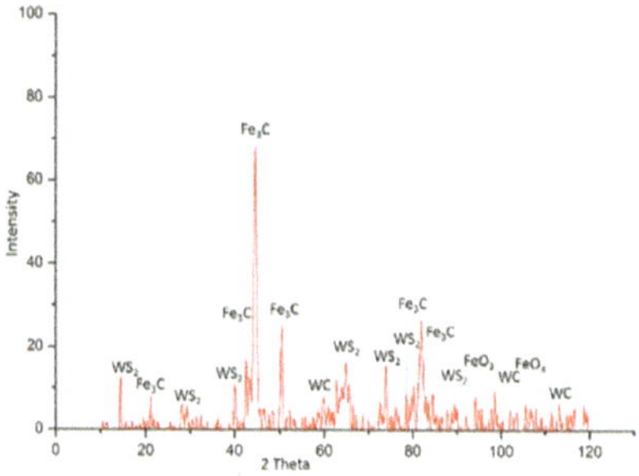

Fig. 6. XRD plot for coated specimen at 60 V, 45% duty factor and 10 min of machining duration.

4 Conclusions

The coating of HSS tool material was successfully done by the use of tungsten disulphide powders mixed in the dielectric with a tungsten tool electrode. The surface roughness and micro-hardness graphs have been plotted that indicate an improvement of surface properties post coating process. The micro-hardness values showed an increase range of 613.93 HV to 1408.52 HV which is thrice the base material (610.2 HV). The XRD results indicate the deposition of hard layer as recast surface owing to the existence of cementite, tungsten carbide and iron oxide. EDS reports confirmed the elemental diffusion from both the dielectric and tool in to the recast layer. Thus, the miniaturise component with enhanced surface properties are obtained by this process. This may find significant application in areas where sub-millimetre or micron sized surface area is to be modified such as in micro-tools, dies, moulds, indent tip, etc. Extensive study on the wear, corrosion, deposition rate and microstructure needs in-depth analysis.

References

1. Gill, A.S., Kumar, S.: Investigation of micro-hardness in electrical discharge alloying of En31 tool steel with Cu–W powder metallurgy electrode. Arab. J. Sci. Eng. **43**(3), 1499–1510 (2018)
2. Mohanty, S., Kumar, V., Das, A.K., Dixit, A.R.: Surface modification of Ti-alloy by micro-electrical discharge process using tungsten disulphide powder suspension. J. Manuf. Process. **37**, 28–41 (2019)
3. Simão, J., Aspinwall, D., El-Menshawy, F., Meadows, K.: Surface alloying using PM composite electrode materials when electrical discharge texturing hardened AISI D2. J. Mater. Process. Technol. **127**(2), 211–216 (2002)
4. Rahul, Datta, S., Masanta, M., Biswal, B.B., Mahapatra, S.S.: Analysis on surface characteristics of electro-discharge machined Inconel 718. Int. J. Mater. Product Technol. **56**(1–2), 135–168 (2018)
5. Siddique, A.R., Mohanty, S., Das, A.K.: Micro-electrical discharge coating of Titanium alloy using WS2 and Brass P/M electrode. Mater. Manuf. Process. **34**(15), 1761–1774 (2019)
6. Mohanty, S., Bhushan, B., Das, A.K., Dixit, A.R.: Production of hard and lubricating surfaces on miniature components through micro-EDM process. Int. J. Adv. Manufa. Technol. **105**(5–6), 1983–2000 (2019)
7. Bai, C.Y., Koo, C.H.: Effects of kerosene or distilled water as dielectric on electrical discharge alloying of superalloy Haynes 230 with Al–Mo composite electrode. Surf. Coat. Technol. **200**(12–13), 4127–4135 (2006)
8. Kumar, R., Singh, I.: Parametric optimisation for micro electric discharge drilling process. Int. J. Addit. Subtractive Mater. Manuf. **1**(3–4), 257–264 (2017)
9. Kumar, A., Maheshwari, S., Sharma, C., Beri, N.: Research developments in additives mixed electrical discharge machining (AEDM): a state of art review. Mater. Manuf. Proces. **25**(10), 1166–1180 (2010)
10. Kuriachen, B., Mathew, J.: Effect of powder mixed dielectric on material removal and surface modification in microelectric discharge machining of Ti-6Al-4V. Mater. Manuf. Process. **31**(4), 439–446 (2016)
11. Furutani, K., Sato, H., Suzuki, M.: Influence of electrical conditions on performance of electrical discharge machining with powder suspended in working oil for titanium carbide deposition process. Int. J. Adv. Manuf. Technol. **40**(11–12), 1093–1101 (2009)
12. Kozak, J., Rozenek, M., Dabrowski, L.: Study of electrical discharge machining using powder-suspended working media. Proc. Inst. Mech. Eng. Part B: J. Eng. Manuf. **217**(11), 1597–1602 (2003)
13. Kumar, S., Singh, R., Singh, T.P., Sethi, B.L.: Comparison of material transfer in electrical discharge machining of AISI H13 die steel. Proc. Inst. Mech. Eng. Part C: J. Mech. Eng. Sci. **223**(7), 1733–1740 (2009)

14. Prakash, V., Shubham, Kumar, P., Singh, P.K., Das, A.K., Chattopadhyaya, S., Mandal, A., Dixit, A.R.: Surface alloying of miniature components by micro-electrical discharge process. Mater. Manuf. Process. **33**(10), 1051–1061 (2018)
15. Batish, A., Bhattacharya, A., Singla, V.K., Singh, G.: Study of material transfer mechanism in die steels using powder mixed electric discharge machining. Mater. Manuf. Process. **27**(4), 449–456 (2012)
16. Bhattacharya, A., Batish, A., Kumar, N.: Surface characterization and material migration during surface modification of die steels with silicon, graphite and tungsten powder in EDM process. J. Mech. Sci. Technol. **27**(1), 133–140 (2013)
17. Kumar, A., Mandal, A., Dixit, A.R., Das, A.K.: Performance evaluation of Al2O3 nano powder mixed dielectric for electric discharge machining of Inconel 825. Mater. Manuf. Process. **33**(9), 986–995 (2018)

Friction Comparison on Tribological Pairs HXLPE/Ti6Al4V Alloy and VE-HXLPE/Ti6Al4V Alloy Under Dry and Physiological Lubricated Conditions

Alessandro Ruggiero[1(✉)], Roberto D'Amato[2], Nicolae Ungureanu[3], and Saverio Affatato[4]

[1] Department of Industrial Engineering, University of Salerno, Via Giovanni Paolo II 132, 84084 Fisciano, SA, Italy
ruggiero@unisa.it
[2] Departamento de Ingeniería Mecánica, Química y Diseño Industrial, ETSDI, Universidad Politécnica de Madrid (UPM), Madrid, Spain
[3] Faculty of Engineering, Department of Engineering and Technology Management, Technical University of Cluj-Napoca, Baia Mare, Romania
[4] Laboratorio di Tecnologia Medica, IRCCS Istituto Ortopedico Rizzoli, Bologna, Italy

Abstract. Friction and wear behavior of new generation Ultra-High Molecular Weight Polyethylene (UHMWPE), Highly Cross-Linked UHMPE (HXLPE) and Vitaminized E XLPE (VE-XLPE), sliding against TiAl6V4 alloy under dry and lubricated conditions were investigated. Tribological tests were conducted by using a reciprocating pin-on-flat tribometer under hip replacement implants conditions. Lubricated sliding tests were carried out with a biological fluid that approximates well the synovial fluid present in the joints with imposed boundary mechanism. The worn samples of the HXLPE and VE-XLPE were analyzed with a gravimetric weighing scale with a sensitivity of 0,1 mg. The evolution of the friction coefficient of the HXLPE/Ti6Al4V and VEXLPE/Ti6Al4V pairs were also compared.

Keywords: Friction · Wear · Reciprocating tribometer · UHMWPE · Ti6Al4V · Vitamin E · Cross-linking

1 Introduction

Tribology can be defined as the discipline that studies friction, wear and lubrication applied to surfaces in relative motion [1, 2]. The understanding and application of these tribological principles applied in the medical field has greatly contributed to the development of prostheses, improving their duration and the quality of life of patients [3, 4]. In the field of orthopedics, prosthesis wear is a significant clinical problem considering that approximately 3 million joint replacements are implanted worldwide every year [5, 6] and the hip prosthesis represent is the most used joint endoprosthesis [7]. In

S. Hloch et al. (Eds.): ICMEM 2021, LNME, pp. 371–385, 2021.
https://doi.org/10.1007/978-3-030-71956-2_31

addition, the increase in the average life of individuals and therefore the increase in the number of elderly people makes that arthroplasty carriers are increasingly younger and with more active lifestyles [8]. This implies the tribological performance optimization of the manufactured implants with greater research of more wear-resistant materials to meet the growing expectations of patients [9]. Until recently, the most commonly adopted coupling was Metal-on-Polyethylene (M-o-P). In these prosthesis the femoral head is usually a Cobalt-Chrome-Molybdenum alloy combined with Ultra-high Molecular Weight Polyethylene (UHMWPE) acetabular cup [10, 11]. This coupling for four decades provided consistent results remaining the standard coupling for wear rate comparisons to other couplings [10, 11]. Despite the great success of these couplings and the introduction of ceramic, the formation of wear debris of Polyethylene, which is associated with osteolysis and subsequent failure of the prosthesis has historically been a limiting factor of longevity of these implants [11, 12]. For this reason, over the years, new generations of UHMWPE (Table 1.) have been introduced that provide greater resistance to wear, such as Highly-Crosslinked UHMWPE (HXLPE), and greater resistance to oxidation, via the addition of vitamin E (VE-HXLPE) [11].

Table 1. The use of UHMWPE for joint protheses [12].

Method/Technology	Name/Example	Reported year of introduction
UHMWPE	RCH-1000, Chirulen	1962
Gamma irradiation sterilization (air)		1968
Carbon fiber reinforcement	Poly-II	1970
First generation highly cross-linked		1972–1978
Higher purity, better consolidation, manufacturing in clean room	Medical grade	1985
Quality without Ca stearate	GUR 402/405	1985
Gamma sterilization in inert gas	Sulene	1986
High-pressure remelted	Hylamer, Hylamer M	1987
Surface heat polishing	PCA	1989
Gamma inert gas sterilization/annealing	Duration	1996
Second generation highly cross-linked, annealed	Crossfire	1998
Second generation highly cross-linked, remelted	Durasul, Longevity, Marathon	1999–2001
Third generation sequentially highly cross-linked, annealed	X3	2005
Third generation Vitamin E doped highly cross-linked, annealed	E1	2007

The cross-linking of a polymer is defined as the union of two independent polymeric molecules with a covalent chemical bond requiring molecule irradiation and, is proportional to the dose of absorbed radiation, on which the crystallinity depends [10]. This in turn influences the mechanical properties, sometimes deteriorating them [10].

Prolonged cross-linking of the Polyethylene (PE), in sheets or layers, is possible with an accumulation of irradiation doses through exposure to gamma radiation or electron beam [13]. Wear resistance increases with the increase in the radiation dosage up to a limit of 100 - 150 kGy, beyond which there is saturation and no further benefits are seen in the reducing wear while the mechanical properties, especially toughness, continue to degrade [12, 13]. UHMWPE components have been shown to become Highly Cross-Linked after irradiation exposure and absorption of $\geq$50 kGy [5].

The new generation of UHMWPE are all irradiated by gamma radiations rather than electron beam because this has more impact on wear resistance. Two methods can be used to eliminate radicals. The first, consists of the irradiation/annealing sequence repeated three consecutive times. The second, incorporates different quantities of vitamin E (alpha-tocopherol) in search of radicals. Neither offer a full solution. Annealing depends on the temperature chosen to preserve the mechanical properties and, therefore, cannot completely eliminate free radicals. Oxidation is then still possible, although lower in value than previously so can be considered a good solution for patients over 60 years old [5].

Only further clinical experience can demonstrate whether HXLPE stabilized with vitamin E will actually maintain long-term expectations. This antioxidant can be mixed with PE through different methods and has been applied in the orthopedic field since it was introduced in 2007. Vitamin E with the polymer powders are sometimes mixed before the consolidation process, then cross-linked and annealed in later. Alternatively, the antioxidant is homogenized after cross-linking. Both sterilizations, with or without irradiation, consume the antioxidant already in cross-linking or in final sterilization [5].

In the past 15 years, HXL-UHMWPE biomaterials have shown great reductions in wear and improved survival of hip prostheses, and also HXLPE with the addition of vitamin E shows promising laboratory results [5].

Recent *in vivo* studies on oxidation have shown that UHMWPE oxidation can be caused not only by radiation-induced free radicals, but also by free radicals generated *in vivo*. These can be caused by cyclic loads and by lipid reactions absorbed by synovial fluid. Therefore, researchers suggest that sufficient Vitamin E activity may be useful and necessary to prevent oxidation. Usually, the typical dosages of Vitamin E are 0.1% in weight [5]. Instead, orthopedic literature does not report on the effects of the concentration of Vitamin E on stability and mechanical behavior in great depth. A main disadvantage of UHMWPE stabilized with Vitamin E is considered to be the greater resistance of the material to the cross-linking process, which depends on the irradiation method as well as on the concentration of antioxidant [5].

One of the most important property of the vitamin E is to hunt the free radicals in order to stabilize UHMWPE at extremely low concentrations. However, this property has a disadvantage at low concentrations, namely the reduction in the efficiency of crosslinking during irradiation, as noted by several authors. This behavior, at low concentrations, of the vitamin E irradiated in the UHMWPE directly affects the wear resistance. Indeed, the development of materials added with vitamin E foresees a potential trade-off between wear resistance and oxidation stability for concentrations up to 3000 ppm [5]. [5]. An alternative method is to stabilize with Vitamin E after irradiation, so that the efficiency of cross-linking is not affected because the additive is not present during irradiation. The last generations of VE-HXLPE show promising improvements in terms of longevity of the HXLPE components, improving fatigue and oxidation resistance, and at the same time preserving wear resistance of irradiated materials that have already shown good results in vivo up to 10 years after implantation. In addition, Vitamin E added to UHMWPE would appear to have no cytotoxic effect. It can act as an effective anti-inflammatory and a lower penetration of the cup into the femoral head has also been hypothesized compared to HXLPE [5, 14, 15]. Since these new generations of UHMWPE have been around for about a decade, data confirming the results in vivo and supporting the use of vitamin E HXLPE is obviously still limited [5, 15]. For these reasons in this work the friction and the wear behavior of new generation Ultra-High Molecular Weight Polyethylene (UHMWPE), Highly Cross-Linked UHMPE and Vitaminized E XLPE, sliding against TiAl6V4 alloy under dry and lubricated conditions, were investigated. The tribological tests were carried out by using a reciprocating pin-on-flat tribometer TR-BIO 282. The tribological tests under lubricated conditions were carried out by using a biological fluid that approximates well the synovial fluid present in the joints with imposed boundary mechanism.

2 Materials and Methods

The materials used for tribological tests, provided by the Center of Research of the Istituto Ortopedico Rizzoli, Bologna (Italy) were Highly Cross-Linked UHMPE (XLPE), Vitaminized E XLPE (VEPE) (Fig. 1).

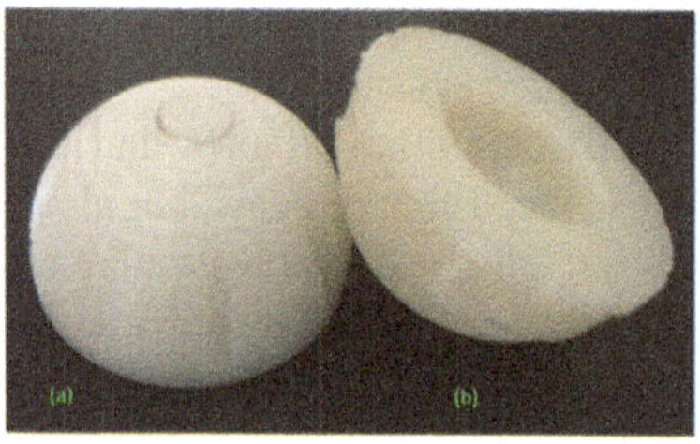

Fig. 1. Images of the acetabular cups a) conventional Highly Cross-linked UHMWPE; b) stabilized with Vitamin E HXL-UHMWPE.

For the tribological tests under lubricated conditions, three to four drops of the same fluid were used, i.e. patient 12.140's blood serum (Fig. 2) containing wear debris deriving from a Metal-on-Polyethylene prosthetic coupling, stored in the freezer at −15 °C.

The composition of the synovial fluid (SF) is similar to a dialysate of blood plasma [16]. During the experimental analysis, it was regarded that the blood serum was therefore similar to the synovial fluid. A fluid viscosity equal to that of the SF for post-arthroplasty patients was considered ($\eta = 0.001$ Pa $\cdot$ s) [17].

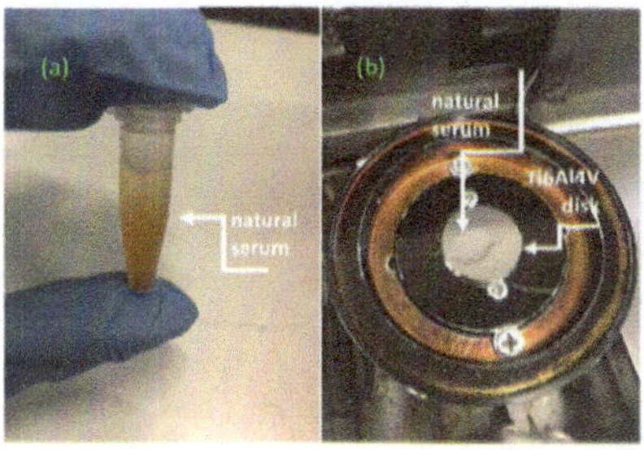

Fig. 2. (a) Bottle containing the serum; (b) illustrative image of the application of the lubricant on the titanium plane.

2.1 Samples Preparation

Table 2 shows the physical properties, for different typical radiation doses, of the XLPE and VEPE specimens used for tribological tests [5, 12].

Table 2. XLPE and VEPE physical properties in relation to radiation doses [18].

Property	Gur 1050	25 kGy	50 kGy	100 kGy	200 kGy
Crystallinity (%)	50.7 ± 0.5	45.4 ± 0.7	46.2 ± 0.7	46.9 ± 0.8	47.7 ± 0.4
Yield stress (MPa)	20.2 ± 1.0	19.0 ± 0.4	19.9 ± 0.8	18.9 ± 0.7	21.2 ± 1.0
Modulus (MPa)	495 ± 56	433 ± 14	412 ± 50	386 ± 23	266 ± 30
True stress at break (MPa)	315.5 ± 31.6	248.8 ± 18	237.6 ± 12.3	185.7 ± 7.5	126.0 ± 14
True strain at break	1.82 ± 0.01	1.74 ± 0.03	1.59 ± 0.01	1.50 ± 0.02	1.37 ± 0.06
Fracture toughness (J_{IC}, kJ/m^2)	2.1	23.8	76.2	$= J_{SS}$	$= J_{SS}$
Steady-state fracture toughness (J_{SS}, kJ/m^2)	116.9 ± 0.1	101.2 ± 0.1	98.5 ± 0.2	87.6 ± 0.1	79.3 ± 1.9

The VEPE and XLPE acetabular cups were perforated with a hollow cutter, with a diameter of 22 cm, mounted on a vertical drill press (Fig. 3a and 3b). Subsequently, once the cylindrical samples with equal radius of curvature ($R = 11$cm) were obtained, it was possible to make two cuts on each sample (Fig. 3c), one parallel to the plane containing the base of the cylinders (α) and other plan perpendicular to it (β), in order to be able to mount the specimens in the tribometer pin (Fig. 4).

During the cutting process, low shock resistance and high toughness of the materials was experienced by the operator.

It was possible to obtained five specimens for both couples of materials (VEPE and XLPE) for the tribological test under dry conditions, and five specimens for both couples of materials (VEPE and XLPE) for the test under lubricated conditions.

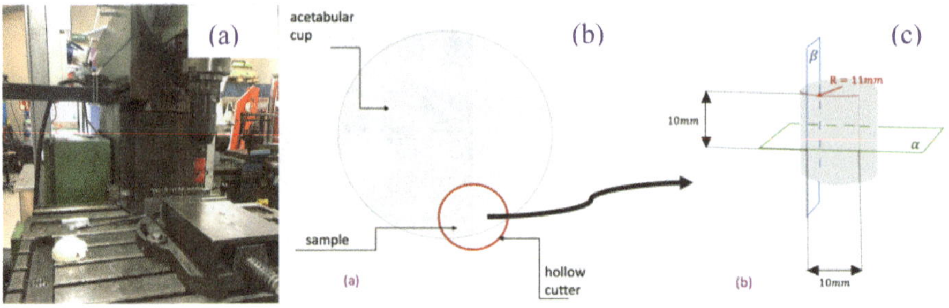

Fig. 3. (a) Machine tool; (b) cutting performed to the acetabular cups; (c) α and β the section planes of the two subsequent cuts.

Fig. 4. Image of a generic XLPE sample inserted in the pin.

The sample labelling nomenclature applied in laboratory and subsequently referred to in this text is listed in Table 3. The VE and XL acronyms stand for the Vitaminized E and traditional Crosslinked, A and B indicate the different two potential sides of the same sample with equal radius of curvature.

Table 3. Sample nomenclature.

	Dry conditions		Lubricated conditions	
Vitaminized E Cross-linked UHMWPE	VE2.A	Sample #1	VE4.B	Sample#11
	VE2.B	Sample #2	VE5.A	Sample #12
	VE3.A	Sample #3	VE5.B	Sample #13
	VE3.B	Sample #4	VE6.A	Sample #14
	VE4.A	Sample #5	VE1.B	Sample #15
Traditional Cross-linked UHMWPE	XL1.A	Sample #6	XL3.B	Sample #16
	XL1.B	Sample #7	XL6.A	Sample #17
	XL2.A	Sample #8	XL6.B	Sample #18
	XL2.B	Sample #9	XL7.A	Sample #19
	XL3.A	Sample #10	XL7.B	Sample #20

A single face of a Ti (grade 5) 6Al4V disk, with a 25 mm diameter and 6.4 mm width, was utilized as counterpart (flat) on which the rubbing of the samples was carried out. The titanium alloy remained stationary throughout the testing. Poisson ratio was considered equivalent to 0.3 and Young modulus was considered to 114 GPa [16].

2.2 Tribological Tests

The pin-on-flat experimental analysis was carried out with the Ducom Reciprocating Tribometer Tr-BIO 282 at the Applied Mechanics Engineering Laboratory of the University of Salerno [19, 20]. The chosen working parameters were calculated with the Hertzian theory and the lubrication mechanism present, in the cases where the biological fluid was interposed, was verified. Through the theory of Hertz, it was possible to evaluate what the average contact pressure of the VEPE-Ti6Al4V coupling was, and the same value was chosen for the XLPE-Ti6Al4V contact. With reference to hip protheses, vertical load N was chosen to be applied during the tribological tests (Fig. 5) in order to reach the contact pressures similar to those typical found in the literature, where the maximum pressure reached in the lubricated human contact varies between 6.7 and 10 MPa during walking [21].

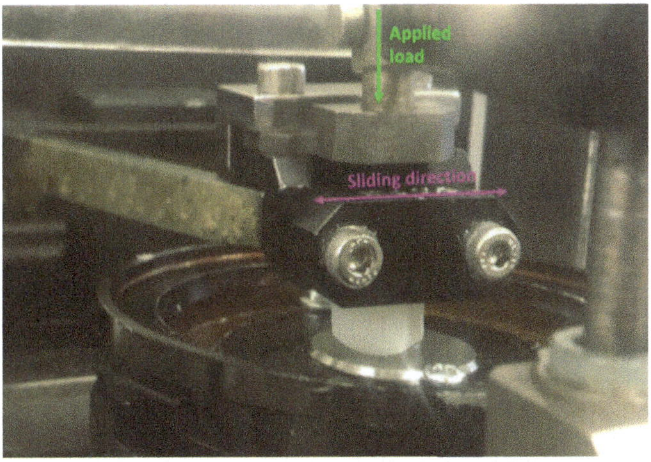

Fig. 5. Tribological contact mounted in the tribometer.

For the calculation of the vertical load N using the Hertz theory, the geometry of the specimens obtained was taken into account and the contact area was determined as cylinder-cylinder coupling, with axes having an inclination angle different from 0° or 90°, and therefore delimited by an ellipse [2]. The values of the contact area, maximum and average contact pressure are provided by the equations shown in Table 4 [2].

Starting from the values shown in Table 5, the contact parameters were calculated for both types of VEPE-Ti6Al4V and XLPE-Ti6Al4V coupling (Table 6), setting a load of 15N, compatible with the tribometer admissible loads and thus avoiding an elastohydrodynamic lubrication mechanism.

Table 4. Elliptic contact Area between two elastic bodies.

Dimension of contact area	Maximum contact pressure	Average contact pressure
$a = k_1\left(\frac{3NR'}{E'}\right)^{1/3}$ $b = k_2\left(\frac{3NR'}{E'}\right)^{1/3}$ $A = \pi ab$	$p_{max} = \frac{3N}{2\pi ab}$	$p_{average} = \frac{N}{\pi ab}$

Table 5. Values of the Poisson ratio, of Young modulus and of curvature radiuses.

	Poisson ratio υ	Young Modulus [GPa]	Rx [mm]	Ry [mm]
VEPE	0,4	0,4	11	∞
XLPE	0,4	0,4	11	∞
Ti6Al4V	0,3	114	∞	∞

Table 6. Hertzian contact parameters of VEPE-Ti6Al4V and XLPE-Ti6Al4V contacts.

	VEPE-Ti6Al4V	XLPE-Ti6Al4V
1/E' [1/Pa]	1,05E-09	1,05E-09
1/R' [1/m]	90,91	90,91
k_0	1	1
k_1	8	8
k_2	0,2	0,2
a [m]	0,0065	0,0065
b [m]	0,00016	0,00016
A [mm^2]	3,258	3,258
N [N]	15	15
Average contact pressure [MPa]	~ 4,60	~ 4,60
Maximum contact pressure [MPa]	~ 6,90	~ 6,90

It was possible to calculate the minimum film thickness in the elastohydrodynamic contacts for the viscous-elastic regime by using the equation provided by Hamrock and Dowson [2, 22] for the tribological contact VEPE-Ti6Al4V and XLPE-Ti6Al4V and as lubricant the patient 12.140's blood serum:

$$\frac{h_{min}}{R'} = 3.63\left(\frac{U\eta_0}{E'R'}\right)^{0.68}\left(\alpha E'\right)^{0.49}\left(\frac{N}{E'R'^2}\right)^{-0.73}\left(1 - e^{-0.68k}\right) \quad (1)$$

Where N is the vertical load, R' is the curvature radius of the VEPE and XLPE sample, U is the sliding speed imposed in the tribometer, α is pressure-viscosity coefficient, k is ellipticity parameter *(a/b)* and E' is define as:

$$E' = \frac{2}{\left(\frac{1-\nu_A^2}{E_A} + \frac{1-\nu_B^2}{E_B}\right)} \tag{2}$$

The tribological tests were carried out with a frequency of 5 Hz and a stroke length of 0.003m. A fluid viscosity of 0,001 Pa · s [18], was then considered. Furthermore, the sliding velocity U_2 of the Ti6Al4V plate was null, while the velocity U_1 of the XLPE and VEPE samples, characterized by an alternative motion, was considered as semi-harmonic function in which the speed is zero at the points of inversion of the movement and reaches a maximum value during the stroke. Therefore, U_1 can be defined as U_{1max}:

$$U_1 = x_{stroke}\omega sin(\omega t) \tag{3}$$

$$U_{1max} = x_{stroke}2\pi fr \tag{4}$$

and the average speed value corresponds to about 47 mm/s, between the typical velocities in the hip joint [17]. Thus, it was possible to calculate the dimensionless groups whose values are reported in Table 7. The minimum thickness is equal to 3,49E^{-09} m and by imposing a roughness of 0,8 μ m equivalent to the milling cutter impression [22, 23], then

$$S = \frac{h_{min}}{R_a} = 0,004954649 < 1,4 \tag{5}$$

Consequently, the elastohydrodynamic lubrication condition was not verified.

Table 7. Values of the dimensionless groups used to calculate the minimum thickness in the elastohydrodynamic contacts.

	VEPE-TI6AL4V	XLPE-TI6AL4V
$H = \frac{h_{min}}{R'}$	3,153E-07	3,153E-07
$U' = \left(\frac{U\eta_0}{E'R'}\right)$	4,52E-12	4,52E-12
$G = (\alpha E')$	5,693	5,693
$W = \left(\frac{N}{E'R'^2}\right)$	0,00013	0,00013
$k = \frac{a}{b}$	40	40

It can be assumed that a boundary lubrication mechanism was present. In fact, it appears that when synovial fluid, bovine serum or artificial fluids are used as lubricants, boundary or mixed lubrication usually take place [24]. Boundary lubrication dominates, indeed, where the loads are high, the sliding speed and the viscosity of the fluid are low and it is potentially favored by the synovial fluid components. For this reason, friction analysis for the respective couplings was then evaluated. The working parameter criteria chosen, summarized below in Table 8, was based on the attempt to reproduce actual hip conditions and on not verifying the condition of elastohydrodynamic lubrication in cases where there was the presence of blood serum. Therefore, the following conditions were set: a low frequency to simulate the low frequencies that occur during normal activities and to prevent generation of heat from friction [25–27], an average tangential speed of between 0–50 mm/s as typical velocities in the hip [17], a load such as to guarantee contact pressure, calculated with the Hertzian theory, similar to that observed in the literature during walking [21], a duration tests as define by the ASTM-G133 standard with reference to the type A procedure [27].

Table 8. Working parameters for the tribological tests.

	VEPE-Ti6Al4V	XLPE-Ti6Al4V	VEPE-Ti6Al4V	XLPE-Ti6Al4V
	Dry Conditions		*Lubricated Conditions*	
Number of samples	5	5	5	5
Load [N]	15	15	15	15
Frequency [s^{-1}]	5	5	5	5
Stroke [mm]	3	3	3	3
Time [min]	20	20	20	20
Sliding distance [m]	36	36	36	36
temperature [°C]	20° ± 1°	20° ± 1°	20° ± 1°	20° ± 1°
Humidity [%]	40 ± 5	40 ± 5	40 ± 5	40 ± 5

After the tribological tests, the wear of the VEPE and XLPE samples was determined using a gravimetric method, by measuring and comparing the weight of the pre-test and post-test samples. The weighing scale available at the Applied Mechanics Engineering Laboratory of the University of Salerno is the Gibertini Model E50S with a readability of 0.1 mg up to 90 g, and repeatability of ± 0.03 mg. After cleaning each sample with acetone, three measurements were carried out for each sample. Following the tribological tests, no difference in pre or post sample weighing was observed.

3 Results and Discussions

Figures 6 and 7 show the friction forces for the VEPE-Ti6Al4V and XLPE-Ti6Al4V tribo-pairs under dry and lubricated conditions. From the analysis of Figs. 6a and 7a, relative to the VEPE-Ti6Al4V tribocouple in dry and lubricated conditions respectively, it is possible to notice a greater variability in the evolution of the friction force (Sample #

5, Fig. 6a, and Sample 14–15 Fig. 7a). On the contrary, in the case of the XLPE-Ti6Al4V tribocouple (Fig. 6b and 7b) both in dry and lubricated conditions, this variability in the evolution of the friction force during tribological tests is significantly reduced. This difference in the behavior in the evolution of friction forces between the two tribocouples studied VEPE-Ti6Al4V and XLPE-Ti6Al4V can be explained by the presence of Vitamin E which, as explained above, is an effective free radical hunter and its ability to stabilizing UHMWPE at extremely low concentrations. For this reason, the development of materials added with vitamin E provides a potential compromise between wear resistance and oxidation stability for concentrations up to 3000 ppm [5].

In the lubricated cases, three to four drops were placed on the contact surface for each test and the presence of wear debris in the fluid was responsible for the irregular aspect of the curve, especially for the last five samples tested and represented in Fig. 9. It was also noted that the XLPE produced in a discontinuous manner scraping noises when scrubbing against the Titanium counterpart during the first 7 to 8 min of testing in conditions of dry friction.

This noise was not manifested in the lubricated case and in the other two cases with the VEPE.

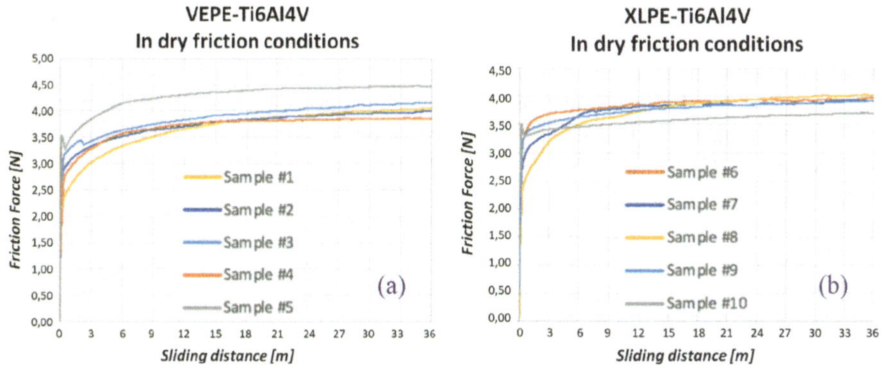

Fig. 6. (a) Friction force evolution for VEPE-Ti6Al4V tribocouple under dry conditions; (b) Friction force evolution for XLPE-Ti6Al4V tribocouple under dry conditions.

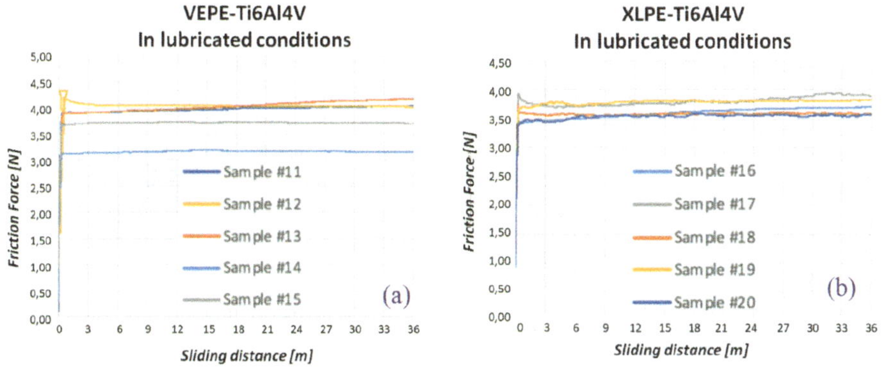

Fig. 7. (a) Friction force evolution for VEPE-Ti6Al4V tribocouple in lubricated conditions; (b) Friction force evolution for XLPE-Ti6Al4V tribocouple under lubricated conditions.

It can also be stated that, in the lubricated cases, for both VEPE-Ti6Al4V and XLPE-Ti6Al4V couplings, the Titanium disk appeared stained at the end of the tests (Fig. 8), probably due to surface-level interactions with the fluid, adsorption of proteins, external agents and also because titanium alloys are subject to corrosion by rubbing [7].

Fig. 8. Ti6Al4V counterpart after tests.

Figure 9a shows t the average friction coefficients evolution for the VEPE-Ti6Al4V and XLPE-Ti6Al4V tribocouples under dry and lubricated conditions. Figure 9b shows the Average friction coefficients with they variabilities for all tribocouples and for both lubricated conditions. The average friction coefficient (COF) was assessed at a steady state, in particular the transient was considered to have died out after 10 min from the start of the tests, and from the results reported in Fig. 9a and Table 9 it can be noted that the COF fluctuates from 0.27 to 0.25 depending on the reference tribosystem. Furthermore, from the analysis of the results of Fig. 9 and of Table 9, both XLPE and VEPE materials exhibit a good tribological behavior, in terms of friction coefficient, even in dry lubrication conditions. The percentage reduction of the non-approximate average value of the friction coefficient can be resumed as:

- 4,5% from VEPE-Ti6Al4V (dry) to XLPE-Ti6Al4V (dry).
- 3% from VEPE-Ti6Al4V (lubricated) to XLPE-Ti6Al4V (lubricated).
- 6% from VEPE-Ti6Al4V (dry) to VEPE-Ti6Al4V (lubricated).
- 4,5% from XLPE-Ti6Al4V (dry) to XLPE-Ti6Al4V (lubricated).

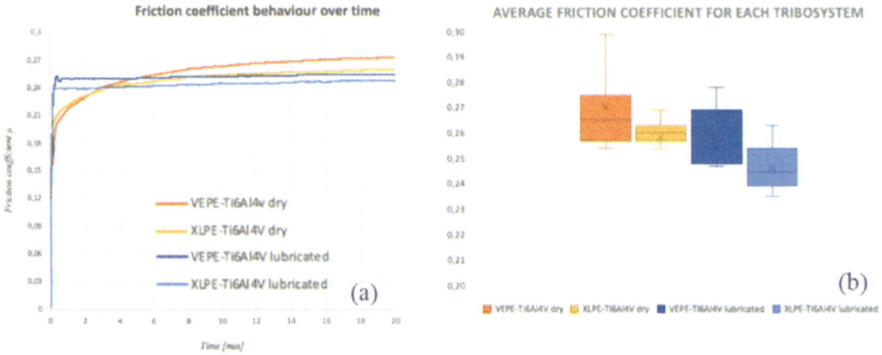

Fig. 9. (a) Evolution of the Average friction coefficient; (b) Average friction coefficients.

Table 9. Friction coefficients of the tribological contacts.

	VEPE-Ti6Al4V	XLPE-Ti6Al4V	VEPE-Ti6Al4V	XLPE-Ti6Al4V
	Dry Conditions		Lubricated conditions	
μ_{mean}	0,27	0,26	0,25	0,25
μ_{max}	0,30	0,27	0,27	0,26
μ_{min}	0,26	0,24	0,21	0,24
STD_r	0,015	0,008	0,023	0,008

The low percentage reduction observed from the dry case to the lubricated case, for each sample, produced results in line with the hypothesis of boundary lubrication, in which relatively high friction coefficients occur due to the close contact of the surfaces. This is also in accordance with the post-arthroplasty synovial fluid properties, in which proteins increase Ph, while viscosity decreases [28]. In general, since the variation between the average COF is lower than 0.1 and the friction force was on average between 3,50 ÷ 4,00 N in the four cases under study, the two types of biopolymers can be deemed equal in performance, as hypothesized pre-test and foreseen in the literature. Indeed, as already outlined, vitamin E plays the role of antioxidant, and on the other hand, it can partially reduce the efficiency of cross-linking if added before irradiation doses. Consequently, in the last-generation VEXL-UHMWPE, with the development and use of new manufacturing techniques, the wear resistance is not influenced by the additive [5]. In this study, a subtle difference in behavior between VEPE and XLPE was verified, both dry and lubricated conditions, confirming that the Cross-linked UHMWPE has, albeit slightly, better tribological properties than the UHMWPE stabilized with Vitamin E.

4 Conclusions

Friction and wear behavior of new generation (UHMWPE), UHMPE (HXLPE) and Vitaminized E XLPE (VE-XLPE), sliding against TiAl6V4 alloy under dry and lubricated conditions were investigated. A natural lubricant, together with the chosen work parameters applied on the Reciprocating Tribometer, allowed for conditions that replicate true hip replacement prothesis. Following the tests on TR-BIO 282 carried out on the five samples of each kind of couple in different lubricant conditions, a total of 20 tests these observations can be noted:

- HXLPE and VE-HXLPE both show a low friction coefficient sliding against Ti6Al4V, under dry conditions.
- The performance in terms of friction coefficient (COF) of HXLPE and VE-HXLPE are similar when paired with Ti6Al4V, especially HXLPE presents a slightly lower COF than VE-HXLPE both under dry and lubricated conditions.
- The boundary lubrication mechanism imposed with the natural synovial fluid has allowed tests to reach a decreasing COF of 4,5% for the tribocouple HXLPE-Ti6Al4V and of 6% for the tribocouple VEXLPE-Ti6Al4V.

- Even in corresponding tribological tests lasting a total of 10 h (I test 2hrs, II test 3hrs, III test 5hrs) with increasing frequencies selected (10 Hz, 15 Hz, 30 Hz), wear rates measurements were unable to be observed due to the technological limitations of the weight scale used.

Future scope: Given the recent introduction of these materials in the past decade and given the limited nature of clinical data describing the outcome of these new HXLPE and VEXLPE, only additional research can help provide further answers to issues in question.

Acknowledgments. The Authors acknowledge B.Eng. Sophie Zizzo for the support during the experimental activity developed in the framework of her Bachelor Thesis. This research was supported by MIUR PRIN 2017 BIONIC.

References

1. Bhushan, B.: Principles and Applications of Tribology, 2nd edn. Wiley, Chichester (2013). ISBN 9781119944546
2. Stachowiak, G.W., Batchelor, A.W.: Engineering Tribology, 4th edn. Butterworth-Heinemann (2013). ISBN 9780123977762
3. Jin, Z.M., Stone, M., Ingham, E., Fisher, J.: (v) Biotribology. In: Proceedings of the Current Orthopaedics. Churchill Livingstone, vol. 20, pp. 32–40 (2006)
4. Zhou, Z.R., Jin, Z.M.: Biotribology: recent progresses and future perspectives. Biosurf. Biotribol. **1**, 3–24 (2015). https://doi.org/10.1016/j.bsbt.2015.03.001
5. Kurtz, S.M.: Ultra-High Molecular Weight Polyethylene in Total Joint Replacement and Medical Devices (2016)
6. Mattei, L., Di Puccio, F., Piccigallo, B., Ciulli, E.: Lubrication and wear modelling of artificial hip joints: a review. Tribol. Int. **44**, 532–549 (2011)
7. Saverio, A., Ruggiero, A., Merola, M.: Advanced biomaterials in hip joint arthroplasty. A review on polymer and ceramics composites as alternative bearings. Compos. Part B Eng. **83,** 276–283 (2015)
8. Ruggiero, A., Merola, M., Affatato, S.: Finite element simulations of hard-on-soft hip joint prosthesis accounting for dynamic loads calculated from a musculoskeletal model during walking. Materials **11**(4), 574 (2018)
9. Di Puccio, F., Mattei, L.: Biotribology of artificial hip joints. World J. Orthop. **6**, 77–94 (2015)
10. Seyyed Hosseinzadeh, H.R., Eajazi, A., Sina, A.: The bearing surfaces in total hip arthroplasty – options, material characteristics and selection. In: Recent Advances in Arthroplasty. InTech (2012)
11. Hu, C.Y., Yoon, T.R.: Recent updates for biomaterials used in total hip arthroplasty. Biomater. Res. **22**, 1–2 (2018)
12. Tribology in Total Hip Arthroplasty. Springer, Heidelberg (2011)
13. Sobieraj, M.C., Rimnac, C.M.: Ultra high molecular weight polyethylene: mechanics, morphology, and clinical behavior. J. Mech. Behav. Biomed. Mater. **2**, 433–443 (2009)
14. Barbour, P.S.M., Barton, D.C., Fisher, J.: The influence of contact stress on the wear of UHMWPE for total replacement hip prostheses. Wear **181–183**, 250–257 (1995)

15. Wyatt, M.C., Roberto, A., Foxall-Smi, M., Beswick, A.D., Kenutsor, S.K., Whitehouse, M.R.: Does vitamin E highly-crosslinked polyethylene convey an advantage in primary total hip replacement? A systematic review and meta-analysis. HIP Int. **30**, 598–608 (2019)
16. Standring, S.: Anatomia del Gray. Le basi per la pratica clinica; Edra (2009)
17. Myant, C., Cann, P: On the matter of synovial fluid lubrication: implications for metal-on-metal hip tribology. J. Mech. Behav. Biomed. Mater. **34**, 338–348 (2014)
18. Pruitt, L.A.: Conventional and cross-linked polyethylene properties. In: Total Knee Arthroplasty: A Guide to Get Better Performance, pp. 353–360. Springer, Heidelberg (2005). ISBN 3540202420
19. Ruggiero, A., D'Amato, R., Gómez, E., Merola, M.: Experimental comparison on tribological pairs UHMWPE/TIAL6V4 alloy, UHMWPE/AISI316L austenitic stainless and UHMWPE/AL2O3ceramic, under dry and lubricated conditions. Tribol. Int. **96**, 349–360 (2016)
20. Ruggiero, A., D'Amato, R., Gómez, E.: Experimental analysis of tribological behavior of UHMWPE against AISI420C and against TiAl6V4 alloy under dry and lubricated conditions. Tribol. Int. **92**, 154–161 (2015)
21. Ruggiero, A., Sicilia, A.: Lubrication modeling and wear calculation in artificial hip joint during the gait. Tribol. Int. **142**, 105993 (2020)
22. Hamrock, B.J., Dowson, D.: Minimum film thickness in elliptical contacts for different regimes of fluid film lubrication (1979)
23. Houpert, L.G., Hamrock, B.: J: Fast approach for calculating film thicknesses and pressures in elastohydrodynamically lubricated contacts at high loads. J. Tribol. **108**, 411–419 (1986)
24. Mishina, H., Kojima, M.: Changes in human serum albumin on arthroplasty frictional surfaces. Wear **265**, 655–663 (2008)
25. Ruggiero, A., Sicilia, A.: A mixed elasto-hydrodynamic lubrication model for wear calculation in artificial hip joints. Lubricants **8**(7), 72 (2020)
26. Sinha, S.K., Briscoe, B.J.: Polymer Tribology. Imperial College Press, London (2009). ISBN 9781848162044
27. Pachi, A., Ji, T.: Frequency and velocity of people walking. Struct. Eng. **83**, 36–40 (2005)
28. Rufaqua, R., Vrbka, M., Choudhury, D., Hemzal, D., Křupka, I., Hartl, M.: A systematic review on correlation between biochemical and mechanical processes of lubricant film formation in joint replacement of the last 10 years. Lubr. Sci. **31**, 85–101 (2019)

Computational Study of Metal Flow Behavior of AA 6061 in Friction Stir Processing with Twin Tools

Vineet Kumar Tiwari[1], Akshoy Ranjan Paul[1], Anuj Jain[1](✉), Somnath Chattopadhyaya[2], Sachindra Shankar[2], and Pedro Vilaca[3]

[1] Motilal Nehru National Institute of Technology Allahabad, Prayagraj 211004, India
anujjain@mnnit.ac.in
[2] Indian Institute of Technology (ISM) Dhanbad, Dhanbad 826004, India
[3] Advanced Manufacturing and Materials Group, Department of Mechanical Engineering, School of Engineering, Aalto University, Espoo, Finland

Abstract. In order to tackle the growing demand of low weight to strength ratio materials in shipbuilding, trains, and aerospace applications, high strength aluminum alloys have shown good promise. Fabrication of parts with aluminum alloys has always been a challenge as conventional welding processes, which are basically fusion welding techniques, are not able to provide enough joint strength against impact. Friction stir welding (FSW) that generates heat from friction between a rotating tool and work-pieces to be welded, has shown promise to provide better strength. One of the important parameters that dictate the strength of joints produced by friction stir welding is the amount of mixing. Research has been carried out to improve mixing by modifying the tool geometry by several scholars. The present work is focused on improving mixing using twin tools instead of a conventional single tool. Numerical analysis is carried out using a finite volume method based computational fluid dynamics (CFD) solver. The result reveals that twin tools in linear alignment in the direction of welding with the rear/follower tool rotating in a direction counter to the front one can improve mixing as well as even distribution of material, leading to stronger joining.

Keywords: Friction stir welding (FSW) · Twin tool · Mixing · Computational fluid dynamics (CFD)

1 Introduction

Friction stir welding (FSW) technique invented by Wayne Thomas at The Welding Institute, U.K. in 1991 is different from conventional fusion welding techniques as it does not involves the melting of either the tool or the workpieces. The maximum temperature attained in this welding process limited to 70% of the melting temperature of the workpiece. At this temperature, the workpieces deform plastically, mixes and join under rotating tool pressure. The heat is generated due to friction between the rotating tool and workpieces. Thus, FSW restricts the formation of defects like hot cracking and

S. Hloch et al. (Eds.): ICMEM 2021, LNME, pp. 386–391, 2021.
https://doi.org/10.1007/978-3-030-71956-2_32

liquation cracking due to the re-solidification of the base material, which makes friction stir welding a potential joining technique of aluminum alloys for the fabrication.

The need of materials with high strength to weight ratio has gone up in the past few years. As the demand for lightweight transport vehicles having high strength has increased due to pollution and safety concerns. Aluminum alloys offer high strength to weight ratio but fabrication of parts using aluminum alloy is still a challenge because it is hard to weld using conventional techniques. The major problems with the fusion welding of aluminum alloys are insufficient fusion, the inclusion of slag, porosity, etc. FSW is found to have the capability to weld aluminum alloys and provide higher resistance against impact.

Because of its potential use, rigorous research is being carried out both experimentally and numerically, to improve the weld quality. It is observed that the mixing of materials being joined is an important parameter that affects the weld quality. Various combinations of tool geometries have been studied to improve stirring. Flow around the bobbin tool has been modeled by Hilgert et al. [1] to predict the shear layer shape around the tool. The numerical study conducted by Mimouni et al. [2] for a threaded tool profile with arbitrary Lagrangian-Eulerian formulation revels different flow patterns in advancing and retreating sides of the tool and also asymmetric temperature distribution has been reported on the work-piece. Numerical analysis to join dissimilar metals has also been carried out by scholars using the Volume of fluids (VOF) method [3, 4]. VOF model has also been used to predict surface formation after welding in a more realistic way considering surrounding air as one phase and metal phase to be the other [5].

In order to improve mixing, experiment using twin tool geometry has been conducted by Kumari et al. [6]. The study revealed that twin tools rotating in counter directions resulted in the better and less defective weld as compared to that produced using double pass with a single tool. This study presents a microstructural analysis of the weld using twin tools. Microstructure depends upon temperature, the flow of material and cooling rate. Thus, the present work is targeted to study computationally the flow behavior of the aluminum alloy during welding to gain insight in the welding process. Work by Soundararajan et al. [7] on thermo-mechanical modeling of FSW has been used to validate the present modeling.

2 Numerical Methodology

2.1 Geometry and Grid Generation

Dimensions of the computational domain are $150 \times 50 \times 3$ mm^3. The tool is having a shoulder of diameter 25 mm and the pin of the tool is 6 mm in diameter and its length being 2.5 mm. The tools have been separated by a distance of 30 mm. The complete flow domain with twin tools is shown in Fig. 1. The sphere of influence method has been used to generate fine mesh near the tool and its surroundings. The final grid has 1,469,793 elements.

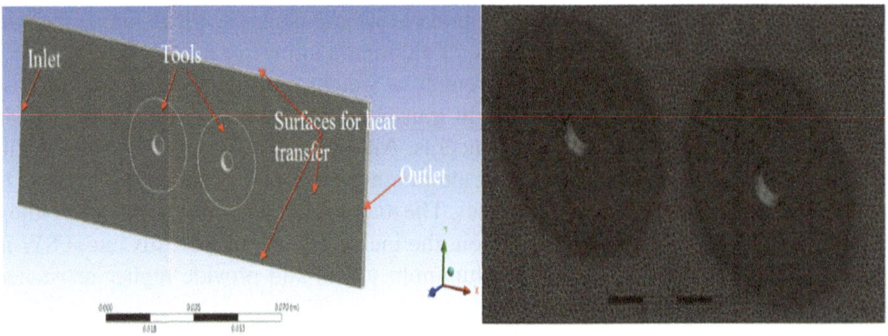

Fig. 1. Flow domain along with mesh near the twin tool geometry.

2.2 Governing Equations

The Eulerian approach has been used to model the flow of aluminum alloy 6061 as non-Newtonian fluid. Equations governing the flow of metal are the mass, momentum, and energy conservation equations as stated below.

Continuity equation:
$$\frac{\partial \rho}{\partial t} + \nabla \cdot (\rho \vec{v}) = 0 \quad (1)$$

Momentum equation:
$$\frac{\partial (\rho \vec{v})}{\partial t} + \nabla \cdot (\rho \vec{v} \vec{v}) = -\nabla p + \nabla \cdot \left(\mu\left(\nabla \vec{v} + \nabla \vec{v}^{T}\right)\right) \quad (2)$$

Energy equation:
$$\frac{\partial (\rho H)}{\partial t} + \nabla \cdot (\rho \vec{v} H) = \nabla \cdot (k \nabla T) \quad (3)$$

Due to the high stirring of metal around the tool, plastic deformation of metal occurs. This results in a non-Newtonian fluid behavior of metal in the vicinity of the tool. Zener-Hollman parameter is used to calculate effective stress and that has been utilized to calculate the non-Newtonian viscosity of the flowing metal as stated below.

Viscosity of metal, μ, is given as

$$\mu = \frac{\sigma}{3\dot{\varepsilon}} \quad (4)$$

where σ is the stress generated, and $\dot{\varepsilon}$ is the strain rate. Stress is calculated using Zener Hollman parameter *Z as*

$$\sigma = \frac{1}{\alpha} \sinh^{-1}\left(\frac{Z}{A}\right)^{\frac{1}{n}} \quad (5)$$

where Z is calculated as

$$Z = \dot{\varepsilon} \exp\left(\frac{Q}{RT}\right) \quad (6)$$

α, A, and n are material constants. The values of these constants for Al 6061alloy are referred from the work by Kadian *et al.* [8]. R is the universal gas constant. The density of the alloy is assigned a constant value of 2700 kg/m^3. Thermal conductivity and heat capacity of the alloy keeps changing with temperature. A piecewise linear function has been used to model these properties [7].

2.3 CFD Solver Settings, Boundary and Initial Conditions

The tools have been set to rotate at constant angular velocities of 344 rpm in opposite directions at a fixed location with walls of the tool having the no-slip condition. And allowed to inlet the flow domain at an inlet velocity of 2.28 mm/s normal to the inlet surface. To simplify, a constant temperature of 800 K is imposed to the tools. On the backide of the aluminum plates higher convective heat transfer coefficient value, 150 W/m^2 has been assigned to model heat transfer through backing plates [2]. For the sidewalls and top surface convective heat transfer coefficient value has been set to 30 W/m^2 [2]. The finite volume method has been used to discretize the governing equations to solve metal flow near the tool. SIMPLE pressure-velocity algorithm with second-order upwinding scheme is used to solve the governing equations. In the CFD solver, viscosity is incorporated using a user-defined function (UDF).

3 Results and Discussion

3.1 Validation of Numerical Model

To ensure the efficacy of the present numerical formulation, temperature variation with time at a location of 2 mm below the surface and 6 mm from the weld line has been plotted and compared to that with the experimental results obtained by Soundrarajan et al. [7]. Inlet velocity of 0.002267 m/s has been provided at the inlet and the tool is allowed to rotate at 344 rpm. Resulting temperature profile is plotted against time and found to be in good agreement with the experimental results (Fig. 2).

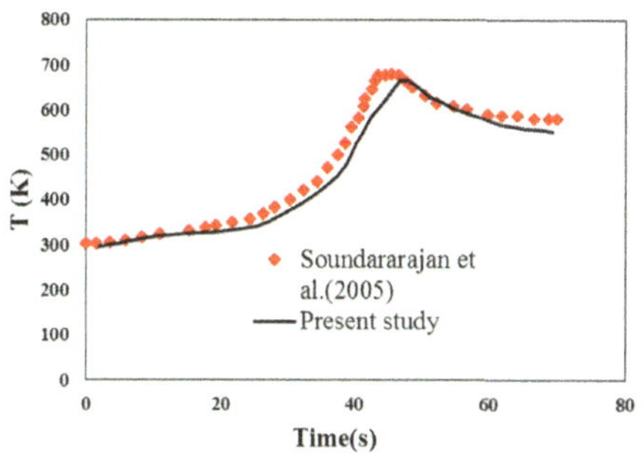

Fig. 2. Temperature variation with time at a distance of 6 mm from the weld line.

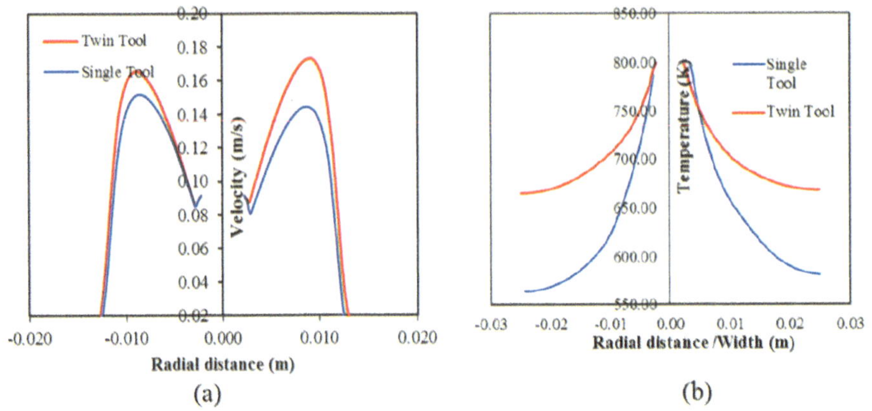

Fig. 3. (a) Variation of velocity along the radial direction at a depth of 2 mm from the top surface (b) Variation of Temperature in the radial direction at the same location.

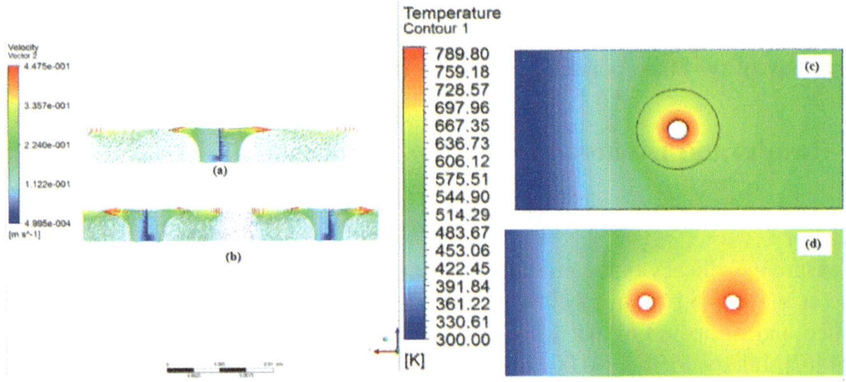

Fig. 4. Velocity vectors of (a) Single tool (b) Twin tool; Temperature contours of (c) Single tool, (d) Twin tool.

3.2 Results and Discussions of a Single Tool and Twin Tool Model

Temperature and velocity variations of the single tool are found to be asymmetric because of the flow of material from retreating to the advancing side of the tool. Also, the maximum velocity reached in the vicinity of the single tool is less (0.15 m/s) and for twin, the tool is higher (0.17 m/s). From the temperature contour plots, it is evident that temperature variation is more symmetric for the twin tool on both advancing and retreating sides. Due to counter-rotation of the backward tool to that with the forward one, relatively even material distribution has taken place which resulted in even temperature distribution on both the side of the tool. As already heated material reaches the second tool temperature drop along the width of the tool is less. This results in lower temperature gradients during the cooling of the material after welding has been completed (Figs. 3 and 4).

4 Conclusion

Numerical modeling using computational fluid dynamics (CFD) is a good tool to gain insight into metal processing. With the Eulerian approach, the aluminum alloy has been successfully modeled as non-Newtonian fluid using Zenner-Hollman parameter. It can be concluded from the present study that the use of twin counter-rotating toolscan help us achieve uniform temperature distribution and flow/mixing of metal on both the advancing and retreating sides of the tool.

References

1. Hilgert, J., Hütsch, L.L., dos Santos, J.F., Huber, N.: Material flow around a bobbin tool for friction stir welding. In: Proceedings of the COMSOL Conference (2010)
2. Mimouni, O., Badji, R., Hadji, M., Kouadr-Davidi, A., Rachid, H., Chekroun, N.: Numerical simulation of temperature distribution and material flow during friction stir welding 2017A aluminum alloys. In: MATEC Web of Conferences, vol. 80, p. 12002 (2016)
3. Padmanaban, R.V.R.K., Kishore, V.R., Balusamy, V.: Numerical simulation of temperature distribution and material flow during friction stir welding of dissimilar aluminum alloys. Procedia Eng. **97**, 854–863 (2014)
4. Aghajani Derazkola, H., Elyasi, M., Hossienzadeh, M.: CFD modelling of friction stir welding of aluminum to steel butt joint. J. Adv. Mater. Process. **3**(3), 47–60 (2015)
5. Hasan, A.F.: CFD modelling of friction stir welding (FSW) process of AZ31 magnesium alloy using volume of fluid method. J. Mater. Res. Technol. **8**(2), 1819–1827 (2019)
6. Kumari, K., Pal, S.K., Singh, S.B.: Friction stir welding by using counter-rotating twin tool. J. Mater. Process. Technol. **215**, 132–141 (2015)
7. Soundararajan, V., Zekovic, S., Kovacevic, R.: Thermo-mechanical model with adaptive boundary conditions for friction stir welding of Al 6061. Int. J. Mach. Tools Manuf **45**(14), 1577–1587 (2005)
8. Kadian, A.K., Puri, G., Das, S., Biswas, P.: Effect of tool geometry and process parameters on the material flow of friction stir welding. In: 5th International and 26th All India Manufacturing Technology, Design and Research Conference (AIMTDR 2014), pp. 12–14 (2014)

Nano Particles as Lubricant Additive: A Concise Review

Roberto D'Amato[1](✉), Alessandro Ruggiero[2], Petr Valašek[3], and Nicolae Ungureanu[4]

[1] Departamento de Ingeniería Mecánica, Química y Diseño Industrial, Universidad Politécnica de Madrid (UPM), Madrid, Spain
r.damato@upm.es

[2] Department of Industrial Engineering, University of Salerno, Via Giovanni Paolo II 132, 84084 Fisciano, SA, Italy

[3] Faculty of Engineering, Department of Material Science and Manufacturing Technology, Czech University of Life Sciences Prague, Prague, Czech Republic

[4] Faculty of Engineering, Department of Engineering and Technology Management, Technical University of Cluj-Napoca, Baia Mare, Romania

Abstract. In order to successfully design the mechanical elements of a machine in relative motion it is necessary to know and understand the tribological principles such as friction and wear. Lubricating oils play an essential role for the proper functioning of these elements. In fact, the durability and economic effectiveness of the machine are related to the quality, performance and reasonable use of the oils. In fact, the basis of all lubrication mechanisms is the knowledge of the physical and chemical interactions between lubricant molecules, surfaces of materials and the environment. The nanoparticles inserted in lubricating oils can improve their properties such as thermal conductivity, anti-wear power and the reduction of the friction coefficient thus reducing emissions and saving fuel. In recent years, many scientific researchers have devoted themselves to the study of lubricant oil additives to improve lubricating properties. This review focused on the effects of nanoparticles on tribological performance in oil lubrication.

Keywords: Lubrication · Friction · Wear · Nanoparticles

1 Introduction

Today, various industries and international societies more than ever pay attention to energy consumption reduction, optimization and saving it. This subject is due to the reduction of fossil fuels sources, rising energy prices and environmental constraints. Although different sciences play an important role in optimizing and reducing energy consumption, such as tribology [1]. In fact, Tribology is the science that investigate the behavior of two surface in contact and in relative motion between them. This word was first introduced in 1966 by Jost [2–4] and it derives from the Greek and it is composed of the words "Tribos" and "logos" which mean "friction" and "study", respectively, therefore literally "the science of friction". Approximately 23% (119 EJ) of the total world's energy is consumed by tribological surfaces in relative motion and this indicates the

S. Hloch et al. (Eds.): ICMEM 2021, LNME, pp. 392–403, 2021.
https://doi.org/10.1007/978-3-030-71956-2_33

key role of this science in saving, optimization and energy consumption. Approximately about 20% of the total world's energy consumption is used for overcoming friction especially in economic sectors as transportation (6%), industry (4%), energy industry (4%) and residential (2%). And about 3% is used for repairing and manufacturing damaged parts by wear. About one-third of the energy consumed in vehicles is used for overcoming friction [5–8]. It means friction can be considered as one of the principal factors in loss and energy consumption. In addition, the energy loss due to the friction causes emissions such as CO_2 and NO_x and consequently environmental constraints and problem. Hence, efforts to reduce friction and wear in order to decrease energy consumption are increasing.

For example, the research of new surface, material and lubrication technologies will be able to reduce energy losses due to friction and wear about 40% in the long term (15 years) and 18% in the short term (8 years). These amount is equal to 1.4% of GDP annually and 8.7% of total energy consumptions in the world in the long term. In addition to the energy topics, the using of advanced tribological technologies can able to reduce the emission of 1460 Mt CO_2 and consequently 450000 M€ cost saving in the short term, and emission 3140 Mt CO_2 and consequently 970000 M€ cost saving in the long term [5–8].

Considering to the interdisciplinary nature of tribology science [9, 10] to achieve optimal tribological system and reducing the amount of friction and wear, it is necessary to use knowledge and new technologies of other sciences as material science, mechanical engineering, chemical engineering and chemistry for the study of new lubricants. The use of lubricants is known as one of the most used and, at the same time, the oldest ways to reduce wear and friction in tribological systems. Lubricants were used to reduce friction and wear even before appearance petroleum and fossil oil. For example, ancient Persians used castor oil as a lubricant in carts and wheels. The non-renewability, the high prices and the environmental pollutions have led to a lot of research to replace fossil base lubricants with renewable and environmentally friendly lubricants. In addition to reducing the friction and wear, lubricants are used "to reduce oxidation, prevent rust, reduce energy consumption, provide insulation in transformer applications, transmit mechanical power in hydraulic fluid power applications and seal against dust, dirt and water" [2, 3, 11, 12].

Fossil base oils, synthetic base oils and bio base oils formed major part of lubricants. In most industrial applications these oils are used with the addition of additives to improve their tribological performance. The use of modern additives has started since 1930 due to the increased use of lubricants and their application diversity, and it grew rapidly [13]. The additives can be classified into two major types that influence chemical and physical properties of base oils and influence the metal surfaces. According to application, additives can be classified as primary and special additive. Primary additives are commonly used in most lubricants while special additives added to special lubricant with special application. The additives will be classified according to their action mechanism in the lubricant to two groups. The first group perform either on a surface and second group perform in the lubricant bulk. In the other classification, lubricants additives can be classified to three group according to performance. 1- Surface additives (protection of the lubricated surface); 2- Performance enhancing additives (improvement of the performance of the lubricant); 3- Fluid protective additives (protection of the lubricant base oil). However, do not exist the exact boundary between these classifications [3, 13, 14].

The appearance of nanoscience and the addition of nanoparticles as additives to lubricants has led to a noticeable improvement in performance in industrial tribological applications. The use of nanoparticles in tribological system, is possible by dispersing them into base oils and embedding them inside a surface coating. The problem of the first method is the colloid instability caused by the aggregation and sedimentation of nanoparticles in the lubricant. While the problem of the second method is the release of nanoparticles by the coating during wear. The use of nanoparticles as lubricants additives have many advantages over organic compound as lubricants additives, which include: 1- The size of nanoparticles allows them to enter the contact area easily, 2- Nanoparticles are often efficient at ambient temperature, 3- Existence of different types of nanoparticles with different properties.

For these reasons, the main purpose of this study is to introduce a review on the effect of Nano Particles, as lubricant additives, on the performance and on tribological properties of the lubricants. This study is summarized in three main sections. 1- Nanofluid, preparation, stability and different types of Nanoparticles. 2- Effect of Nano particles additive on the lubricant performance properties. 3- Effect of Nano particles additive on the lubricants tribological properties such as friction and wear [15].

2 Nano-fluid, Preparation, Stability and Different Types of Nanoparticles

The nanofluid, which was first used by Choi, is created by the dispersing of particles with size less than 100 nm in the fluid [16]. In other words, nanofluids are the colloids that nanoparticles formed the disperse phase of them. In general, there are two methods for nanofluid preparation: one-step method (direct evaporation technique) and two-step method are the two developed methods for the nanofluid production [16]. In a one-step method, the nanoparticles are created and dispersed in the base fluid by physical or chemical method simultaneously. The advantages of this method are to reduce the agglomeration and more stability of the nanofluid. But the disadvantage of this method is the presence of the residual reactants due to the probability of incomplete chemical reaction. In the two-step method, nanoparticles are created as dry powder by physical and chemical methods at the first and then dispersed in the fluid. The two-step method is due to the production of nanoparticles at an industrial scale is more economical than the one-step method. Like another solid-liquid two phase systems, instability is the most important disadvantage of nanofluid that consist of nanoparticles or disperse phase and liquid phase or continuous phase. The most common methods to investigate the stability of nanofluids are measuring of Zeta potential, the effects of sedimentation and centrifugation, and thermal conductivity. The term "Zeta potential" is refer to electro-kinetic potential in colloidal dispersions, and it can be measured by using the UV-vis spectrum method or the dynamic light scattering (DLS) in order to evaluate the ultrasonic mixing time on UV absorbency and on the stability of nanofluids. For the preparation of nanofluids stability are used physical, chemical methods or the combination of these two one

[17]. In the physical methods, magnetic force agitation, ultrasonic agitation, high-shear mixing, homogenizing are used for stability of nano fluid. While in chemical methods surfactant and pH adjustment are used for the stability of nano fluid [17–21]. In terms of chemical characterization, nanoparticles can be divided into seven groups: carbon and its derivatives, metals, metal oxide, sulfides, rare earth compounds, nanocomposites and others. In terms of morphology, nanoparticles can be divided into five groups: granular, onion, sheet, spherical and tube. Approximately in the all of nano-lubricants studies, a two-step method has been used to create nano-lubricant [22, 23].

3 Effect of Nano Particles Additive on the Lubricant Performance Properties

The lubricants performance is strongly influenced by their properties. As mentioned, lubricants additives are added to improve the base oil properties according the working conditions and achieving the best performance in the specified conditions. Lubricants properties can be classified in three general group taking in account the performance properties, environmental properties and long lifetime properties. These properties are determined according to application and operation condition.

3.1 Effect of Nano Particles Additive on the Lubricant Thermal Conductivity

Heat transfer and cooling are one of the most important lubricants applications. For example, in internal combustion engines the primary use of the lubricants is to reduce friction wear, however it also plays an important role for heat transfer. The same happens in the transformers and hydraulic systems in which the lubricants are used for cooling effect.

However, the change of viscosity, thermal oxidation stability, thermal decomposition and similar constraints prompted lubricants have a less application in the cooling system compare water but their thermal properties are still very important. Thermal conductivity is one of the lubricant properties to be kept under control, for this reason numerous studies have been carried out in order to improve the performance in terms of heat transfer [3, 11, 24, 25].

Evidence indicate that thermal conductivity of the lubricant fluids can be improved by adding solid particles to them [26, 27].

For this reason, with the advent of the nano science it has been possible to add nano particles to lubricant fluids to improve the thermal properties such as thermal conductivity. In fact, thanks to the size of nanoparticles it is possible to obtain a better stability compared to micro particles.

3.1.1 Theoretical Investigation

In recent years, many scientific research has attempted to develop mathematical models to study and improve the thermal conductivity of suspensions, colloids and nanofluids. The models presented today can be divided into two categories: classic models and developed models. The effective parameters on the behavior of thermal conductivity in classical modeling are: The thermal conductivity of the base fluid, the thermal conductivity of solid particles, the concentration of solid particles and the shape of solid particles. One of the theoretical limitations of these studies is the very low volumetric concentration value of solid particles considered. This did not allow to take due account of some physical phenomena such as the interaction between particles. The scientist, Maxwell, J [28, 29] was the first to develop a theory on the effective thermal conductivity, k, of two-phase mixtures containing powders that have particle diameters in the order of millimeters or micrometers. "Maxwell's effective medium theory (EMT) for dilute suspensions of spherical particles predicts a 3% increase in thermal conductivity for each 1% volume concentration added to the host fluid. Furthermore, there are, however, other classical theories of effective nanofluid media. For example, the Hamilton–Crosser (HC) model and Hashin–Shtriman (HS) model consider the shapes of nanoparticles, those which are spherical and non-spherical" [28]. Instead, the Maxwell–Garnett (MG) model accounts for spherical nanoparticles, together with a dependence on both the thermal conductivity of nanoparticles and base fluids, and the volume fraction of nanoparticles. As mentioned above, the classical models due to not pay attention to real condition such as Brownian motion, surface chemical properties, interaction between solid particles and similar item, are not able to correctly determine the suspensions thermal conductivity. Therefore, studies in this field have led to the development of developed models. Developed models are classified based on the consideration interfacial layer between solid and liquid or considering particles Brownian motion.

In the following, some of the most important developed models will be mentioned. Jeffrey et al. have developed the M-G model for non-uniform distribution particle size and high volume fraction [30]. Yu et al. have proposed a model according to the effect of the absorbed layer on the nanoparticle surface on the thermal conductivity of the nanofluid [31, 32]. In fact, the above model has been achieved with considering the interfacial layer between solid particles and liquid. Based on this hypothesis (interfacial layer) by Yang et al. [33], Xue [34], Rizvi et al. [35], Leong et al. [36] and Xie et al. [37] have been presented different models. Xue [38] has presented models for determining the thermal conductivity of nanofluids contains nanotubes with considering the nanotubes orientation distribution. Jiang et al. [39] and Murshed et al. [40] and Yamada et al. [41], have gained models by considering the interfacial layer between nanotube and liquid for the nanofluids contains nanotube. Xuan has gained semi-experimental model for determining the thermal conductivity of nanofluids with considering convection heat transfer in micro-scale, the effect of adsorption on the surface of the nanoparticles, the size of the nanoparticles, the volume fraction of nanoparticles, the temperature and energy transfer by particles and the base fluid [42]. Yang et al., has gained new model considering effect of surfactants and particle shape of nanoparticles on the thermal conductivity [43].

Bhattacharya et al. [44], Xiao et al. [45], Shukla et al. [46], Xiao et al. [47] and Felderhof [48] have gained models for computing nanofluid thermal conductivity based on the nanoparticle Brownian motion and the convection heat transfer. Evans et al. has gained model for computing nanofluid thermal conductivity based on the effect of aggregation and interfacial thermal resistance of nano particles [49]. Wang et al. has gained a model for prediction of nanofluid thermal conductivity base on aggregation of nanoparticles in the fluid, effect of adsorption on the surface of the nanoparticles and the distribution of particle size [50].

Cui et al. have gained a modified model for spherical nanoparticles suspension thermal conductivity considering two static and dynamic mechanism [51]. Murshed et al. [52] and Kumar et al. [53] have gained models for nanofluid thermal conductivity with static and dynamic mechanism. Yang et al. have gained new model for thermal conductivity of nanofluids with cylindrical nanoparticles by anisotropy analysis [54]. Tso et al. have gained the thermal conductivity model for the nanofluids as a function of the interfacial layer thickness, particle size, volume concentration and thermal conductivity of the nanoparticles and base fluid [55]. Timofeeva et al. [56] and Xuan et al. [57] have gained the nanofluid effective thermal conductivity models by considering the effect of nanoparticles aggregation. The Anomalous increasing of nanofluid thermal conductivity with decreasing nano-particle size has been studied by Keblinski et al. by assuming four possible mechanisms:1-nanoparticles Brownian motion, 2-interfacial layer, 3-effect of nanoparticles clustering, 4-the nature of heat transport in the nanoparticles [58]. Koo et al. have investigated the nanofluid thermal conductivity model by assuming that the thermal conductivity is divided into two parts, the thermal conductivity of the static part and the thermal conductivity of the moving part with a Brownian motion [59].

4 Effect of Nanoparticles Additive on the Lubricants Tribological Properties

Adding nanoparticles to the lubricants, in addition to effecting on the performance properties such as viscosity and thermal conductivity, have a direct effect on tribological properties like including coefficient of friction and wear. In the following, we will mention to the experimental and theoretical investigation and the mechanisms presented in this field.

4.1 Anti-friction and Anti-wear Mechanism of Nanoparticles

In general, the review of the studies shows the positive effect of increasing nanoparticles as additive on tribological properties such as reducing wear and friction. In this part, will be mentioned to the most important causes and mechanisms of this positive effect on the tribological properties. Due to that tribology science is so complicate and influenced by different parameters, results interpretation and the mechanism presentation for tribological properties such as wear rate and coefficient of friction are very difficult. However, researchers have presented different mechanisms to explain the cause of reducing friction and wear due to increasing nanoparticles. In general, nanoparticles as additives have a direct and indirect effects on the tribology of contact [60]. The indirect effect of nanoparticles on tribology of contact is reducing real contact area between surfaces. But the direct effects of nanoparticles on the tribology of surface is due to the interaction between the nanoparticles and the surface that will be mentioned in the following.

Rolling effect or ball bearing effect is one of the mechanisms for reducing friction and wear due to the increases of nanoparticles as lubricant additive, especially for spherical nanoparticles. In fact, nanoparticles reduce friction and wear by converting slipping to rolling friction. This mechanism depends strongly to the nanoparticles size and the lubrication regime (the thickness of the lubricant film). In fact, this mechanism, when is effective, depends on the size or diameter of the nano-particles is larger or equal to the lubricant film between the tribological surfaces [61–66]. Tribofilm formation on tribological surface is another mechanism for interpreting the causes of reducing friction and wear by adding nanoparticles as lubricant additives. This film is formed due to penetration and interaction of the nanoparticles with tribological surfaces. This mechanism is strongly dependent on the nanoparticles diffusion and surface energy to react with the tribology surface. Positive wear can justified in this mechanism by creating film of nanoparticles on the tribology surfaces [67, 68]. In some studies, is mentioned to protective film or protective layer formed by nanoparticles on the tribological surface as a mechanism for reducing friction and wear. This mechanism is similar to the tribofilm mechanism [23, 67–70]. Nanoparticles as a third component prevents from direct contact between tribology surface asperities. This mechanism is named third body effect. The third body also can be considered as a mixture of lubricants, nanoparticles and abrasive particles [63–65]. Tribosintering is another mechanism that improves tribological properties by forming nanoparticle film on tribological surfaces and reducing direct contact between tribology surfaces. If fine metal nanoparticles are used at high contact pressure, the sintering process starts as soon as the temperature increases above ambient temperature [65, 68, 71]. The nanoparticles are deposited on the surface and filling the grooves of the worn surface and creating a smoother surface that cause improve the tribological properties. This mechanism is known as the mending effect [63, 72]. Polishing the tribology surfaces by nanoparticles, known as polishing effect, is another mechanism that has positive effect on tribological properties of the lubricants. Initially this mechanism, due to polishing the surface, may be increase the coefficient of friction, but over time create a smooth surface and reduce friction and wear. The rate of surface polishing by nanoparticles depends on the size of the nanoparticles and increase by increasing the size of the nanoparticles [73, 74]. Nanoparticles also improve tribological properties by melting, welding and forming the film on the tribology surfaces. This phenomena are known as melting and mending effect. Due to the high melting temperature of the most nanoparticles, the possibility occurrence of this mechanism is less than other mechanisms [69, 75]. Hardening of tribology surfaces by incorporation of nanoparticles with a tribology surface, especially for hard nanoparticles, such as diamond nanoparticles, also can improve tribological properties [62]. The colloidal effect, size effect, transfer film and antioxidant effect for oil at high temperature are other mechanisms that mentioned for improve tribological properties due to using nanoparticles in some references [23, 62, 76]. Although used various mechanisms with different names for explain positive effect of increasing nanoparticles as lubricants additive on tribological properties, but some of these mechanisms overlap and do not have a precise boundary between them.

4.2 Theoretical Investigation and Modelling for Prediction Nano-lubricants CoF and Wear Rate

The complexity of the tribology science and the dependence of this science to various parameters, have influenced on theoretical investigation and modeling of friction coefficient and wear rate. Review of the studies show that investigations done on the prediction of tribological properties in the absence of nanoparticles are more than investigations done in the presence of nanoparticles. For example the prediction equation proposed for Coefficient of friction of gears by Carlos et al. [77], the prediction model proposed for steady state coefficient of friction in the lubricated contact by Mehdizadeh et al. [78], the prediction model proposed for coefficient of friction in mixes lubrication regime by Akchurin et al. [79], the prediction model proposed for sliding coefficient of friction "based on a novel hybrid molecular-mechanical model" by Zhang et al. [80], the prediction model proposed for coefficient of friction and wear in spiral bevel gears by Zhang et al. [81] and the prediction model proposed for steady-state wear rate in spur gears by Masjedi et al. [82] are some of the newest studies done about modelling of tribological properties in the absence of nanoparticles. The analytical investigations done by Ghaednia et al. about the effect of nanoparticles on tribological properties, have led to the development of the most important model presented in this regard. This model was obtained based on the reduction of the real contact area due to the presence of nanoparticles between tribological surfaces as a third body [83, 84]. This model only obtained based on one of the nanoparticles effect mechanisms of reducing the wear and friction provided in the previous section, namely, reducing the real contact area. Therefore, in this model the effect of other mechanisms such as rolling effect, mending effect and etc. on the tribological properties was ignored. In order to investigate about this model two scales, micron scale for contact between surface asperities and smaller scale for nanoparticles contact are considered. According statistical particle sub-model in this study assumed, surfaces inside the contact region are effectively flat compared to the forms and size nanoparticles.

5 Conclusions

Several factors have been found that influence tribological performance such as the chemical composition of the nanoparticles or their preparation. An important role in the anti-wear mechanism of lubricating oils with the addition of nanoparticles is certainly played by the chemical composition, while it has had a subtle effect on the friction performance. In fact, from the studies reported in this review it was seen that the wear resistance was closely related to the properties of the tribofilm in the boundary lubrication regime.

The dimensions of the nanoparticles have shown visible effects on both friction and wear, while their morphology plays a fundamental role in reducing friction.

The mechanisms of friction reduction and anti-wear of nanoparticles in the lubricant can be studied as colloidal effect, rolling effect, protective film and third body.

The use of lubricating oils with addition of nanoparticles can improve tribological properties such as:

- thermal conductivity
- the anti-wear effects
- the reduction of friction thus improving the efficiency and durability of the operating machines.

Acknowledgments. The Authors acknowledge the assistance in the bibliographic research of M.Sc. M. E. Soltani from the Department of chemical engineering, Isfahan University of Technology, Isfahan, Iran during his PhD period at the Department of Industrial Engineering of the University of Salerno.

References

1. Ruggiero, A., et al.: Tribological characterization of vegetal lubricants: comparative experimental investigation on Jatropha curcas L. oil, Rapeseed Methyl Ester oil, Hydrotreated Rapeseed oil. Tribol. Int. **109**, 529–540 (2017)
2. Mang, T., Dresel, W.: Lubricants and Lubrication. Wiley (2007)
3. Bart, J.C.J., Cavallaro, S.: Biolubricants: Science and Technology. Elsevier Science & Technology (2018)
4. Blau, P.J.: Friction Science and Technology: From Concepts to Applications, 2nd edn. CRC Press (2008)
5. Holmberg, K., Erdemir, A.: Influence of tribology on global energy consumption, costs and emissions. Friction **5**(3), 263–284 (2017)
6. Holmberg, K., Andersson, P., Erdemir, A.: Global energy consumption due to friction in passenger cars. Tribol. Int. **47**, 221–234 (2012)
7. Holmberg, K., et al.: Global energy consumption due to friction in trucks and buses. Tribol. Int. **78**, 94–114 (2014)
8. Fox, I.E.: Numerical evaluation of the potential for fuel economy improvement due to boundary friction reduction within heavy-duty diesel engines. Tribol. Int. **38**(3), 265–275 (2005)
9. Ruggiero, A., D'Amato, R., Gomez, E.: Experimental analysis of tribological behavior of UHMWPE against AISI420C and against TiAl6V4 alloy under dry and lubricated conditions. Tribol. Int. **92**, 154–161 (2015)
10. Ruggiero, A., et al.: Experimental comparison on tribological pairs UHMWPE/TIAL6V4 alloy, UHMWPE/AISI316L austenitic stainless and UHMWPE/AL(2)O(3) ceramic, under dry and lubricated conditions. Tribol. Int. **96**, 349–360 (2016)
11. Honary, L., Richter, E.: Biobased Lubricants and Greases: Technology and Products. Wiley (2011)
12. Seireg, S.: Friction and Lubrication in Mechanical Design. CRC Press (1998)
13. Stepina, V., Vesely, V.: Lubricants and Special Fluids. Elsevier Science (1992)
14. Torbacke, M., Rudolphi, Å.K., Kassfeldt, E.: Lubricants: Introduction to Properties and Performance. Wiley (2014)
15. Martin, J.M., Ohmae, N.: Nanolubricants. Wiley (2008)
16. Chol, S., Estman, J.: Enhancing thermal conductivity of fluids with nanoparticles. ASME-Publications-Fed **231**, 99–106 (1995)
17. Yu, W., Xie, H.: A review on nanofluids: preparation, stability mechanisms, and applications. J. Nanomater. **2012**, 1 (2012)
18. Drzazga, M., et al.: Preparation of metal oxide-water nanofluids by two-step method. Inż. Ap. Chem **51**, 213–215 (2012)
19. Mukherjee, S., Paria, S.: Preparation and stability of nanofluids-a review. IOSR J. Mech. Civil Eng. **9**(2), 63–69 (2013)
20. Chamsa-ard, W., et al.: Nanofluid types, their synthesis, properties and incorporation in direct solar thermal collectors: a review. Nanomaterials **7**(6), 131 (2017)
21. Yazid, M.N.A.W.M., et al.: A review of the impact of preparation on stability of carbon nanotube nanofluids. Int. Commun. Heat Mass Transfer **78**, 253–263 (2016)

22. Dai, W., et al.: Roles of nanoparticles in oil lubrication. Tribol. Int. **102**, 88–98 (2016)
23. Shahnazar, S., Bagheri, S., Hamid, S.B.A.: Enhancing lubricant properties by nanoparticle additives. Int. J. Hydrogen Energy **41**(4), 3153–3170 (2016)
24. Pirro, D.M., Daschner, E., Wessol, A.A.: Lubrication Fundamentals, 2nd edn. CRC Press (2001)
25. Astarita, A., Rubino, F., Carlone, P., Ruggiero, A., Leone, C., Genna, S., Merola, M., Squillace, A.: On the improvement of AA2024 wear properties through the deposition of a cold-sprayed titanium coating. Met. **6**(8), 185 (2016)
26. Fricke, H.: A mathematical treatment of the electric conductivity and capacity of disperse systems I. The electric conductivity of a suspension of homogeneous spheroids. Phys. Rev. **24**(5), 575 (1924)
27. Levin, M., Miller, M.: Maxwell a treatise on electricity and magnetism. Uspekhi Fizicheskikh Nauk **135**(3), 425–440 (1981)
28. Hamilton, R.L., Crosser, O.: Thermal conductivity of heterogeneous two-component systems. Ind. Eng. Chem. Fundam. **1**(3), 187–191 (1962)
29. Maxwell, J.: A Treatise on Electricity and Magnetism. Oxford University Press , Cambridge (1904)
30. Jeffrey, D.J.: Conduction through a random suspension of spheres. Proc. R. Soc. Lond. A **335**(1602), 355–367 (1973)
31. Yu, W., Choi, S.: The role of interfacial layers in the enhanced thermal conductivity of nanofluids: a renovated Maxwell model. J. Nanopart. Res. **5**(1–2), 167–171 (2003)
32. Yu, W., Choi, S.: The role of interfacial layers in the enhanced thermal conductivity of nanofluids: a renovated Hamilton-Crosser model. J. Nanopart. Res. **6**(4), 355–361 (2004)
33. Yang, L., et al.: A new thermal conductivity model for nanorod-based nanofluids. Appl. Therm. Eng. **114**, 287–299 (2017)
34. Xue, Q.-Z.: Model for effective thermal conductivity of nanofluids. Phys. Lett. A **307**(5–6), 313–317 (2003)
35. Rizvi, I.H., et al.: Mathematical modelling of thermal conductivity for nanofluid considering interfacial nano-layer. Heat Mass Transf. **49**(4), 595–600 (2013)
36. Leong, K., Yang, C., Murshed, S.: A model for the thermal conductivity of nanofluids–the effect of interfacial layer. J. Nanopart. Res. **8**(2), 245–254 (2006)
37. Xie, H., Fujii, M., Zhang, X.: Effect of interfacial nanolayer on the effective thermal conductivity of nanoparticle-fluid mixture. Int. J. Heat Mass Transf. **48**(14), 2926–2932 (2005)
38. Xue, Q.: Model for thermal conductivity of carbon nanotube-based composites. Physica B **368**(1–4), 302–307 (2005)
39. Jiang, H., et al.: The role of interfacial nanolayer in the enhanced thermal conductivity of carbon nanotube-based nanofluids. Appl. Phys. A **118**(1), 197–205 (2015)
40. Murshed, S., Leong, K., Yang, C.: Investigations of thermal conductivity and viscosity of nanofluids. Int. J. Therm. Sci. **47**(5), 560–568 (2008)
41. Yamada, E., Ota, T.: Effective thermal conductivity of dispersed materialsEffektive Wärmeleitfähigkeit in dispersen Systemen. Wärme-und Stoffübertragung **13**(1–2), 27–37 (1980)
42. Xuan, Y., et al.: Stochastic thermal transport of nanoparticle suspensions. J. Appl. Phys. **100**(4), 043507 (2006)
43. Yang, L., et al.: Roles of surfactants and particle shape in the enhanced thermal conductivity of TiO2 nanofluids. AIP Adv. **6**(9), 095104 (2016)
44. Bhattacharya, P., et al.: Brownian dynamics simulation to determine the effective thermal conductivity of nanofluids. J. Appl. Phys. **95**(11), 6492–6494 (2004)
45. Xiao, B., et al.: A fractal model for heat transfer of nanofluids by convection in a pool. Phys. Lett. A **373**(45), 4178–4181 (2009)

46. Shukla, K.N., et al.: Effective thermal conductivity of nanofluids–a new model taking into consideration Brownian motion. Int. J. Heat Mass Transf. **99**, 532–540 (2016)
47. Xiao, B., Yang, Y., Chen, L.: Developing a novel form of thermal conductivity of nanofluids with Brownian motion effect by means of fractal geometry. Powder Technol. **239**, 409–414 (2013)
48. Felderhof, B.: The effect of Brownian motion on the transport properties of a suspension of spherical particles. Physica A **118**(1–3), 69–78 (1983)
49. Evans, W., et al.: Effect of aggregation and interfacial thermal resistance on thermal conductivity of nanocomposites and colloidal nanofluids. Int. J. Heat Mass Transf. **51**(5–6), 1431–1438 (2008)
50. Wang, B.-X., Zhou, L.-P., Peng, X.-F.: A fractal model for predicting the effective thermal conductivity of liquid with suspension of nanoparticles. Int. J. Heat Mass Transf. **46**(14), 2665–2672 (2003)
51. Cui, W., et al.: Modified prediction model for thermal conductivity of spherical nanoparticle suspensions (nanofluids) by introducing static and dynamic mechanisms. Ind. Eng. Chem. Res. **53**(46), 18071–18080 (2014)
52. Murshed, S., Leong, K., Yang, C.: A combined model for the effective thermal conductivity of nanofluids. Appl. Therm. Eng. **29**(11–12), 2477–2483 (2009)
53. Kumar, D.H., et al.: Model for heat conduction in nanofluids. Phys. Rev. Lett. **93**(14), 144301 (2004)
54. Yang, L., Du, K., Zhang, X.: A theoretical investigation of thermal conductivity of nanofluids with particles in cylindrical shape by anisotropy analysis. Powder Technol. **314**, 328–338 (2017)
55. Tso, C.Y., Fu, S.C., Chao, C.Y.: A semi-analytical model for the thermal conductivity of nanofluids and determination of the nanolayer thickness. Int. J. Heat Mass Transf. **70**, 202–214 (2014)
56. Timofeeva, E.V., et al.: Thermal conductivity and particle agglomeration in alumina nanofluids: experiment and theory. Phys. Rev. E **76**(6), 061203 (2007)
57. Xuan, Y., Li, Q., Hu, W.: Aggregation structure and thermal conductivity of nanofluids. AIChE J. **49**(4), 1038–1043 (2003)
58. Keblinski, P., et al.: Mechanisms of heat flow in suspensions of nano-sized particles (nanofluids). Int. J. Heat Mass Transf. **45**(4), 855–863 (2002)
59. Koo, J., Kleinstreuer, C.: A new thermal conductivity model for nanofluids. J. Nanopart. Res. **6**(6), 577–588 (2004)
60. Ghaednia, H., Jackson, R.L.: The role of nanoparticles in lubricants; performing lubricated and dry friction tests. Tribol. Lubr. Technol. **71**(7), 20–24 (2015)
61. Chinas-Castillo, F., Spikes, H.: Mechanism of action of colloidal solid dispersions. J. Tribol. **125**(3), 552–557 (2003)
62. Puzyr, A.P., et al.: Modified nanodiamonds as antiwear additives to commercial oils. Tribol. Trans. **55**(1), 149–154 (2012)
63. Jatti, V.S., Singh, T.P.: Copper oxide nano-particles as friction-reduction and anti-wear additives in lubricating oil. J. Mech. Sci. Technol. **29**(2), 793–798 (2015)
64. Müller, M., Valášek, P., Ruggiero, A.: Strength characteristics of untreated short-fiber composites from the plant ensete ventricosum. BioResources **12**(1), 255–269 (2017)
65. Battez, A.H., et al.: Friction reduction properties of a CuO nanolubricant used as lubricant for a NiCrBSi coating. Wear **268**(1–2), 325–328 (2010)
66. St. Dennis, J., et al.: Carbon microspheres as ball bearings in aqueous-based lubrication. ACS Appl. Mater. Interfaces **3**(7), 2215–2218 (2011)
67. Li, B., et al.: Tribochemistry and antiwear mechanism of organic–inorganic nanoparticles as lubricant additives. Tribol. Lett. **22**(1), 79–84 (2006)

68. Kato, H., Komai, K.: Tribofilm formation and mild wear by tribo-sintering of nanometer-sized oxide particles on rubbing steel surfaces. Wear **262**(1–2), 36–41 (2007)
69. Hernández Battez, A., et al.: CuO, ZrO2 and ZnO nanoparticles as antiwear additive in oil lubricants. Wear **265**(3–4), 422–428 (2008)
70. Padgurskas, J., et al.: Tribological properties of lubricant additives of Fe Cu and Co nanoparticles. Tribol. Int. **60**, 224–232 (2013)
71. Zhou, Y.-H., Harmelin, M., Bigot, J.: Sintering behaviour of ultra-fine Fe, Ni and Fe-25wt% Ni powders. Scr. Metall. **23**(8), 1391–1396 (1989)
72. Liu, G., et al.: Investigation of the mending effect and mechanism of copper nano-particles on a tribologically stressed surface. Tribol. Lett. **17**(4), 961–966 (2004)
73. Xu, T., Zhao, J., Xu, K.: The ball-bearing effect of diamond nanoparticles as an oil additive. J. Phys. D Appl. Phys. **29**(11), 2932 (1996)
74. Zeng, T., Sun, T.: Size effect of nanoparticles in chemical mechanical polishing-a transient model. IEEE Trans. Semicond. Manuf. **18**(4), 655–663 (2005)
75. Alves, S., et al.: Nanolubricants developed from tiny CuO nanoparticles. Tribol. Int. **100**, 263–271 (2016)
76. Greenberg, R., et al.: The effect of WS2 nanoparticles on friction reduction in various lubrication regimes. Tribol. Lett. **17**(2), 179–186 (2004)
77. Fernandes, C.M., Martins, R.C., Seabra, J.H.: Coefficient of friction equation for gears based on a modified Hersey parameter. Tribol. Int. **101**, 204–217 (2016)
78. Mehdizadeh, M., Akbarzadeh, S.: Experimental investigation and a model to predict the steady-state friction coefficient in the lubricated contact. STLE, Vegas (2016)
79. Akchurin, A., et al.: On a model for the prediction of the friction coefficient in mixed lubrication based on a load-sharing concept with measured surface roughness. Tribol. Lett. **59**(1), 19 (2015)
80. Zhang, X., et al.: Prediction of sliding friction coefficient based on a novel hybrid molecular-mechanical model. J. Nanosci. Nanotechnol. **18**(8), 5551–5557 (2018)
81. Zhang, J.-G., Liu, S.-J., Fang, T.: On the prediction of friction coefficient and wear in spiral bevel gears with mixed TEHL. Tribol. Int. **115**, 535–545 (2017)
82. Masjedi, M., Khonsari, M.: On the prediction of steady-state wear rate in spur gears. Wear **342**, 234–243 (2015)
83. Ghaednia, H., Jackson, R.L.: The effect of nanoparticles on the real area of contact, friction, and wear. J. Tribol. **135**(4), 041603 (2013)
84. Ghaednia, H.: An analytical and experimental investigation of nanoparticle lubricants (2014)

Automatic Generation of the Machining Processes for Turned Components in CAD/CAM System

Tomáš Dodok[1(✉)], Nadežda Čuboňová[1], Miroslav Císar[1], and Vitalii Ivanov[2]

[1] Faculty of Mechanical Engineering, Department of Automation and Production Systems, University of Žilina, Univerzitná 1, 010 26 Žilina, Slovakia
tomas.dodok@fstroj.uniza.sk

[2] Faculty of Technical Systems and Energy Efficient Technologies, Department of Manufacturing Engineering, Machines and Tools, Sumy State University, Rymskogo-Korsakova St. 2, Sumy 40007, Ukraine

Abstract. At present, some more complex CAD/CAM systems provide advanced tools focused on optimization and automation of the NC program preparation. Previous researches was focused on the issue of optimization and automatic generation of NC programs for components with milled features. The article deal mainly with the automated generation of turning processes in CAD/CAM system Edgecam 2021. Experiments are divided into two parts. The first part focuses on the identification and recognition of features on the turned components. The second part is focused on application of the experimental NC strategies on the selected components.

Keywords: NC strategies · CAD/CAM system · Edgecam 2020

1 Introduction

Modern CAD/CAM systems contain a number of the integrated models to optimize their usage, for example by eliminating repetitive tasks. This optimization uses NC strategies that allow users to automatically generate parts of the machining process as well as generate the entire machining processes.

In the past, several studies have been published with focus on the possibility of automatic feature recognition from CAD models and design of automated systems for generating of machining processes based on the feature properties [1, 2]. At present, some CAD/CAM systems already have integrated different methods and algorithms for automatic feature recognition [3]. The authors have dealt with this issue in the past. Part of the research was the identification of suitable features and the application of NC strategies in the creation of machining processes for milled components, especially parts of injection molds. Based on achieved the results, the main advantages of using these integrated modules have been formulated. Deficiencies in the recognition of individual features as well as deficiencies in the application of the proposed experimental NC strategies were also defined [4].

S. Hloch et al. (Eds.): ICMEM 2021, LNME, pp. 404–410, 2021.
https://doi.org/10.1007/978-3-030-71956-2_34

The article is focused on the area of automatic feature recognition on turned components and their subsequent use in the design and application of experimental NC strategies.

2 Feature Recognition on Turned Components

Geometric feature recognition is performed after the CAD model has been created. Recognition is realized directly in CAD/CAM system Edgecam 2020.0, user can choose the type of recognized features. Feature recognition is realized without user intervention based on predefined templates. In the CAD/CAM system Edgecam 2020 allows to recognize several groups of features (Fig. 1).

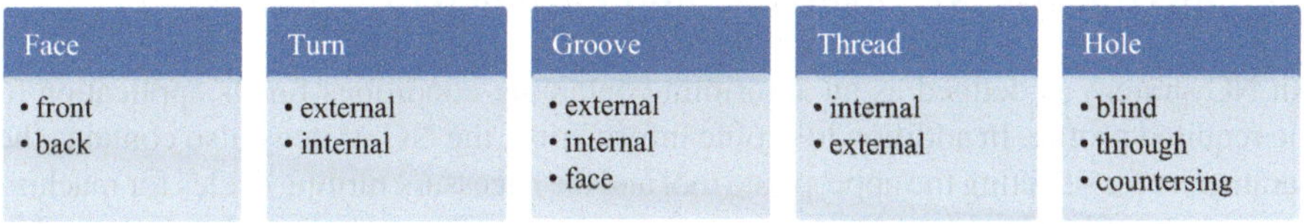

Fig. 1. Type of features in CAD/CAM system Edgecam 2020.0.

The properties of the recognized feature can be divided into several groups:

1. Basic - define the type and subtype of the recognized profile for example, the basic features are face, turn, hole and subtypes are grooves, profiles (internal and external).
2. Geometry - contains geometric information about the feature (turning profile), its start and end points (in X and Z coordinates), profile length, information about angles (TurnMinimumCutAngle) and describes the elements of which it is composed, individual lines or arcs (start and end points, length of element and angles).

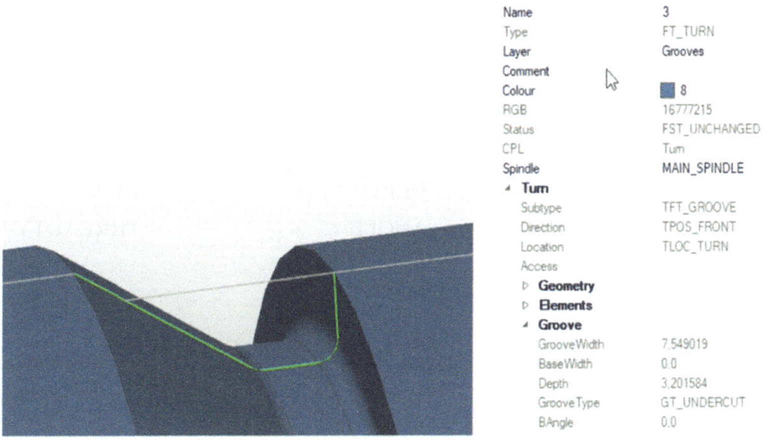

Fig. 2. Feature type Groove with properties.

3. Subtype properties - describes parameters of a recognized feature e.g. basic properties of groove are width, base width, depth, chamfer or radius dimensions (Fig. 2).

In addition to automatic recognition of the profile CAD/CAM system Edgecam also enables user profile creation. This is done either by selecting the surface of which it will be composed or by profile where the user selects the start and end elements of the profile. The main difference between the user-created turning features and user-created milling features are their properties. The user-created turning profiles have the same parameters as the automatically recognized profiles and the same NC strategies can be applied to them.

3 Experimental NC Strategies and Parameters

An NC strategy is defined as an algorithm containing conditions for its application to the required profile. In addition to profile information, the NC strategy also contains the parameters for selecting the appropriate tool and the necessary turning cycles for machining of the profile. Groups of NC strategies and type specifications were implemented in the past [7–9]. Where the strategies have been divided according to the type of information required for their correct application to the feature into automatic, semi-automatic and manual. In this case, fully automated experimental strategies were designed [5, 6]. The input information for the strategies was determined as follows:

1. Feature definition - Based on the profile recognition analysis, five experimental NC strategies were proposed – face-turning, profile-turning, groove-turning, thread and hole drilling.
2. Parameters for tool selection - for each strategy several properties of features for tool selection were selected (Table 1).

Table 1. Features properties used for proper selection of tools.

Type of feature	Parametre 1	Parametre 2	Parametre 3
PROFILE	CUT ANGLE	RADIUS (MAX - MIN)	
GROOVE	BASEWIDTH	RADIUS (MAX - MIN)	DEPTH
HOLE	DIAMETER	DEPTH	TYPE
THREAD	DIAMETER	ANGLE	THREADTYPE

The following types of experimental NC strategies were proposed for the experiments (Table 2 and 3):

Table 2. Proposed NC strategies for turning.

Face turning	Profile turning	Groove turning
FACE roughing	PROFILE external roughing	GROOVE roughing
FACE finishing	PROFILE external finishing	GROOVE finishing
FACE groove roughing	PROFILE internal roughing	
FACE groove finishing	PROFILE internal finishing	

Table 3. Proposed NC strategies for hole making.

Hole drilling			Thread cutting
HOLE drilling	HOLE chamfering	HOLE tapping	THREAD external

Figure 3a) shows the strategy for drilling holes and Fig. 3b) shows the NC strategy for the profile roughing and finishing.

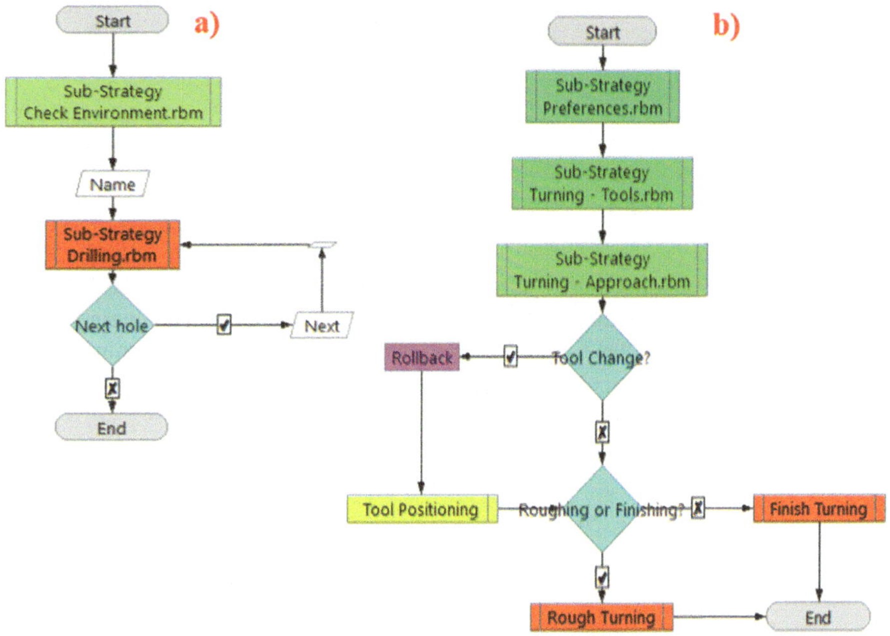

Fig. 3. Experimental NC strategies for a) drilling holes and b) profile turning.

4 Experiments

Several components were selected for verification of the proper functionality of the user-defined NC strategies and their impact on the creation of machining process (Fig. 4). The components have different dimensions and contain different types of features with different properties. The time required to automatically generate the machining process was selected as the evaluation parameter. Three measurements were made for each component with usage of user-defined NC strategies and the resulting average value was compared with the time required to create machining processes manually.

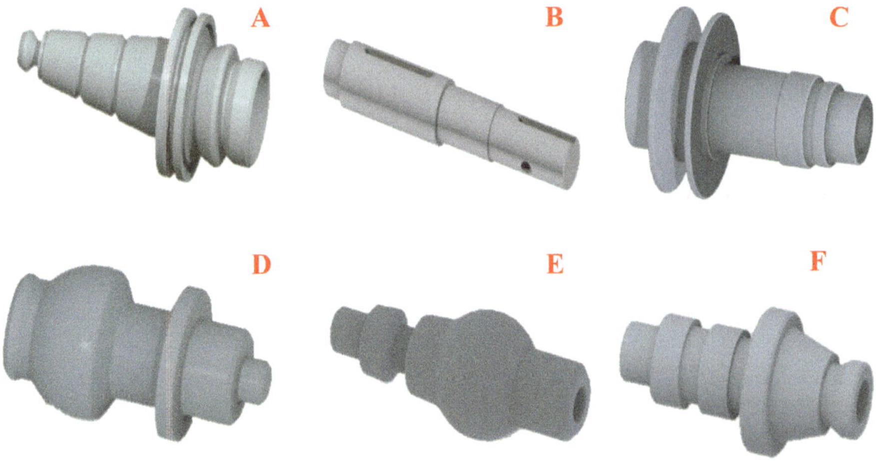

Fig. 4. Selected components for experiments.

The time measurement started after inserting the component into the CAD/CAM system Edgecam 2020 and ended after the creation of the complete machining process. Each machining process has to contain all cycles (for roughing and finishing) necessary for complex machining of company and individual commands for tool change and updating of the stock. Results of experiments are shown on Fig. 5.

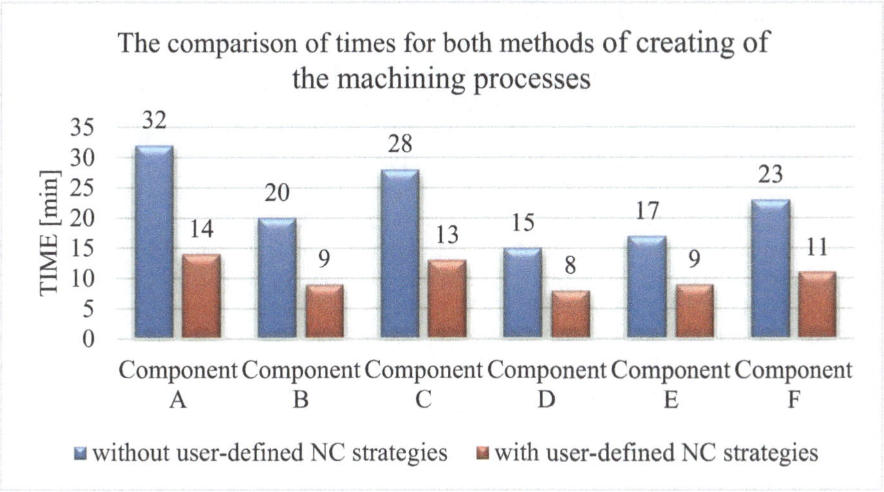

Fig. 5. Comparison of the time required to create machining processes.

5 Conclusion

Based on the achieved results and findings in the study of the NC strategy issues, it is possible to formulate the following conclusions:

1. The combination of the NC strategies together with the automatic recognition of machining features and profiles allows to eliminate repetitive tasks of the programmer and reduce the time required for the creation of new NC programs.
2. Appropriately designed NC strategies can be used as a tool for generating entire machining processes or only for the generation of selected parts of the processes, such as operations for the machining of grooves and threads.
3. The CAD/CAM system Edgecam 2020 does not allow to modify or create templates or algorithms for recognizing new feature. During feature recognition, errors occurred in the recognized turning profiles, which contained grooves.
4. The proposed NC strategies or the proposed procedures for their creation can also be used when working with other CAD/CAM systems with similar possibilities for creating of NC strategies.

Acknowledgments. This article was made under the support of KEGA project No. 006ŽU-4/2019 Quality improving and innovation the subject's content continuity of Automated Production Systems study program to bachelor study programs of Mechanical Engineering Area.

References

1. Lee, J.Y., Kim, K.: A feature-based approach to extracting machining features. Comput. Aided Design **30**, 1019–1035 (1998)

2. Magill, W.R., Mcleod, A.J.: Automated generation of NC part programs from a feature-based component description. Int. J. Comput. Integr. Manuf. **2**, 194–204 (1989)
3. Babic, B., Nesic, N., Miljkovic, Z.: A review of automated feature recongnition with rule-based pattern recognition. Comput. Ind. **59**(4), 321–337 (2008)
4. Dodok, T., Cubonova, N., Cisar, M.: Utilization of strategies to generate and optimize machining sequences in CAD/CAM. Proc. Eng. **192**, 113–118 (2017)
5. Samardžiová, M., Pokorný, P., Zvončan, M.: Influence of cutting parameters on achieved geometric tolerances in rotary ultrasonic drilling of Al2O3. In: Applied Mechanics and Materials: 2012 International Conference on Sensors, Measurement and Intelligent Materials (ICSMIM 2012), 26–27 December 2012, Guilin, China, vol. 303–306, pp. 2699–2703 (2013)
6. Peterka, J., Pokorný, P., Václav, Š.: CAM strategies and surfaces accuracy. In: Annals of DAAAM and Proceedings of DAAAM Symposium 2008: Vol. 19, No. 1: Proceedings of the 19th International DAAAM Symposium "Intelligent Manufacturing & Automation: Focus on Next Generation of Intelligent Systems and Solutions", 22–25th October 2008, Trnava, Slovakia, pp. 1061–1062 (2008)
7. Košinár, M., Kuric, I.: Monitoring possibilities of CNC machin tools accuracy. In: 1st International Conference on Quality and Innovation in Engineering and Management (QIEM), Cluj Napoca, 17–19 March 2011, pp. 115–118 (2011)
8. Kuric, I.: New methods and trends in product development and planning. In: 1st International Conference on Quality and Innovation in Engineering and Management (QIEM), Cluj Napoca, 17–19 March 2011, pp. 453–456 (2011)
9. Tlach, V., Cisár, M., Kuric, I.: Determination of the industrial robot positioning performance. In: Modern Technologies in Manufacturing (MTEM 2017 - AMATUC) Book Series: MATEC Web of Conferences, vol. 137 (2017)

Author Index

A
Affatato, Saverio, 371

B
Babu, Avinash Babu Ramesh, 120
Barosh, Mykola, 249
Bhattacharyya, Partha, 361
Bock, Silke, 3
Botko, Frantisek, 75
Botkova, Dominika, 75
Burkhard, Roger, 3

C
Chan, Chun, 212
Chattopadhyaya, Somnath, 155, 386
Chinnampalli, Yogesh, 270
Chlupová, Alice, 63
Císar, Miroslav, 404
Čuboňová, Nadežda, 404

D
D'Amato, Roberto, 199, 343, 371, 392
Das, Alok Kumar, 361
Dixit, Amit Rai, 100, 120, 361
Djapan, Marko, 199
Dodok, Tomáš, 404

E
Erb, Nico, 3

F
Feiler, Stefanie, 3
Fiedeń, Mateusz, 226

G
Gilgen, Lilian, 3
Gopi, Suresh, 270

H
Habrat, Witold, 226
Hloch, Sergej, 63, 100, 120, 142
Hölscher, Meike, 3
Hromasová, Monika, 87, 111, 120
Hussain, Shaik Tappa Nayab, 270

I
Ivanov, Vitalii, 404

J
Jain, Anuj, 355, 386
Jayakumar, Arunkumar, 270

K
Klich, Jiri, 41, 75
Klichova, Dagmar, 41, 75
Kodácsy, János, 236
Kolář, Viktor, 87
Kovács, Zsolt Ferenc, 236, 332
Krishna, Ram, 249, 270
Kruml, Tomáš, 63
Krupa, Krzysztof, 226
Kuchtová, Gabriela, 258
Kun, Krisztián, 282, 290, 332
Kuppusamy, Balaji, 132

L
Liguori, Consolatina, 324
Líska, János, 290, 332

S. Hloch et al. (Eds.): ICMEM 2021, LNME, pp. 411–412, 2021.
https://doi.org/10.1007/978-3-030-71956-2

Lisowicz, Joanna, 226
Luther, Anna, 3
Lyutyk, Mykola, 249

M
Maccarini, Giancarlo, 23
Madheswaran, Dinesh Kumar, 270
Mahalingam, Sivakumar, 132
Mallin, Levi, 301
Maurya, Anil Kumar, 355
Milosevic, Marko, 199
Mohan, Sreenivasan, 270
Mohanty, Shalini, 361
Mondala, Amitava, 155
Monka, Peter Pavol, 111
Monková, Katarína, 111
Mughilvalavan, Mohan, 183
Mülken, Oliver, 3
Muller, Miroslav, 87
Muruganandhan, Radhakrishnan, 183
Musial, Wojciech, 50

N
Nag, Akash, 100, 120, 142
Natarajan, Yuvaraj, 132
Novotný, Jan, 87

P
Paul, Akshoy Ranjan, 355, 386
Pellegrini, Giuseppe, 23
Peng, Guoyi, 41
Perec, Andrzej, 50
Poloprudský, Jakub, 63
Pude, Frank, 3

R
Raj, Piush, 155
Rajendran, Sounder Rajan, 132
Rausenberger, Julia, 3
Ravasio, Chiara, 23
Ravichandiran, Ashokkumar, 111
Rudawska, Anna, 87
Ruggiero, Alessandro, 199, 324, 343, 371, 392
Russo, Domenico, 324, 343

S
Shamli, Chinnakulanthai Sampath, 183
Shankar, Sachindra, 386
Simkulet, Vladimir, 75
Sitek, Libor, 41
Sommella, Paolo, 324
Sreenivasan, Mohan, 249
Stolárik, Gabriel, 142
Svehla, Brano, 142
Svobodova, Jaroslava, 12

T
Thangamuthu, Mohanraj, 270
Tichý, Martin, 87
Tiwari, Vineet Kumar, 386

U
Ungureanu, Nicolae, 199, 343, 371, 392

V
Valašek, Petr, 392
Viharos, Zsolt János, 236
Vilaca, Pedro, 386

W
Walton, Karl, 212
Weltsch, Zoltan, 282, 290

Y
Yuvaraj, Natarajan, 183

The manufacturer's authorised representative in the EU is Springer Nature Customer Service Centre GmbH, Europaplatz 3, 69115 Heidelberg, Germany. If you have any concerns regarding our products, please contact ProductSafety@springernature.com

Printed and bound by CPI Group (UK) Ltd, Croydon, CR0 4YY
30/09/2026
02225251-0012